P9-DEP-228

Digital Transmission Systems

DAVID R. SMITH

Chief, Transmission System Development Division
Defense Communications Engineering Center

Associate Professorial Lecturer
Department of Electrical Engineering and Computer Science
George Washington University

LIFETIME LEARNING PUBLICATIONS

Production Management: Bernie Scheier & Associates
Designer: Rick Chafiar
Copy Editor: Don Yoder
Illustrator: John Foster
Compositor: Science Typographers, Inc.

Printed in the United States of America

Published by Van Nostrand Reinhold Company Inc.
135 West 50th Street
New York, New York 10020

Van Nostrand Reinhold Company Limited
Molly Millars Lane
Wokingham, Berkshire RG11 2PY, England

Van Nostrand Reinhold
480 Latrobe Street
Melbourne, Victoria 3000, Australia

Macmillan of Canada
Division of Gage Publishing Limited
164 Commander Boulevard
Agincourt, Ontario M1S 3C7, Canada

1 2 3 4 5 6 7 8 9 10—88 87 86 85

Library of Congress Cataloging in Publication Data
Smith, David R. (David Russell), 1945–
Digital transmission systems.

Includes index.
1. Digital communications. I. Title.
TK5103.7.S65 1985 621.38 84-27049
ISBN 0-534-03382-2

To my wife, Carolyn, and our children
Christy, Stacy, Allison, and Andrew.

Preface

In the development of telecommunication networks throughout the world, digital transmission has now replaced analog transmission as the predominant choice for new transmission facilities. This trend began in the early 1960s when the American Telephone and Telegraph Company first introduced pulse code modulation as a means of increasing capacity in their cable plant. Since that time, digital transmission applications have grown dramatically, notably in the United States, Canada, Japan, and Western Europe. With the rapidity of digital transmission developments and implementation, however, there has been a surprising lack of textbooks written on the subject.

This book grew out of my work, research, and teaching in digital transmission systems. My objective is to provide an overview of the subject. To accomplish this end, theory has been blended with practice in order to illustrate how one applies theoretical principles to actual design and implementation. The book has abundant design examples and references to actual systems. These examples have been drawn from common carriers, manufacturers, and my own experience. Considerable effort has been made to include up-to-date standards, such as those published by the CCITT and CCIR, and to interpret their recommendations in the context of present-day digital transmission systems.

The intended audience of this book includes engineers involved in development and design, system operators involved in testing, operations, and maintenance, managers involved in system engineering and system planning, and instructors of digital communications courses. For engineers, managers, and operators, the book should prove to be a valuable reference because of its practical approach and the insight it provides into state-of-the-art digital transmission systems. In the classroom the book will allow the student to relate theory and practice and to study material not covered by other textbooks. Theoretical treatments have been kept to a minimum by summarizing derivations or simply stating the final result. Moreover, by consulting the extensive list of references given with each chapter the interested reader can explore the theory behind a particular subject in greater detail. Even so, to gain from this book the reader should have a fundamental knowledge of communication systems and probability theory.

The organization of this book follows a natural sequence of topics:

- Chapter 1: Historical background and perspective on worldwide digital transmission systems
- Chapter 2: System design including services to be provided, parameters used to characterize performance, and means of allocating performance as a function of the service, media, and transmission equipment
- Chapter 3: Analog-to-digital conversion techniques, starting with the most commonly used technique, pulse code modulation, and ending with specialized voice coders
- Chapter 4: Time-division multiplexing using both asynchronous and synchronous techniques
- Chapter 5: Baseband transmission including coding, filtering, equalization, and scrambling techniques
- Chapter 6: Digital modulation, starting with binary (two-level) forms and generalizing to M-ary forms, showing comparisons of error performance, bandwidth, and implementation complexity
- Chapter 7: Digital transmission over the telephone network, from voice-channel modems for low data rates to wideband modems for high data rates
- Chapter 8: Digital cable systems for twisted-pair, coaxial, and optical fibers, showing design and performance
- Chapter 9: Digital radio systems including propagation and interference effects, radio equipment design, and link calculations
- Chapter 10: Digital network timing and synchronization, describing fundamentals of time and frequency and means of network synchronization
- Chapter 11: Testing, monitoring, and control techniques used to measure and maintain system performance
- Chapter 12: Assessment of the future of digital transmission services and technology, together with a look at the emerging integrated services digital network (ISDN)

The material presented in Chapters 3 through 8 has been used for a graduate course in digital communications at George Washington University. The contents of the entire book form the basis for a short course on digital transmission, also offered by George Washington University.

David R. Smith

Acknowledgment

I first wish to acknowledge the support of my colleagues at the Defense Communications Agency, especially K. R. Belford, J. J. Cormack, W. A. Cybrowski, B. S. McAlpine, T. L. McCrickard, J. L. Osterholz, M. J. Prisutti, D. O. Savoye, D. O. Schultz, P. S. Selvaggi, S. Soonachan, and H. A. Stover.

My graduate students at George Washington University used early versions of chapters of this book and made many helpful contributions. Participants in numerous short courses offered encouragement and suggested improvements. I am indebted to all those who have taken my courses and at the same time helped me with the writing of this book.

I would like to pay special thanks to Dr. Heinz Schreiber of Grumman Aerospace Corporation and the State University of New York, who carefully reviewed the entire manuscript and provided many valuable comments.

In writing a book of this size, I have made considerable use of the open literature, especially the *IEEE Transactions on Communications*, the *Bell System Technical Journal*, and proceedings of various technical conferences. In describing standards applicable to digital transmission systems, I have extracted from the reports and recommendations of the CCIR and CCITT.*

Finally, I would like to express my thanks to Linda Thomas who diligently typed the original and revised versions of the manuscript.

*The reproduction in this book of material taken from the publications of the International Telecommunications Union (ITU), Place des Nations, 1211 Geneva 20, Switzerland, has been authorized by the ITU.

Contents

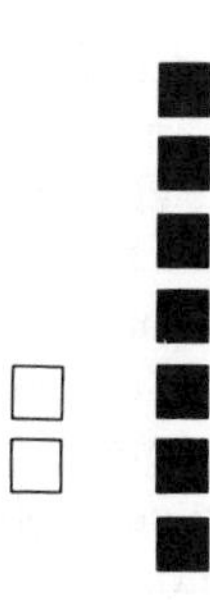

Introduction to Digital Transmission Systems

OBJECTIVES

- Describes the development of digital communications from the ancient Greeks to the global telecommunications networks of today
- Discusses the present-day use of digital transmission in the United States and abroad
- Explains national and international standards that have been established for digital transmission
- Compares the advantages and disadvantages of digital versus analog transmission
- Defines the fundamental components of a digital transmission system

1.1 HISTORICAL BACKGROUND

From the biblical passage:

> But let your communication
> be yea, yea, nay, nay.
> Matthew 5:37

one could argue that digital (actually, binary with redundancy!) transmission was divined two thousand years ago. In fact, early written history tells us that the Greeks used a form of digital transmission based on an array of torches in 300 B.C. and the Roman armies made use of semaphore signaling. The modern history of digital transmission, however, begins with the invention of the telegraph. Of the sixty or more different methods that had been proposed for telegraphy by 1850, the electromagnetic telegraph of Samuel Morse emerged as the most universally accepted one [1]. The Morse telegraph consisted of a battery, a sending key, a receiver in the form of an electromagnet, and the connecting wire. Pressure on the key activated the armature of the electromagnet, which produced a clicking sound. Although the original intention was to make a paper recording of the received signals, operators found that they could read the code faster by listening to the clicks. Morse's original code was based on a list of words in which individual numbers were assigned to all words likely to be used. Each number was represented by a succession of electrical pulses—for example, a sequence of 10 pulses signified the tenth word in the code. This cumbersome code was replaced by the now familiar dots and dashes with the assistance of Alfred Vail. Morse and Vail constructed a variable-length code in which the code length was matched to the frequency of occurrence of letters in newspaper English. Vail estimated letter probabilities by counting the numbers of letters in type boxes at the office of a local newspaper. This idea of assigning the shortest codes to the letters most frequently used was an early application of information theory.

Multiplexing techniques were also introduced in early telegraphy to allow simultaneous transmission of multiple signals over the same line. Time-division multiplexing was used in which a rotary switch connected the various signals to the transmission line. A basic problem then (and now) was the synchronizing of the transmitting and receiving commutators. A practical solution introduced by Baudot in 1874 involved the insertion of a fixed synchronizing signal once during each revolution of the switch. Baudot also introduced a five-unit code in place of the Morse type. Up to six operators could work on a single line; each operator had a keyboard with five keys. For each character to be transmitted the operator would depress those keys corresponding to marks in the code. The keys remained locked in place until connection was made with the rotary switch and the signals were transmitted down the line. The keys were then unlocked and an audible signal given to the operator to allow resetting of the keys for the next character transmission. Practical limitation for this multiplexing scheme was about 90 bits per second (b/s).

The telephone was born in 1876 when Alexander Graham Bell transmitted the first sentence to his assistant: "Mr. Watson, come here, I want you." The first telephones were primarily a transmitter or receiver, but not both. Thus initial use was limited to providing news and music, much in the fashion of radio in the twentieth century. Even when refinements provided the telephone with both transmitting and receiving capabilities, there was the problem of educating customers about its use. Read one Bell circular in 1877: "Conversation can be easily carried on after slight practice and with occasional repetition of a word or sentence." But by 1880 the telephone had begun to dominate the development of communication networks, and digital transmission in the form of telegraphy had to fit with the characteristics of analog transmission channels. Today telephony remains dominant in the world's communication systems, and thus transmission systems are still largely analog. As for Bell, he continued research in speech after the telephone became an economic success in the early twentieth century. But in his study there was a conspicuous absence: no telephones. Bell did not like interruptions [2].

Alexander Graham Bell is also credited as the inventor of modern lightwave communications. In 1880, Bell demonstrated the transmission of speech over a beam of light using a device he called the "photophone." One version of this invention used sunlight as a carrier, reflected from a voice-modulated thin mirror and received by a light-sensitive selenium detector. Although the photophone was Bell's proudest invention, the practical application of lightwave communication was made possible only after the invention of reliable light sources (such as the laser, invented in 1958) and the use of low-loss glass fiber. In 1966, Kao first proposed the use of low-loss glass fibers as a practical optical waveguide. In the early 1970s, Corning developed and introduced practical fiber cables having a loss of only a few decibels per kilometer. Field demonstrations of digital telephony transmission over fiber optic cable began in the mid-1970s. Since that time, the use of fiber optics for communications has grown rapidly, notably in the United States, Canada, Europe, and Japan.

Although many researchers contributed to the introduction of "wireless" transmission, Marconi is usually designated as the inventor of radio. He began his experiments in Italy in 1895. In 1899 he sent a wireless message across the English Channel, and in 1901 across the Atlantic Ocean between St. John's, Newfoundland, and Cornwall, England. Marconi's radiotelegraph produced code signals by on-off keying similar to that used for land telegraphy. Pressure on the key closed an electrical circuit, which caused a spark to jump an air gap between two metal balls separated by a fraction of an inch. The spark gave off radio waves that went through the antenna and over the air to the receiver. Still, radio transmission had limited use until the invention and perfection of the vacuum tube. Used for both amplification and detection, the vacuum tube allowed longer transmission distances and better reception. With the introduction of modulation, radiotelephony was possible and became the dominant mode of radio transmission by the 1920s. Digital radio for telephony was still a long way off, however, and did not become available until the 1970s.

With the advent of power amplifiers, limitations in basic noise levels were discovered and, in 1928, Nyquist described noise theory in a classic paper [3]. At the same time, the relationship between bandwidth and sampling rate was being investigated, and in 1928 Nyquist published his famous account of digital transmission theory [4]. Nyquist showed mathematically that a minimum sampling rate of $2W$ samples per second is required to reconstruct a band-limited analog signal of bandwidth W. Twenty years later, Shannon linked noise theory with digital transmission theory, which led to his famous expression for the information capacity C of a channel with noise [5]:

$$C = W \log_2(1 + S/N) \text{ bits/second} \tag{1.1}$$

where W is the bandwidth, S is the signal power, and N is the noise power. This expression implies that data transmission at any rate is possible for an arbitrary bandwidth provided an adequate S/N can be achieved, the actual data rate R is less than C, and a sufficiently involved encoder/decoder is used.

Digital transmission of analog signals was made practical with the invention of pulse code modulation (PCM) by A. Reeves in 1936–1937. Reeves first discovered pulse-position modulation and then recognized the noise immunity possible by quantizing the amplitude and expressing the result in binary form. The first fully operational PCM system did not emerge until ten years later in the United States [6]. PCM systems in the late 1940s were large, unreliable, and inefficient in power consumption. The key invention responsible for the success of PCM was of course the transistor, which was invented in 1948 and in practical use by the early 1950s. Initial applications of PCM systems involved twisted-pair cable, whose characteristics limited analog multichannel carrier operation but were acceptable for PCM. Today PCM is the universal standard for digital transmission of telephony and is used with terrestrial microwave radio, satellites, and fiber optics as well as metallic cable.

The history of satellite communications began with Echo, a passive satellite experiment conducted in 1960. Telstar, launched in 1962 for use in the Bell System, was the first satellite to use active repeaters, or transponders. These early satellites were placed in low orbits that required complex earth stations to track the motion of the satellite. Beginning with the Early Bird (INTELSAT I) satellite in 1965, communication satellite orbits have since been geostationary—that is, stationary above a fixed point on the earth's surface. Using geostationary satellites, complete earth coverage is possible with only three or four satellites. The international telecommunications satellite consortium known as INTELSAT was organized in 1964 and by 1970 had established a global telecommunications network. Today there exist a number of domestic and international satellite carriers offering various services including voice, data, and broadcast television. Present-day satellites contain a number of transponders, each providing a separate service or leased separately to a customer.

1.2 PRESENT-DAY DIGITAL TRANSMISSION

In 1962, the Bell System introduced the T1 cable carrier system in the United States [7]. This carrier system has a capacity of 24 PCM voice channels and a transmission rate of 1.544 megabits per second (Mb/s). This 24-channel PCM system is now the standard PCM multiplexer in North America and Japan. By 1982 the Bell System had over 100 million circuit miles of T1 carrier systems. These systems, operating over twisted-pair cable, already dominate the U.S. metropolitan networks. The prime application of PCM over twisted-pair cable was to take advantage of the local telephone companies' investment in buried copper cable. Frequency-division multiplex (FDM) equipment was expensive and did not provide economic advantage over the cost of copper pairs for individual voice channels over short distances between telephone offices (typically less than 10 mi). PCM carrier equipment reduced hardware costs so that T1 carrier became more economical than individual copper pairs for short distances. The ability to regenerate digital signals also allowed better performance than analog transmission.

To illustrate the relative use of digital transmission in the Bell System (now AT & T and the Bell Operating Companies), Figure 1.1 shows cumulative circuit miles for metro and short haul (up to about 100 mi) transmission projected to 1989. Digital radio has been used to connect metropolitan islands of T carrier and to interconnect digital switches. Since 1977, digital

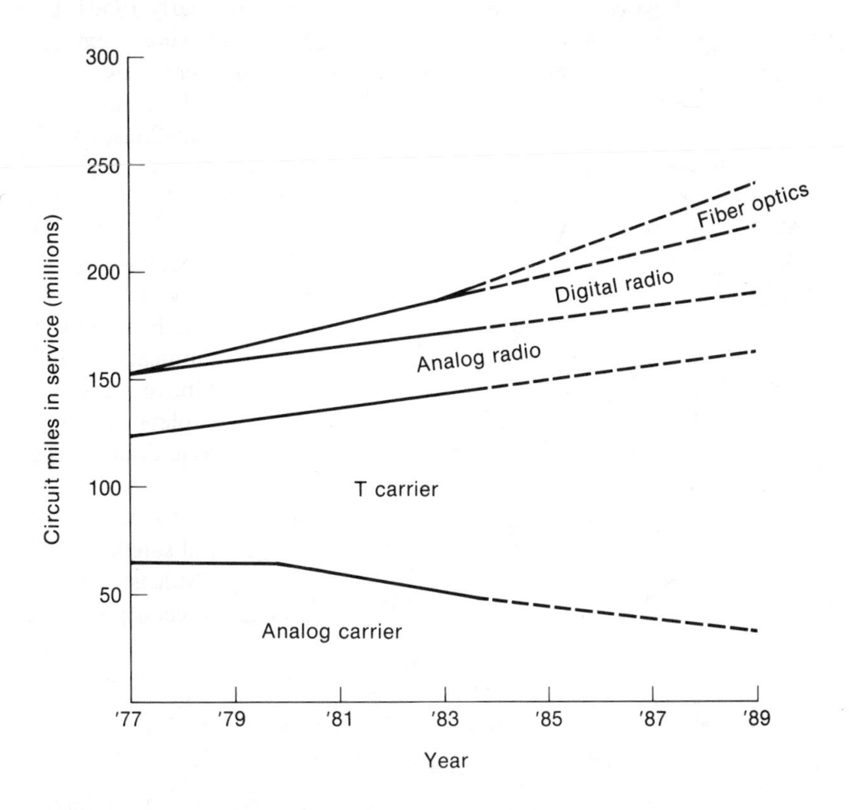

Figure 1.1 Cumulative Short Haul and Metro Circuit Miles by Transmission Type in the AT & T and Bell Operating Companies' Networks

radio for short haul has experienced rapid growth and accounted for nearly 5 million circuit miles in 1979. Fiber optic transmission is another rapidly growing transmission technology in the United States. The first standard fiber optic system was introduced by the Bell System in 1980. It operates at the standard rate 44.736 Mb/s and provides 672 voice channels on a fiber pair.

Figure 1.1 indicates that nearly all future growth in the metro and short-haul areas will be digital. In contrast, the long-haul transmission plant of AT & T is predominantly analog and is projected to remain that way through the 1980s, as indicated in Figure 1.2. Since the long-haul plant carries primarily voice circuits, AT & T employs analog radio techniques, such as single sideband, which are more bandwidth efficient than digital radio. However, digital transmission has penetrated the long-haul network in two ways. First, many analog radio and coaxial cable facilities can be adapted to carry voice and data simultaneously, which is called **hybrid transmission**. Second, fiber optic systems operating at 90 Mb/s are being installed with initial deployment in 1983 to link metropolitan areas along the East and West coasts of the United States. As shown in Figure 1.2, the AT & T network in 1990 is projected to be 23 percent coaxial cable, 58 percent analog radio, 3 percent satellite, and 16 percent fiber optics.

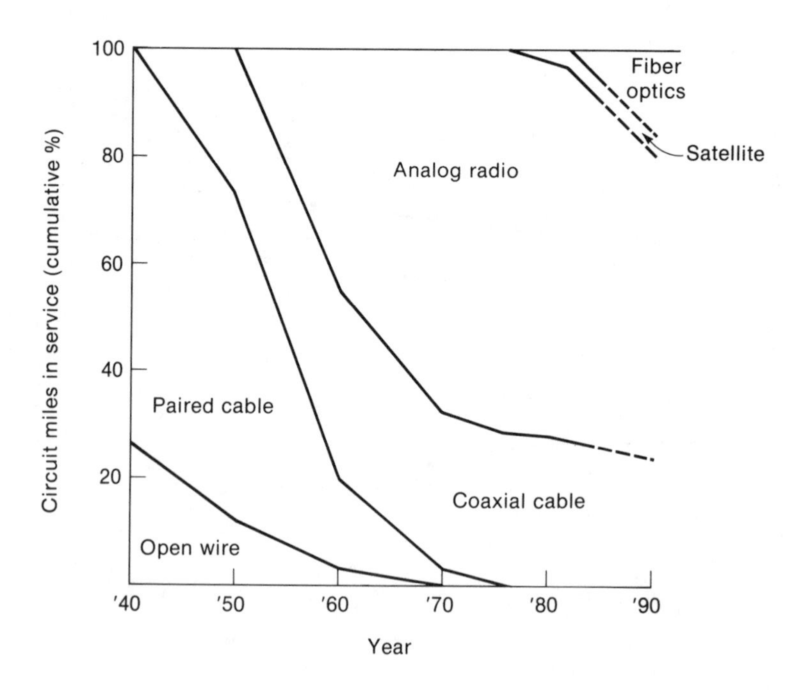

Figure 1.2 Trends in Long-Haul Transmission Within the AT & T Network

Outside AT & T and the Bell Operating Companies, satellite and common carriers have emerged in the United States to provide digital services—primarily by satellite and fiber optics for long haul and by digital radio and cable for local areas. The first satellite systems provided voice, data, and television signals by means of analog transmission, as was done, for example, by Western Union with their Westar satellites beginning in 1974. Many present-day satellite systems, however, are using digital transmission. The first all-digital satellite system was developed by Satellite Business Systems (SBS) to provide digital voice, data, high-speed facsimile, and video teleconferencing. Other U.S. carriers, such as MCI Communications Corp., have selected fiber optics, installed along railway rights-of-way, as their major long-haul medium. Military communication networks are also undergoing conversion from analog to digital transmission for both terrestrial and satellite media. Within the U.S. Defense Communications System, both voice (AUTOVON) and data (AUTODIN) networks are being upgraded with digital transmission and digital switching technology.

Although the United States may have been the early leader in developing digital transmission technology, other countries, notably Japan, Canada, and several European nations, have been much more energetic in exploiting digital transmission. From the mid-1960s to early 1970s, T1 carrier systems were introduced in many national systems and led to considerable growth in digital transmission throughout the world. In Japan, Nippon Telegraph and Telephone (NTT) has developed and applied higher-order PCM multiplex for coaxial cable, digital radio, and fiber optics systems. The NTT program for network digitization indicates that 90 percent of transmission facilities in Japan will be digital by the turn of the century [8]. Bell Canada's progress in moving to a digital network has also been dramatic. Local facilities are already 50 percent digital and are projected to grow to 65 percent by 1990; intertoll facilities were 17 percent digital in 1980 and expected to rise to 65 percent by 1990 [9]. Although the United States, Japan, and Canada have based their digital hierarchy on the original 24-channel, 1.544 Mb/s PCM standard, European countries have since developed another standard based on a 30-channel grouping and a transmission rate of 2.048 Mb/s. This second PCM standard now forms the basis for the European digital hierarchy as defined by the CEPT (European Conference of Post and Telecommunications Administrations). Italy was the first European nation to use PCM, beginning in 1965 with the 24-channel standard. Since 1973, only 30-channel systems have been installed—principally by the Societa Italiana per l'Esercizio Telefonico (SIP), which is the major Italian operating company. In 1979, some 25 percent of Italian carrier systems were digital; by 1985 about 45 percent of installed carrier channels will be digital [10]. In the United Kingdom, the British Post Office (BPO) introduced 24-channel PCM in the late 1960s, but since 1978 only 30-channel systems have been installed [11]. France has also been a leader in the use of PCM transmission facilities in conjunction with digital switching.

1.3 DIGITAL TRANSMISSION STANDARDS

Standards play an important role in establishing compatibility among communications facilities and equipment that are designed by different manufacturers and used by the many communications carriers throughout the world. These standards are developed by various national and international groups. The organization that has the greatest effect on digital transmission standards is the International Telecommunications Union (ITU), which was founded in 1865 and in 1982 had 157 member countries. There are three main organizations within the ITU: the International Frequency Registration Board, which registers and standardizes radio frequency assignments; the International Radio Consultative Committee (CCIR), which deals with standards for radio communications; and the International Telegraph and Telephone Consultative Committee (CCITT). Both the CCIR and CCITT are divided into a number of study groups that make recommendations on various aspects of telephony, telegraphy, and RF communications. These study groups report their results at the CCITT and

Table 1.1 CCITT and CCIR Study Groups for Digital Transmission Topics

Designation	*Title*	*Recommendations and Reports*
CCITT Study Group VII	Data Communication Networks	CCITT 1981 Vol. VIII.2 and VIII.3 (X Series Rec.)
CCITT Study Group XV	Transmission Systems	CCITT 1981 Vol. III.1 (G Series Rec.) CCITT 1981 Vol. III.4 (H, J Series Rec.)
CCITT Study Group XVII	Data Communication Over the Telephone Network	CCITT 1981 Vol. VIII.1 (V Series Rec.)
CCITT Study Group XVIII	Digital Networks	CCITT 1981 Vol. III.3 (G Series Rec.)
CCIR Study Group 4	Fixed Satellite Service	CCIR 1982 Vol. IV — Pt. 1
CCIR Study Group 5	Propagation in Nonionized Media	CCIR 1982 Vol. V
CCIR Study Group 7	Standard Frequencies and Time Signals	CCIR 1982 Vol. VII
CCIR Study Group 9	Fixed Service Using Radio-Relay Systems	CCIR 1982 Vol. IX — Pt. 1
CCIR Study Group 11	Broadcasting Service (Television)	CCIR 1982 Vol. XI — Pt. 1
Joint CCIR / CCITT Study Group CMTT	Transmission of Sound Broadcasting and Television Signals Over Long Distances	CCIR 1982 Vol. XII

CCIR plenary assemblies held every three or four years. Here recommendations are formally adopted and work is organized for the next study period. Table 1.1 indicates the study groups that are mostly concerned with digital transmission. Examples of CCITT and CCIR standards will be given throughout this book.

Another important source of data transmission standards is the International Organization for Standardization (ISO). The ISO was organized in 1926 with the objective of developing and publishing international standards. Currently 71 countries are represented in the ISO; the American National Standards Institute (ANSI) is the U.S. member. The ISO is divided into more than 150 technical committees. The technical committee responsible for data transmission standards is TC-97, Computers and Information Processing. Within ANSI, X3S3 is the technical committee on data communications. Another organization playing a significant role in data transmission standards in the United States is the Electronic Industries Association (EIA), established in 1924 to assist U.S. industry in legislation, regulation, and standards. Examples of ISO and EIA standards will be given in data transmission sections of the book. The interrelationship of these various standards organizations is shown in Figure 1.3.

Various military organizations also play an important role in establishing standards for digital transmission. In the United States, the Defense Communications Agency (DCA), the joint tactical communications office, and individual military departments are collectively responsible for developing and promulgating military standards in the MIL-STD-188 series for

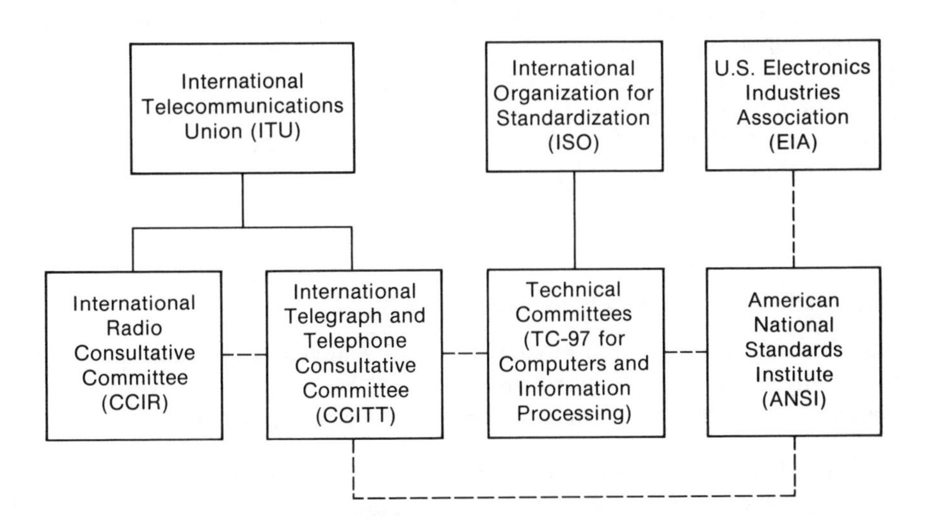

Figure 1.3 Relationships Among International and U.S. National Standards Organizations

Table 1.2 Transmission Rates for Digital Hierarchy of North America, Japan, and Europe

Multiplex Level	*North America (Mb / s)*	*Japan (Mb / s)*	*Europe (Mb / s)*
1	1.544	1.544	2.048
2	6.312	6.312	8.448
3	44.736	32.064	34.368
4	274.176	97.728	139.264
5		397.200	

strategic and tactical communications systems. Specifically, U.S. MIL-STD-188-323 provides design and performance criteria for digital transmission systems [12], and DCA Circular 300-175-9 provides operations and maintenance performance criteria for circuits and links [13]. These U.S. military standards generally adopt national and international standards unless unique military requirements dictate otherwise. Foreign military organizations follow procedures similar to those of the United States. Within NATO, the United States, Canada, and European member nations form standardization agreements (STANAGs) as a means of ensuring compatible military communication systems among these various nations.

The worldwide evolution of digital transmission networks has led to three digital hierarchies in use today. These are referred to as North American, Japanese, and European (CEPT) and have been recognized by international standards organizations, most importantly the CCITT. The transmission rate structure of the three hierarchies is shown in Table 1.2. A more detailed comparison is deferred to Chapter 4.

1.4 DIGITAL VS. ANALOG TRANSMISSION

Since most of today's transmission systems are analog, an obvious question is "why digital?" To answer this question, we must first consider the advantages of digital transmission versus analog transmission:

- Digital performance is superior. The use of regenerative repeaters eliminates the noise accumulation of analog systems. Performance is nearly independent of the number of repeaters, system length, or network topology.
- Sources of digital data, such as teletypewriters or computers, can be more efficiently accommodated. Present analog systems use individual modems

(modulator-demodulators), one per data source, making inefficient use of bandwidth.

- With the advent of space and computer technologies, advances in digital logic and use of medium to large-scale integration have led to superior reliability and maintainability of digital transmission equipment.
- The cost of digital transmission equipment has effectively decreased over the last 10 years. Previously the cost and poor reliability of vacuum tube and discrete transistor technology made digital transmission expensive. Today PCM and digital multiplex equipment is generally less costly than frequency-division multiplex (FDM) equipment.
- Digital radio systems are less susceptible to interference than are analog radio systems. Time-division multiplex signals are tolerant of single-frequency interference, since that interference is spread over all channels. Frequency-division multiplex, however, is susceptible to the single-frequency interferer since the interference is translated directly to a few baseband channels. Digital radio systems commonly employ scramblers to smooth the power spectrum and use modulation schemes that suppress the carrier; hence the overall radio system's performance is immune to single-frequency interference.
- The use of digital transmission greatly facilitates the protection of communications against eavesdropping. The simple addition of **cryptographic equipment**, a scrambling device based on digital logic, provides privacy to the users. In military applications, digital transmission has led to extensive use of **bulk encryption**, a process in which multiple signals are encrypted by a single cryptographic apparatus, and **secure voice**, in which voice communications are encrypted by cryptographic equipment. Encryption has always been desirable for military communications, but it is becoming more important in other communication systems for protection of electronic banking, industrial secrets, and the like.

Now let us consider the disadvantages of using digital transmission:

- Current communication systems worldwide use primarily analog transmission and analog switches. With these large analog assets, there is no economic justification to precipitously convert all-analog systems to digital. Instead, this conversion is an evolutionary process in which analog equipment is replaced as it becomes obsolete.
- The universal acceptance of an FDM standard hierarchy has not led to an analogous standardization of the TDM hierarchy. As shown in Table 1.2, there are three digital hierarchies in existence today. With the exception of the first two multiplex levels in the North American and Japanese hierarchies, the multiplex equipment is all incompatible.
- Present digital microwave systems are not as bandwidth efficient as analog microwave systems for the transmission of analog signals—that is, voice and video. Advances in digital modulation and analog-to-digital conversion techniques are expected to narrow the gap.

1.5 A SIMPLIFIED DIGITAL TRANSMISSION SYSTEM

To introduce the terminology for the remaining chapters in this book, consider the block diagram of a digital transmission system shown in Figure 1.4. The fundamental parts of the system are the *transmitter*, the *medium* over which the information is transmitted, and the *receiver*, which estimates the original information from the received signal. At the transmitter the information source may take several forms—for example, a computer, teletypewriter, telemetry system, voice signal, or video signal. The source *encoder* formats data signals—for example, by creating alphanumeric characters for computer output—or digitizes analog signals such as voice and video. The *modulator* then interfaces the digital signal with the transmission medium by varying certain parameters (such as frequency, phase, or amplitude) of a particular square wave or sinusoidal wave called the *carrier*. A transmission medium such as cable or radio provides the *channel*, a single path that connects the transmitter and receiver. Transmitted signals are corrupted by noise, interference, or distortion inherent in the transmission medium. The receiver must then demodulate and decode the degraded signal to make an estimate of the original information.

The block diagram of Figure 1.4 represents a simplified digital transmission system. The direction of transmission is indicated as one-way, or *simplex*, between two points. Transmission systems usually operate *duplex*, in which simultaneous two-way information is passed point to point. The electrical path between the two endpoints, called a *circuit*, may be dedicated for exclusive use by two users or may be switched to provide temporary connection of the two users. Further, a circuit may be composed of several types of transmission medium, or channels. A long-haul circuit, for example, may require a mix of satellite, radio, and cable media.

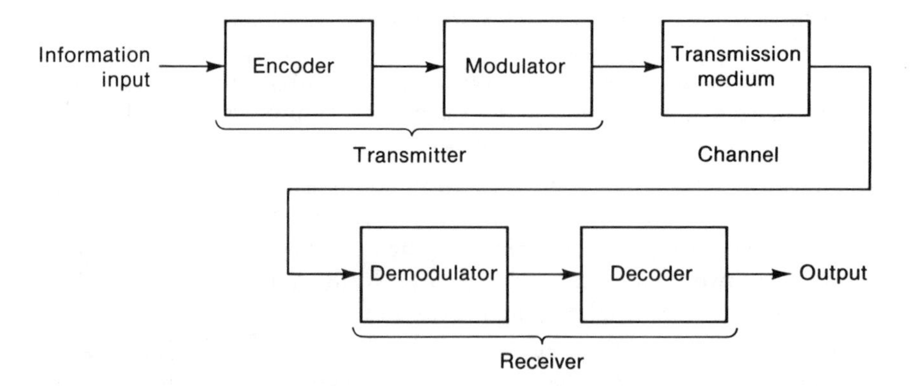

Figure 1.4 Block Diagram of Digital Transmission System

Transmission systems may be devised for single circuits or for multiple circuits through the use of multiplexing techniques. Transmission systems may also interconnect many users to provide a *network* for communication. As the size of a network grows, it becomes impractical to provide dedicated transmission channels for each circuit. Hence most networks employ switches that provide a temporary circuit at the request of one of the connected users.

The theme of this book is transmission systems used in digital communication systems. Emphasis is placed on long-haul, multichannel, terrestrial transmission systems, although the theory and practices presented here are also applicable to local transmission systems such as mobile radio and subscriber loops and to satellite communication systems. Inevitably, certain digital communication topics had to be omitted to keep the book to a manageable size. A description of error correction codes has been omitted because of the philosophy of most digital transmission system designers. Transmission systems are typically designed to provide a certain level of error performance without the use of error correction codes. Thus a user who requires better service must resort to error correction codes built into the encoder/decoder. Switching systems have been discussed only to the extent that they influence transmission system design and performance. The growth in data communication requirements has led to the development of new networks that employ packet switching, a topic beyond the scope of this book. Moreover, certain topics relevant to military communications are not covered in detail, including the use of spread spectrum for protection against electronic countermeasures and the use of cryptography for privacy. For information on these peripheral topics, the reader is asked to make use of the references [14, 15, 16].

1.6 SUMMARY

The use of digital transmission dates back to early written history, but the modern history of digital transmission barely stretches over the last hundred years. The major events in the development of digital transmission, summarized in Figure 1.5, have culminated in worldwide applications, led by the United States, Canada, Japan, and several European nations. Today these developments in digital transmission are facilitated by standards and hierarchies established by the ITU, ISO, and various national standards organizations. Advantages of digital transmission—performance, bandwidth efficiency for data, reliability, maintainability, equipment cost, and ease of encryption—in most applications will outweigh its disadvantages—pervasiveness of existing analog systems, lack of a single universal digital hierarchy, and bandwidth inefficiency for voice and video. The remaining chapters of this book describe the theory and practices of digital transmission systems and explain in more detail the terminology, standards, and examples given in this first chapter.

1850 — Invention of telegraph by Morse

1874 — Invention of time-division multiplexing by Baudot
1876 — Invention of telephone by Bell
1880 — Invention of photophone by Bell

1899 — Invention of radio by Marconi

1928 — Development of sampling theory by Nyquist

1936 — Invention of pulse code modulation by Reeves

1948 — Development of channel capacity theory by Shannon
Invention of transistor by Bell Laboratories
1962 — First 1.544-Mb / s T1 cable carrier system by Bell System
Telstar, first communications satellite by Bell System
1965 — Early Bird, first geostationary communications satellite by INTELSAT
1966 — Low-loss optical fiber proposed by Kao
1970 —
1980 — } Development of worldwide applications of digital transmission
1990 —

Figure 1.5 Significant Events in the Modern History of Digital Transmission

REFERENCES

1. W. R. Bennett and J. R. Davey, *Data Transmission* (New York: McGraw-Hill, 1965).
2. J. Brooks, *Telephone: The First Hundred Years* (New York: Harper & Row, 1976).
3. H. Nyquist, "Thermal Agitation of Electricity in Conductors," *Phys. Rev.* 32(1928):110–113.
4. H. Nyquist, "Certain Topics in Telegraph Transmission Theory," *Trans. AIEE* 47(1928):617–644.
5. C. E. Shannon, "A Mathematical Theory of Communication," *Bell System Technical Journal* 27(July and October 1948):379–423 and 623–656.

6. H. S. Black and J. O. Edson, "Pulse Code Modulation," *Trans. AIEE* 66(1947):895–899.
7. K. E. Fultz and D. B. Penick, "The T1 Carrier System," *Bell System Technical Journal* 44(September 1965):1405–1451.
8. T. Mirrami, T. Murakami, T. Ichikawa, "An Overview of the Digital Transmission Network in Japan," *International Communications Conference*, 1978, pp. 11.1.1–11.1.5.
9. J. A. Harvey and J. R. Barry, "Evolution and Exploitation of Bell Canada's Integrated Digital Network," *International Communications Conference*, 1981, pp. 17.4.1–17.4.6.
10. M. R. Aaron, "Digital Communications—The Silent (R)evolution," *IEEE Comm. Mag.*, January 1979, pp. 16–26.
11. J. F. Boag, "The End of the First Pulse Code Modulation Era in the U.K.," *Post Office Elec. Eng. J.* 71(April 1978):2.
12. Draft MIL-STD-188-323, "System Design and Engineering Standards for Long Haul Digital Transmission System Performance," U.S. Dept. of Defense, Washington, D.C., November 1984.
13. Defense Communications Agency Circular 300-175-9, *DCS Operating-Maintenance Electrical Performance Standards*, U.S. Dept. of Defense, Washington, D.C., August 1982.
14. W. W. Peterson and E. J. Weldon, Jr., *Error Correcting Codes* (Cambridge, Mass.: MIT Press, 1972).
15. R. D. Rosner, *Packet Switching: Tomorrow's Communication Today* (Belmont, Calif.: Lifetime Learning Publications, 1981).
16. R. H. Pettit, *ECM and ECCM Techniques for Digital Communication Systems* (Belmont, Calif.: Lifetime Learning Publications, 1982).

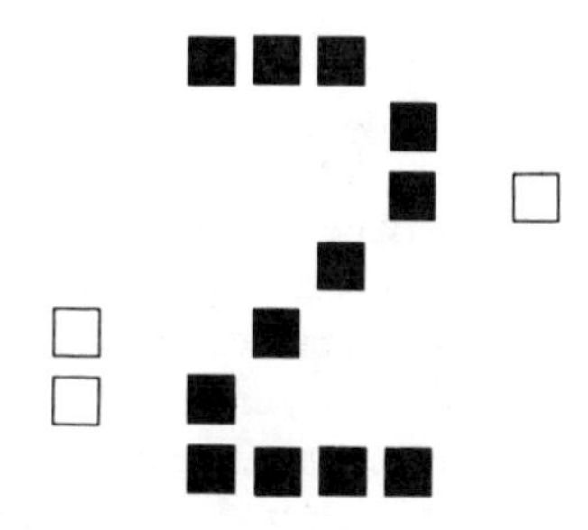

Principles of System Design

OBJECTIVES

- Describes how performance objectives are established for a digital transmission system
- Discusses the three main services provided by a digital communications system: voice, data, and video
- Explains how the hypothetical reference circuit is used to determine overall performance requirements
- Discusses the two main criteria for system performance: availability and quality

2.1 GENERAL PLAN

The first step in transmission system design is to establish performance objectives. This process begins by stating the overall performance from circuit end to circuit end. The requirements of the user depend on the service being provided. For human communications such as voice and video, satisfactory performance is largely a subjective judgment; for data users, performance can be better quantified through measures such as throughput, efficiency, and error rate. For any type of service, the measure used to indicate performance must be based on statistics, since performance will vary with time due to changes in equipment configuration and operating conditions. Once these end-to-end objectives have been stated, the allocation of objectives to individual subsystems and equipment may take place.

The parameters used to characterize performance are based on the service and the type of transmission media and equipment. Here we will use two basic parameters for all cases: availability and quality. Circuit availability indicates the probability or percentage of time that a circuit is usable. During periods of circuit availability, other quality-related parameters such as error rate express the soundness of the circuit. These quality parameters are selected according to the service and transmission system. In many cases, however, performance objectives are based on telephony service because of its dominance in today's communication systems. Other services provided in the same transmission network may demand greater performance calling for special procedures.

The allocation process can be done for each combination of transmission equipment unique to a particular circuit, but this approach is unnecessarily redundant and clearly tedious. Instead, hypothetical reference circuits are used to represent the majority of practical cases and to reduce the system designer's job to the allocation of performance for just this single circuit. Once this hypothetical circuit has been formulated, the overall

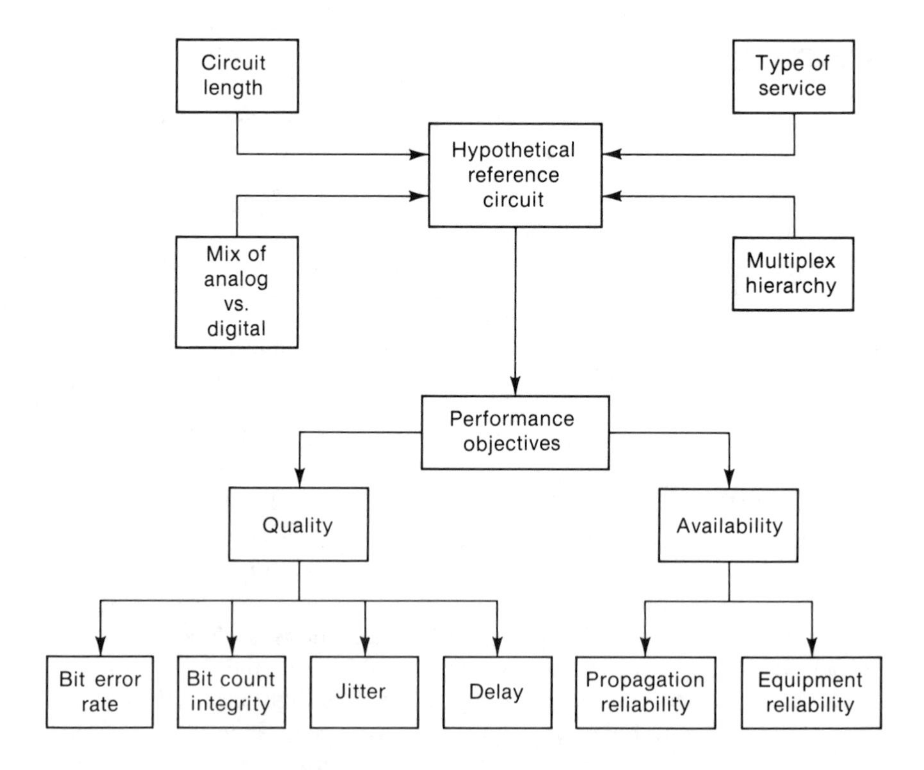

Figure 2.1 Process for Performance Allocation of a Digital Transmission System

(end-to-end) performance objective can be allocated to the individual subsystems and equipment. Numerous factors must be considered in constructing the hypothetical reference circuit, such as type of service, mix of transmission media, mix of analog and digital transmission, and multiplex hierarchy. Thus, because of diverse requirements in the type of service or transmission system, more than one reference circuit may be required to establish performance objectives for all cases. Figure 2.1 summarizes the general process for establishing performance allocations in a digital transmission system.

2.2 TRANSMISSION SERVICES

The main services provided by a digital communications system are voice, data, and video. Each of these services is described in the following sections from the standpoint of how digital transmission systems accommodate the service. Emerging digital services are described in Chapter 12.

2.2.1 Voice

Voice signals are the major service provided by today's communication systems. For digital transmission applications, the voice signal must first be converted to a digital representation by the process of analog-to-digital (A/D) conversion. The principles of A/D conversion are illustrated in Figure 2.2. The input voice signal is first band-limited by a low-pass filter and then periodically sampled to generate a time-sampled, band-limited signal. These samples are then made discrete in amplitude by the process of quantization in which the continuous sample value is replaced by the closest discrete value from a set of quantization levels. The coder formats the quantized signal for transmission or multiplexing with other digital signals. **Pulse code modulation** (PCM) represents each quantization level with a binary code. Thus if there are 256 quantization levels, an 8-bit PCM code is required for each sample. Rather than code the absolute value of each voice sample as is done in PCM, the difference or delta between samples can be coded in an attempt to eliminate the redundancy in speech and in the

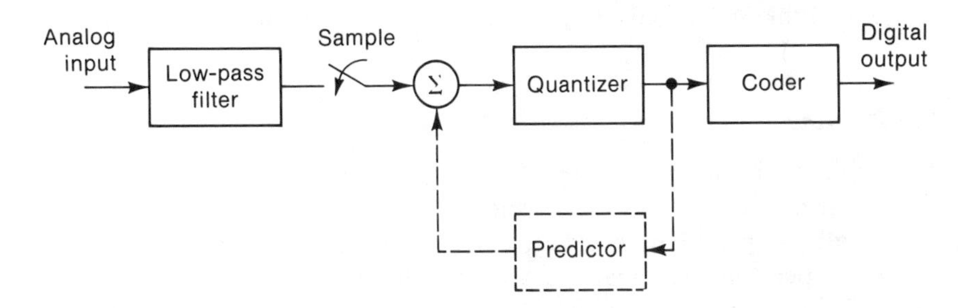

Figure 2.2 Block Diagram of Analog-to-Digital Converter

process lower the transmission bit rate. As shown in the dashed lines of Figure 2.2, the quantized sample can be fed back through a prediction circuit and subtracted from the next sample to generate a difference signal. This form of A/D conversion is known as **differential PCM** or **delta modulation**. A detailed description of these and other A/D techniques is given in Chapter 3.

The universal standard for voice digitization is PCM at a transmission rate of 64 kb/s. This rate results from a sampling rate of 8000 samples per second and 8 bits of quantization per sample. For multichannel digital transmission of voice, digitized signals are time-division multiplexed. The resulting composite transmission bit rate depends on the coding rate per voice signal, the number of voice signals to be multiplexed, and the overhead required for multiplex synchronization. In the North American standard PCM multiplex, for example, there are 24 voice channels, each coded at 64 kb/s, plus 8 kb/s of overhead for synchronization, with a total transmission rate of 1.544 Mb/s. In the European CEPT standard, there are 30 voice channels, each at 64 kb/s, plus two 64 kb/s overhead channels used for signaling and synchronization, with a total transmission rate of 2.048 Mb/s.

The transmission quality of digitized voice channels is determined primarily by the choice of A/D technique, the number of tandem A/D conversions traversed by the circuit, and the characteristics of transmission impairment. For a choice of 64-kb/s PCM as the voice coder, the transfer characteristics of the voice channel are essentially the same as the analog channel being replaced. A/D conversion, however, introduces **quantization noise**, caused by the error of approximation in the quantizing process. With successive A/D conversions in the circuit, quantization noise will accumulate, although certain encoding techniques such as 64-kb/s PCM provide tolerable performance for a large number of A/D conversions. Regarding the effects of transmission impairment on voice quality, certain bit errors and timing slips will cause audible "clicks" and jitter will cause distortion in the speech signal recovered from 64-kb/s PCM [1]. Effects of a 10^{-4} error rate are considered tolerable and a 10^{-6} error rate negligible for 64-kb/s PCM. Depending on the technique employed, low-bit-rate voice (LBRV) coders may be more susceptible to degradation from greater quantization noise, multiple A/D conversions, and transmission impairment (Chapter 3).

2.2.2 Data

Data communication requirements are a fast-growing service, due largely to the emergence of networks that connect computer-based systems to geographically dispersed users. Most current data services are provided by modems operating over analog channels within a frequency-division multiplex (FDM) hierarchy as described in Chapter 7. Rates up to 9600 b/s can be operated over a single voice channel. Higher data rate modems require

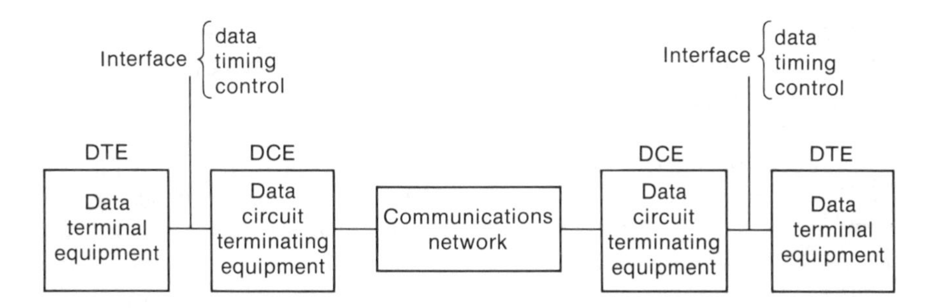

Figure 2.3 Data Transmission Network Interfaces

proportionally larger FDM bandwidths; for example, 56-kb/s modems have been developed for a 48-kHz bandwidth and 1.544-Mb/s modems for a 480-kHz bandwidth. Digital transmission facilities can be designed to handle data requirements more efficiently. However, early implementations of PCM for digital telephony did not include direct data interfaces. Thus modems were still required to provide the necessary conversion for interface of data with a 64-kb/s PCM voice channel. PCM channels can provide satisfactory performance for modems at rates up to 9.6 kb/s (see Chapter 3 for details), but at the expense of wasted bandwidth. To be more efficient, data channels can be directly multiplexed with PCM channels. In this case, a single 64-kb/s PCM channel might be replaced with, say, six channels of 9.6-kb/s data. Submultiplexing of 64-kb/s channels is now common in digital transmission facilities, as evidenced by the number of low-speed TDM equipment available today.

To understand the organization of emerging data transmission networks, it is necessary to define some specialized terminology and review existing interface standards. Figure 2.3 illustrates the basic interface between the **data terminal equipment** (DTE) and **data circuit terminating equipment** (DCE) of a data transmission network. The DTE is the user device, typically a computer, while the DCE is the communication equipment, say modem or multiplex, that connects the user to the network. The DTE/DCE interface is characterized by three types of signals—data, timing, and control—and by several levels of organization necessary for network applications.

Timing Signals

The use of the timing signal is determined by the type of DTE. **Asynchronous** terminals transmit data characters independently, separated by start-stop pulses. The use of unique start and stop pulses allows the receiving terminal to identify the exact beginning and end of each character. The stop pulse is made longer than the data bits to compensate for differences in clock frequency at the transmitter and receiver. With this arrangement, no timing line is required between the DTE and DCE. With **synchronous** transmission, characters are transmitted continuously in large blocks without separation by start or stop pulses. The transmitting and receiving terminals must be operating at the same speed if the receiver is to know which bit is first in a block of data. Likewise, the speed of the DTE must be matched to the communication line. The various possible arrangements of the timing lines for synchronous transmission are shown in Figure 2.4. With **codirectional timing**, the DTE supplies the timing line that synchronizes the DCE frequency source (oscillator) to the DTE. Conversely, with **contradirectional timing**, the DTE is slaved to the timing of the DCE. A third configuration, **buffered timing**, is often used when the communication line interfaces with several different terminals as in the case of time-division multiplexing. Here buffers allow each DTE to operate from its own timing source by compensating for small differences in data rate between each DTE and the DCE. These buffers are typically built into the DTE or DCE and are sized to permit slip-free operation for some minimum period of time. (See Chapter 10 for calculation of the buffer slip rate.)

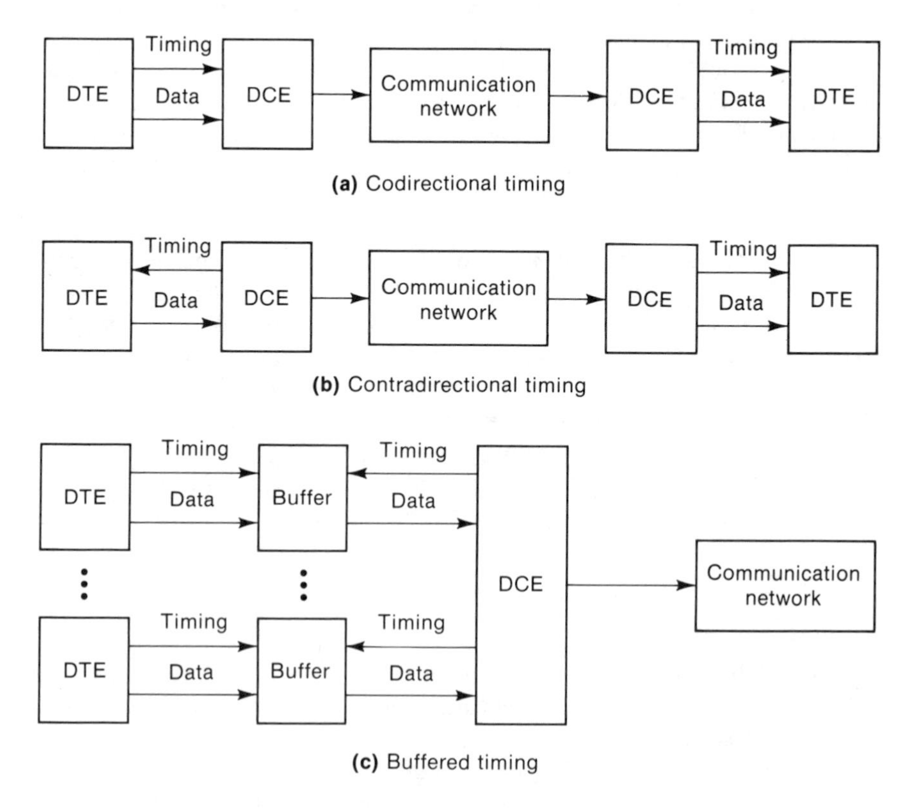

Figure 2.4 Synchronous Data Transmission Configurations

Interface Standards

The various levels of interface can be best described by examining the standards that exist for data transmission. Only a brief summary of the topic is given here; a detailed account is beyond the scope of this book and has been given ample treatment elsewhere [2]. The basic architecture that has emerged identifies seven interface levels (Figure 2.5):

- Level 1: the physical, electrical, functional, and procedural level used to establish, maintain, and disconnect the physical link between the DTE and the DCE
- Level 2: the link control level for interchange of data between the DTE and the network in order to provide synchronization control and error detection/correction functions
- Level 3: the network control level that defines the formatting of messages or packets and the control procedures for establishing end-to-end connections and transferring user data through the network
- Higher levels: system control and user protocols that provide data exchange on a totally end-to-end basis

The development of these standards has been the responsibility of different organizations. Level 1 standards have been developed by the ISO and CCITT and in the United States by the ANSI and EIA. At level 2, the ISO has had responsibility internationally and ANSI in the U.S. Level 3 standards have been developed by the ISO, CCITT, and ANSI. These levels

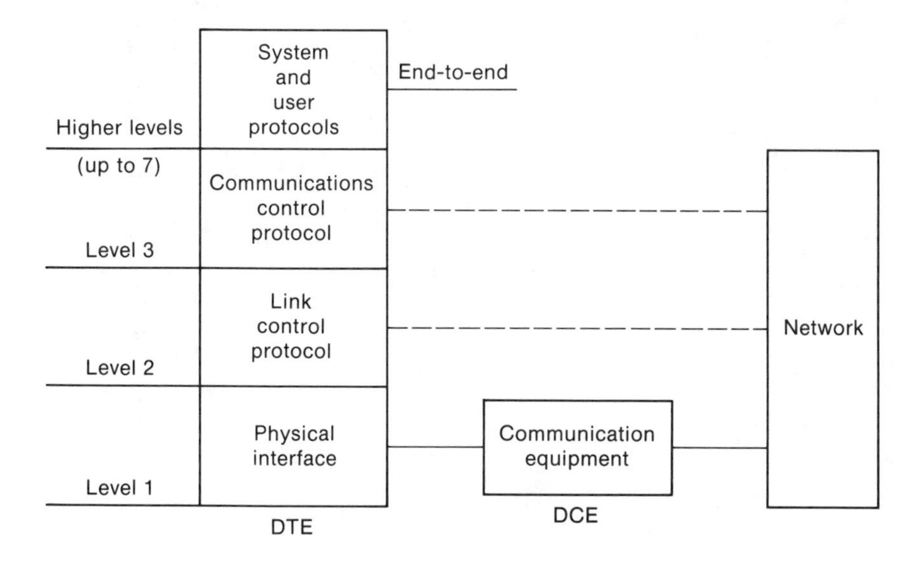

Figure 2.5 Architecture of Data Transmission Networks

Table 2.1 Level 1 Interface Standards for Data Transmission

	First Generation		*Second Generation*	
Type	*Standard*	*Characteristics*	*Standard*	*Characteristics*
Electrical	EIA RS-232C CCITT V.28	Unbalanced interface 0 – 20 kb/s data rate 50 ft maximum distance	EIA RS-422 CCITT X.27 (V.11)	Balanced interface 0 – 10 Mb/s data rate Interoperable with RS-423 and X.26 (V.10) Distance limitation: 4000 ft at 100 kb/s, 40 ft at 10 Mb/s
			EIA RS-423 CCITT X.26 (V.10)	Unbalanced interface 0 – 300 kb/s data rate Interoperable with RS-232C, V.28, RS-422, and X.27 (V.11) Distance limitation: 4000 ft at 3 kb/s, 40 ft at 300 kb/s
Functional	EIA RS-232C CCITT V.24	Definitions of data, timing, and control signals	EIA RS-449 CCITT V.24	Definitions of data, timing, and control signals
Mechanical	EIA RS-232C ISO 2110	25-pin connector	EIA RS-449 ISO 4902	37-pin connector interoperable with RS-232C

are not independent of each other. Interface at a particular level requires that all subordinate levels also be specified.

The interface of digital transmission systems with data transmission networks can be characterized primarily by the physical interface level. The standards for this level 1 interface have undergone two generations of development, which are summarized in Table 2.1. The first generation was based on the EIA "Recommended Standard" RS-232C and its international counterpart, CCITT V.24 for functional characteristics and V.28 for electrical characteristics. These standards were limited to data transmission over telephone networks using modems at rates up to 20 kb/s. The new, second-generation standards apply both to modems and to the new public data networks offering circuit-switching, packet-switching, and leased-circuit services. These new standards are the EIA RS-422 for balanced circuits (using a differential generator with both outputs aboveground) and RS-423 for unbalanced circuits (using a single-ended generator with one side grounded) and their international equivalents, CCITT X.27 (V.11) and X.26 (V.10). The CCITT X and V standards are equivalent; the X version is for public data network interfaces and the V version for modems operating in a telephone network. Standard RS-449 specifies the mechanical connector design and functional pin assignments for implementation of RS-422 and RS-423. Salient characteristics of these level 1 standards are listed in Table 2.1. Note the interoperability specification for second-generation standards, which allows continued use of older equipment in modern data networks.

Performance

Data transmission quality is measured by response time and block error rate. Digital transmission impairments that affect data quality include delay, slips, and bit errors. Most data users employ some form of error detection/correction such as automatic request for repetition (ARQ) or forward error correcting (FEC) codes to augment the performance provided by a transmission channel. FEC techniques are based on the use of overhead bits added at the transmitter to locate and correct erroneous bits at the receiver. ARQ systems detect errors at the receiver and, after a request for repeat, automatically retransmit the original data. These error-correcting schemes act on blocks of data. Hence the data user is concerned with block error rate, where the length of a block is determined either by the basic unit of traffic, such as an alphanumeric character, or by standard ARQ or FEC coding schemes.

2.2.3 Video

Digital transmission of television signals is a rapidly developing technology [3] where standardization is still ongoing. A joint CCITT/CCIR committee, the Television Transmission Joint Committee (CMTT), has been established to deal with transmission of sound and color television signals over long distances. Recommendations for digital coding in television studios have been developed, but the choice of coding technique for digital transmission is still open [4, 5]. Studio standards lead to bit rates between 200 and 300 Mb/s using uniform PCM with 8 bits per sample. Such bit rates, however, are not economical for digital transmission. The trend is toward image bandwidth compression techniques that can reduce the transmission rate to well below 100 Mb/s. Systems operating at the European standard rate of 34.368 Mb/s and the North American standard of 44.736 Mb/s have been developed. Typical characteristics of such systems are the use of differential PCM, interframe or intraframe coding for reduction of image redundancy, and built-in error detection and correction.

Digital transmission impairments that affect television performance include jitter, slips, and bit errors. Depending on the choice of A/D conversion method, another impairment is the distortion caused by the sampling and filtering process, amplitude quantization, and bit rate reduction coding. Error performance requirements for broadcast-quality television are usually more stringent than those for voice and data because of the disturbing effects of loss of TV synchronization. Permissible error rates are in the range between 1×10^{-4} and 1×10^{-9}, depending on the choice of A/D conversion and the use of forward error correction techniques [5].

2.3 HYPOTHETICAL REFERENCE CIRCUITS

The task of assigning performance allocations to a digital transmission system is greatly facilitated by use of a hypothetical reference circuit composed of interconnected segments that are representative of the media and equipment to be used. Reference circuits are a standard tool used by communication system designers, including the CCITT, CCIR, and AT & T, to provide the basis for development of:

- Overall user-to-user performance requirements for satisfactory service
- Performance specifications of typical segments of the overall circuit that when connected together will meet the overall performance objectives

The hypothetical reference circuit should represent a majority of all circuits but is not necessarily the worst-case design. The resulting design should provide satisfactory service to a majority of users. The numerous factors to be considered in constructing a reference circuit are discussed in the following paragraphs.

Path Length

The dominant factor in formulating the reference circuit is the path length. International circuits may span the globe and use several media types, including terrestrial radio, satellite, fiber optic cable, and metallic cable. National reference circuits are considerably shorter in length and use primarily cable and terrestrial radio. A national reference circuit may be connected to an international reference circuit—as in the case of the longest reference circuit described by the CCITT, shown in Figure 2.6a. This circuit

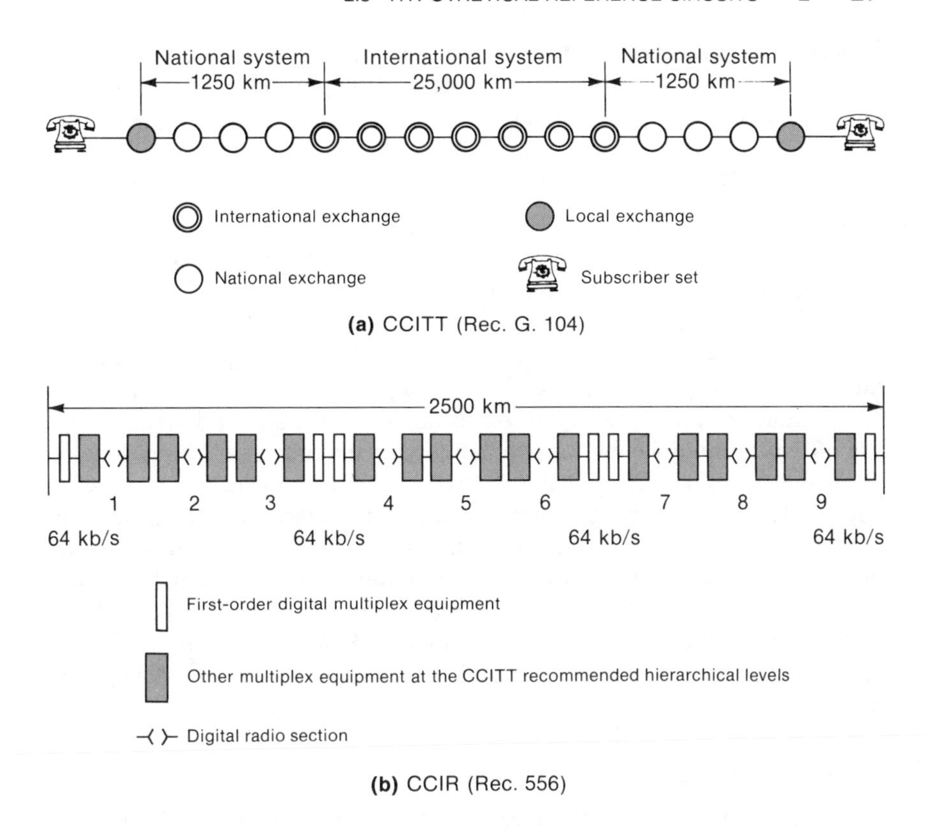

(a) CCITT (Rec. G. 104)

(b) CCIR (Rec. 556)

Figure 2.6 Hypothetical Reference Circuits for 64-kb/s Digital Transmission [6, 7] (Courtesy CCITT and CCIR)

has a 25,000-km international connection with two 1250-km national systems. Separate reference circuits have also been developed by the CCITT and CCIR for specific transmission media including coaxial cable (CCITT Rec. G.332), line-of-sight radio relay (CCIR Rec. 556), tropospheric scatter radio relay (CCIR Rec. 396-1), and communication satellite systems (CCIR Rec. 521-1).

Multiplex Hierarchy

Another factor in structuring the reference circuit is the number of multiplex stages and their location in the network. The reference circuit should represent the tandem multiplex and demultiplex functions that occur from one end of the circuit to the other. If the service is telephony, the circuit may also undergo a number of tandem analog-to-digital conversions. The reference circuit should show the relative composition of multiplex and A/D functions at appropriate locations. Typically, higher-level multiplexers appear more often than lower-level multiplexers. The A/D conversion required for digital telephony would occur at the lowest level in the hierarchy and at analog switching or patching locations. The CCIR's hypothetical reference circuit for digital radio-relay systems shown in Figure 2.6b includes nine sets of digital multiplex equipment at the hierarchical levels recommended by the CCITT.

Analog vs. Digital

Since most existing transmission systems are analog, reference circuits in the past have been based on the use of analog equipment—for example, a choice of frequency-division multiplex for telephony. As analog transmission networks incorporate and evolve toward digital facilities, the reference circuit must show a representative mix of analog and digital transmission facilities. Interfaces of digital and analog equipment also must be reflected in the choice of reference circuit. An example of such a hybrid system is the use of digital multiplexing for telephony but with circuit switching done on an analog basis, requiring that A/D conversion equipment (such as PCM) appear at each switching center. Such hybrid reference circuits are important in planning the transition strategy for exchange of analog with digital transmission. The only all-digital reference circuits that have been developed by the CCITT or CCIR are shown in Figures 2.6 and 2.7. These circuits are limited principally to telephony. Reference circuits for other services and other media using digital transmission are under development by both organizations.

Type of Service

The composition of the reference circuit may be affected by the type of service being provided. The hypothetical reference circuits shown in Figure

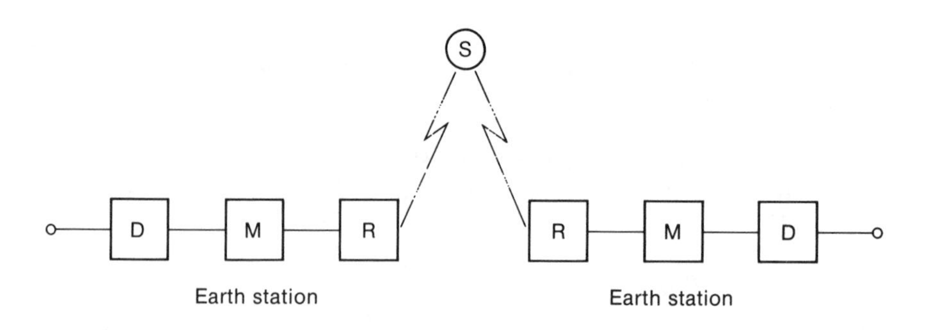

S: Space station
D: Direct digital interface equipment
M: Modem equipment
R: IF/RF equipment

Figure 2.7 Hypothetical Reference Circuit for Digital Transmission Systems Using Satellite Communications (CCIR Rec. 521-1) [8] (Courtesy CCIR)

2.6 are based on telephony. Should the service be data instead, the multiplex hierarchy would likely be affected—showing either additional levels of multiplex (below the 64 kb/s level) for low-speed data or fewer levels of multiplex for high-speed data. Television transmission generally requires an entire terrestrial radio or satellite bandwidth allocation, eliminating multiplex equipment from the hypothetical reference circuit—as shown, for example, by the CCIR's hypothetical reference circuit for television (Rec. 567-1) [5]. The most significant difference in these various services, as seen by the designer of the hypothetical reference circuit, is the method of describing performance. Each service—data, voice, and video—may have a different set of parameters and values. If the transmission system is to carry only a single type of service, then a single reference circuit and single set of performance criteria will suffice. If more than one service is to be supplied, however, the system designer is faced with two approaches: (1) Set performance objectives to satisfy the largest class of users, which is usually telephony, and force other users to adopt special measures for better performance, or (2) set objectives according to the service with the most stringent set of performance criteria. The latter approach provides satisfactory performance for all services, whereas the former approach accounts for the majority of users.

Example 2.1

Figure 2.8 shows a typical global reference circuit spanning two continents and one ocean. The transoceanic circuit consists of two satellite hops,

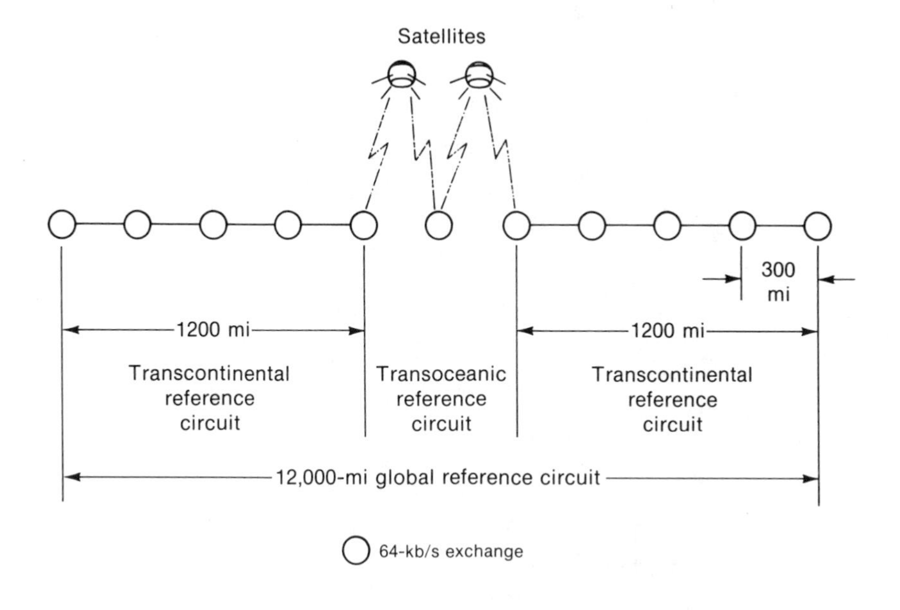

Figure 2.8 Example of Global Reference Circuit

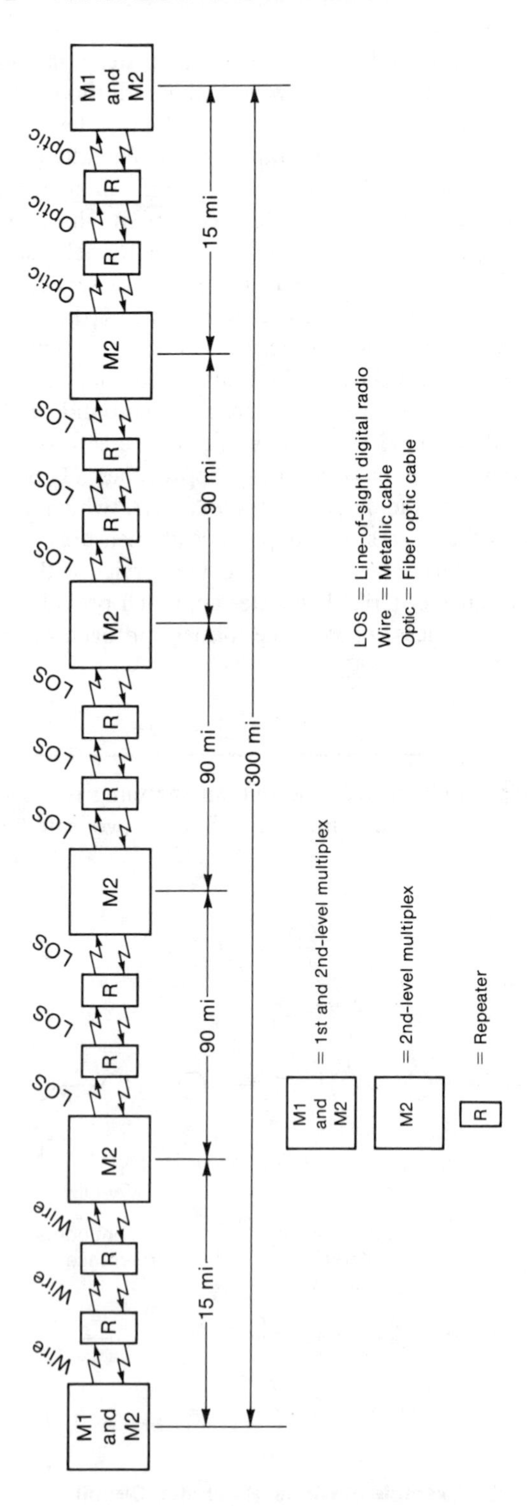

Figure 2.9 Example of Terrestrial Reference Circuit

although one or both of these hops could well be transoceanic cable. Each transcontinental circuit is 1200 mi in length and is composed of four terrestrial reference circuits, which are each 300 mi in length. Figure 2.9 shows the terrestrial reference circuit to be composed of line-of-sight (LOS) digital radio along the backbone and digital cable, both metallic and fiber optic, at the tails. The repeater sites provide regeneration of the digital signal but do not include multiplexers. The sites labeled M2 utilize second-level multiplexers to interconnect 1.544-Mb/s digital signals. Finally, the sites at the end of the terrestrial reference circuit will use first and second levels of multiplex equipment to derive individual 64-kb/s voice and data channels.

2.4 PERFORMANCE OBJECTIVES

Performance objectives for a digital transmission system may be stated with two characteristics: availability and quality. **Availability** is defined as the probability or fraction of time that circuit continuity is maintained and usable service is provided. The quality of transmission, then, is considered only when the circuit is available. For digital transmission, quality parameters commonly used include bit error rate, timing slips, jitter, and delay. These parameters apply to all types of services, although parametric values may differ and additional parameters may be called out that are related to service. Here we will consider availability and the quality parameters just cited.

Before proceeding, a distinction must be made between design objectives discussed here and operation and maintenance (O & M) standards of performance. In operation, the performance of a service may deteriorate for various reasons: aging, excessive environmental conditions, human error, and so forth. Therefore design objectives should provide a suitable margin to assure that the service in operation meets O & M performance standards.

2.4.1 Availability Objectives

Periods of time when a circuit is not usable, or **unavailable**, may be categorized as long term or short term. This distinction is made because of the relative impact of long-term and short-term outages on the user. A voice user subjected to a long-term outage (say, greater than a few seconds in length) will be automatically disconnected from the multiplexer or switch or will impatiently hang up his instrument. A data user subjected to a long-term outage will have messages corrupted, since the outage length will exceed the ability of coding or requested retransmission to provide error correcting. Short-term outages (less than a few seconds) will be a disturbance to voice users but probably will not cause termination of the connection. Likewise, short-term outages in the data user's channel may be compensated by error correction or retransmission techniques. Multipath fading in radio transmission, resynchronization in time-division multiplexers, and automatic switching actions for restoral in redundant equipment are all sources of short-term outages. Long-term outages consist primarily of catastrophic failure of equipment such as radio, antennas, or power supply, although certain propagation conditions such as precipitation attenuation and power fading also produce long-term outages.

There are two basic approaches to allocating availability: (1) Lump all outages together, long and short term, or (2) deal only with long-term outages. The second approach places short-term outages in the category of circuit quality and not circuit availability. In either case, an end-to-end availability is levied that takes on typical values of 0.9999 to 0.99. Further suballocation is then made to individual segments and finally equipment within the hypothetical reference circuit.

Following the definition given by the CCITT and CCIR (shown later in Tables 2.3 and 2.4), we will consider a circuit to be unavailable when the service is degraded (for example, bit error rate is worse than a specified threshold) for periods exceeding n seconds. Typically the values of n that have been used or proposed are in the range of a few seconds to a minute. Thus this definition tends to exclude short-term outages and includes only long-term outages caused principally by equipment failure.

Definitions

To allocate performance, the system designer must understand and apply the science of statistics as used in the prediction and measurement of equipment availability. The following definitions are essential to this understanding.

Failure refers to any departure from specified performance, whether catastrophic or gradual degradation. For prediction and system design purposes, the types of failure considered are limited to those that can be modeled statistically, which excludes such instances as natural disasters or human error. **Failure rate** (λ) refers to the average rate at which failures can be expected to occur throughout the useful life of the equipment. Equipment failure rates can be calculated by using component failure rate data or obtained from laboratory and field tests.

Reliability, $R(t)$, refers to the probability that equipment will perform without failure for a period of time t given by

$$R(t) = e^{-\lambda t} \tag{2.1}$$

Studies of failure rates in electronic equipment have shown that the reliability of components decays exponentially with time. It is known, however, that electronic equipment may contain parts subject to early failure, or "infant mortality." It is also known that the rate of failures increases as the device nears the end of its life. During the useful life of the equipment, however, the rate of failure tends to remain constant.

Mean time between failures (MTBF) refers to the ratio of total operating time divided by the number of failures in the same period:

$$\text{MTBF} = \int_0^\infty e^{-\lambda t}\, dt = \frac{1}{\lambda} \tag{2.2}$$

Mean time to repair (MTTR) is defined as the average time to repair failed equipment, including fault isolation, equipment replacement or repair, and test time, but excluding administrative time for travel and locating spare modules or equipment. **Mean time to service restoral (MTSR)** is the average time to restore service, including repair and administrative time.

Availability (A) is the probability that equipment will be operable at any given point in time or the fraction of time that the equipment performs the required function over a stated period of time. It is also defined as the ratio of uptime to total time:

$$A = \frac{\text{MTBF}}{\text{MTBF} + \text{MTSR}} \tag{2.3}$$

Unavailability (U) is the complement of availability:

$$\begin{aligned} U &= 1 - A \\ &= \frac{\text{MTSR}}{\text{MTBF} + \text{MTSR}} \end{aligned} \tag{2.4}$$

For large MTBF and small MTSR (usually the case)

$$U \approx \frac{\text{MTSR}}{\text{MTBF}} \tag{2.5}$$

Outage (O) is the condition whereby the user is deprived of service due to failure within the communication system. For redundant equipment, a single failed unit should not cause an outage but should cause switchover to a backup unit. **Mean time between outages (MTBO)** refers to the ratio of total operating time divided by the number of outages in the same period.

Redundant Equipment

In equipment designed with redundancy, an outage can occur either from simultaneous failure of all redundant units or from a failure to properly recognize and switch out a failed unit. If the equipment consists of n identical units, each with the same MTBF, for which at least r must be operable for the system to be operable, then the first type of outage can be characterized by

$$\text{MTBO}_1 = \frac{\text{MTBF}(\text{MTBF}/\text{MTTR})^{n-r}}{n(n-1)!/(r-1)!(n-r)!} \tag{2.6}$$

where the sensing of a failure and resulting switchover is itself assumed to occur without failure. Expression (2.6) thus accounts for the situation where the backup units all fail within the MTTR for the first failed unit. As an example, consider two identical units, one used on-line and the other used as a redundant standby. For this case, $n = 2$, $r = 1$, and

$$\text{MTBO}_1 = \frac{\text{MTBF}^2}{2(\text{MTTR})} \tag{2.7}$$

The monitoring and control necessary for redundancy switching has some probability P_s of successfully detecting failure in the on-line unit and completing the switching action to remove the failed unit. This second type of outage for redundant equipment has an MTBO given by

$$\text{MTBO}_2 = \frac{\text{MTBF}}{1 - P_s} \tag{2.8}$$

The total MTBO of the redundant equipment is then

$$\frac{1}{\text{MTBO}} = \frac{1}{\text{MTBO}_1} + \frac{1}{\text{MTBO}_2} \tag{2.9}$$

With double redundancy, for example, the total MTBO is found from (2.7) and (2.8):

$$\text{MTBO} = \frac{\text{MTBF}^2}{2(\text{MTTR}) + \text{MTBF}(1 - P_s)} \tag{2.10}$$

Availability for redundant equipment can be expressed by using the basic definition given in (2.3) with MTBO used in place of MTBF:

$$A = \frac{\text{MTBO}}{\text{MTBO} + \text{MTSR}} \tag{2.11}$$

whereas the unavailability can be expressed from (2.5) as

$$U \approx \frac{\text{MTSR}}{\text{MTBO}} \tag{2.12}$$

Provision for redundancy in equipment design allows uninterrupted service when a single unit fails. For redundancy to be effective, the system designer must be able to guarantee that:

1. Failures in the active (on-line) unit are recognized and successfully switched out. This requires a sophisticated monitoring and control capability (see Chapter 11). Both on-line and off-line units should be continuously monitored and compared with some performance factor such as error rate.
2. A prescribed MTBO can be met with the design proposed. This requires a failure analysis to demonstrate that a certain percentage of all failures will be recognized and removed by automatic switchover.
3. Failure sensing and switching times are short enough to be classified as a short-term outage, so that they are not counted as a contribution to system unavailability.

Example 2.2

A doubly redundant system has the following characteristics:

$$\text{MTBF} = 3200 \text{ hr}$$
$$\text{MTTR} = \text{MTSR} = 15 \text{ min}$$
$$P_s = 0.97$$

Find the total MTBO and unavailability. What effect will $P_s = 0.9$ have on the system?

Solution

From (2.10) we have, for a doubly redundant system,

$$\text{MTBO} = \frac{(3200)^2}{2(0.25) + 3200(1 - 0.97)}$$
$$= 106{,}114 \text{ hr}$$

The unavailability is given by (2.12) as

$$U = \frac{0.25}{106{,}114} = 2.4 \times 10^{-6}$$

For $P_s = 0.9$, the corresponding MTBO and U are

$$\text{MTBO} = 31{,}950 \text{ hr}$$
$$U = 7.8 \times 10^{-6}$$

System Availability

To calculate the availability of a system composed of interconnected equipment, one needs only to apply the basic laws of probability to the definitions and equations given here. Failures can be assumed to be independent from one piece of equipment to another. Availability expressions are given here for typical system configurations.

Series Combination. For a series combination (Figure 2.10) the system availability is the product of individual availabilities, and system failure rate is the sum of individual failure rates:

$$A_N = \prod_{i=1}^{N} A_i \tag{2.13}$$

$$\lambda_N = \sum_{i=1}^{N} \lambda_i \tag{2.14}$$

The unavailability is then

$$U_N = 1 - \prod_{i=1}^{N} A_i \tag{2.15}$$

It can be shown that for small outages system unavailability is determined

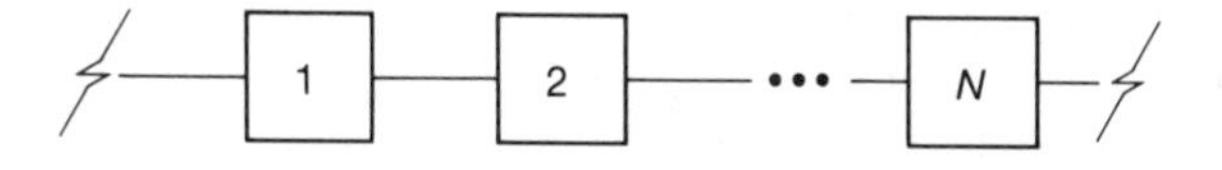

Figure 2.10 Series Combination

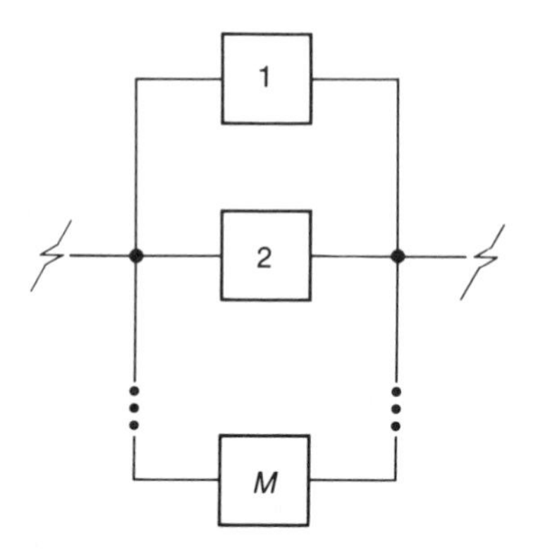

Figure 2.11 Parallel Combination

by summing the outages of individual equipment. Therefore

$$A_N \approx 1 - \sum_{i=1}^{N} U_i \tag{2.16}$$

Parallel Combination. If only one piece of equipment must be available for the whole system to be available, then for the parallel configuration of Figure 2.11

$$A_M = 1 - \prod_{i=1}^{M} U_i \tag{2.17}$$

where the product of individual unavailabilities yields the probability that all M units are unavailable.

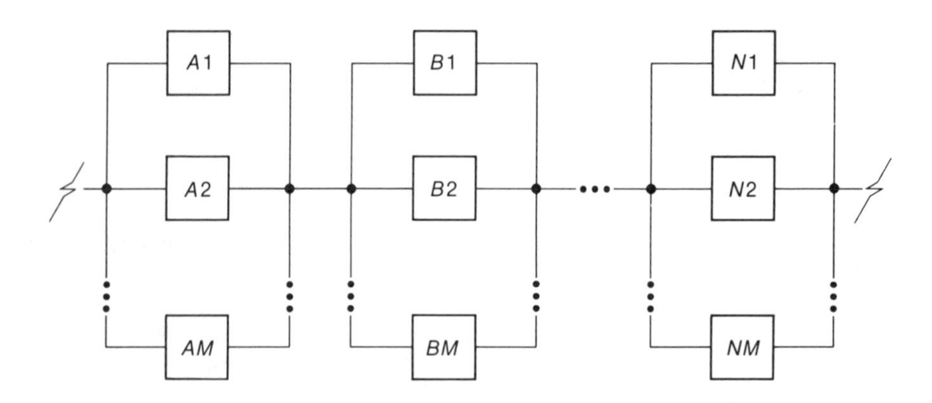

Figure 2.12 Series-Parallel Combination

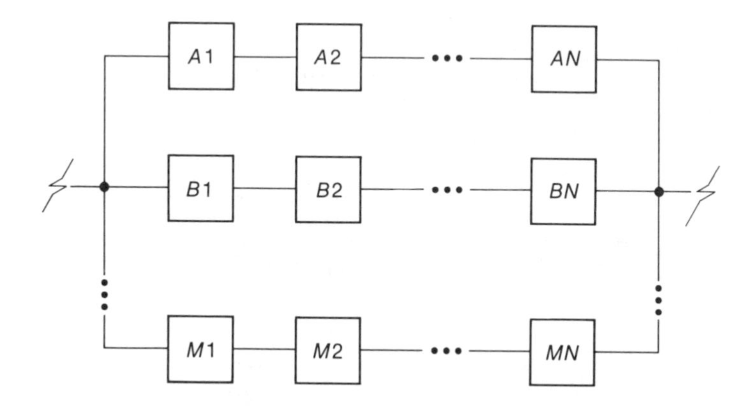

Figure 2.13 Parallel-Series Combination

Series-Parallel Combination. Using (2.16) and (2.17), the availability for a series-parallel combination (Figure 2.12) can be written as

$$A_{NM} = \prod_{i=1}^{N} \left(1 - \prod_{j=1}^{M} U_{ij} \right) \tag{2.18}$$

Parallel-Series Combination. Using (2.16) and (2.17), the availability for a parallel-series combination (Figure 2.13) can be written as

$$A_{MN} = 1 - \prod_{i=1}^{M} \left(\prod_{j=1}^{N} A_{ij} \right) \tag{2.19}$$

Effect of Maintenance and Logistics Practices

Maintenance and logistics practices have a strong impact on system availability because of potential outage extension beyond the MTTR. This outage extension is influenced by:

- *The sparing of modules and equipment*: Without a complete supply of replacement parts, failed equipment may go unrepaired for a time much in excess of the MTTR. Logistics support must maintain a readily available supply of spare parts to avoid jeopardizing system availability.
- *Travel time*: Outages at unmanned sites are largely determined by the time required for a maintenance team to arrive on site. Unavailability of spares at the unmanned site is another potential cause of outage. Moreover, failures at such sites must be recognized and isolated by a fault reporting system before any repair action can be taken.
- *Periodic inspection and test*: Although digital transmission equipment generally provides extensive self-diagnostics, periodic inspection and testing can prevent later outages. Suppose, for example, the off-line unit of a

redundant piece of equipment fails without the system operator's knowledge. Periodic manual switching between redundant units would allow recognition and repair of this failure, thus preventing later outages.

Example 2.3

Based on an end-to-end availability objective of 0.99 for the global reference circuit of Example 2.1, the following unavailability allocations are proposed:

Transcontinental 1	0.004
Satellite 1	0.001
Satellite 2	0.001
Transcontinental 2	0.004
Total unavailability	0.01

Since the two transcontinental circuits are each composed of four identical 300-mi terrestrial reference circuits in tandem, each terrestrial reference circuit has an unavailability allocation of 0.001. To proceed with further suballocation some assumptions must be stated about the terrestrial reference circuit of Example 2.1 and the unavailability allocations shown in Table 2.2:

1. The equipment's MTBO is based on current design practices in digital communications. Redundancy with automatic switchover is employed in radio and cable systems and with higher-level multiplex (level 2 and

Table 2.2 Unavailability Allocation for Hypothetical Reference Circuit Shown in Figure 2.9

Equipment	*Quantity*	*MTSR (hr)*	*MTBO (hr)*	*Unavailability ($\times 10^4$)*
Digital LOS radio	18	2.16	125,000	3.12
Digital fiber optic cable system	6	2.16	125,000	1.04
Digital metallic cable system	6	2.16	50,000	2.60
Second-level multiplex				
Common equipment	10	0.5	200,000	0.25
Channel equipment	10	0.5	220,000	0.27
First-level multiplex				
Common equipment	2	0.5	7,000	1.43
Channel equipment	2	0.5	170,000	0.06
Station power	16	2.06	500,000	0.66
Total				9.43

above) due to the number of circuits that are affected when this equipment fails.

2. The MTSR values shown in Table 2.2 are $\frac{1}{2}$ hr at an attended location and 3 hr at an unattended site. The MTSR values for radio, cable, and station power equipment are averages and result from the ratio of attended locations to unattended locations.
3. All repeater sites are unattended. Sites with multiplex equipment are all attended.

Based on these assumptions, the unavailability contribution from each type of equipment was calculated using Equation (2.12). Unavailabilities were then summed according to the quantity of each piece of equipment and entered in the last column of Table 2.2. The total unavailability is then simply the sum of that column's entries. The resulting value of 0.000943 for this example is less than 0.001, thus meeting the objective for unavailability of this terrestrial reference circuit.

2.4.2 Error Performance

The basic quality measure for a digital transmission system is its error performance. Transmission errors occur with varying statistical characteristics, depending on the media. On digital satellite links, in the absence of forward error correcting, errors tend to occur at random, suggesting that average bit error rate (BER) can be satisfactorily used for error allocation in satellite transmission. For cable and terrestrial radio media, the error performance is characterized by long periods with no or few errors interspersed with short periods of high error rate. For such media, error allocation is best done by the percentage of time spent below a specified error threshold. The choice of the error parameter, be it average BER or percentage of time below threshold, is also determined by the service's requirements (voice, data, or video). One service may be tolerant of randomly distributed errors but intolerant of error bursts whereas another service may perform better under the opposite conditions. Finally, the system designer should also be motivated to select error parameters that are easy to allocate and apply to link design, easy to verify by measurement, and easy for the user to understand and apply to the performance criteria.

Error Parameters

The probability of error in a transmitted bit is a statistical property. The measurement or prediction of errors can be expressed in various statistical ways, but four main parameters have been traditionally used:

- *Bit error rate (or ratio) (BER)*: ratio of errored bits to the total transmitted bits in some measurement interval
- *Error-free seconds (EFS) or error seconds (ES)*: percentage or probability of one-second measurement intervals that are error free (EFS) or in error (ES)

- *Percentage of time* (T_1) *that the BER does not exceed a given threshold value*: percentage of specified measurement intervals (say, 1 min) that do not exceed a given BER threshold (say, 10^{-6})
- *Error-free blocks* (*EFB*): percentage or probability of data blocks that are error free

These parameters all attempt to characterize the same performance characteristic—bit errors—but yield different results for the same link and measurement interval. Before examining specific applications of each parameter, we will establish the relationships among them. First, for statistically independent errors with an average probability of error p = BER, a binomial distribution of errors can be assumed, which leads to the relationship

$$\%\ \text{EFS} = 100(1 - p)^R \tag{2.20}$$

for a bit rate R. This probability may also be expressed in terms of the Poisson distribution when p is small and the number of transmitted bits is large. Then the probability of observing x errors in n transmitted bits is given by

$$P(x) = \frac{e^{-\mu}\mu^x}{x!} \qquad \text{for } p \ll 1, n \gg 1 \tag{2.21}$$

where $\mu = np$.

The probability of no errors is given by

$$P(0) = e^{-\mu} \tag{2.22}$$

For a time period of 1 s, the Poisson distribution may be related to percentage of error-free seconds by

$$\%\ \text{EFS} = 100e^{-pR} \tag{2.23}$$

This relationship is shown in Figure 2.14 for a rate R = 64 kb/s. As an example, a mean probability of error of 1.3×10^{-6} is required to meet an objective of 92 percent error-free seconds for a 64-kb/s channel rate.

The percentage of time T_1 that the BER does not exceed a given threshold value may be related to long-term average error rate, p, by use of the Poisson distribution:

$$\begin{aligned} \%\ T_1(\text{BER} \leq \text{threshold}) &= 100P(x \leq T_0Rp) \\ &= 100 \sum_{x=0}^{T_0Rp} \frac{(T_0Rp)^x}{x!} e^{-(T_0Rp)} \end{aligned} \tag{2.24}$$

where
x = number of errors in time T_0
R = bit rate
T_0 = measurement interval
T_0Rp = error threshold (integer value)

To take an example, assume the BER threshold to be 1×10^{-6} averaged over 1-min intervals for a 64-kb/s bit rate. The error threshold to meet this

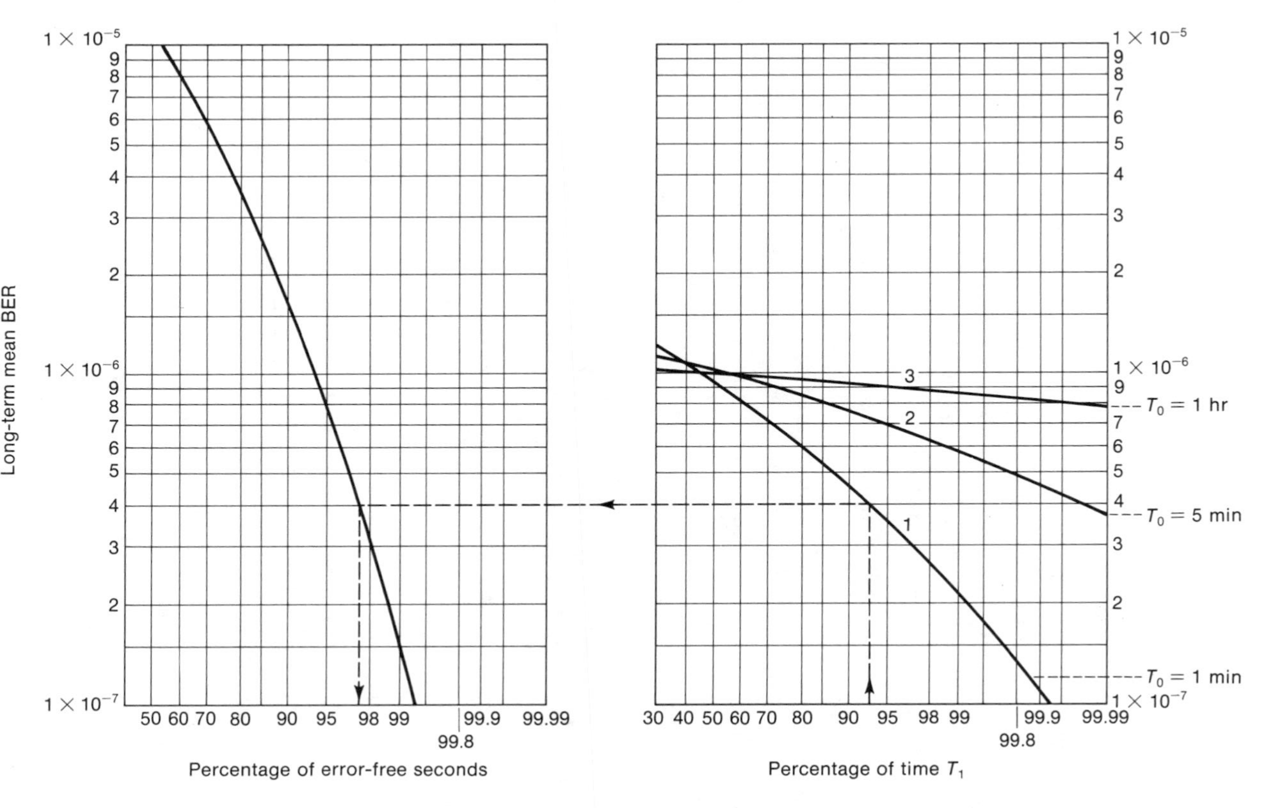

Figure 2.14 Relationship Between Percentage of Error-Free Seconds and Long-Term Mean BER for 64 kb/s [9] (Reprinted by Permission of Bell-Northern Research)

Figure 2.15 Relationship Between Percentage of Time T_1 and Long-Term Mean BER for Mean BER of 1×10^{-6} of Various Averaging Periods T_0 for 64 kb/s [9] (Reprinted by Permission of Bell-Northern Research)

BER averaged over 1-min intervals is $T_0 Rp = (60)(64 \times 10^3)(10^{-6}) = 3.84$; therefore the 64-kb/s connection must have three or fewer errors, where

$$P(x \leq 3) = \sum_{x=0}^{3} \frac{(60Rp)^x}{x!} e^{-(60Rp)} \tag{2.25}$$

This relationship is plotted as curve 1 in Figure 2.15. To illustrate the use of Figure 2.15, consider the following two examples:

1. If 90 percent of 1-min intervals must have a BER better than 10^{-6}, a long-term mean BER of $\leq 4.5 \times 10^{-7}$ would be required.
2. A long-term mean BER of 10^{-6} would yield only 46 percent of the 1-min intervals meeting the 10^{-6} BER threshold.

Note by comparison of Figure 2.15 with Figure 2.14 that the percentage of time spent below BER threshold can be related to the percentage of EFS through a common parameter: long-term average BER.

For statistically independent errors with average probability of error p, the probability of block error is given by the binomial distribution

$$P(\text{block error}) = \sum_{k=1}^{n} \binom{n}{k} p^k (1-p)^{n-k} = 1 - (1-p)^n \tag{2.26}$$

where n = block length

$\binom{n}{k}$ = binomial coefficient $= \dfrac{n!}{k!(n-k)!}$

Conversely, the probability of error-free blocks (EFB) is simply $(1-p)^n$. Expression (2.26) may be approximated for two cases that are commonly assumed:

$$P(\text{block error}) \approx 1 - e^{-np} \qquad \text{for } p \ll 1, n \gg 1 \tag{2.27a}$$

$$P(\text{block error}) \approx np \qquad \text{for } np \ll 1 \tag{2.27b}$$

The relationship of block error probability to bit error probability and error-second probability is illustrated in Figure 2.16 and summarized here:

1. For a block of length 1, the bit error probability is equal to the block error probability.
2. For a block length equal to the data rate, the error-second probability is equal to the block error probability.
3. As the block length increases, the probability of block error also increases, approaching 1 asymptotically.

The error parameters and their interrelationships have been described here under the assumption that errors occur at random. Most transmission systems experience errors in bursts or clusters, however, which invalidates the use of the Poisson (random) error model. To characterize this clustering

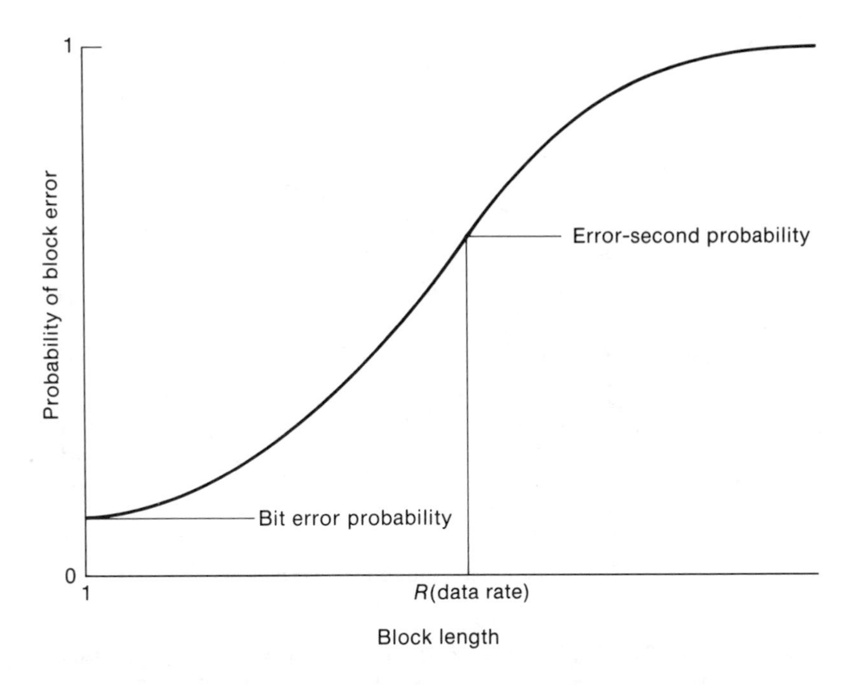

Figure 2.16 Probability of Block Error vs. Block Length

effect, various models for channels with memory have been developed, among them the Neyman Type A contagious distribution [10, 11]. This distribution is a compound Poisson model in which error clusters have a Poisson distribution and errors within a cluster also have a Poisson distribution. Each distribution can be described by its mean:

M_1 = mean number of clusters per sample size of transmitted bits
M_2 = mean number of errors per cluster

The probability of a sample containing exactly r errors is given by

$$P(r) = \frac{M_2^r}{r!} e^{-M_1} \sum_{j=0}^{\infty} \frac{z^j j^r}{j!} \tag{2.28a}$$

where $z = M_1 e^{-M_2}$. The probability of a sample containing no errors is given by

$$P(0) = \exp\{-M_1(1 - e^{-M_2})\} \tag{2.28b}$$

The mean of the distribution is $M_1 M_2$; this is equal to np, where n is the number of bits in the sample and p is the long-term mean bit error probability. The variance of the distribution is $M_1 M_2(1 + M_2)$. For a given error distribution, defined in terms of M_1 and M_2, it is possible to calculate

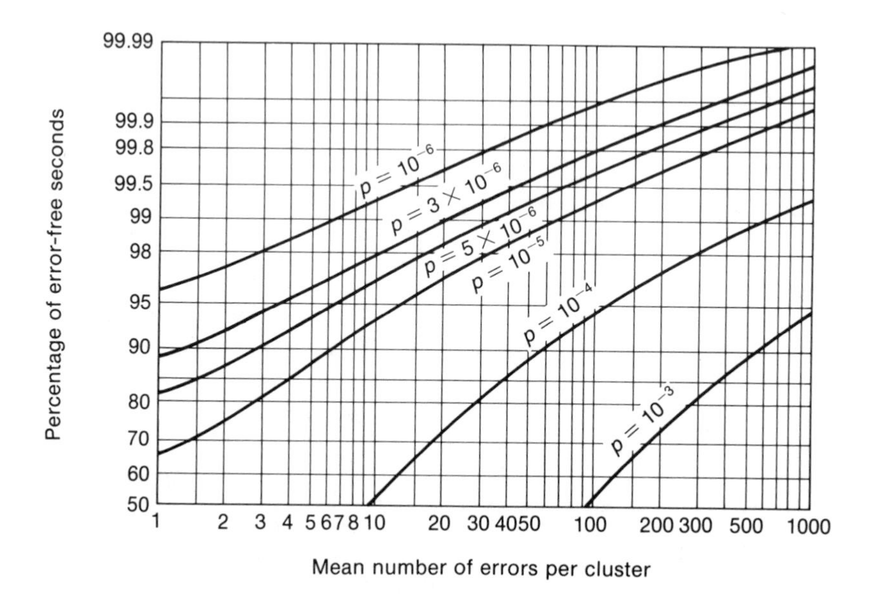

Figure 2.17 The Variation with Degree of Clustering of the Percentage of Seconds of 64-kb/s Transmission That Are Error-Free for Different Values of Long-Term Mean Error Ratio (*p*) [9] (Reprinted by Permission of IT & T)

the percentage of:

- Error-free seconds (Figure 2.17)
- Minutes during which the error rate is better than a specified threshold, say 10^{-5} or 10^{-6} (Figure 2.18)

Figure 2.17 indicates that the random error model represents worst case results for percentage of error-free seconds. As the errors become more clustered, the performance expressed in terms of EFS is significantly better than would be observed for a random error channel. Figure 2.18a also indicates that the random error model represents worst case results for percentage of minutes below a 10^{-6} BER. For the same measurement interval of one minute but a BER threshold of 10^{-5}, Figure 2.18b reveals that error clustering up to a mean value of errors per cluster (M_2) of about 50 actually reduces the percentage of acceptable minutes for a given long-term mean error rate. For M_2 greater than 50, the percentage of acceptable minutes increases monotonically. The difference in the curves of Figure 2.18a versus Figure 2.18b is explained by the maximum acceptable number of errors per minute, which is 38 for a 10^{-5} BER threshold but only three for a 10^{-6} BER threshold.

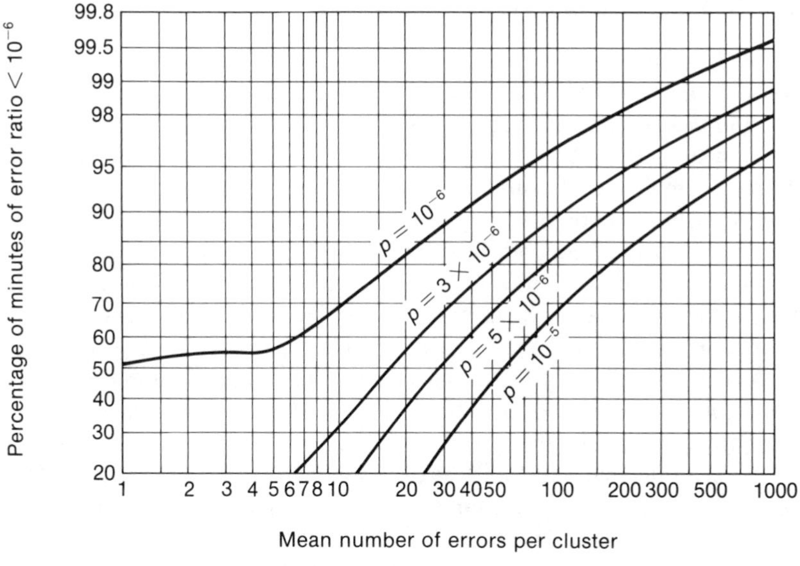

(a) BER threshold = 10^{-6}

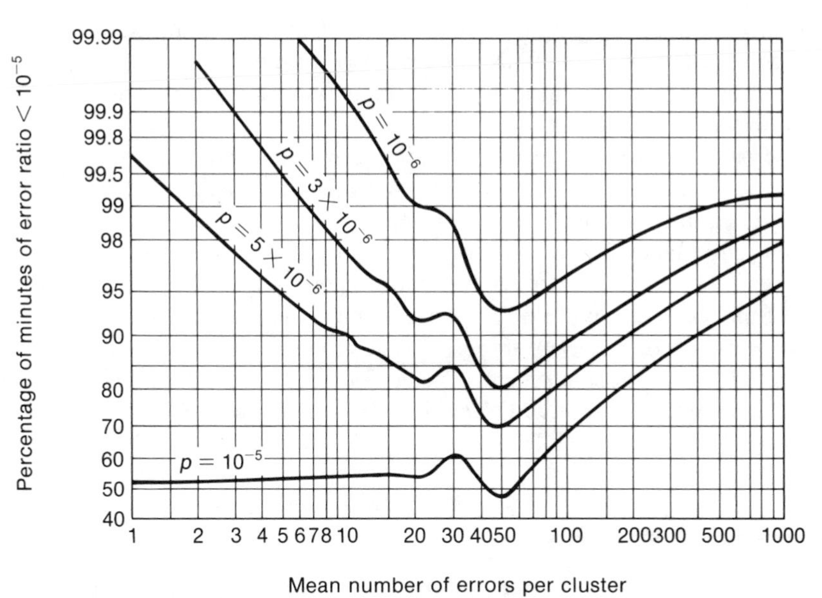

(b) BER threshold = 10^{-5}

Figure 2.18 The Variation with Degree of Clustering of the Percentage of Minutes of 64-kb/s Transmission Having an Error Ratio Less Than Threshold for Different Values of Long-Term Mean Error Ratio (p) [11] (Reprinted by Permission of IT & T)

Service Requirements

As noted earlier, the choice of error parameter is determined largely by the effects of errors on the user, which are different for various services—voice, data, and video. For the voice user, a suitable parameter is the percentage of time in which the BER does not exceed a specified threshold. Based on informal listening tests, subjective evaluations of disturbances in 64-kb/s PCM speech indicate that the total outage time is the most important parameter, not the frequency or duration of outages [1].

For typical data applications, EFS or EFB is better for evaluating the effect of errors. Data are transmitted in block or message form, often accompanied by some form of parity to allow a check on data integrity at the receiver. For this application the number of errors is unimportant, since a single error requires retransmission of the whole block or message. If an error-correcting code is employed in a data application, the effects of single errors are minimized. However, error correction coding can only cope with error bursts of finite length. Hence the data user employing error-correcting codes will be interested in the probability of error bursts exceeding a certain length. In video applications, with the use of differential coding and various image compression techniques, much of the redundancy in video is usually removed. As the transmission rate is reduced, the effects of errors become more pronounced. A single error may persist through many frames. Hence the choice of error characterization by the video user is similar to that of the data user and is influenced by whether or not error-correcting codes are employed.

For transmission systems providing more than one service, say voice, video, and data, the choice of appropriate error parameters for system design requires further comparison. Since voice and video signals contain much redundancy and because of the tolerance to errors by the human eye and ear, the data user often has the greatest requirement for error-free performance. Yet the use of error-correcting codes or retransmissions upon error makes a data system more tolerant of error. Further, with the use of bandwidth compression schemes, voice and video systems may become the most demanding service with regard to error performance.

International Standards

The CCITT has defined error performance objectives using two parameters reflecting the requirements of voice and data services. In practice, it is expected that both parts of the objective should be met simultaneously. These performance objectives are stated for a 64-kb/s connection operating over the CCITT hypothetical reference circuit identified in Figure 2.6*a*. For voice services the parameter is based on an averaging time of 1 min and a BER threshold of 1×10^{-6}. For data services the parameter is based on error-free seconds. These objectives apply to the period when the connection is available. A connection is considered to be unavailable and not included in the error performance objectives when the BER is worse than 1×10^{-3} for periods exceeding n seconds. Values of n in the range 1 to 10 s had been proposed that led to a choice of 10 s. At present no objective has been proposed for the allowed percentage of unavailable time, but a value on the order of 0.5 percent has been mentioned [12]. Table 2.3 summarizes these CCITT error and availability performance objectives. The allocation of performance to segments of the CCITT hypothetical reference circuit has been tentatively proposed as:

- 50 to 60 percent to the international portion together with both national systems
- 40 to 50 percent to the sum of both local portions

Table 2.3 CCITT Performance Objectives for Digital Transmission (Rec. G.821) [13]

Reference Channel

64-kb/s channel traversing a 27,500-km hypothetical reference connection (Rec. G.104)

User Types

1. Voice
2. Data

Availability

A connection is considered to be unavailable when the BER is worse than 1×10^{-3} for periods exceeding 10 seconds.[a]

Error Performance[bc]

BER in 1 Min	*Percentage of Available Minutes*	*BER in 1 Sec*	*Percentage of Available Seconds*
Worse than 10^{-6}	< 10%	Worse than 10^{-3}	< 0.2%
Better than 10^{-6}	> 90%	0	> 92% (% EFS)

[a] The value of 10 s was adopted at the 1984 CCITT plenary assembly.

[b] Measurement time for error rate is unspecified. A period of 1 month is suggested.

[c] These BER thresholds were adopted at the 1984 CCITT plenary assembly.

Although the CCITT has proposed a dual error specification, the goal of using a single method for describing error performance has been recognized. There is also controversy over how to reconcile the conflicts between the two forms. Some contributors have pointed out that circuits meeting the CCITT's 10^{-6} error rate specification may or may not meet its specification on error-free seconds [9, 14]. Since the relationship between these two parameters cannot be stated without some assumptions about the manner in which errors are clustered, CCITT participants have analyzed this problem in terms of both the Poisson distribution and the Neyman Type A contagious distribution. CCITT Study Group XVIII has requested its members to institute measurement programs for digital cable, radio relay, and satellite systems in order to establish their performance characteristics for future standardization work. Recommendations of error performance of higher-bit-rate services, above 64 kb/s, are also under study with emphasis on the need for digital TV performance objectives.

For digital radio systems, the CCIR has defined its error performance objectives using three BER thresholds and measurement averaging times. These objectives are based on a 64-kb/s connection operating over the 2500-km hypothetical reference circuit identified in Figure 2.6*b*. The error performance objectives apply only when the circuit is available. The CCIR error and availability performance objectives are summarized in Table 2.4.

Table 2.4 CCIR Performance Objectives for Digital Radio Systems [7]

Reference Channel

64-kb/s channel traversing a 2500-km hypothetical reference path (Rec. 556)

Bit Error Rate (CCIR Rec. 594 as modified in May 1984)

$\geq 10^{-6}$:	not more than 0.4% of any month averaged over 1-min intervals.
$\geq 10^{-3}$:	not more than 0.054% of any month averaged over 1-s intervals.
> 0:	not more than 0.32% of any month for 1-s intervals.

Availability (CCIR Rec. 557)

The objective for a hypothetical reference digital path is 99.7% with a possible range of 99.5 to 99.9%.

Unavailability (CCIR Rec. 557)

The concept of unavailability of a hypothetical reference digital path should be as follows. In at least one direction of transmission, one or both of the two following conditions occur for at least 10 consecutive seconds:

1. The digital signal is interrupted (alignment or timing is lost).
2. The error rate is greater than 10^{-3}.

Under Study

Burst errors [Rep. 930]

Outages < 10 s [Rec. 557, note 8]

Table 2.5 CCIR Performance Objectives for Digital Transmission on Fixed Satellite Systems [8]

Reference Channel

64-kb/s PCM channel traversing hypothetical reference path (Rec. 521-1).

Bit Error Rate (CCIR Rec. 552-1)

$\geq 10^{-6}$: not more than 20% of any month averaged over 10-min intervals
$\geq 10^{-4}$: not more than 0.3% of any month averaged over 1-min intervals
$\geq 10^{-3}$: not more than 0.01% of any year averaged over 1-s intervals

Availability (CCIR Rep. 706-1)

The provisional objective is $\geq$ 99.8%.

Unavailability (CCIR Rep. 706-1)

The reference channel is considered unavailable if one or both of the two following conditions occur for more than 10 consecutive seconds:

1. The bit error rate exceeds 10^{-3}.
2. The digital signal is interrupted (alignment or timing is lost).

For each of the three error rate thresholds, performance is stated as a percentage of time in which the error rate should not be exceeded. The availability objective is considered provisional with an allowed range of 99.5 to 99.9 percent. It should be noted that this availability figure accounts only for radio-related outages and does not include multiplex equipment. For transmission of PCM telephony via satellite, the CCIR has defined error and availability performance using a similar approach to digital radio objectives. As summarized in Table 2.5, three error rate thresholds have been stated with measurement intervals and allowed percentage of time in which the BER is not exceeded. Only the effects of random errors are considered in this recommendation, but interference, atmospheric absorption, and rain effects are included. Areas under study by the CCIR include burst error characterization and short-term (< 10 s) outages for digital radio and satellite systems.

Error Allocation

Once the end-to-end objectives have been established for the hypothetical reference circuit, the allocation of error performance to individual segments and links can be accomplished. Several methods have been used. The CCITT has proposed a bit error rate objective of 1×10^{-10} per kilometer for the 25,000-km hypothetical reference circuit. This results in an end-to-end objective of 2.5×10^{-6} BER. This allocation procedure is possible only when the hypothetical reference circuit is homogeneous—that is, where each kilometer has the same expected performance. When the hypothetical reference circuit consists of various media and equipment producing different error characteristics, error objectives are best allocated to reference segments and reference links.

To prescribe link performance, a model of error accumulation is needed for n tandem links. If p is the probability of bit error per link, and assuming statistical independence of errors from link to link, then [15]

$$P_n(e) = P[\text{error in } n \text{ links}] = \tfrac{1}{2}\left[1 - (1 - 2p)^n\right] \qquad (2.29)$$

For small enough p, the probability of multiple errors on the same bit is negligible. Then (2.29) simplifies to a sum of individual link error probabilities or

$$P_n(e) \approx np \qquad (np \ll 1) \qquad (2.30)$$

These results can be extended to any of the error parameters discussed earlier: bit error rate, error-free seconds, percentage of time below error threshold, and block error rate. For example, if the percentage EFS and percentage T_1 are known for a particular type of link, then for n such links the percentage of error-free seconds is

$$\%\ \text{EFS}_n \approx 100 - n(100 - \%\ \text{EFS}) \qquad (n)(100 - \%\ \text{EFS}) \ll 1 \qquad (2.31)$$

and the percentage of time where the BER does not exceed threshold is

$$\%\ T_{1n} \approx 100 - n(100 - \%\ T_1) \qquad (n)(100 - \%\ T_1) \ll 1 \qquad (2.32)$$

Again note that Expressions (2.30) to (2.32) assume that multiple errors in the same bit, second, or measurement interval (T_0) do not occur from link to link. For large probability of error or for a large number of links, however, this assumption becomes invalid and forces the system designer to use an exact expression as in (2.29).

This model is easily extended to combinations of links with different error characteristics. Designating link types by numbers $(1, 2, \ldots, m)$, the probability of error expression for a hypothetical reference circuit (hrc) is

$$\begin{aligned} P[\text{error in hrc}] &= \tfrac{1}{2}\left[1 - \prod_{i=1}^{m} (1 - 2p_i)^{n_i}\right] \\ &\approx \sum_{i=1}^{m} n_i p_i \qquad (n_i p_i \ll 1 \text{ for all } i) \end{aligned} \qquad (2.33)$$

This result is directly applicable to other error parameters. For example, if $\% \text{ EFS}_1, \% \text{ EFS}_2, \ldots, \% \text{ EFS}_m$ are the percentages of error-free seconds for each link type, then the percentage of error-free seconds for the hypothetical reference circuit is

$$\begin{aligned} \% \text{ EFS[hrc]} &\approx 100 - \sum_{i=1}^{m} n_i(100 - \% \text{ EFS}_i) \\ (n_i)(100 - \% \text{ EFS}_i) &\ll 1 \text{ for all } i \end{aligned} \tag{2.34}$$

and similarly

$$\begin{aligned} \% \ T_1[\text{hrc}] &\approx 100 - \sum_{i=1}^{m} n_i(100 - \% \ T_{1i}) \\ (n_i)(100 - \% \ T_{1i}) &\ll 1 \text{ for all } i \end{aligned} \tag{2.35}$$

Example 2.4

For the global reference circuit of Example 2.1, let

$$\begin{aligned} \% \text{ ES}_L &= \%[\text{error seconds for LOS radio reference link}] \\ \% \text{ ES}_F &= \%[\text{error seconds for fiber optic cable reference link}] \\ \% \text{ ES}_M &= \%[\text{error seconds for metallic cable reference link}] \\ \% \text{ ES}_S &= \%[\text{error seconds for satellite reference link}] \end{aligned}$$

Assume the relative error performance of these different links to be given by

$$\% \text{ ES}_S = (10)(\% \text{ ES}_M) = (100)(\% \text{ ES}_L) = (100)(\% \text{ ES}_F)$$

(This relationship is somewhat arbitrary; the actual case would be determined by link error statistics, perhaps based on experimental data.) That is, the satellite link is allocated 10 times the error-second percentage of metallic cable and 100 times that of line-of-sight radio and fiber optic cable reference links. Further assume an end-to-end objective of 99 percent EFS. From Example 2.1, the numbers of each medium contained in the hypothetical reference circuit are

$$\begin{aligned} n_S &= \text{number of satellite links} = 2 \\ n_M &= \text{number of metallic cable links} = 24 \\ n_F &= \text{number of fiber optic cable links} = 24 \\ n_L &= \text{number of LOS radio links} = 72 \end{aligned}$$

From Equation (2.34) the following link allocations then result for percentage of EFS:

Fiber optic cable reference link:	99.998%
LOS radio reference link:	99.998%
Metallic cable reference link:	99.98%
Satellite reference link:	99.8%

2.4.3 Bit Count Integrity Performance

Bit count integrity is defined as the preservation of the precise number of bits (or characters or frames) that are originated in a message or unit of time. Given two enumerable bits in a transmitted bit stream that are separated by n bits, bit count integrity is maintained when the same two bits are separated by n intervals in the received bit stream. Losses of bit count integrity (BCI) cause a short-term outage that may be extended if equipment resynchronization (such as multiplex reframe) is required.

When a digital signal is shifted (or slipped) by some number of bits without loss of alignment, the resulting loss of BCI is known as a **slip**. Slips may be controlled or uncontrolled. If controlled, the slip is limited to repetition or deletion of a single bit, single character, or single frame, which then limits the effect on the user. As an example, for 64-kb/s PCM the slippage can be controlled to 8-bit characters (octets) in order to maintain character synchronization. For multichannel PCM, the slippage must be confined to whole frames in order to avoid loss of frame alignment. The effect of such controlled slips on PCM speech is very slight, but it can be significant for other signal types such as secure (encrypted) voice, digital data, and high-speed data via modems on telephone channels. Table 2.6 presents slip rate objectives for various services that were submitted by AT & T to the CCITT. For PCM voice, a rate of 300 slips per hour would result in an audible "click" every 5 min since only one slip in 25 will be audible. In the case of secure voice, each slip is audible since the encryption device will lose alignment and must be resynchronized. To meet an objective of one audible click every 5 min, the allowed slip rate is 12 slips per hour for secure voice. For data services, an objective of 0.1 percent time lost results in slip rates on the order of a few slips per hour. Table 2.7 gives the CCITT controlled slip rate objectives for the 25,000-km 64-kb/s international connection. These objectives apply to both telephone and nontelephone services at 64 kb/s.

Uncontrolled slips or high error rates can cause loss of frame alignment in multiplexing equipment. For these cases, the outage caused by loss of bit count integrity is extended by the time required to regain frame synchroni-

Table 2.6 AT & T Suggested Slip Rates for Various Services [16]

Service	*Objectives*	*Slips / hr*
PCM voice	1 audible click / 5 min	300
Secure voice	1 audible click / 5 min	12
Digital data on 64-kb/s carrier	0.1% time lost	Fixed block length: 6 Variable block length: 0.6
Voice-band data	0.1% time lost	Fixed block length: 7.2 Variable block length: 0.6

zation. The effect on the user is much more evident than in the case of a controlled slip. In most PCM multiplexers, for example, speech outputs are suppressed during periods of frame loss. Frame resynchronization times are typically several milliseconds for PCM multiplex, thus affecting on the order of 100 frames. The CCITT has included a specification on the percentage of short-term outages for BER greater than 10^{-3} as a safeguard against excessive short interruptions such as loss of frame alignment. A 1-s measurement time was selected to distinguish between an unacceptable performance

Table 2.7 CCITT Performance Objectives for Slip Rate on 64-kb/s International Connections (Rec. G.822) [17]

Performance Classification	*Mean Slip Rate Thresholds*	*Measurement Averaging Period*	*Network Objectives as Percentage of Total Time*
Unacceptable	$\geq$ 1 slip in 2 min	1 hr	< 0.1%
Degraded	< 1 slip in 2 min and > 1 slip in 5 hr	24 hr	< 1.0%
Acceptable	$\leq$ 1 slip in 5 hr	24 hr	$\geq$ 99%

Notes:
1. These objectives are primarily addressed to unencrypted transmission.
2. Rec. G.822 notes that further study is required to confirm that these values are compatible with other objectives such as the error performance objectives.

(≥ 1 s) and a period of unavailability (≥ 10 s), both of which have the same BER threshold but different measurement periods. As shown in Table 2.3, the specification within the CCITT is that less than 0.2 percent of available seconds should have a BER worse than 10^{-3} for the 64-kb/s hypothetical reference circuit [13].

After determining the end-to-end objectives for bit count integrity, allocations are established for the segments and links of the hypothetical reference circuit. First we must develop the model for loss of BCI in n tandem links. Since this parameter is probabilistic, we can associate a probability q that each received bit (or character or frame) leads to a loss of BCI. Assuming statistical independence of the parameter q from link to link, then for n links each with the same q,

$$P_n(\text{LBCI}) = P[\text{loss of BCI in } n \text{ links}] = \tfrac{1}{2}\left[1 - (1 - 2q)^n\right] \tag{2.36}$$

For small enough q and n, Equation (2.36) reduces to

$$P_n(\text{LBCI}) \approx nq \qquad (nq \ll 1) \tag{2.37}$$

This model is easily extended to combinations of links with different BCI characteristics. For link types designated $(1, 2, \ldots, m)$ in a hypothetical reference circuit,

$$P_{\text{hrc}}(\text{LBCI}) = \tfrac{1}{2}\left[1 - \prod_{i=1}^{m} (1 - 2q_i)^{n_i}\right] \tag{2.38}$$

which simplifies to

$$P_{\text{hrc}}(\text{LBCI}) \approx \sum_{i=1}^{m} n_i q_i \qquad (n_i q_i \ll 1 \text{ for all } i) \tag{2.39}$$

With the assumption of statistically independent losses of BCI, the number of losses of BCI has a binomial distribution. Therefore, the expected number of losses of BCI per unit time is

$$E[\text{LBCI}] = qR \tag{2.40}$$

where R represents bit, character, or frame rate, as determined by the definition of q. The mean time to loss of BCI, $\bar{X}$, is then the reciprocal of (2.40), or

$$\bar{X} = \frac{1}{qR} \tag{2.41}$$

The rate of BCI losses given in Equation (2.40) and mean time to loss of

BCI given in (2.41) are the two parameters commonly used in allocating BCI performance. When they are applied to the hypothetical reference circuit, we obtain

$$\begin{aligned} \mathrm{E}[\mathrm{LBCI}_{\mathrm{hrc}}] &= \frac{R}{2}\left[1 - \prod_{i=1}^{m}(1 - 2q_i)^{n_i}\right] \\ &\approx R\sum_{i=1}^{m} n_i q_i \qquad (n_i q_i \ll 1 \text{ for all } i) \end{aligned} \tag{2.42}$$

and

$$\overline{X}_{\mathrm{hrc}} = \frac{1}{E[\mathrm{LBCI}_{\mathrm{hrc}}]} \tag{2.43}$$

A more convenient form of (2.43) is given by

$$\frac{1}{\overline{X}_{\mathrm{hrc}}} \approx R\sum_{i=1}^{m} n_i q_i = \sum_{i=1}^{m} \frac{n_i}{\overline{X}_i} \qquad (n_i q_i \ll 1 \text{ for all } i) \tag{2.44}$$

where $\overline{X}_i = (q_i R)^{-1}$.

From link allocations of BCI, performance requirements can be determined for those functions that can cause loss of BCI, such as:

- Bit synchronization in a timing recovery loop
- Buffer lengths and clock accuracy required for network synchronization
- Frame and pulse stuffing synchronization in a digital multiplexer
- Protection switching in redundant equipment

Example 2.5

For the global reference circuit of Example 2.1, let

$\overline{X}_{\mathrm{L}}$ = mean time to loss of BCI on LOS radio reference link
$\overline{X}_{\mathrm{F}}$ = mean time to loss of BCI on fiber optic cable reference link
$\overline{X}_{\mathrm{M}}$ = mean time to loss of BCI on metallic cable reference link
$\overline{X}_{\mathrm{S}}$ = mean time to loss of BCI on satellite reference link

Assume the relative BCI performance of these different links to be given by

$$\overline{X}_{\mathrm{L}} = \overline{X}_{\mathrm{F}} = 10\overline{X}_{\mathrm{M}} = 100\overline{X}_{\mathrm{S}}$$

That is, the line-of-sight radio and fiber optic cable links are allocated a mean time to loss of BCI that is 10 times that of metallic cable and 100 times that of satellite reference links. Further assume an end-to-end objective of 1 hr mean time to loss of BCI. From Example 2.1, the numbers of each medium contained in the hypothetical reference circuit are: $n_{\mathrm{S}} = 2$, $n_{\mathrm{M}} = 24$, $n_{\mathrm{F}} = 24$, and $n_{\mathrm{L}} = 72$. The following link allocations of mean time to loss of BCI result from application of Equation (2.44):

Fiber optic cable reference link:	536 hr
LOS radio reference link:	536 hr
Metallic cable reference link:	53.6 hr
Satellite reference link:	5.36 hr

2.4.4 Jitter Performance

Jitter is defined as a short-term variation of the sampling instant from its intended position in time or phase. Longer-term variation of the sampling instant is sometimes called **wander** or **drift**. Jitter causes transmission impairment in three ways:

1. Displacement of the ideal sampling instant leads to a reduction in the noise margin and a degradation in system error rate performance.
2. Slips in timing recovery circuits occur when excessive jitter causes a loss of phase lock or cycle slip.
3. Irregular spacing of the decoded samples of digitally encoded analog signals introduces distortion in the recovered analog signal.

Wander causes buffers used for frame or clock alignment to fill or empty, resulting in a repetition or deletion (slip) of a frame or data bit.

Contributions to *jitter* arise primarily from two sources: regenerative repeaters and digital multiplexers. These two sources of jitter are briefly discussed here and described in more detail in Chapters 4 and 8. Regenerative repeaters must derive a train of regularly spaced clock pulses in order to properly sample the digital signal. Jitter is introduced in the clock recovery process due to the following:

- Mistuning of a resonant circuit causes accumulation of phase error, especially during the absence of data transitions when the circuit drifts toward its natural frequency.
- Misalignment of the threshold detector from ideal (zero amplitude) causes variation in the clock triggering time for any amplitude variation of the signal.
- Imperfect equalization results in **intersymbol interference** in which the pulses may be skewed, shifting the signal from the ideal transition time and causing a phase shift in the clock triggering time.
- Certain repetitive data patterns produce a characteristic phase shift in the recovered clock. A change in pattern can cause a relative change in clock timing resulting in a form of timing jitter.

Jitter associated with digital multiplexing operation stems from the process of buffering input signals prior to multiplexing and the inverse process for demultiplexing. The insertion of overhead bits for synchronization causes phase differences between the input message traffic and the composite transmitted bit stream. Jitter is created when deriving the timing signal for the message traffic at the receiving demultiplexer. When pulse stuffing is used in a multiplexer, an additional form of jitter called **stuffing jitter** can arise. This jitter results from the periodic insertion of stuff bits in the multiplexer and their removal at the demultiplexer.

Contributions to *wander* arise primarily from two sources: oscillator instability and propagation delay variation. These two effects are important considerations in the design of network synchronization and are described in detail in Chapter 10. The magnitude of wander determines the buffer sizes required at each node in a network.

The specification of jitter must indicate the upper limits of allowed output jitter and tolerable input jitter. For amounts exceeding the jitter specification, bit errors or loss of BCI would be expected. The standard parameter used in jitter specifications is *maximum peak-to-peak*, since this value has the greatest effect on transmission performance. Because jitter amplitude is a function of the measurement bandwidth, it is necessary to specify both a frequency and amplitude of jitter. An example of such a specification is the mask of tolerable input jitter and wander shown in Figure 2.19, which has been proposed by the CCITT for national networks or international links [18]. The CCITT wander specification, given by amplitude A_0 and frequency f_0, accounts for effects due to transmission facilities (cable and repeaters) and network synchronization. Jitter at each hierarchical level is based on the CCITT G.700 series recommendations for multiplex, which specify amplitudes A_1 and A_2 and frequency f_1 to f_4. Figure 2.19 thus provides the envelope of lower limits on the maximum tolerable jitter for any combination of equipment, and it imposes a limit on the jitter generated and accumulated in a network.

The allocation of jitter performance in a reference circuit requires a model of the jitter characteristics of the components in the circuit and a model of jitter accumulation from component to component. A complete jitter characterization of components in the reference circuit can be made with knowledge of:

- Tolerable input jitter
- Output jitter in the absence of input jitter
- Jitter transfer function, defined as the ratio of component output jitter to applied input jitter

Jitter accumulation depends on the structure of the reference path and the sources of jitter considered. Certain jitter sources are systematic and lead to accumulation in a predetermined manner. Sources of jitter associated with repeatered cable systems tend to be systematic and therefore accumulate with a known characteristic. Multiplexers tend to contribute nonsystematic jitter and work as jitter reducers. Therefore a reference circuit that has frequent multiplex points tends to have less jitter accumulation than the case for no multiplex points. Without the jitter reducing effect of demultiplexers, means of reducing jitter (called *dejitterizers*) may be neces-

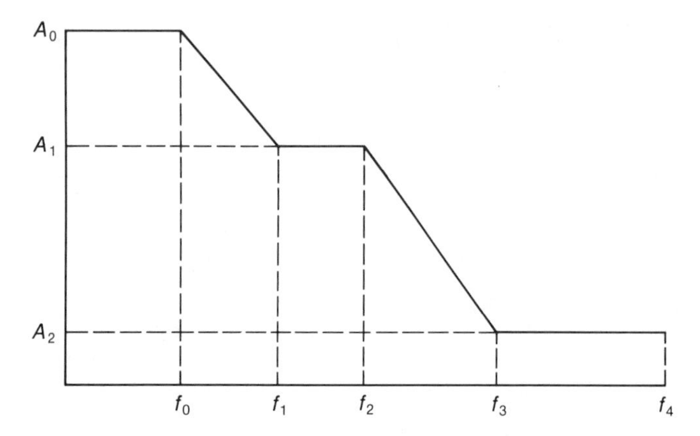

(a) Mask of tolerable sinusoidal jitter and wander

	2048 kb/s	8448 kb/s	1544 kb/s
A_0 (μs)	17.2	17.2	17.2
A_1 (UI)	1.5	1.5	2.0
A_2 (UI)	0.2	0.2	0.05
f_0 (Hz)	3×10^{-8}	3×10^{-8}	3×10^{-8}
f_1 (Hz)	20	20	10
f_2 (Hz)	2.4×10^3	400	200
f_3 (Hz)	18×10^3	3×10^3	8×10^3
f_4 (Hz)	100×10^3	400×10^3	40×10^3

Note: UI = unit interval. For 1544-kb/s systems, 1 UI = 648 ns; for 2048-kb/s systems, 1 UI = 488 ns; for 8448-kb/s systems, 1 UI = 118 ns.

(b) Values for the mask of tolerable jitter and wander

Figure 2.19 CCITT Proposed Draft Recommendation G.8YY for Tolerable Jitter and Wander in a Digital Network [18] (Reprinted by Permission of Nippon Telegraph and Telephone)

sary to meet reference circuit performance objectives. The mask of Figure 2.19*a* is an example of the maximum jitter that may accumulate in a reference circuit and still be tolerated by the end equipment.

2.4.5 Delay

Absolute delay in transmission affects any interactive service, such as voice and certain types of data circuits. Delay in voice transmission imparts an unnatural quality to speech, while data circuits using some form of automatic response are affected if the transmission delay exceeds the allowed response time. Sources of transmission delay are propagation time and equipment processing. Propagation delays are a function of distance and are independent of the bit rate. Equipment delays are due to buffering—as used in digital processing (multiplexing; error correction coding; packet, message, or circuit switching) and in network synchronization—and are inversely proportional to bit rate. The total delay is therefore a complex function of the geographical, media, and equipment configuration. Delay in a satellite link is by far the largest contributor to propagation time, however; delays in terrestrial links usually can be considered negligible. Excessive delay can be avoided by eliminating or minimizing the use of satellite links or by using commercially available delay compensation units designed for interactive data circuits.

The specification of allowed end-to-end delay is determined by user tolerance. The allocation of delay to transmission segments and links of a reference circuit then depends on the media (satellite versus terrestrial), multiplex hierarchy, use of error control techniques, type of network synchronization, and type of switching. Although the CCITT has not yet addressed delay allocations for the hypothetical reference circuits, delay in circuit-switching and packet-switching networks is the subject of current Q series and draft X series recommendations [12].

2.5 SUMMARY

Digital transmission system planning is based on the following steps:

1. Identify services and their performance requirements.
2. Formulate a hypothetical reference circuit.
3. Specify performance objectives for the end-to-end circuit.
4. Allocate performance objectives to individual segments, links, and equipment.

The performance requirements of various services are a complex function involving human perception for voice and video and transmission efficiency and response time for data. Once these requirements are established, the system designer can assign performance objectives to end-to-end service. Hypothetical reference circuits are a valuable tool used by the system designer in allocating performance to representative portions of the circuit. These circuits can be constructed with different media and equipment in order to reflect an appropriate combination for the transmission system being designed.

The performance parameters used in communication systems can be divided into two areas: availability and quality. Availability is defined as the probability or percentage of time that a connection is usable. A typical criterion for digital transmission is to define a circuit as unavailable if the bit error rate exceeds a specified threshold for some stated period of time. This definition distinguishes short-term interruptions that affect quality from long-term interruptions that are periods of unusable time. For periods of available service, various parameters are used to indicate the quality of service. Although these parameters and their value depend on the type of service, four quality parameters play a dominant role in digital transmission: error rate, bit count integrity, jitter, and delay.

Error rate can be stated in several different ways, such as average bit error rate (BER), percentage of error-free seconds, percentage of time the BER does not exceed a given threshold value, and percentage of error-free blocks. Each of these methods has application to certain types of service or media, although different error performance readings may result under the same test conditions, depending on the distribution of bit errors. Standard error performance criteria and allocation techniques are proposed for various media and services.

Bit count integrity (BCI) is defined as the preservation of the precise number of bits that are originated in a message or unit of time. Losses of BCI are due to loss of bit synchronization, buffer slips, loss of multiplex synchronization, and protection switching in redundant equipment. The result of BCI losses is a short interruption due to deletion or repetition of a bit, character, or frame, and in some cases this interruption is extended due to the time required to resynchronize equipment. The effect on a voice user is minimal, but data and secure (encrypted) voice users suffer more serious effects due to retransmission and resynchronization. BCI performance is most commonly expressed as the rate of BCI losses or the time between losses of BCI, either of which can be stated for end-to-end connections and individual links in a reference circuit.

Jitter is defined as a short-term variation of the sampling instant from its intended position. The effects of jitter include degradation in error rate performance, slips in timing recovery circuits, and distortion in analog signals recovered from digital (PCM) representations. The principal sources of jitter are regenerative repeaters and digital multiplexers. Long-term variation of the sampling instant, called wander or drift, is caused by oscillator instability and propagation delay variations. The specification of jitter should include allowed output jitter and tolerable input jitter. From knowledge of the jitter transfer function for components of a reference circuit, a jitter specification can be developed for the end-to-end connection and for individual equipment.

Absolute delays in transmission affect all interactive service, but primarily the data user. Since most interactive data transmission systems require automatic responses within a specified time, excessive delays will affect such data users. Delay is due to equipment processing and propagation time. The most significant contributor is the delay inherent in satellite transmission. Allocation of delay performance thus depends on the media and equipment of the reference circuit.

REFERENCES

1. J. Gruber, R. Vickers, and D. Cuddy, "Impact of Errors and Slips on Voice Service," *1982 International Conference on Communications*, pp. 2D.5.1–2D.5.7.
2. H. C. Folts (ed.), *McGraw-Hill's Compilation of Data Communications Standards*, 2nd ed. (New York: McGraw-Hill, 1982).
3. J. O. Limb, C. B. Rubinstein, and J. E. Thompson, "Digital Coding of Color Video Signals—A Review," *IEEE Trans. on Comm.*, vol. COM-25, no. 11, November 1977, pp. 1349–1385.
4. CCIR XVth Plenary Assembly, vol. XI, pt. 1, *Broadcasting Service (Television)*, "Section IIF: Digital Methods of Transmitting Television Information" (Geneva: ITU, 1982).
5. CCIR XVth Plenary Assembly, vol. XII, *Transmission of Sound Broadcasting and Television Signals over Long Distances (CMTT)*, "Section CMTT A—Television Transmission Standards and Performance Objectives" (Geneva: ITU, 1982).
6. CCITT Yellow Book, vol. III.1, *General Characteristics of International Telephone Connections and Circuits* (Geneva: ITU, 1981).
7. CCIR XVth Plenary Assembly, vol. IX, pt. 1, *Fixed Service Using Radio-Relay Systems* (Geneva: ITU, 1982).
8. CCIR XVth Plenary Assembly, vol. IV, pt. 1, *Fixed-Satellite Service* (Geneva: ITU, 1982).
9. Bell Northern Research, *Considerations on the Relationship Between Mean Bit Error Ratio, Averaging Periods, Percentage of Time and Percent Error-Free Seconds*, COM XVIII no. 1-E (Geneva: CCITT, 1981), pp. 73–75.
10. J. Neyman, "On a New Class of Contagious Distribution, Applicable in Entomology and Bacteriology," *Ann. Math. Statist.* 10(35)(1939).
11. International Telephone and Telegraph Corp., *Error Performance Objectives for Integrated Services Digital Network (ISDN)*, COM XVIII, no. 1-E (Geneva: CCITT, 1981), pp. 89–95.
12. M. Decina and U. deJulio, "Performance of Integrated Digital Networks: International Standards," *1982 International Conference on Communications*, pp. 2D.1.1–2D.1.6.
13. *Error Performance of an International Digital Connection Forming Part of an Integrated Services Digital Network*, COM XVIII, no. 95-E (Geneva: CCITT, 1984), pp. 155–163.
14. American Telephone and Telegraph Company, *Relation Between Error Measures*, COM XVIII, no. 1-E (Geneva: CCITT, 1981), p. 95.
15. H. D. Goldman and R. C. Sommer, "An Analysis of Cascaded Binary Communication Links," *IRE Trans. on Comm. Systems*, vol. CS-10, no. 3, September 1962, pp. 291–299.
16. American Telephone and Telegraph Company, "Effects of Synchronization Slips," CCITT Contribution COM SpD-TD, no. 32, Geneva, November 1969.
17. CCITT Yellow Book, vol. III.3, *Digital Networks—Transmission Systems and Multiplexing Equipment* (Geneva: ITU, 1981).
18. Nippon Telegraph and Telephone Public Corp., *Jitter and Wander in a Digital Network*, COM XVIII, no. 1-E (Geneva: CCITT, 1981), pp. 49–56.

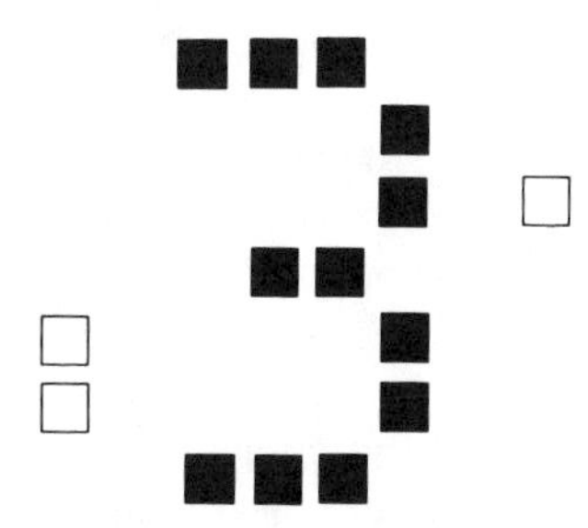

Analog-to-Digital Conversion Techniques

OBJECTIVES

- Explains the overall design, performance, and applications of analog-to-digital conversion techniques
- Discusses sampling, quantizing, and coding—the basis for pulse code modulation
- Covers linear and logarithmic PCM, with emphasis on international standards for voice transmission
- Describes techniques such as differential PCM and delta modulation that apply to highly correlated signals (speech or video)
- Compares the performance of pulse code modulation, delta modulation, and differential PCM—including the effects of transmission error rate and tandeming (multiple A/D coding)
- Describes the basic principles of vocoding techniques
- Compares speech coding techniques with respect to transmission bit rate, quality, and complexity (cost)

3.1 INTRODUCTION

Most communication systems are dominated by signals that originate in analog form, such as voice, music, or video. For such analog sources, the initial process in a digital transmission system is the conversion of the analog source to a digital signal. Numerous analog-to-digital (A/D) conversion techniques have been developed. Some have widespread use and are the subject of standards, such as pulse code modulation in voice networks, while others have very specialized and limited use, such as vocoders for secure voice. The type of A/D converter selected by the system designer thus depends on the application and the required level of performance.

Performance of A/D converters can be characterized by both objective and subjective means. The most significant parameter of objective assessment is the **signal-to-distortion (S/D) ratio.** Typical sources of distortion are quantization distortion and slope or amplitude overload distortion, both of which can be characterized mathematically in a straightforward manner. Subjective evaluation is more difficult to quantify because it involves human hearing or sight. In speech transmission, several tests have been devised for quantifying speech quality. They require listener juries who judge such factors as speech quality, word intelligibility, and speaker recognition.

Analog-to-digital coders can be divided into two classes: waveform coders and source coders [1]. Waveform coders are designed to reproduce the input signal waveform without regard to the statistics of the input. In speech coding, for example, pulse code modulation can provide adequate performance not only for speech but also for nonspeech signals such as signaling tones or voice-band data, which may appear on 3-kHz telephone circuits. Moreover, waveform coders provide uniform performance over a wide range of input signal level (dynamic range) and tolerate various sources of degradation (robustness). These advantages are offset somewhat by poor economies in transmission bit rates. The second class of A/D coders makes use of a priori knowledge about the source. If one can use certain physical constraints to eliminate source redundancy, then improved transmission efficiency can be realized. Human speech, for example, is known to have an information content of a few tens of hertz but requires a transmission bandwidth of a few thousand hertz. Source coders for speech, known as vocoders, are nearly able to match the transmission bandwidth to the information bandwidth. Vocoders tend to lack robustness, however, and produce unnatural sounding speech.

3.2 PULSE CODE MODULATION

Pulse code modulation (PCM) converts an analog signal to digital format by three separate processes: sampling, quantizing, and coding. The analog signal is first sampled to obtain an instantaneous value of signal amplitude at regularly spaced intervals; the sample frequency is determined by the Nyquist sampling theorem. Each sampled amplitude is then approximated by the nearest level from a discrete set of quantization levels. The coding process converts the selected quantization level into a binary code. If 256 ($= 2^8$) quantization levels are used, for example, then an 8-bit binary code is required to represent each amplitude sample.

The following step-by-step description of PCM is based on the illustration in Figure 3.1:

1. An analog input signal $s(t)$ is band-limited by a low-pass filter to F hertz.
2. The band-limited signal is sampled at a rate f_s that must equal or exceed the Nyquist frequency ($f_s \geq 2F$).
3. The sampled signal $s(iT)$ is held in a sample-and-hold circuit between two sampling instants (T seconds).
4. During this interval the sample is quantized into one of N levels. Quantization thus converts the sample into discrete amplitudes and in the process produces an error equal to the difference between input and quantized output. The greater the number of levels in the quantizer, the smaller the quantizing error introduced by the quantizer.

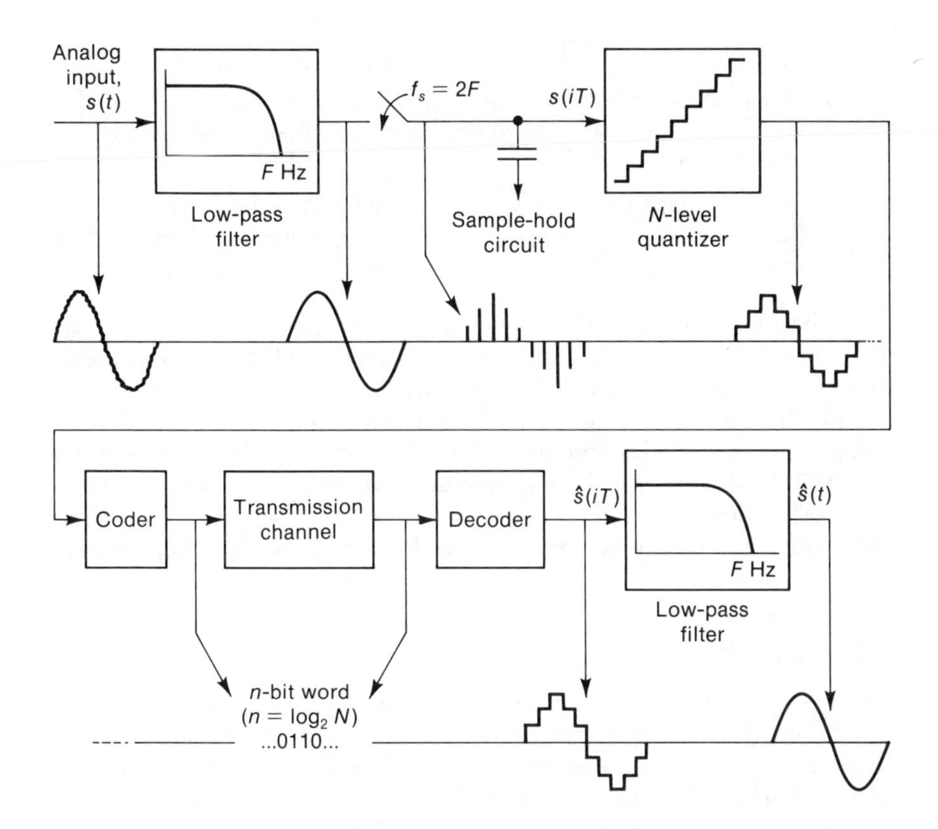

Figure 3.1 Block Diagram of Pulse Code Modulation

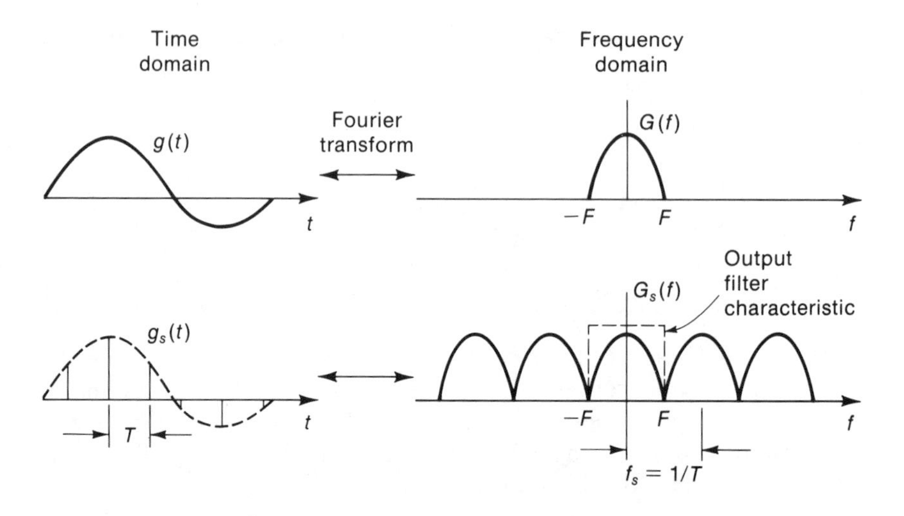

Figure 3.2 Illustration of Sampling Theorem

5. The coder maps the amplitude level selected by the quantizer into an $n(n = \log_2 N)$ bit code designated $b_1, b_2, \ldots, b_n$, where $b_n = 1$ or 0. The format of the code is selected to facilitate transmission over a communication channel.
6. The decoder maps the PCM words back into amplitude levels, and the amplitude samples are low-pass filtered by a filter having a bandwidth of F hertz, which results in an estimate $\hat{s}(t)$ of the original sample.

The importance of PCM in digital transmission is based on its high level of performance for a variety of applications and its universal acceptance as a standard, especially for voice digitization. Pulse code modulation at 64 kb/s is an international standard for digital voice, based on a sampling rate of 8 kHz and an 8-bit code per sample. Two standards for 64-kb/s PCM voice have been developed, and these will be examined and compared. First, however, we will consider the processes of sampling, quantizing, and coding in more detail, with emphasis on voice applications.

3.2.1 Sampling

The basis for PCM begins with the sampling theorem, which states that a band-limited signal can be represented by samples taken at a rate f_s that is at least twice the highest frequency f_m in the message signal. An illustration of the sampling theorem is shown in Figure 3.2. A summarized proof of the sampling theorem is presented here, followed by a discussion of the application of sampling to a practical communication system.

Proof of Sampling Theorem [2]

Consider a band-limited signal $g(t)$ that has no spectral components above F hertz. Suppose this signal is sampled at regular intervals T with an ideal impulse function $\delta_T(t)$ having infinitesimal width. The sampled signal $g_s(t)$ may be written as a sequence of impulses multiplied by the original signal $g(t)$:

$$g_s(t) = \sum_{k=-\infty}^{\infty} g(t)\,\delta_T(t - kT) \tag{3.1}$$

Applying the Fourier transform to both sides of (3.1) we obtain

$$G_s(f) = G(f) * \delta_{f_s}(f) \tag{3.2}$$

where * denotes the convolutional integral, $\delta_{f_s}(f)$ is a sequence of impulse functions separated by f_s hertz, and $f_s = 1/T$. Carrying out the convolution indicated in (3.2) we obtain

$$G_s(f) = \sum_{k=-\infty}^{\infty} G(f - \frac{k}{T}) \tag{3.3}$$

The sampled spectrum is thus a series of spectra of the original signal separated by frequency intervals of $f_s = 1/T$. Note that $G(f)$ repeats periodically without overlap as long as

$$f_s \geq 2F$$

or equivalently

$$F \leq \frac{f_s}{2}$$

Therefore, as long as $g(t)$ is sampled at regular intervals $T \leq 1/2F$, the spectrum of $g_s(t)$ will be a periodic replica of $G(f)$ and will contain all the information of $g(t)$. We can recover the original signal $G(f)$ by means of a low-pass filter with a cutoff at F hertz. The ideal filter characteristic required to recover $g(t)$ from $g_s(t)$ is shown in Figure 3.2.

Proof of the sampling theorem is contingent upon the assumptions that the input signal is band-limited to F hertz, that the samples are taken with impulses of infinitesimal width, and that the low-pass filter used to recover the signal is ideal. In practice, these assumptions do not hold and therefore certain types of error are introduced. Because an ideal (rectangular) low-pass filter characteristic is not feasible, the spectrum of the "band-limited" analog signal contains frequency components beyond $f_s/2$ as shown in Figure 3.3*a*. The resulting adjacent spectra after sampling will overlap, causing an error in the reconstructed signal, as illustrated in Figure 3.3*c*. This error is called **aliasing** or **foldover distortion** (Figure 3.3*d*). The filtered spectrum is distorted due to the addition of tails from higher harmonics and the loss of tails for $|f| > f_s/2$. Design limitations on filter performance make it necessary to leave a guard band near the half sampling rate to minimize aliasing effects. For example, sampling a voice signal at 8 kHz provides a usable bandwidth of approximately 3.4 kHz. In this case the original signal can be recovered at the receiver by use of a low-pass filter with a gradual rolloff characteristic and a 3-dB cutoff around 3.4 kHz.

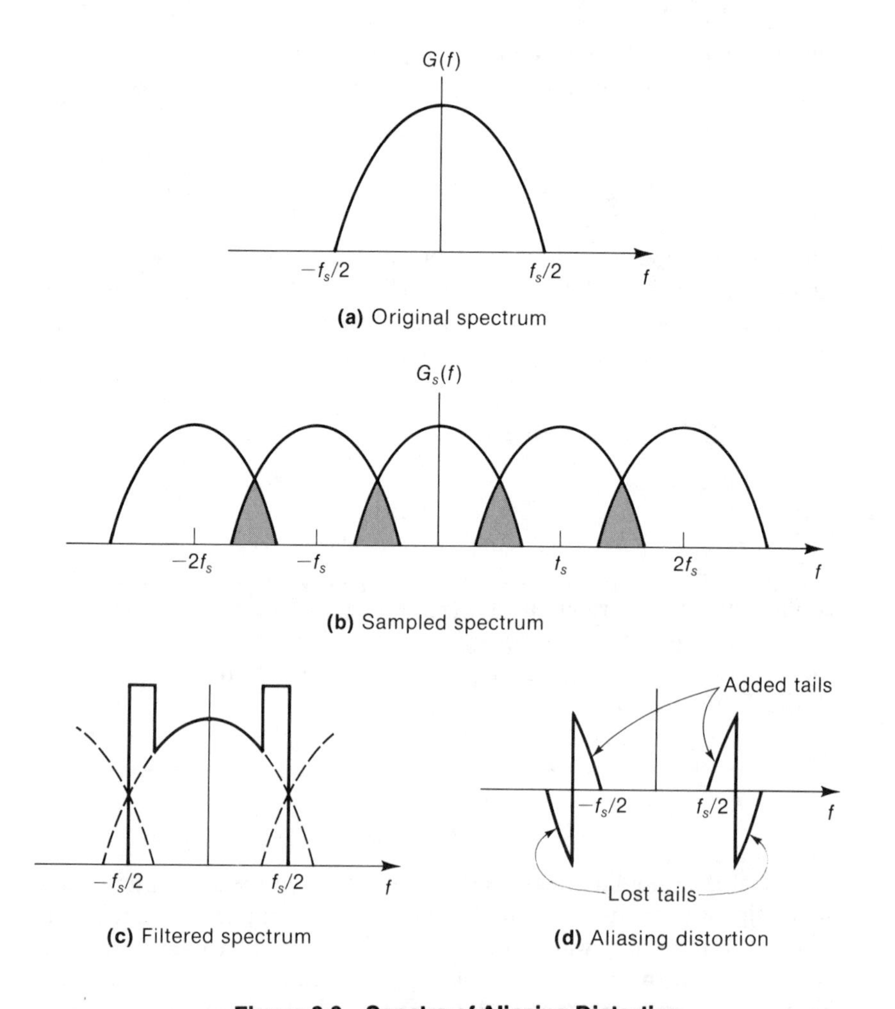

Figure 3.3 Spectra of Aliasing Distortion

3.2.2 Quantization

In contrast to the sampling process, which converts a continuous signal to one that is discrete in time, **quantizing** is the process of converting a continuous signal to one that is discrete in amplitude. As shown in Figure 3.4, the amplitude range of the analog input is converted into discrete steps. All samples of the input falling into a particular quantizing interval are replaced by a single value. An N-step quantizer may be defined by specifying a set of $N+1$ decision thresholds $x_0, x_1, \ldots, x_N$ and a set of output values $y_1, y_2, \ldots, y_N$. When the value x of the input sample lies in the jth quantizing region, that is,

$$R_j = (x_{j-1} < x < x_j) \tag{3.4}$$

the quantizer produces the output y_j. The end thresholds x_0 and x_N are given values equal to the minimum and maximum values, respectively,

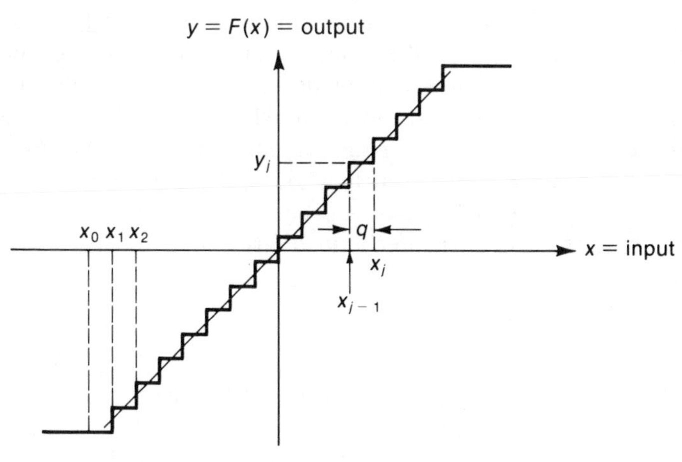

(a) Linear quantizer characteristic

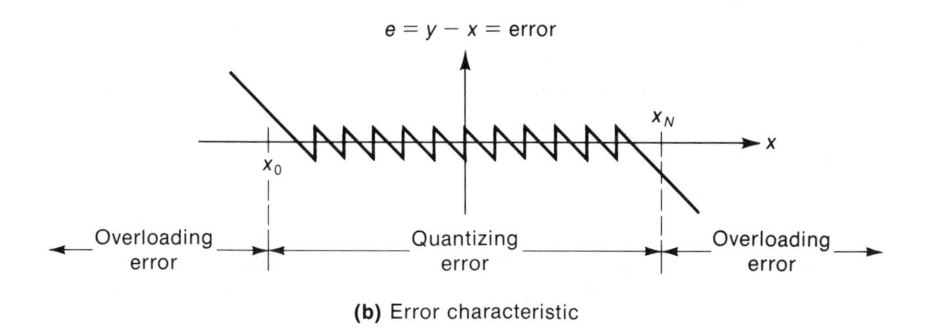

(b) Error characteristic

Figure 3.4 Linear Quantization

expected of the input signal. A distinct n-bit binary word can then be associated with each output value if the set of output values contains $N = 2^n$ members—hence the term *n-bit quantizer*.

The input/output characteristic $F(x)$ of a quantizer has a staircase form. For the simplest type of quantizer, the quantizing steps are all equal in size—hence the name *uniform* or *linear* quantization. If the allowed input range is bounded by $\pm V$, then the uniform step size q is

$$q = \frac{2V}{2^n} = \frac{2V}{N} \tag{3.5}$$

for an n-bit quantizer. The error introduced by the quantizing process is categorized into two types: **quantizing distortion** and **overloading** (or **clipping**), illustrated in Figure 3.4*b*. As long as the input signal lies within the quantizer-permitted range of $-V$ to $+V$, the only form of error introduced is quantizing distortion,* limited to a maximum or $\pm q/2$ for the linear quantizer. If the input signal exceeds this allowed range, the quantizer output will remain at the maximum allowed level ($\pm Nq/2$ for the linear quantizer), resulting in the input signal being clipped. Quantizing error is bounded in magnitude and tends to be sawtooth-shaped except at the maxima and at points of inflection, while overload noise is unbounded. A properly designed quantizer should match the particular input signal statistics (to the extent such statistics are known). In particular, the choice of overload thresholds x_0 and x_N controls a tradeoff between the relative amounts of quantizing and overload distortion.

Quantizing Distortion

Performance of a quantizer can be described by a signal-to-distortion ratio that takes both quantizing and overload distortion into account. Distortion performance is stated in statistical terms, usually by mean square error, which will be used here. If we define the error due to the quantization process as

$$e = F(x) - x \tag{3.6}$$

then the mean square error can be computed as

$$\overline{e^2} = \int_{-\infty}^{\infty} [F(x) - x]^2 p(x)\,dx \tag{3.7}$$

where $p(x)$ is the probability density function (pdf) of the input signal x. To compute $\overline{e^2}$ for a particular type of quantizer (linear or otherwise) we must divide the region of integration according to the decision thresholds

*The terms *noise* and *distortion* are used interchangeably here when describing quantization performance. It should be noted that quantizing and overload distortion are solely dependent on the particular quantization process and the statistics of the input signal.

$x_0, x_1, \ldots, x_N$ that define the quantizing regions:

$$\overline{e^2} = \sum_{j=1}^{N} \int_{x_{j-1}}^{x_j} (y_j - x)^2 p(x)\,dx + \int_{-\infty}^{x_o} (y_1 - x)^2 p(x)\,dx + \int_{x_N}^{\infty} (y_N - x)^2 p(x)\,dx \tag{3.8}$$

where $F(x) = y_j$ when x is in the region R_j. Note that the first term of (3.8) corresponds to quantizing distortion and the last two terms correspond to overload distortion. The regions R_j can be made arbitrarily small with large N, with the exception of the overload regions R_1 and R_N, which are unbounded. Then $p(x)$ can be assumed to be uniform over the inner regions and replaced by a constant $p(y_j)$. Further assuming that $p(x) \approx 0$ for x in the overload regions, Equation (3.8) simplifies to

$$\begin{aligned} \overline{e^2} &= \sum_{j=1}^{N} \frac{(x_j - x_{j-1})^2}{12}(x_j - x_{j-1})p(y_j) \\ &= \frac{1}{12}\sum_{j}(\delta e_j)^2\left[p(y_j)\delta e_j\right] \end{aligned} \tag{3.9}$$

where $\delta e_j = x_j - x_{j-1}$.

For the special case of linear quantization, the decision thresholds are equally spaced so that the quantizing steps $(x_j - x_{j-1})$ are of constant length and equal to the step size q, as illustrated in Figure 3.4*a*. The mean square error then becomes

$$\begin{aligned} \overline{e^2} &= \frac{q^2}{12}\sum_{j=1}^{N} p(y_j)q \\ &= \frac{q^2}{12} \end{aligned} \tag{3.10}$$

since $p(y_j)q$ describes the probability that the signal amplitude is in the region x_{j-1} to x_j and the sum of probabilities over index j equals 1. Thus the mean square distortion of a linear quantizer increases as the square of the step size. This result may also be obtained from the expression

$$\overline{e^2} = \int_{-\infty}^{\infty} e^2 p(e)\,de \tag{3.11}$$

by assuming the quantizing error pdf to be uniform over each quantization interval $(-q/2, +q/2)$ and neglecting overload distortion, so that

$$\overline{e^2} = \int_{-q/2}^{q/2} e^2 p(e)\,de = \int_{-q/2}^{q/2} \frac{e^2}{q}\,de = \frac{q^2}{12} \tag{3.12}$$

To compute the ratio of signal to quantizing distortion S/D_q, the input signal characteristics (such as power) must also be specified. Quite often performance for a quantizer is based on sinusoidal inputs, because S/D_q for speech and sinusoidal inputs compare favorably and use of sinusoidal inputs facilitates measurement and calculation of S/D_q [3]. For the case of a full range $(-V, V)$ sinusoidal input that has zero overload error, the average signal power is

$$S = \frac{V^2}{2} \tag{3.13}$$

From (3.2) we know that the peak-to-peak $(-V, V)$ range of the linear quantizer is $2^n q$. Therefore from (3.12)

$$\overline{e^2} = \frac{2}{3} \frac{V^2/2}{2^{2n}} \tag{3.14}$$

so that for the linear quantizer the signal to quantizing distortion ratio is

$$\frac{S}{D_q} = \frac{V^2/2}{(2/3)\left[(V^2/2)/2^{2n}\right]} = \left(\frac{3}{2}\right)2^{2n} \tag{3.15}$$

or, expressed in decibels,

$$\left(\frac{S}{D_q}\right)_{\text{dB}} = 6n + 1.8 \text{ dB} \tag{3.16}$$

This expression indicates that each additional quantization bit adds 6 dB to the S/D_q ratio.

For random input signals with root mean square (rms) amplitude σ, a commonly used rule for minimizing overload distortion is to select a suitable **loading factor** α, defined as $\alpha = V/\sigma$. A common choice is the so-called 4σ loading ($\alpha = 4$), in which case the total amplitude range of the quantizer is 8σ. The linear quantizing step then becomes

$$q = \frac{8\sigma}{2^n} \tag{3.17}$$

The S/D_q can then be expressed as

$$\frac{S}{D_q} = \frac{\sigma^2}{(2\alpha\sigma/2^n)^2/12} = \frac{\sigma^2}{(8\sigma/2^n)^2/12} = \frac{3}{16}(2^{2n}) \tag{3.18}$$

or, in decibels,

$$\left(\frac{S}{D_q}\right)_{\text{dB}} = 6n - 7.2 \text{ dB} \tag{3.19}$$

which is 9dB less than the expression given in (3.16) because of the difference in average signal power.

Example 3.1

A sinusoid with maximum voltage $V/2$ is to be encoded with a PCM coder having a range of $\pm V$ volts. Derive an expression for the signal-to-quantization distortion and determine the number of quantization bits required to provide an S/D_q of 40 dB.

Solution

For the given sinusoid, we know that

$$\frac{S}{D_q} = \frac{V^2/8}{\overline{e^2}}$$

From (3.14) we have

$$\frac{S}{D_q} = \frac{V^2/8}{(2/3)\left[(V^2/2)/2^{2n}\right]} = \frac{3}{8}(2^{2n})$$

or, in decibels,

$$\left(\frac{S}{D_q}\right)_{\text{dB}} = 6n - 4.3 \text{ dB}$$

The number of bits required to yield an S/D_q of 40 dB is

$$n = \frac{40 + 4.3}{6}$$
$$= 8 \text{ bits}$$

Overload Distortion

Overload distortion results when the input signal exceeds the outermost quantizer levels $(-V, V)$. The distortion power due to quantizer overload has been previously defined by the last two terms of (3.8). To compute the mean square error due to overload distortion (D_0), the input signal pdf must be specified. First, let us assume the pdf to be symmetric so that the last two terms of (3.8) can be written

$$D_0 = 2\int_V^{\infty} (V - x)^2 \, p(x)\,dx \tag{3.20}$$

If the input signal has a noiselike characteristic, then a gaussian pdf can be assumed, described by

$$p(x) = \frac{1}{\sigma\sqrt{2\pi}} e^{-x^2/2\sigma^2} \tag{3.21}$$

An important example of signals that are closely described by a gaussian pdf, by virtue of the law of large numbers, is an FDM multichannel signal [4]. Speech statistics are often modeled after the laplacian pdf [5], given by

$$p(x) = \frac{1}{\sigma\sqrt{2}} e^{-\sqrt{2}|x|/\sigma} \tag{3.22}$$

where σ in (3.21) and (3.22) is the rms value of the input signal and σ^2 is the average signal power. Substituting the gaussian and laplacian pdf's into

(3.20) we obtain, respectively,

$$D_0 = \sqrt{\frac{2}{\pi}}\,(V^2 + \sigma^2)\int_{V/\sigma}^{\infty} e^{-x^2/2}\,dx - \sqrt{\frac{2}{\pi}}\,V\sigma e^{-V^2/2\sigma^2} \qquad \text{(gaussian input)} \tag{3.23a}$$

and

$$D_0 = \sigma^2 e^{-\sqrt{2}\,V/\sigma} \qquad \text{(laplacian input)} \tag{3.23b}$$

Figure 3.5 plots the S/D ratio for linear PCM versus the loading factor α for a gaussian input signal. For small α, performance is limited by an asymptotic bound due to amplitude overload given by (3.23a). Similarly for large α, performance is bounded by a quantization noise asymptote given by (3.18). The peak of each curve indicates an optimum choice of loading factor, which maximizes the S/D ratio by balancing amplitude overload and quantization noise. Code lengths of $n = 2, 4, 6, 8$, and 10 are shown. These curves also illustrate that by increasing the code length by 1 bit, the S/D ratio is improved by 6 dB.

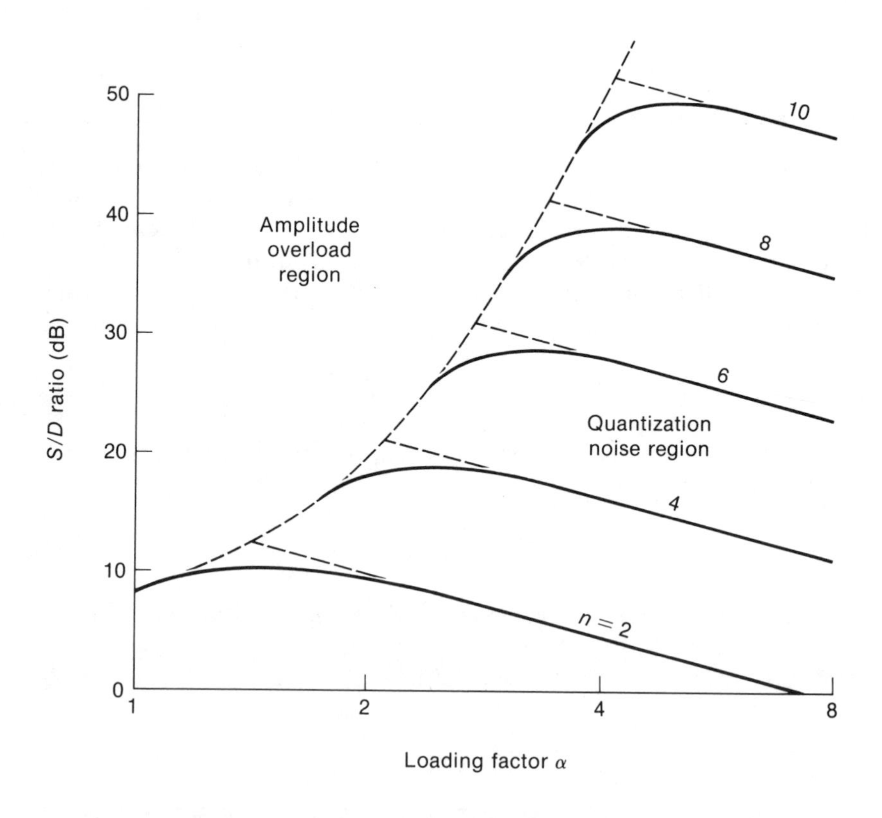

Figure 3.5 Linear PCM S/D Performance for Gaussian Signal Input

Effects of Transmission Errors

Transmission errors cause inversion of received bits resulting in another form of distortion in the decoded analog signal. The distortion power due to decoded bit errors can be computed on the basis of mean square value of the decoded errors. Assuming the use of linear quantization, each sample $s(iT)$ is transmitted by a weighted binary code that can be written

$$s(iT) = \sum_{j=1}^{n} qb_{ij}2^{j-1} \tag{3.24}$$

and the reconstructed sample can be written

$$\hat{s}(iT) = \sum_{j=1}^{n} qc_{ij}2^{j-1} \tag{3.25}$$

where n represents the number of bits used in the code and b_{ij}, c_{ij} are the transmitted and received bits, respectively, where the c_{ij} may have errors. An error in the jth bit will cause an error in the decoded sample value of $e_j = 2^{j-1}q$. If each error occurs with probability p_j, then the mean square value is

$$\begin{aligned} \overline{e^2} &= \sum_{j=1}^{n} e_j^2 p_j \\ &= \sum_{j=1}^{n} 4^{j-1}q^2 p_j \end{aligned} \tag{3.26}$$

If bit errors are assumed independent, then $p_j = p_e$ for all bits in the code so that

$$\begin{aligned} \overline{e^2} &= q^2 p_e \sum_{j=1}^{n} 4^{j-1} \\ &= q^2 p_e \frac{4^n - 1}{3} \end{aligned} \tag{3.27}$$

In computing the signal to bit error distortion ratio, S/D_e, two cases are calculated here: gaussian input with average signal power σ^2 and sinusoidal input with peak power V^2:

$$\frac{S}{D_e} \approx \frac{3\sigma^2}{4^n q^2 p_e} \quad \text{(gaussian)} \tag{3.28}$$

and

$$\frac{S}{D_e} \approx \frac{3}{4p_e} \quad \text{(sinusoidal)} \tag{3.29}$$

To compare the effects of transmission error versus quantization error, consider the ratio of signal to total distortion, $S/(D_q + D_e)$. For a sinusoidal input signal with peak power V^2, we have, using (3.15) and (3.29),

$$\frac{S}{D_q + D_e} = \frac{3}{4p_e + 2^{-2n}} \tag{3.30}$$

Rearranging terms, we may rewrite (3.30) as

$$\frac{S}{D_q + D_e} = \frac{(3)2^{2n}}{4p_e 2^{2n} + 1} \tag{3.31}$$

Two specific cases are of interest. First, for large p_e we have $4p_e 2^{2n} \gg 1$ and Equation (3.31) is dominated by transmission errors, in which case

$$\frac{S}{D_q + D_e} \approx \frac{3}{4p_e} \qquad \text{(large } p_e\text{)} \tag{3.32}$$

Second, for small p_e we have $1 \gg 4p_e 2^{2n}$ and Equation (3.31) is dominated by quantization distortion, in which case

$$\frac{S}{D_q + D_e} \approx (3)2^{2n} \qquad \text{(small } p_e\text{)} \tag{3.33}$$

3.2.3 Companding

For most types of signals such as speech and video, a linear quantizer is not the optimum choice in the sense of minimizing mean square error. Strictly speaking, the linear quantizer provides minimum distortion only for signals

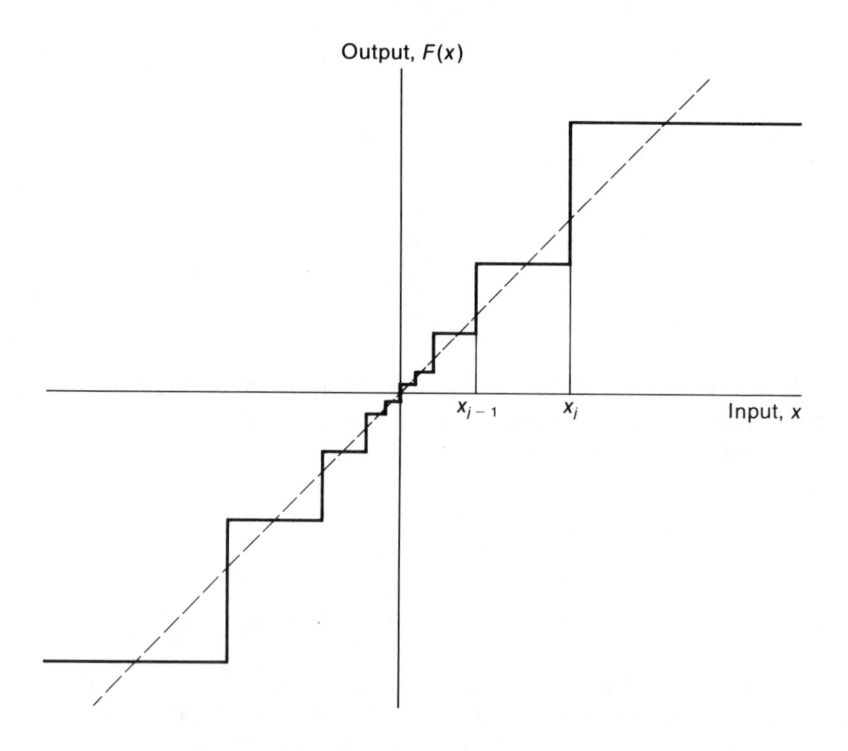

Figure 3.6 Nonuniform Quantizer Characteristic

with a uniform probability density function and performs according to predictions only for stationary signals. Speech signals, however, exhibit nonuniform statistics with a wide dynamic range (up to 40 dB) and smaller amplitudes are more likely than larger amplitudes. Here a choice of linear quantization would result in a poorer S/D (up to 40 dB worse) for weak signals. An alternative approach is to divide the input amplitude range into nonuniform steps by increasing the number of quantization steps in the region around zero and correspondingly decreasing the number around the extremes of the input range. The result of this nonuniform code is an input/output characteristic that is a staircase with N steps of unequal width, as shown in Figure 3.6.

As indicated in Figure 3.7, nonuniform quantization can be achieved by first compressing the samples of the input signal and then linearly quantizing the compressed signals; at the receiver a linear decoder is followed by an expander that provides the inverse characteristic of the compressor. This technique is called **companding**.

The input signal is compressed according to the characteristic $F(x)$, illustrated in Figure 3.8 for positive values of x. The characteristic of the linear quantizer is once again a staircase but now preceded by the compressor, which provides N steps of unequal width as shown in Figure 3.6. The input signal x is recovered by applying the inverse characteristic $F^{-1}(x)$ to produce an estimate $\hat{x}$.

An approximate expression for the mean square error of a nonuniform quantizer can be derived by starting with (3.9). First assume a compression characteristic $y = F(x)$, with N nonuniform input intervals and N uniform output intervals each of step size q. Next define the compression slope to be $F'(x) = dF(x)/dx$. For large N, this curve of $F(x)$ in the jth quantizing interval can be approximated by

$$F'(x_j) \approx \frac{q}{x_j - x_{j-1}} \tag{3.34}$$

Substituting (3.34) into (3.9) and approximating the resulting sum by an

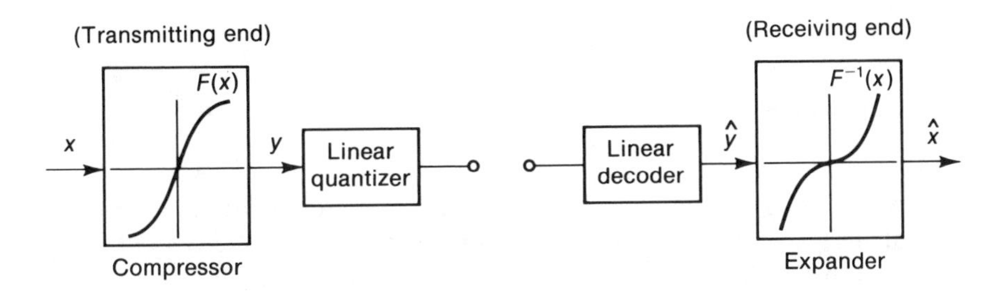

Figure 3.7 Companding Technique

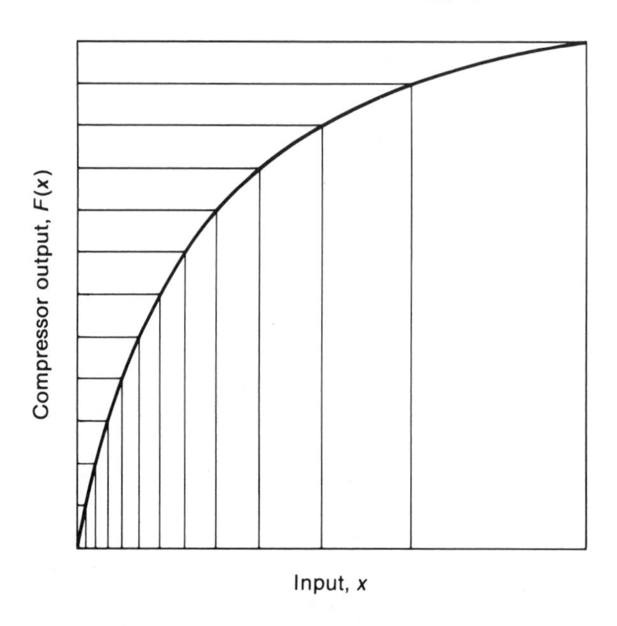

Figure 3.8 Compressor Characteristic (Positive Values Only)

integral (good for $N \gg 1$) yields

$$\begin{aligned}\overline{e^2} &= \frac{1}{12}\sum_{j=1}^{N}\frac{q^2}{\left[F'(x_j)\right]^2}(x_j - x_{j-1})p(y_j)\\ &\approx \frac{q^2}{12}\int\frac{p(x)}{\left[F'(x)\right]^2}\,dx\end{aligned} \tag{3.35}$$

The improvement in performance of the nonuniform quantizer over the uniform quantizer is expressed by the ratio of (3.10) to (3.35), which yields a factor C_I called the companding improvement:

$$C_I = \frac{1}{\displaystyle\int\frac{p(x)}{\left[F'(x)\right]^2}\,dx} \tag{3.36}$$

Logarithmic Companding

The choice of compression characteristic $F(x)$ is made to provide constant performance (S/D) over the dynamic range of input amplitudes for a given signal. For speech, a logarithmic compression characteristic is used, since the resulting S/D is independent of input signal statistics [6]. In practice, an approximate logarithmic shape must be used since a truly logarithmic representation would require an infinite code set. One such logarithmic compressor curve widely used for speech digitization is the μ-law curve given by

$$F(x) = \text{sgn}(x) V \frac{\ln\left(1 + \frac{\mu|x|}{V}\right)}{\ln(1 + \mu)} \qquad 0 \leq |x| \leq V \tag{3.37}$$

where sgn(x) is the polarity of x. As seen in Figure 3.9, $F(x)$ approaches a linear function for small μ and a logarithmic function for large μ. Specifically, by applying l'Hôpital's rule to (3.37), the case for $\mu = 0$ can be seen to correspond to no companding and therefore reduces to linear quantization. For $\mu \gg 1$ and $\mu x \gg V$, $F(x)$ approximates a true logarithmic form. The mean square error for the logarithmic compandor can be found by first differentiating (3.37) to obtain the compressor slope

$$\frac{dF}{dx} = \frac{\mu}{\ln(1 + \mu)} \left[1 + \frac{\mu|x|}{V}\right]^{-1} \tag{3.38}$$

Substituting (3.38) into the expression for mean square error (3.35) then yields

$$D_q = \frac{q^2[\ln(1 + \mu)]^2}{12\mu^2} \left(1 + \frac{2\mu E[|x|]}{V} + \frac{\mu^2\sigma_x^2}{V^2}\right) \tag{3.39}$$

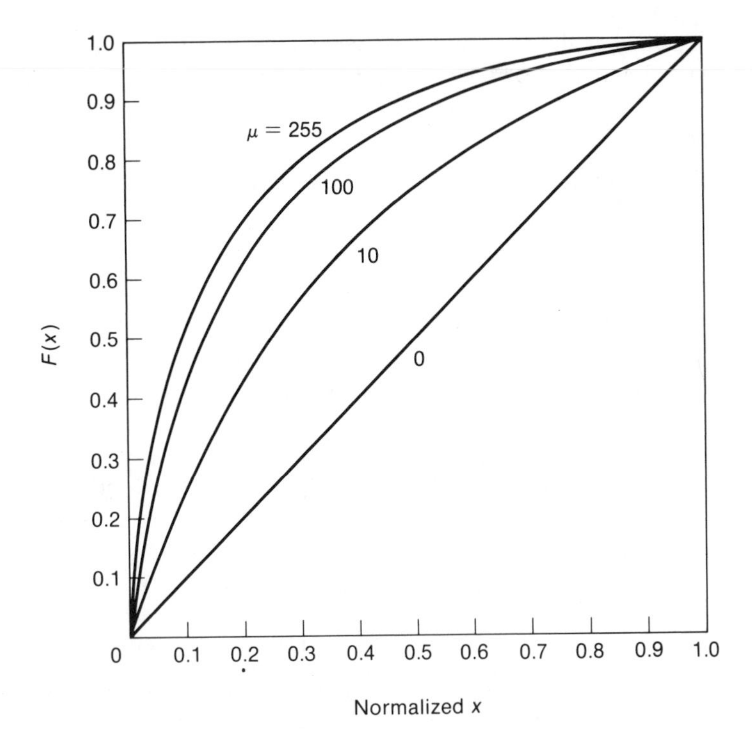

Figure 3.9 Normalized μ-Law Compressor Curve (Positive Values Only)

where the quantity $E[|x|]$ is the rms mean absolute value of the input signal and where σ_x is the rms value of the input signal. Now since $q = 2V/N$, Equation (3.39) can be simplified to

$$D_q = \frac{\ln^2(1+\mu)}{3N^2\mu^2}\left(V^2 + 2V\mu E[|x|] + \mu^2\sigma_x^2\right) \tag{3.40}$$

The signal to quantizing distortion ratio then becomes

$$\frac{S}{D_q} = \frac{3N^2}{\ln^2(1+\mu)} \frac{1}{1 + 2BC/\mu + C^2/\mu^2} \tag{3.41}$$

where we define

$$B = \frac{E[|x|]}{\sigma_x} = \frac{\text{mean absolute input}}{\text{rms input}}$$

$$C = \frac{V}{\sigma_x} = \frac{\text{compressor overload voltage}}{\text{rms input}}$$

Example 3.2

Find an expression for S/D_q for the case of a speech input to a $\mu = 255$ companded PCM coder. Assume that speech may be modeled with laplacian statistics and that the rms amplitude of the input is equal to the compressor overload voltage. Repeat the calculation for $\mu = 100$.

Solution

To calculate S/D_q for μ-law companding, we will use (3.41). But first we note that from (3.22) for a laplacian pdf,

$$B = \frac{1}{\sqrt{2}}$$

and from the assumptions given

$$C = 1$$

Substituting these values into (3.41), we find

$$\left(\frac{S}{D_q}\right)_{\text{dB}} = 10\log\left[\frac{3N^2}{\ln^2(1+\mu)} \frac{1}{1 + 1/\mu^2 + \sqrt{2}/\mu}\right] \tag{3.42}$$

For $\mu = 255$,

$$\begin{aligned}\left(\frac{S}{D_q}\right)_{\text{dB}} &= 20\log N - 10.1\text{ dB}\\ &= 6n - 10.1\text{ dB}\end{aligned} \tag{3.43}$$

where $N = 2^n$ so that n represents the number of bits required to code N levels. For $\mu = 100$ a similar calculation yields

$$\left(\frac{S}{D_q}\right)_{\text{dB}} = 6n - 8.5\text{ dB} \tag{3.44}$$

Now let us consider the performance of μ-law companding for the frequently used case of sinusoidal inputs. The parameter C, which indicates the range of the input, can be related to a full-load sine wave by noting that

$$10\log\left[\frac{V^2/2}{\sigma_x^2}\right] = 10\log\left[\frac{C^2}{2}\right] = 20\log C - 3\text{ dB} \tag{3.45}$$

We can then plot the S/D_q ratio given by (3.41) versus input signal power referenced to a full-load sinusoid. Figure 3.10 is such a plot for $N = 256$ (8-bit quantizer) and for $\mu = 0, 10, 40, 100, 255$, and 400. The S/D_q ratio for $\mu = 0$ (linear quantization) is found by applying l'Hôpital's rule to (3.41) to obtain

$$\left(\frac{S}{D_q}\right)_{\text{dB}} = 10\log\left(\frac{3N^2}{C^2}\right) \tag{3.46}$$

The results of Figure 3.10 indicate that $\mu \geq 100$ is required in order to obtain a relatively flat S/D_q ratio over a 40-dB dynamic range. In practice

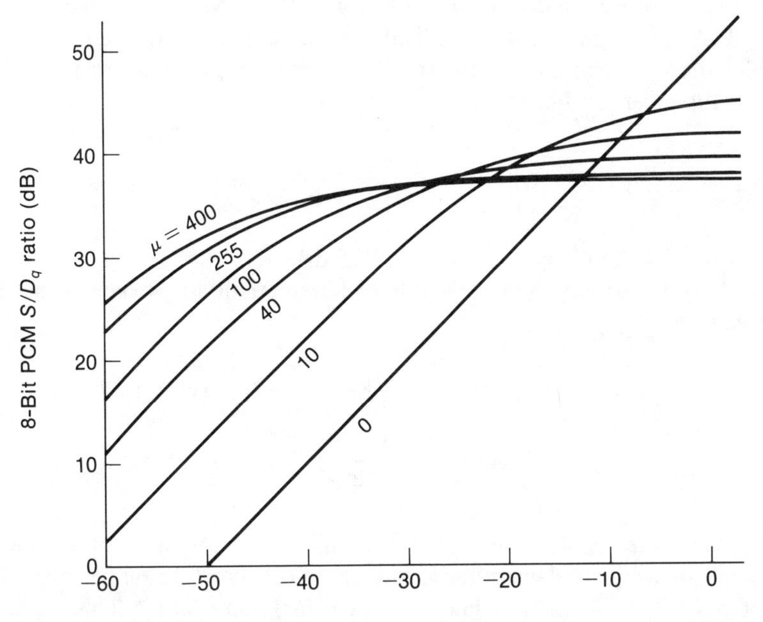

Figure 3.10 Quantizing Distortion Curves for μ-Law Compandor with Sinusoidal Input

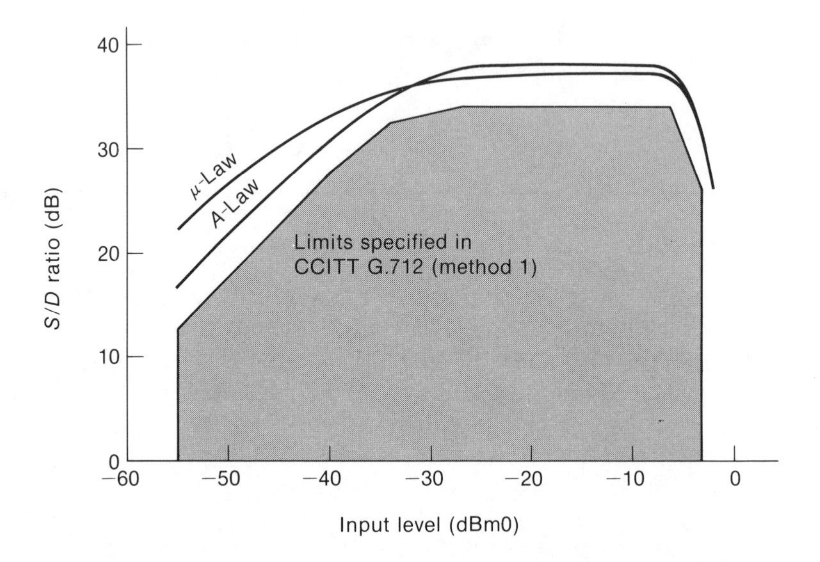

Figure 3.11 Quantizing Distortion Curves with Noiselike (Gaussian) Input for 8-Bit PCM

the usual choices made are $\mu = 100$ for 7-bit PCM and $\mu = 255$ for 8-bit PCM. Another performance indicator is the companding improvement C_I, which for weakest signals (where the companding characteristic is nearly linear) is given by Equation (3.36):

$$C_I = [F'(0)]^2 = \left[\frac{\mu}{\ln(1 + \mu)}\right]^2 \tag{3.47}$$

Then for $\mu = 255$, we obtain $C_I = 33.25$ dB.

Another widely used logarithmic characteristic for speech is the A-law curve where

$$F(x) = \begin{cases} \operatorname{sgn}(x)\dfrac{A|x|}{1 + \ln A} & 0 \le |x| \le 1/A \\ \operatorname{sgn}(x)\dfrac{1 + \ln(A|x|)}{1 + \ln A} & 1/A \le |x| \le 1 \end{cases} \tag{3.48}$$

The parameter A determines the dynamic range. A value of $A = 87.6$ is typical for 8-bit implementation. Over a 40-dB dynamic range, μ-law has a flatter S/D_q ratio than A-law, as shown in Figure 3.11.* This figure also reproduces the recommended CCITT specification [7]. Comparison of these

*The term dBm is the absolute power level in decibels referred to 1 milliwatt; dBm0 is the absolute power level in dBm referred to a point of zero relative level.

curves indicates that 8-bit A-law and μ-law both meet the recommended limits. A similar comparison of 7-bit S/D_q ratios would show that neither A-law nor μ-law meets the recommended limits.

Piecewise Linear Segment Companding

In practice, logarithmic companding laws are approximated by nonlinear devices such as diodes or by piecewise linear segments. The D1 channel bank, for example, uses 7-bit, μ-law companding implemented with diodes [8]; however, because of characteristic diode variation and temperature dependency, experience has indicated difficulty in maintaining matching characteristics at the compressor and expander [9, 10]. In later PCM systems, such as the D2 channel bank, segment implementations have been more successfully used. Two of the segment law families have emerged as international (CCITT) standards—the 13-segment A-law and the 15-segment μ-law. The significance of the segment laws resides in the digital linearization feature, by which the segment intervals can be conveniently coded in digital form. Digitally linearizable compression also facilitates a variety of digital processing functions commonly found in digital switches such as call conferencing, filtering, gain adjustment, and companding law conversion [11].

As an example, consider Figure 3.12, which shows the piecewise linear approximation using 15-segment $\mu = 255$ PCM. The vertices of the connected segments lie on the logarithmic curve being approximated. For this μ-law curve, there are eight segments on each side of zero. The two segments about the origin are collinear and counted as a single segment, however, thus yielding a total of 15 segments. The encoding of the 8 bits—B1 (most significant bit) through B8 (least significant bit)—of each PCM word is accomplished as follows. The B1 position is the polarity bit encoded as a 0 if positive or 1 if negative. The encoding of bits B2, B3, and B4 is determined by the segment into which the sample value falls. Examination of the curve in Figure 3.12 shows that the linear segments have a binary progression (that is, segment 2 represents twice as much input range as segment 1, and segment 3 twice as much as segment 2, and so on) which provides the overall nonlinear characteristic. The remaining 4 bits are encoded using linear quantization with 16 equal-size intervals within each segment.

The corresponding curve for an 8-bit A-law characteristic with $A = 87.6$ uses 13 segments, where the center four segments around zero are made collinear and the other remaining segments are related by binary progression. The 15-segment $\mu = 255$ characteristic and 13-segment $A = 87.6$ characteristic have been adopted by the CCITT as two PCM standards; μ-law is used principally in North America and Japan and A-law primarily in Europe [7]. The S/D_q curve for the 15-segment and 13-segment laws are essentially as shown in Figure 3.11, although the curves would show a scalloping effect that is due to the use of segmented rather than continuous companding.

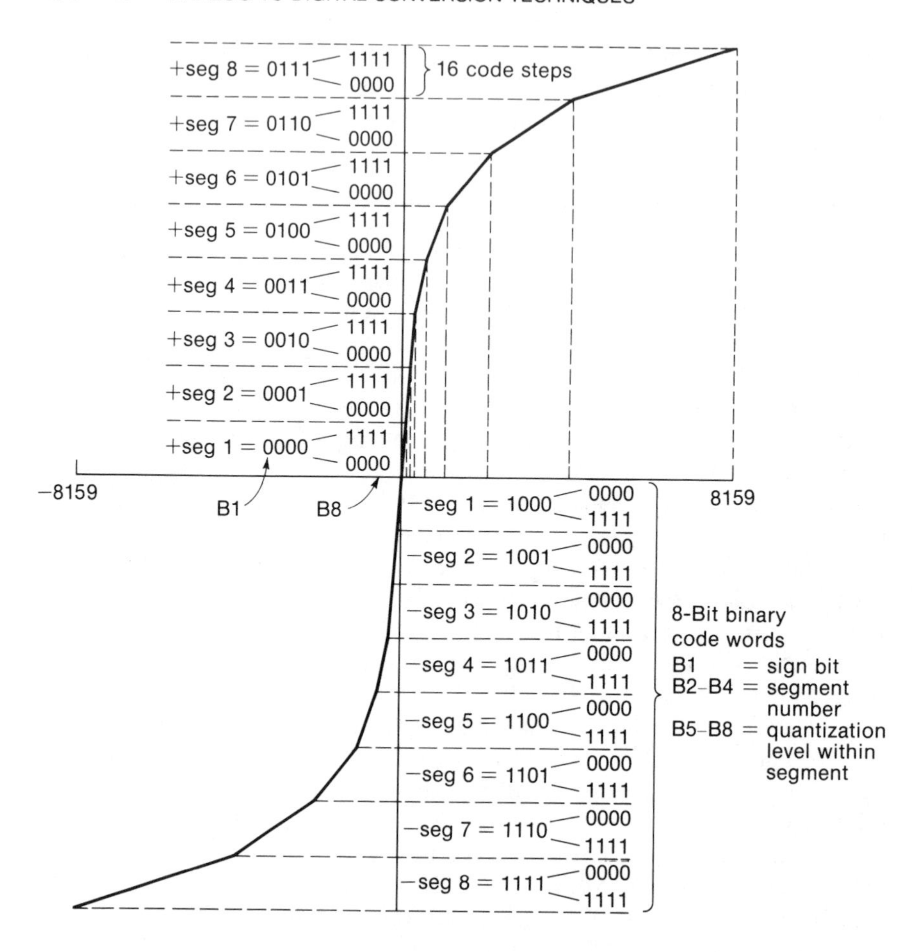

Figure 3.12 Piecewise Linear Segment Approximation to $\mu = 225$ Logarithmic Compandor

3.2.4 Coding

Table 3.1 lists the PCM code words corresponding to the magnitude bits (B2–B8) of Figure 3.12. The input amplitude range has been normalized to a maximum amplitude of 8159 so that all magnitudes may be represented by integer values. Note that the quantization step size doubles with each successive segment, a feature that generates the logarithmic compression characteristic. With 128 quantization intervals for positive inputs and likewise 128 levels for negative inputs, there are 256 code words that correspond to the relative input amplitude range of -8159 to $+8159$. The first and most significant bit indicates the sign and the remaining bits indicate the magnitude. Each quantization interval is assigned a separate code word that is decoded as the midpoint of the interval.

Table 3.1 PCM Coding for μ = 255 Segment Companding

Segment Number	*Segment Endpoint*	*Step Size*	*Amplitude Input Range*[a] (x_n to x_{n+1})	*Code Value* (n)	*Coder Output* Segment Code (B2 – B4)	Quantization Code (B5 – B8)	*Decoder*[b] *Output* (y_n)
	0		0 – 1	0		0000	0
			1 – 3	1		0001	2
			3 – 5	2	000	0010	4
1		2	⋮	⋮		⋮	⋮
	31		29 – 31	15		1111	30
	31		31 – 35	16		0000	33
2		4	⋮	⋮	001	⋮	⋮
	95		91 – 95	31		1111	93
	95		95 – 103	32		0000	99
3		8	⋮	⋮	010	⋮	⋮
	223		215 – 223	47		1111	219
	223		223 – 239	48		0000	231
4		16	⋮	⋮	011	⋮	⋮
	479		463 – 479	63		1111	471
	479		479 – 511	64		0000	495
5		32	⋮	⋮	100	⋮	⋮
	991		959 – 991	79		1111	975
	991		991 – 1055	80		0000	1023
6		64	⋮	⋮	101	⋮	⋮
	2015		1951 – 2015	95		1111	1983
	2015		2015 – 2143	96		0000	2079
7		128	⋮	⋮	110	⋮	⋮
	4063		3935 – 4063	111		1111	3999
	4063		4063 – 4319	112		0000	4191
8		256	⋮	⋮	111	⋮	⋮
	8159		7903 – 8159	127		1111	8031

[a]Magnitude only; normalized to full-scale value of 8159. Sign bit B1 = 0 for positive and 1 for negative.
[b]Decoder output is $y_0 = x_0 = 0$ and $y_n = (x_n + x_{n+1})/2$ for $n = 1, 2, \ldots, 127$.

Example 3.3

For $\mu = 255$, 15-segment companded PCM, determine the code word that represents a 5-volt signal if the encoder is designed for a ± 10-volt input range. What output voltage will be observed at the PCM decoder, and what is the resulting quantizing error?

Solution

Since a 5-volt signal is half the allowed maximum input, the corresponding PCM amplitude is represented by

$$\left(\tfrac{1}{2}\right)(8159) = 4080$$

From Table 3.1, the code word is found to be

$$01110000 \qquad \text{(decimal 112)}$$

The corresponding decoder output is

$$y_{112} = \frac{x_{112} + x_{113}}{2}$$
$$= \frac{4063 + 4319}{2} = 4191$$

The voltage associated with this output value is

$$\left(\frac{4191}{8159}\right)(10 \text{ volts}) = 5.14 \text{ volts}$$

Therefore the quantizing error is $5.14 - 5.0 = 0.14$ volt.

The code set shown in Table 3.1, known as a *signed* or *folded binary* code, is superior in performance to ordinary binary codes in the presence of transmission errors. With a folded binary code, errors in the sign or most significant bit cause an output error equal to twice the signal magnitude. Since speech signals have a high probability of being equal to or near zero, however, the output error due to a transmission error also tends to be small. For ordinary binary codes, where negative signals are expressed as the complement of the corresponding positive signal, an error in the first or most significant bit always causes an output error of half the amplitude range.

Table 3.1 also reveals that choice of a natural folded binary code results in a high density of zeros due to the most probable speech input amplitudes. When using bipolar coding in repeatered line applications, this code set would result in poor synchronization performance, since good clock recovery from a bipolar signal is dependent on a high density of 1's. (See Chapter 5 for more details.) By complementing the folded binary code, a predominance of 1's is created, which provides good synchronization performance for bipolar coding. The transmitted PCM code words for the inverted folded binary code are shown in Table 3.2, which also shows suppression of the all-zero code. When a sample encodes at this value (-127), the all-zero code is replaced by 00000010, which corresponds to the value -125. This **zero code suppression** then guarantees that no more than 13 consecutive zeros can appear in two consecutive PCM code words, which further enhances clock recovery in repeatered line transmission. Since the maximum amplitude represented by the all-zero code occurs infrequently, the degradation introduced by zero code suppression is insignificant. The inverted folded binary code with zero code suppression shown in Table 3.2 is exactly the set of codes specified by CCITT for μ-law PCM in Rec. G.711. For the A-law coder prescribed by CCITT Rec. G.711, only the *even* bits

Table 3.2 PCM Code Words as Prescribed by CCITT for μ-Law Companding

256 Transmission Code Words

1	0	0	0	0	0	0	0	(127)
1	0	0	0	0	0	0	1	(126)
1	0	0	0	0	0	1	0	(125)
1	1	1	1	1	0	1	0	(5)
1	1	1	1	1	0	1	1	(4)
1	1	1	1	1	1	0	0	(3)
1	1	1	1	1	1	0	1	(2)
1	1	1	1	1	1	1	0	(1)
1	1	1	1	1	1	1	1	(+ 0)
0	1	1	1	1	1	1	1	(− 0)
0	1	1	1	1	1	1	0	(− 1)
0	1	1	1	1	1	0	1	(− 2)
0	1	1	1	1	1	0	0	(− 3)
0	1	1	1	1	0	1	1	(− 4)
0	1	1	1	1	0	1	0	(− 5)
0	0	0	0	0	0	1	0	(− 125)
0	0	0	0	0	0	0	1	(− 126)
0	0	0	0	0	0	0	0	(− 127)[a]

[a] The all-zero code is not used and instead is replaced by the code word 00000010.

(B2, B4, B6, and B8) are inverted from the natural folded binary sequence and zero code suppression is not used [7].

As a result of these differences between A-law and μ-law coders, conversion is required when interfacing these two coders as with intercountry gateways [12]. The CCITT has prescribed that any necessary conversion will be done by the countries using μ-law. A conversion algorithm prescribed by the CCITT (Rec. G.711, tables 3 and 4 [7]) provides direct digital conversion via a "look-up table" that is realizable in the form of read-only memory. The S/D performance of the converters is displayed in Figure 3.13 for an input with laplacian distribution to simulate speech. For purposes of comparison, the S/D performance for ideal A-law and μ-law (curves 1 and 2) plus the S/D performance for nonoptimum $\mu \rightarrow A$ and $A \rightarrow \mu$ conversion (curves 3 and 4) are also shown. Curves 3 and 4 show the

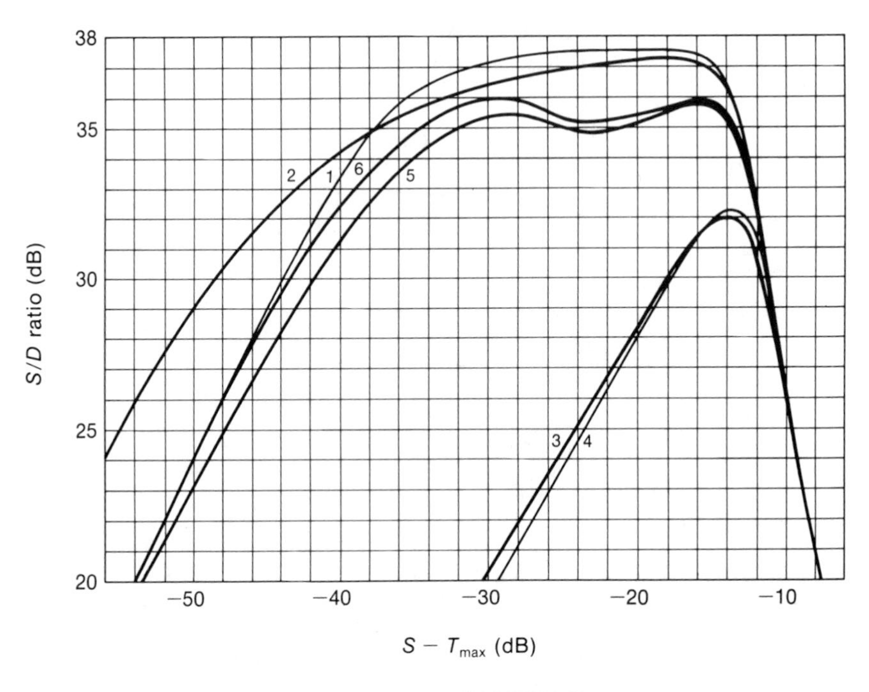

Source: Reprinted by permission of ITU-CCITT [12].
Curve 1: 13-segment A-law compandor
Curve 2: 15-segment μ-law compandor
Curve 3: 15-segment μ-law compressor with 13-segment A-law expander
Curve 4: 13-segment A-law compressor with 15-segment μ-law expander
Curve 5: same as curve 3 but with recoding
Curve 6: same as curve 4 but with recoding
$T_{max} = \begin{cases} 3.14 \text{ dBm0 for } A\text{-law} \\ 3.17 \text{ dBm0 for } \mu\text{-law} \end{cases}$

Figure 3.13 Signal-to-Distortion Ratio (S/D) as a Function of the Load $S - T_{max}$ for 8-Bit Coding with a Simulated (Laplacian) Speech Signal

effects of using a μ-law compressor with an A-law expander, and vice versa, assuming the use of a natural folded binary code. Because of the difference in assignment of signal level to PCM codes between A-law and μ-law, curves 3 and 4 indicate much worse degradation than curves 1 and 2. For high input signal level the S/D is only slightly degraded, due to the fact that high input levels are represented by nearly identical PCM words for μ-law and A-law. For lower input signal levels the S/D degrades rapidly, due to the fact that lower signal levels are represented by significantly different PCM words for μ-law versus A-law. Comparing curves 5 and 6 with curves 1 and 2, we see that the degradation attendant upon optimum conversion is only about 3 dB. Hence the optimized code conversions (curves 5 and 6) via the look-up table must be used to provide acceptable S/D performance.

3.3 DIFFERENTIAL PCM AND DELTA MODULATION

For highly correlated signals such as speech or video, the signal value changes slowly from one Nyquist sample to the next. This makes it possible to predict a sample value from preceding samples and to transmit the difference between the predicted and actual samples. Since the variation of this difference signal is less than that of the input signal, the amount of information to be transmitted is reduced. This technique is generally known as *differential PCM encoding.*

An implementation of differential pulse code modulation (DPCM) is shown in Figure 3.14. The input signal is first low-pass filtered to limit its bandwidth to one-half (or less) of the sampling rate f_s. The predicted input value is then subtracted from the analog input $s(t)$ and the difference is sampled and quantized. The predicted input signal is generated in a feedback loop that uses an integrator to sum past differences as the decoded sample estimate. The receiver contains a decoder identical to that used in the transmitter.

The simplest form of DPCM is the delta modulator, which provides 1-bit quantization of the difference signal. The output bits then represent only the polarity of the difference signal. If the difference signal is positive, a 1 is generated; if it is negative, a 0 is generated. The local decoder generates steps with amplitude $+\Delta$ or $-\Delta$ in accordance with the outputs 1 and 0.

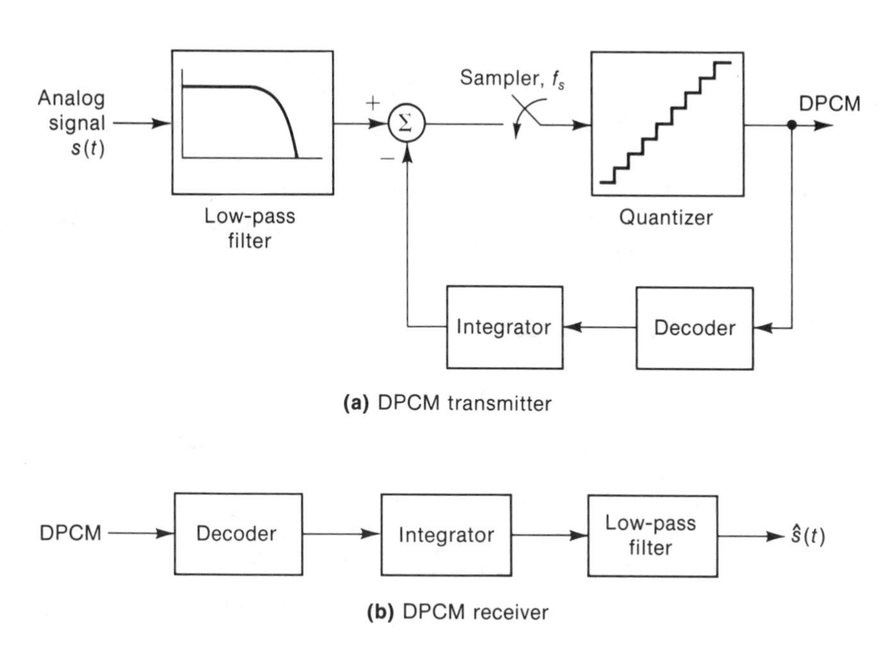

Figure 3.14 DPCM Implementation

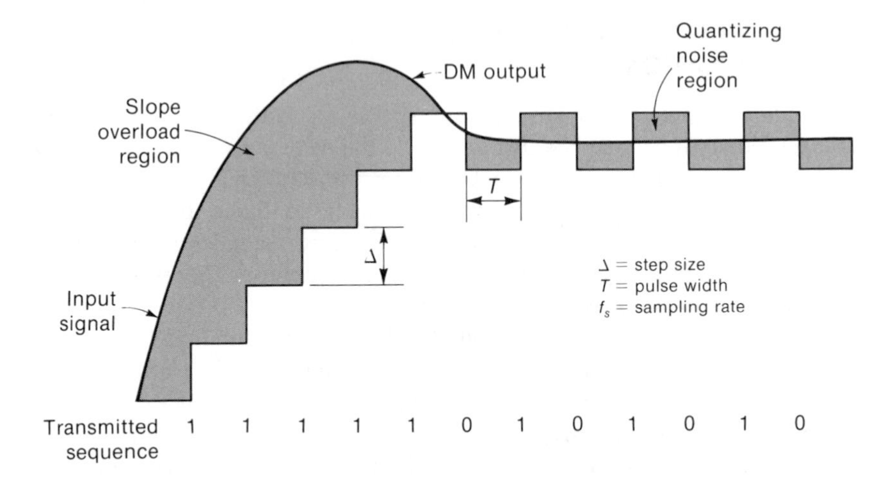

Figure 3.15 Delta Modulator Waveform

The decoded unfiltered signal is then a staircase waveform with uniform step size Δ, as illustrated in Figure 3.15. The integrator in the receiver is periodically reset to limit the effect of transmission errors on the output, and the low-pass filter smoothes the staircase signal to recover the analog signal.

Figure 3.15 also illustrates the effects of slope overload distortion and quantizing distortion. If the slope of the input signal exceeds the slope of the staircase (Δ/T), the resulting error is known as **slope overload distortion**. For an input slope less than Δ/T, the errors are a form of quantizing distortion. For a fixed sampling rate, optimum S/D performance is obtained by selecting the step size to minimize the sum of slope overload and quantizing distortion. With a choice of small step size, slope overload distortion will dominate; with a choice of large step size, quantizing distortion will dominate.

3.3.1 Quantizing Distortion

Delta Modulation

Consider first the simplest measure of performance: quantizing distortion with no slope overload. Assume the input to be sinusoidal with the following characteristics:

$$\text{Input:} \qquad s(t) = A \sin 2\pi f_0 t$$

$$\text{Slope:} \qquad \frac{ds}{dt} = 2\pi f_0 A \cos 2\pi f_0 t$$

$$\text{Maximum slope of input:} \qquad \text{Slope max} = 2\pi f_0 A$$

To prevent slope overload, the slope of the delta modulator must be greater than or equal to the maximum slope of the input, or

$$\Delta f_s \geq 2\pi f_0 A \tag{3.49}$$

The output signal power S can then be deduced from (3.49) as

$$S = \frac{A^2}{2} = \frac{1}{2}\left(\frac{\Delta f_s}{2\pi f_0}\right)^2 \tag{3.50}$$

The maximum quantizing error is $\pm\Delta$; assuming the quantizing error to be uniformly distributed (good approximation for small Δ), the mean square quantizing error is

$$\overline{e^2} = \int_{-\Delta}^{\Delta} e^2 p(e)\, de = \frac{1}{2\Delta}\int_{-\Delta}^{\Delta} e^2\, de = \frac{\Delta^2}{3} \tag{3.51}$$

It is worth noting that for a PCM linear quantizer with $q = 2\Delta$, the same result as (3.51) is obtained for quantizing distortion. For a sinusoidal input the S/D_q ratio is given by

$$\frac{S}{D_q} = \frac{3A^2}{2\Delta^2} \tag{3.52}$$

The maximum S/D_q ratio for no slope overload is obtained from using the equality sign in (3.49):

$$\frac{S}{D_q} = \frac{3}{2}\frac{f_s^2}{(2\pi f_0)^2} \tag{3.53}$$

which indicates that doubling the sampling frequency (and hence doubling the bit rate) improves the S/D_q ratio by 6 dB.

The preceding calculation of the S/D_q ratio assumes no filtering of the output samples. Since the input bandwidth is less than the sampling frequency, we can enhance the output S/D_q ratio by proper filtering at the receiver. Assume the noise (error) power at the receiver input to be uniformly distributed over frequencies $(0, f_s)$. If an ideal low-pass or bandpass filter of bandwidth f_m is used at the receiver, the mean square output error becomes

$$\overline{e^2} = \frac{\Delta^2}{3}\left(\frac{f_m}{f_s}\right) \tag{3.54}$$

Assuming a sinusoidal input, the output S/D_q ratio is then

$$\frac{S}{D_q} = \frac{3A^2 f_s}{2\Delta^2 f_m} \tag{3.55}$$

The maximum S/D_q for no slope overload is obtained from (3.55) and (3.49):

$$\frac{S}{D_q} = \frac{3}{2}\frac{f_s^3}{f_m (2\pi f_0)^2} \tag{3.56}$$

or

$$\left(\frac{S}{D_q}\right)_{\text{dB}} = 10\log\left[\frac{f_s^3}{f_m f_0^2}\right] - 14\,\text{dB} \tag{3.57}$$

Notice that the low-pass filter yields an S/D_q ratio proportional to f_s^3 rather than to f_s^2 as found in (3.53), indicating a 9-dB improvement with doubling of the sample frequency.

Additional improvement in the S/D_q ratio is realized by increasing the order of integration in the delta modulator feedback loop. When double integration is incorporated, for example, the maximum S/D_q for a sinusoidal input has been derived [13]:

$$\left(\frac{S}{D_q}\right)_{\text{dB}} = 10\log\left[\frac{f_s^5}{f_m^2 f_0^2}\right] - 32\,\text{dB} \tag{3.58}$$

Differential PCM (DPCM)

The maximum positive slope for l-bit DPCM is

$$\text{Slope max} = (2^l - 1)\,\Delta f_s \tag{3.59}$$

where the 2^l levels are equispaced and separated by Δ. The slope of the DPCM quantizer must be greater than or equal to the maximum slope of the input to prevent slope overload. For a sinusoidal input, this means that

$$(2^l - 1)\,\Delta f_s \geq 2\pi f_0 A \tag{3.60}$$

so that the output signal power S can be given as

$$S = \frac{A^2}{2} = \frac{1}{2}\left[\frac{(2^l - 1)\,\Delta f_s}{2\pi f_0}\right]^2 \tag{3.61}$$

The quantization distortion for filtered DPCM is identical to (3.54), so that the S/D_q ratio can be given as

$$\frac{S}{D_q} = \frac{3}{2}\,\frac{(2^l - 1)^2 f_s^3}{f_m (2\pi f_0)^2} \tag{3.62}$$

This yields an increase of $(2^l - 1)^2$ relative to the S/D_q ratio for two-level quantization (delta modulation). However, l bits per sample are transmitted with DPCM rather than 1 bit for delta modulation (DM). As shown in Table 3.3, a comparison of l-bit DPCM and DM, normalized to the same bit rate, indicates that DPCM provides significant improvement for $l \geq 3$.

Expression (3.62) for peak S/D_q ratio of DPCM applies only to sinusoidal inputs. The main effect of a gaussian or speech input signal compared to a sinusoidal input is to reduce the peak S/D_q ratio. For example, Van de Weg [14] has derived an expression for S/D_q ratio with

Table 3.3 Comparison of Quantization Distortion for DPCM vs. DM (Normalized to the Same Bit Rate)

Sample Frequency (xf_s)	DM D_q Improvement Factor x^3	DM D_q Improvement Factor $10 \log x^3$	Bits / Sample (l)	DPCM D_q Improvement Factor $(2^l - 1)^2$	DPCM D_q Improvement Factor $10 \log(2^l - 1)^2$
$1f_s$	1	0	1	1	0
$2f_s$	8	9	2	9	9.5
$3f_s$	27	14	3	49	17
$4f_s$	64	18	4	225	23

gaussian input statistics:

$$\frac{S}{D_q} \approx \left(\frac{3}{4\pi}\right)^2 \left(\frac{f_s}{f_m}\right)^3 \frac{(2^l - 1)^2}{l^3} \qquad l \geq 3 \tag{3.63}$$

Notice that the output S/D_q ratio is proportional to f_s^3 as obtained in (3.62). For DM ($l = 1$) systems the S/D_q ratio for gaussian input statistics is given by

$$\frac{S}{D_q} \approx 2\left(\frac{3}{4\pi}\right)^2 \left(\frac{f_s}{f_m}\right)^3 \tag{3.64}$$

We can compare the case for a gaussian input with sinusoidal inputs by rearranging (3.56) to read

$$\frac{S}{D_q} = \frac{3}{2(2\pi)^2} \left(\frac{f_s}{f_m}\right)^3 \left(\frac{f_m}{f_0}\right)^2 \tag{3.65}$$

Obviously this result for sinusoidal inputs exceeds the result for a gaussian input for low-frequency input sinusoids with $f_0 \ll f_m$.

3.3.2 Slope Overload Distortion in Delta Modulation

Here we will outline a classic derivation of slope overload distortion, credited to S. O. Rice [15], and then interpret that result for a simple example. Using an analysis similar to the study of fading statistics in radio transmission, Rice first calculated the average noise energy N_b for a single slope overload event and then calculated the expected number of slope overload events in 1 second, r_b. The product $N_b \cdot r_b$ is then an approximation of the average noise power due to slope overload, D_{SO}, given as

$$D_{SO} = N_b \cdot r_b \approx \frac{1}{4\sqrt{2\pi}}\left(\frac{b_0^2}{b_2}\right)\left(\frac{3b_0^{1/2}}{\Delta f_s}\right)^5 \exp\left[\frac{-(\Delta f_s)^2}{2b_0}\right] \tag{3.66}$$

where

$$\begin{aligned}
\Delta &= \text{quantizer step size} \\
f_s &= \text{sampling frequency} \\
b_0 &= \text{variance of } f'(t) \\
b_2 &= \text{variance of } f''(t) \\
b_n &= \int_0^\infty (2\pi f)^{n+2} F(f)\, df,\ n = 0, 2 \qquad (3.67) \\
F(f) &= \text{power spectrum of } f(t) \\
f(t) &= \text{gaussian input signal}
\end{aligned}$$

Example 3.4

Let the gaussian input signal to a delta modulator be band-limited with spectrum

$$F(f) = \begin{cases} \dfrac{1}{f_m} & 0 < f < f_m \\ 0 & \text{otherwise} \end{cases}$$

For this case, find the expression for slope overload distortion.

Solution

From (3.67) the expressions for b_0 and b_2 are calculated as

$$b_0 = \frac{(2\pi f_m)^2}{3}$$

$$b_2 = \frac{(2\pi f_m)^4}{5}$$

These expressions allow calculation of the slope overload noise from (3.66), which becomes

$$D_{SO} = \frac{5}{4}\sqrt{\frac{3}{2\pi}}\left(\frac{2\pi f_m}{\Delta f_s}\right)^5 \exp -\left[\frac{3}{2}\left(\frac{\Delta f_s}{2\pi f_m}\right)^2\right] \tag{3.68}$$

A plot of S/D ratio versus normalized step size shows a tendency to approach a slope overload asymptote for small step size and a quantization asymptote for large step size. To illustrate this effect, assume the total distortion power D to be approximated by the sum of D_{SO} and D_q. This approximation is valid since quantizing noise does not occur due to a slope overload condition, and vice versa. Now if we normalize the rms value of the input to unity, the approximation for S/D is

$$\frac{S}{D} \approx \frac{1}{D_{\text{SO}} + D_q} \tag{3.69}$$

Figure 3.16 plots the S/D ratio for a flat band-limited gaussian input signal. These theoretical curves were obtained by substituting expressions for slope overload noise (3.68) and quantization noise (3.64) into (3.69). These S/D ratios are plotted as a function of normalized step size, $\Delta f_s/f_m$. Sampling rates of 4, 8, 16, and 32 times the signal bandwidth are shown. For step sizes to the left of the peaks, slope overload noise dominates while to the right of the peaks quantization noise dominates. The peaks indicate an optimum value of normalized step size that maximizes the S/D ratio. These curves also demonstrate that by doubling the sampling frequency, the S/D ratio is improved by 9 dB in the quantization noise region.

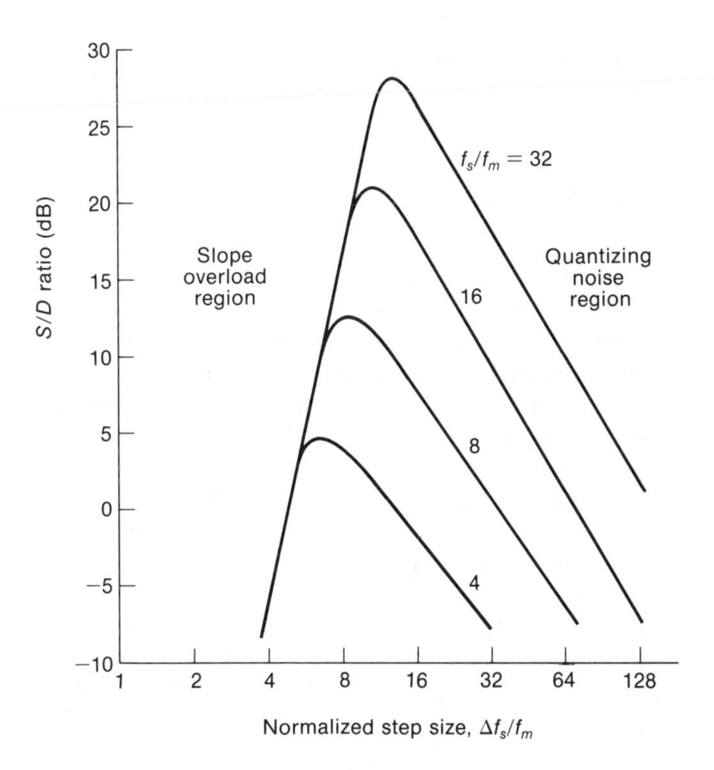

Figure 3.16 Delta Modulation Performance for a Flat Band-Limited Gaussian Input

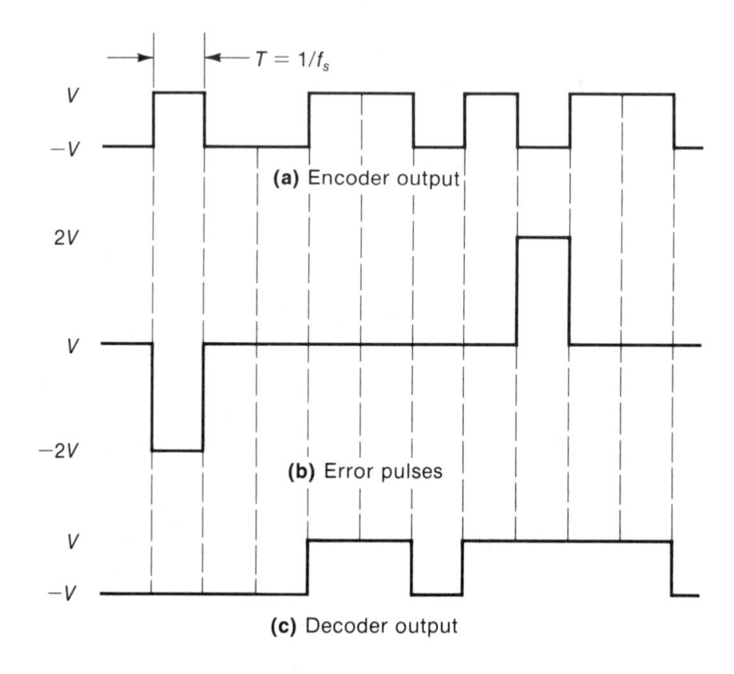

Figure 3.17 Transmission Error Effects on Delta Modulator

3.3.3 Effects of Transmission Errors on Delta Modulation

A delta modulator used with a digital transmission link suffers degradation from transmission errors. To characterize this measure of performance, consider Figure 3.17*a*, which shows an arbitrary two-level waveform produced at the transmitter. Due to channel disturbances, the recovered waveform is observed to contain errors. These errors are assumed to occur randomly with p being the probability of digit error. The resulting error pulses have duration equal to one sampling period (T) and an amplitude which is double that of the transmitted signal, as shown in Figure 3.17*b*.

The receiver shown in Figure 3.18 is composed of a binary detector, an RC circuit that approximates an integrator,* and a bandpass filter. The effect of digit errors on the receiver output is determined by first computing the noise power due to the error pulses and then considering the effect of the integrator and bandpass filters. For sufficiently small p, the autocorrelation function $R_e(\tau)$ of the error pulses is maximum at $\tau = 0$ and effectively zero for $|\tau| > T$, as shown in Figure 3.19*a*. The power spectral density $S_e(f)$ is

*A perfect integrator cannot be realized in a delta modulator and is therefore replaced by an RC circuit. Since the "integration" of a constant input results in an output that follows an exponential curve, this configuration is called an **exponential delta modulator** [16].

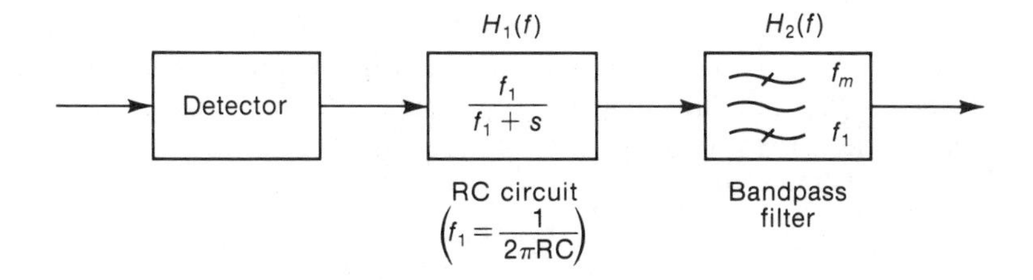

Figure 3.18 Receiver for Exponential Delta Modulator

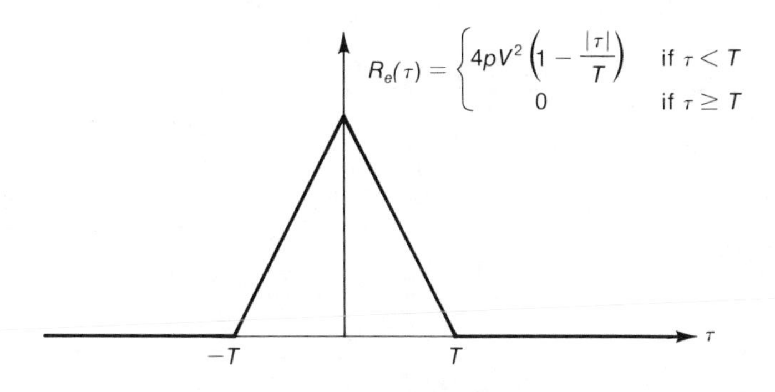

(a) Autocorrelation function

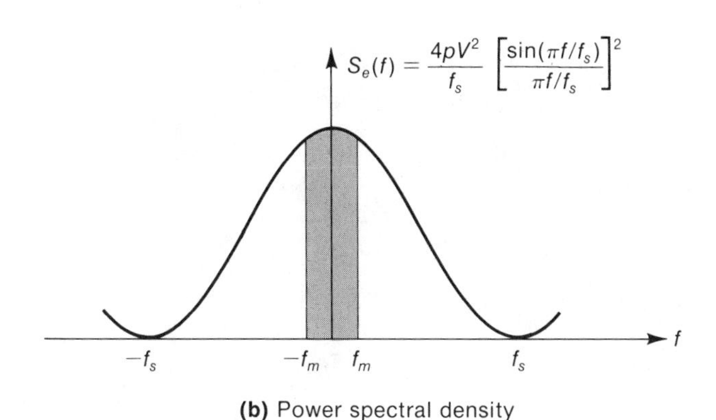

(b) Power spectral density

Figure 3.19 Error Waveform Characteristics for Exponential Delta Modulation

then of the form $(\sin x/x)^2$ with the first null at f_s, as shown in Figure 3.19*b*. If the maximum frequency f_m of the original analog signal is small compared to the sampling frequency f_s (that is, $f_m \ll f_s$), then the power density of the error noise spectrum is essentially constant over the range $(-f_m, f_m)$ and equal to

$$S_e(0) = \frac{4pV^2}{f_s} \tag{3.70}$$

If the RC integrator and bandpass filter have transfer function $H_1(f)$ and $H_2(f)$, respectively, then the spectral density at the receiver output is

$$S_0(f) = S_e(f)|H_1(f)|^2|H_2(f)|^2 \tag{3.71}$$

If $H_2(f)$ is an ideal filter with cutoff at f_1 and f_m, the mean square distortion due to transmission errors is

$$\begin{aligned} D_e &= 2\int_{f_1}^{f_m}\left(\frac{4pV^2}{f_s}\right)\left(\frac{f_1^2}{f^2+f_1^2}\right)df \\ &= \frac{8pV^2f_1}{f_s}\left[\tan^{-1}\left(\frac{f_m}{f_1}\right) - \frac{\pi}{4}\right] \end{aligned} \tag{3.72}$$

For the usual case of $f_m \gg f_1$,

$$D_e \simeq \frac{2\pi pV^2f_1}{f_s} \tag{3.73}$$

For a sinusoidal input $A\sin\omega_0 t$, Johnson [16] has shown that for the exponential delta modulator the overload condition is given by the relationship

$$A = \frac{V}{\sqrt{1+(2\pi f_0/2\pi f_1)^2}} \tag{3.74}$$

The mean square value of the output signal is therefore

$$S = \frac{A^2}{2} = \frac{V^2}{2\left[1+(2\pi f_0/2\pi f_1)^2\right]} \tag{3.75}$$

or for the usual case of $f_0 \gg f_1$,

$$S \approx \frac{V^2}{2}\frac{(2\pi f_1)^2}{(2\pi f_0)^2} \tag{3.76}$$

We can then write the signal to transmission error distortion ratio using (3.76) and (3.73):

$$\frac{S}{D_e} = \frac{f_s f_1}{4\pi f_0^2 p} \tag{3.77}$$

To consider the relative effects of transmission errors, compare (3.77) with the expression for the signal to quantizing error ratio (3.56). Solving for

Table 3.4 Values of Transmission Error Rate p' for $S/D_q = S/D_e$ in Delta Modulator Applied to Typical Voice Channel

Bit Rate (f_s)	p'
64 kb/s	4.6×10^{-4}
32 kb/s	1.8×10^{-3}
16 kb/s	7.3×10^{-3}
8 kb/s	2.9×10^{-2}

the value of p' for which $S/D_q = S/D_e$, we obtain

$$p' = \frac{2\pi f_1 f_m}{3 f_s^2} \tag{3.78}$$

For a typical voice channel bandwidth of $f_m = 3000$ Hz with low-end cutoff frequency of $f_1 = 300$ Hz, Table 3.4 indicates values of p' for various bit rates (f_s). These figures indicate that when $f_s = 16$ kb/s and with an error rate as high as 0.73 percent, the noise produced at the decoder output is of the same magnitude as the quantization noise. Delta modulation systems therefore offer good performance in the presence of a high transmission error rate.

3.3.4 Adaptive Techniques Applied to DM and DPCM

As in PCM, it is possible to improve the dynamic range of DPCM and DM coders by the use of companding. For DM coders, the most suitable form of companding consists of adapting the quantizer step size according to the input signal. Unlike PCM where the companding characteristic is fixed, in companded DM systems the variable step sizes are derived dynamically; step sizes increase during steep segments of the input and decrease with slowly varying segments.

Linear (nonadaptive) DM and DPCM have the undesirable characteristic that there is only one input level which maximizes the signal-to-distortion ratio. Companding can greatly enhance the dynamic range of these coders, rendering them useful for widely varying signals such as speech. Whereas performance characterization for linear DM and DPCM could be done by straightforward derivation of quantization and slope overload distortion formulas, such derivations are not readily accomplished for adaptive techniques. Companding optimization has been done largely by experiment and computer simulation [17, 18] rather than by theory. Several versions of companded DM have been developed and given special names by their inventors. Here we will discuss a few representative examples and show applications.

In general, there are both *discrete* and *continuous* methods of adapting the DM system to changes in the slope of the input signal. In the discrete adaptive system, illustrated in Figure 3.20, the slope is controlled by a logic circuit that operates on the sequence of bits produced by the quantizer. When slope overload occurs, for example, the quantizer output tends toward all 1's or all 0's. In response to these bits of the same polarity, the logic selects a larger step size, continuing to do so until the largest discrete allowed value is reached. Conversely, the step size incrementally decreases when polarity reversals occur. The decoder senses the same bit sequences and controls slope values using logic identical to that in the coder, thus producing the same step sizes. Since step size changes are made at a rate equal to the sampling rate, this form of discrete adaptive delta modulation is known as **instantaneous companding**.

To illustrate one example, consider the set of weights $\{1, 2, 4, 6, 8, 16\}$ for the allowed step sizes and the logic given in Table 3.5. The step sizes to be used are determined by the last three transmitted bits. The direction (positive or negative) is controlled by the current bit. The logic rules have been chosen to minimize slope overload by continuing to increase the step size for runs of three consecutive like bits. Figure 3.21 illustrates the adaptive DM output generated for a sample input using the logic of Table 3.5. The biggest design problems for discrete ADM are the choice of step sizes—both the number of allowed sizes and the gain per size—and the

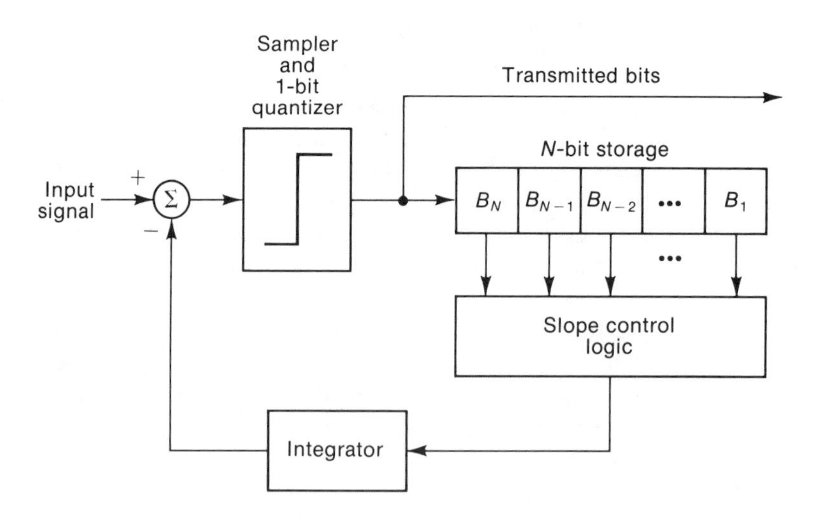

Figure 3.20 Block Diagram of Adaptive Delta Modulator

Table 3.5 Example of Slope Control Logic for Adaptive Delta Modulation

Current Bit	*Previous Bit*	*Pre-Previous Bit*	*Slope Magnitude Relative to Previous Slope Magnitude*
0	0	0	Take next larger magnitude[a]
0	0	1	Slope magnitude unchanged
0	1	0	Take next smaller magnitude[b]
0	1	1	Take next smaller magnitude[b]
1	0	0	Take next smaller magnitude[b]
1	0	1	Take next smaller magnitude[b]
1	1	0	Slope magnitude unchanged
1	1	1	Take next larger magnitude[a]

[a]If slope magnitude is already at the maximum, this instruction is not obeyed.
[b]If slope magnitude is at the minimum, this instruction is not obeyed.

logic required to control changes in step sizes. Abate [17] has investigated this effect of the choice of step sizes on S/D performance of discrete ADM.

Winkler [19] proposed one of the earliest companded DM schemes in a system he termed **high-information delta modulation** (HIDM), so named because the system contained more information per pulse than linear DM. His method consisted of doubling the step size whenever identical consecutive bits are produced at the coder output. Step sizes are divided in half

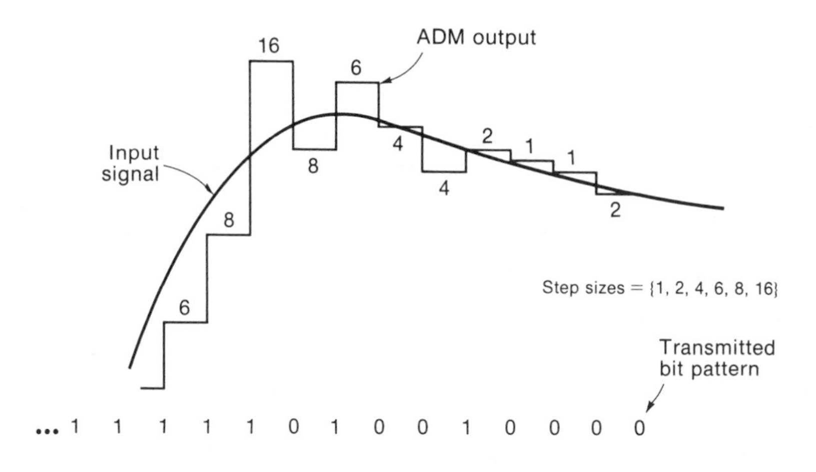

Figure 3.21 Sample Output of Adaptive Delta Modulator Using Logic of Table 3.5

after each polarity reversal. This scheme was applied to video signals and significantly improved the dynamic range over linear DM [20].

Bosworth and Candy [21] described a companded coder specifically designed for picture-phone transmission. It differs from HIDM in that its step response is more damped. The adaptation algorithm was designed to promptly increase the step size for transients (busy areas of picture and at edges) and to promptly decrease step size after a transient in order to minimize overload distortion and quantization noise, respectively. Subjective tests on picture-phone signals were used to pick the optimum weighting sequence (1, 1, 2, 3, 5), in which each weight is equal to the sum of the two previous weights. This sequence minimizes overshoot by converging to the smallest increment after a transient. Specifically, when three or more consecutive 1's or 0's are produced, the weighting logic increases the step size by 1, 1, 2, 3, 5, . . . , 5. Eventually a bit of opposite polarity is observed, and the step size returns to 1. The direction of any step is determined by the polarity of the current coder bit.

Jayant [22] has described an ADM design applicable to speech coding. Here the step size Δ_i at time i is related to the previous step size by

$$\Delta_i = \Delta_{i-1} \cdot P^{b_i b_{i-1}} \tag{3.79}$$

where b_i and b_{i-1} are the present and previous bits, respectively. Variation in step size is determined by the factor P, which for speech has an optimum value $1 < P < 2$ [22]. Note that a choice of $P = 1$ equates to linear delta modulation (LDM). For a choice of $P = 1.5$, a sampling rate of 60 kHz, and a speech signal band-limited to 3.3 kHz, computer simulation showed a 10-dB S/D improvement over LDM.

At low bit rates for digitized speech, instantaneously companded DM systems are seriously degraded by excessive quantization noise. One means of improving speech quality for low-rate coders is the use of **syllabic companding**, wherein the step sizes are adapted more smoothly in time, controlled by the syllabic rate of human speech. A syllabic filter has a typical bandwidth of 100 Hz, so that the step sizes are adapted with a time constant on the order of 10 ms. The slow adaptation of syllabic companding improves upon quantization noise at the expense of increased slope overload distortion. Even so, for speech bit rates below 25 kb/s, syllabic companding is preferable [23]. Moreover, syllabic companding provides resistance to bit errors by its slow-changing nature.

Continuously adaptive delta modulation, first described by Greefkes and DeJager [24], is another version of DM that can improve performance (that is, increase dynamic range) or, alternatively, reduce the bit rate required for digitized speech. This approach derives its name from the way the slope of the coder output varies in an almost continuous fashion, as opposed to discrete ADM where the step sizes are limited to a discrete set. Current implementations of this technique have been generally called **continuously variable slope delta modulation** (CVSD). A typical CVSD encoder/decoder is shown in Figure 3.22. Step size is controlled in a manner similar to discrete ADM, but with unlimited step size values and

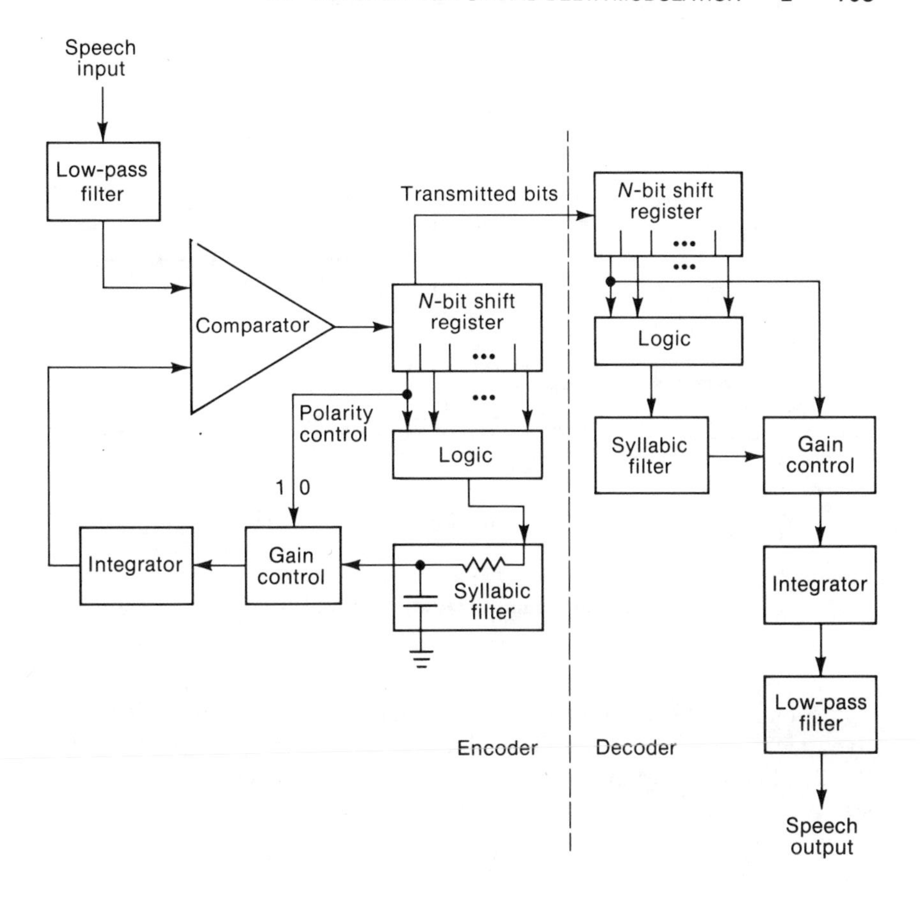

Figure 3.22 Block Diagram of Typical CVSD Modulator

usually syllabic control to slow the rate of change of step size. For typical CVSD speech coders (32 kb/s) a dynamic range of 30–40 dB is possible, which represents approximately 10 dB improvement over discrete ADM coders.

To illustrate the operation of a CVSD coder, assume a 3-bit shift register in Figure 3.22 with the following logic applied:

- Let the smallest step size be Δ.
- For 3-bit combinations 000,111: Increase step size by 3Δ.
- For present bit reversals $0 \rightarrow 1$, $1 \rightarrow 0$: Decrease step size by 3Δ.
- For other 3-bit combinations: Leave slope unchanged.
- The direction of slope is determined by the present bit.

Figure 3.23 shows the resulting CVSD coder output for a sample analog input, assuming instantaneous companding (that is, no syllabic control).

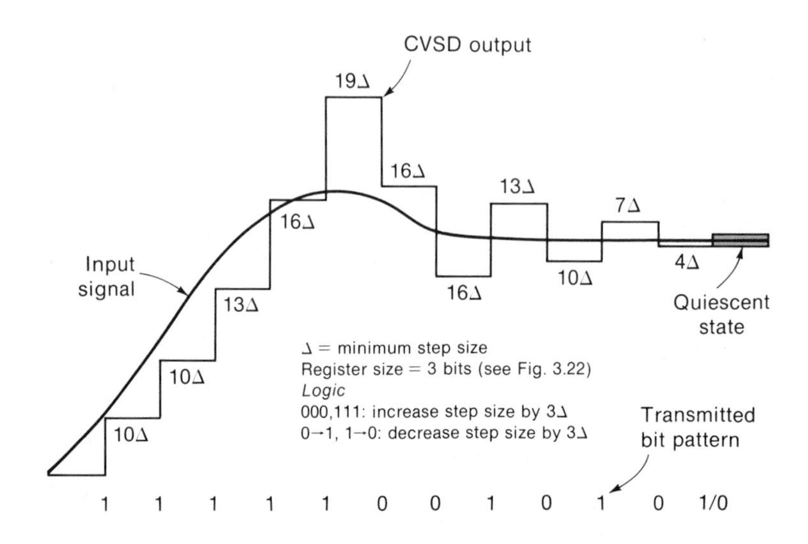

Figure 3.23 Sample Output of CVSD Coder

Similar to ADM, DPCM step sizes can be adapted to provide increased dynamic range. Coders with these adaptive features have been referred to as ADPCM coders. Jayant [25] has described one such coder in which the step size Δ_i at time i is related to the previous step size Δ_{i-1} by

$$\Delta_i = \Delta_{i-1} M(|P_{i-1}|) \tag{3.80}$$

The adaptation factor M is a time-invariant function of the previous code word magnitude $|P_{i-1}|$. According to (3.80), the quantizer range is compressed when $M < 1$ and expanded when $M > 1$. Within a given range, the step sizes are uniform. The preceding algorithm has been used with 3-bit and 4-bit ADPCM speech coding at bit rates of 20 to 32 kb/s [26]. A more recent version of 4-bit ADPCM at 32 kb/s has been developed for telephone network applications; this coder uses a dynamic locking quantizer (DLQ) that is unlocked for speech and locked for voice-band data and signaling tones [27]. Because of its robustness, versatility, and transmission efficiency, 32 kb/s ADPCM is emerging as a digital transmission standard for telephone networks. (See Chapter 12 for a discussion of this 32 kb/s ADPCM standard.)

3.4 COMPARISON OF DM, DPCM, AND PCM

Because DM, DPCM, and PCM have emerged as leading standards for digital transmission of analog information, a comparison of their performance is given here to assist the system engineer in the choice of applicable techniques. Since PCM has been shown to provide adequate performance for a variety of analog sources, the kernel of this comparison lies in the potential for reduced bit rate capabilities via some alternative form of coding. In light of this objective, issues of coder performance (quantizing and overload distortion), transmission performance (effects of error rate and multiple A/D–D/A conversion and ability to handle voice-band and nonspeech signals), and implementation complexity are examined using the example of speech coding.

3.4.1 Quantizing Distortion

Figure 3.24 shows the output S/D_q ratio for logarithmic (μ-law) PCM, linear PCM, and delta modulation as a function of input power level. A sinusoidal input of frequency $f_0 = 800$ Hz is assumed. The results for linear and logarithmic PCM were obtained from (3.41) by setting $\mu = 0$ and $\mu = 255$, respectively; $A = 1$; $N = 256$ for a transmission bit rate of 64 kb/s; and by varying the parameter C according to (3.45). The corresponding results for delta modulation were obtained from (3.55) and (3.56)

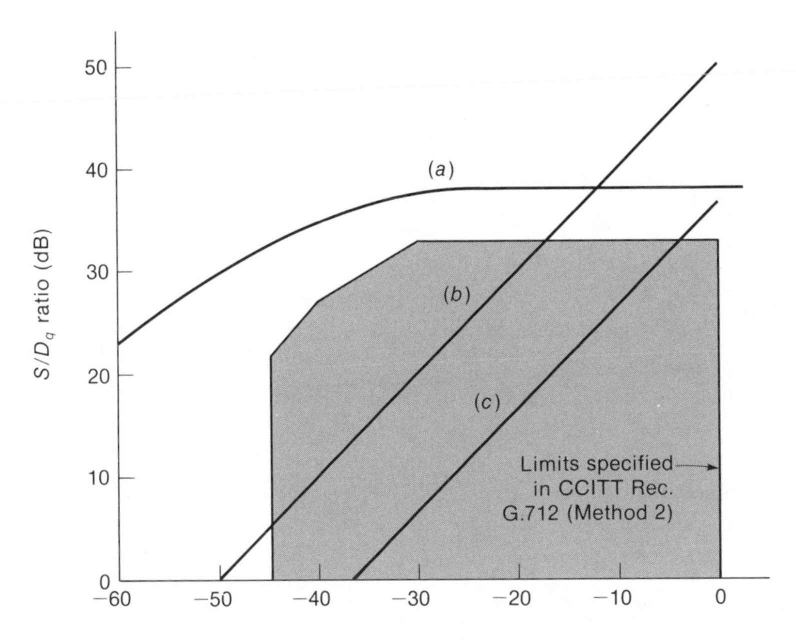

Figure 3.24 S/D_q **Versus Input Signal Power: (*a*) μ-Law PCM with $\mu = 255$ and $n = 8$ Bits; (*b*) Linear 8-Bit PCM; (*c*) Linear DM with $f_s = 64$ kb/s, $f_m = 3400$ Hz, $f = 800$ Hz**

by letting $f_s = 64$ kb/s and $f_m = 3400$ Hz. For all three curves, the input is normalized to a full-load sinusoid that exactly matches the quantizer range so that no overload distortion is introduced. Note that only μ-law PCM meets CCITT Rec. G.712.

A comparison of the three curves in Figure 3.24 indicates the limited dynamic range of the delta modulator as compared to linear or logarithmic PCM. A similar comparison for gaussian or speech input signals would yield curves having the same general shapes as those in Figure 3.24 [28, 23].

Figure 3.25 shows the variation in S/D_q versus bit rate for μ-law PCM and delta modulation. Again a sinusoidal input is assumed. The curve for μ-law PCM was obtained from (3.41), where $\mu = 255$, $A = C = 1$, and N is varied according to the selected bit rate. The maximum S/D_q curves for delta modulation were obtained from (3.56) with $f_m = 3400$ Hz; $f_0 = 800$, 1600, and 2600 Hz, respectively, for curves a, b, and c; and f_s varied according to the bit rate. Curves a, b, and c illustrate the characteristic of delta modulation that the peak S/D_q decreases by 6 dB for every octave increase in f_0 and increases by 9 dB for every doubling of the sampling rate (bit rate) f_s. Curve d illustrates the PCM characteristic that the S/D_q improves by 6 dB for each additional bit of coding. A comparison of PCM

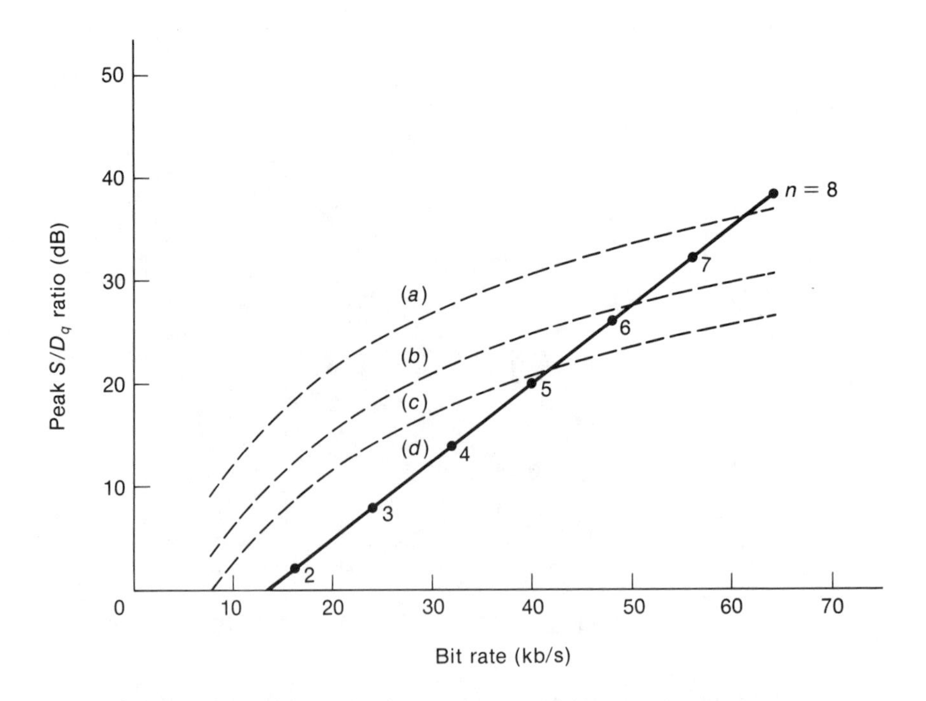

Figure 3.25 Peak S/D_q Versus Bit Rate for Sinusoidal Inputs: (*a*), (*b*), (*c*) Are DM with f_m = 800, 1600, and 2600 Hz Respectively; (*d*) Is μ-Law PCM with μ = 255

versus DM performance curves shown in Figure 3.25 indicates that at high bit rates PCM outperforms DM while the reverse holds true for low bit rates.

3.4.2 Overload Distortion

Delta modulation overload noise results when the input signal *slope* exceeds the maximum slope capability of the DM quantizer; PCM overload noise results when the input signal *amplitude* exceeds the maximum levels of the PCM quantizer. Analytic expressions for both DM overload noise (3.68) and PCM overload noise (3.23*a*) have been derived for gaussian input signals. These two expressions can be conveniently compared as a function of a bandwidth expansion factor B, defined as the ratio of the bandwidth of the transmission channel to that of the signal. For DM systems, B is simply one-half the ratio of sampling rate to signal bandwidth, or $f_s/2f_m$. For PCM systems, B is identical to the number of coding bits. A comparison can now be made by plotting peak performance (that is, maximum S/D ratio obtained from Figures 3.5 and 3.16) versus the bandwidth expansion factor. It is clear from Figure 3.26 that for gaussian signals, linear PCM provides superior performance to that of linear DM. A comparison of the results of

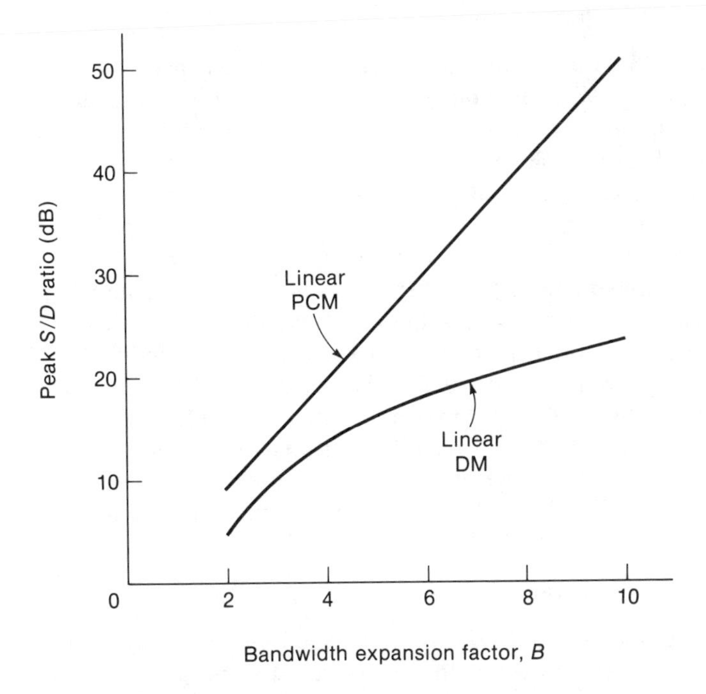

Figure 3.26 Peak S/D Versus Bandwidth Expansion Factor for Gaussian Inputs

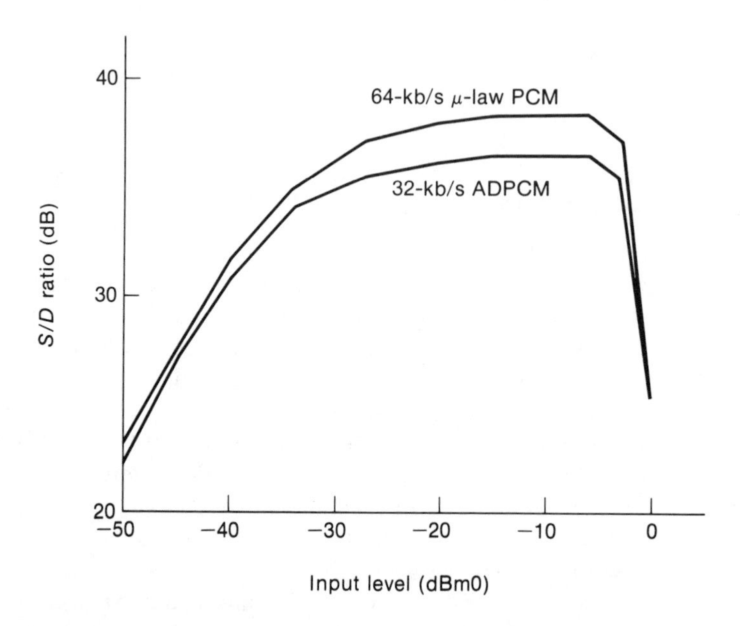

Figure 3.27 *S / D* Ratio for PCM and ADPCM with Gaussian Input Signal [27] (Reprinted by Permission of AT & T)

Figure 3.25 with Figure 3.26 indicates the significance of slope overload in DM systems. With only quantization noise taken into account, Figure 3.25 indicates that DM has superior performance over μ-law PCM for low rate coders. With *both* slope overload and quantization noise taken into account, however, Figure 3.26 indicates that linear PCM provides superior performance at all bit rates.

A similar comparison involving adaptive delta modulation would show that the maximum S/D (where D is the sum of quantizing and slope overload distortion) remains approximately the same as that of linear DM, thus allowing us to extend the results of Figures 3.25 and 3.26 to ADM [17]. The companding improvement inherent in ADPCM and ADM does increase the dynamic range, however, so that the results of Figure 3.24 are not applicable to ADPCM or ADM. Measured results of log PCM and ADPCM S/D performance are shown in Figure 3.27 to illustrate the relative dynamic ranges achievable with these two companding techniques.

3.4.3 Transmission Errors

Delta modulation and DPCM are affected differently by bit errors than is PCM. The feedback action in DPCM leads to a propagation of errors in the decoder output. The cumulative effect of this error propagation becomes a significant problem for error bursts, producing a "streaking" effect in video transmission. For speech transmission, however, the natural gaps and pauses

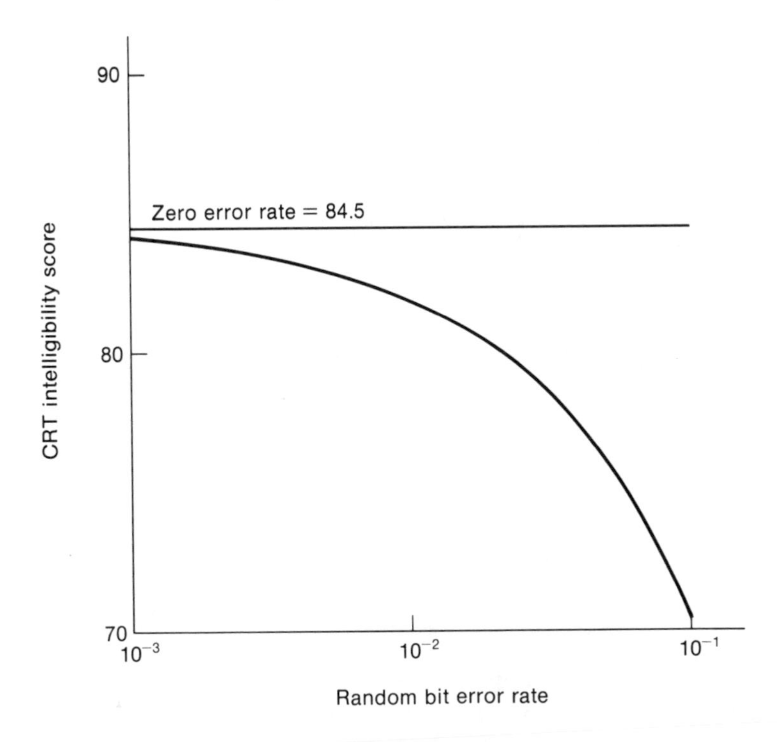

Figure 3.28 Intelligibility vs. Random Bit Error Rate for CVSD at 16 kb / s [29]

in speech effectively mitigate this effect. Subjective evaluations show that DPCM is more tolerant of bit errors than PCM for voice transmission systems. A bit error in PCM transmission can cause an error spike that has magnitude on the order of the peak-to-peak value of the input signal. Corresponding error spikes in a DPCM transmission systems are of smaller magnitude, since DPCM operates on the difference signal rather than absolute magnitude of the input signal. Therefore, even though errors do not propagate in a PCM system, the greater magnitude of error makes PCM more vulnerable to bit errors.

Speech intelligibility scores indicate that CVSD suffers negligible loss at an error rate of 10^{-3} as compared to error-free transmission. This finding is indicated in Figure 3.28, which plots intelligibility versus bit error rates, scored by use of the Consonant Recognition Test.* Similarly, at error rates

*The Consonant Recognition Test (CRT) assesses intelligibility on the basis of a listener jury's ability to perceive the initial consonants of spoken monosyllabic words correctly. The CRT test material typically consists of 360 words arranged in 36 lists of 10 words. Each group of 10 words differs only in their initial consonants. A jury listens to a recording of spoken test words and attempts to write the initial consonants on an answer sheet. A computer analysis of the responses of the listeners then produces an intelligibility score [29].

in the vicinity of 1×10^{-3}, delta modulation suffers only a loss that is comparable to quantization noise, as shown in Table 3.4. By comparison, a subjective evaluation of logarithmic PCM at 64 kb/s indicates significant degradation at an error rate of 10^{-3} [1].

3.4.4 Effect of Multiple A / D and D / A Conversions

An analog signal passed through multiple analog-to-digital and digital-to-analog conversions is degraded because of the accumulation of quantization noise and slope overload distortion. Jayant and Shipley [30] have investigated this effect for DM and PCM and have expressed results in terms of S/D degradation after M successive conversions. For a 60-kb/s delta modulation system a computer simulation using real speech was used to obtain the following empirical relationships [22, 30]:

$$\text{LDM:} \quad \left(\frac{S}{D}\right)_M = 30 - 10\log(3.15 + 0.35M^2) \tag{3.81}$$

$$\text{ADM:} \quad \left(\frac{S}{D}\right)_M = 40 - 10\log(0.9 + 3.5M + 1.4M^2) \tag{3.82}$$

Additionally, by assuming that successive PCM conversions have the effect of adding equal amounts of quantization noise, the following recursive formula results for S/D in a PCM system

$$\text{PCM:} \quad \left(\frac{S}{D}\right)_M = \left(\frac{S}{D}\right)_1 - 10\log(M) \tag{3.83}$$

Inspection of (3.83) and (3.82) indicates that S/D degradation in a PCM and ADM system decreases monotonically with M, while (3.81) indicates that LDM exhibits a constant rate of degradation, as illustrated in Table 3.6. As a consequence, PCM and ADM systems provide superior performance over LDM for networks employing multiple conversions.

The CCITT has used the quantizing distortion produced by a single 8-bit PCM conversion (A-law or μ-law) as a basic reference, assigning it a

Table 3.6 Degradation in S / D Due to Multiple (M) Analog-To-Digital and Digital-to-Analog Conversions

	Values of $\|(S/D)_M - (S/D)_{M+1}\|$ *(dB)*				
M	*1*	*2*	*4*	*6*	*8*
PCM	3.0	1.8	1.0	0.7	0.5
ADM	3.7	2.4	1.6	1.1	0.9
LDM	1.1	1.3	1.4	1.1	0.9

Source: After Jayant and Shipley [30].

Table 3.7 Quantizing Distortion for Various A / D Conversions [27, 31]

A / D Conversion	*Quantizing Distortion (qd units)*	*S / D Degradation in dB Relative to 8-bit PCM (10 log qd)*
8-bit PCM	1	0
7-bit PCM	4	6
A / μ law or μ / A law conversion	1	0
32-kb / s ADPCM	4	6
International telephone connection (CCITT Rec. G113)	14	11.5

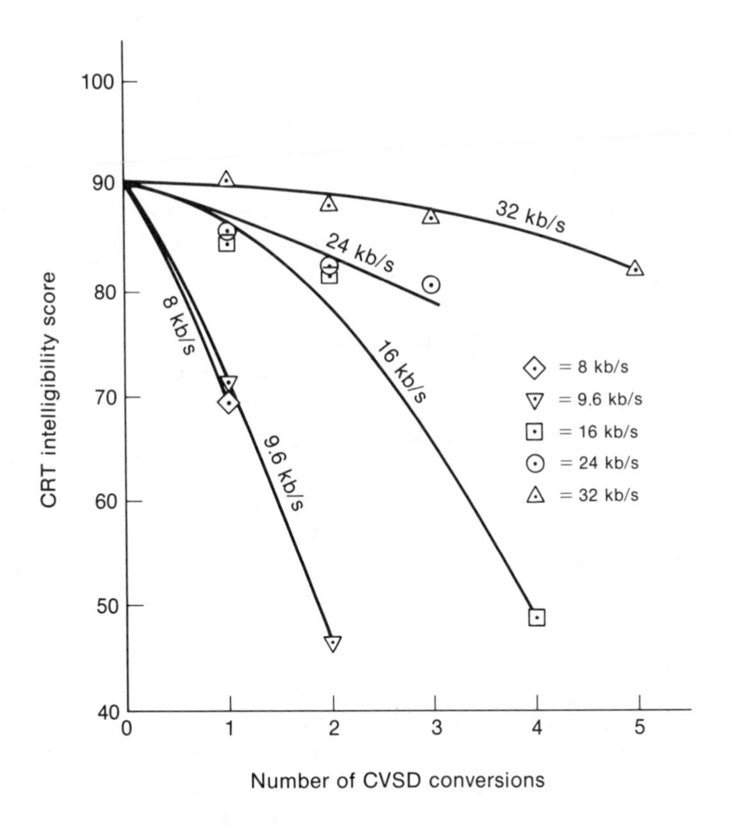

Figure 3.29 Intelligibility vs. Number of CVSD Conversions at Various Bit Rates [29]

value of 1 qd (quantizing distortion) unit. Other A/D converters are assigned values referenced to 8-bit PCM, as shown in Table 3.7. For example, 7-bit PCM is known to produce 6 dB more quantizing distortion and is therefore assigned 4 qd units. Since quantizing distortion is uncorrelated from one converter to the next, individual qd units can be added algebraically to yield the overall distortion for an end-to-end connection. CCITT Rec. G.113 provisionally assigns a limit of 14 qd units for an international telephone connection [31]. This value corresponds to the quantizing distortion produced by 14 consecutive PCM conversions and to a S/D degradation of 11.5 dB relative to a single PCM conversion.

The effect of multiple conversions is strongly dependent on the sampling rate of the coder. For example, Figure 3.29 shows intelligibility scores for several CVSD coder rates from 32 to 8 kb/s. Whereas 32-kb/s CVSD provides acceptable performance through five conversions, 9.6-kb/s CVSD performance becomes unacceptable with the second conversion. In general, 64-kb/s PCM maintains acceptable quality with up to 14 conversions, while coders at $\leq$ 32 kb/s including ADM, CVSD, and ADPCM are not nearly as robust with multiple conversions.

3.4.5 Transmission of Voice-Band Nonspeech Signals

When DM and PCM coders are designed for transmission of speech-generated analog signals, there may also exist requirements to pass quasi-analog signals (voice-band data modems) or other nonspeech signals (signaling information). Logarithmic PCM used in telephony provides the wide dynamic range required for speech but leads to nonoptimum performance for voice-band data. The quantization noise of logarithmic PCM is largest at the peak of the data signal envelope, the point at which data detection is done. Hence the average quantization noise cannot be used as a parameter to predict voice-band data performance accurately. Similarly, the adaptation of step size used in ADM enhances the dynamic range for speech but is nonoptimum for modem signals. The average power level of modem signals remains fairly constant so that companding is not required if the proper (fixed) step size is selected. For DM systems to work properly with data signals, the adaptive feature must be disabled.

Performance of modem signals transmitted via various voice coders has been reported in analytic studies and simulation [32–34], but empirical results are not as widely published. Notably, May and colleagues have experimentally studied performance of five different modems transmitted through tandem connections of both DM and PCM [33]. Significant results are summarized in Table 3.8. PCM and DM are seen to provide comparable modem BER performance for a 64-kb/s channel rate. With errors inserted into the transmission line, however, the BER performance of a modem was superior for DM channels vis-à-vis PCM channels; both were operated at a

Table 3.8 Modem BER Performance over Multilink PCM and DM Coders

Coder Type and Channel Bit Rate	*Modem Bit Error Rates*											
	1200 b / s			*2400 b / s*			*4800 b / s*			*9600 b / s*		
	1 Link	*2 Links*	*4 Links*	*1 Link*	*2 Links*	*4 Links*	*1 Link*	*2 Links*	*4 Links*	*1 Link*	*2 Links*	*4 Links*
DM												
32 kb / s	O	O	X	O	O	X	X	X	X	X	X	X
48 kb / s	O	O	O	O	O	O	O	X	X	X	X	X
64 kb / s	O	O	O	O	O	O	O	O	X	X	X	X
PCM												
64 kb / s	O	O	O	O	O	O	O	O	X	X	X	X

Note: O = error free; X = error-free operation not possible.

Source: After May and others [33].

channel rate of 64 kb/s. Results indicated that the modem BER falls off exponentially as the DM transmission BER decreases, whereas it decreases linearly as the PCM transmission BER decreases. Table 3.8 indicates that 32-kb/s DM will support modem data rates up to 2400 b/s, a conclusion reached by similar experiments [29, 35, 36] and analyses [32]. Moreover, analysis [32] and experiment [29] have shown that 16-kb/s DM will not support modem data rates of 1200 b/s and above. Other test results indicate that 32-kb/s ADPCM-DLQ will support modem data rates at 4800 b/s and below [27].

For switched telephone networks, in-band signaling information may be transmitted through the voice coder. Testing of single-frequency (SF) and dual-tone multifrequency (DTMF) signaling transmitted over CVSD, PCM, and ADPCM channels indicated the following [27, 29, 37]:

1. For SF signaling at 2.6 kHz, 64-kb/s CVSD provided acceptable performance; 16-kb/s and 32-kb/s CVSD did not.
2. For DTMF signaling at frequencies from 697 to 1477 Hz, 32-kb/s CVSD provided acceptable performance but 16 kb/s CVSD did not.
3. Log PCM at 64 kb/s and ADPCM-DLQ at 32 kb/s provided adequate performance for transmission of SF or DTMF signaling.

3.4.6 Implementation Complexity

Coders

Coder implementation for DM and DPCM is potentially much simpler than for PCM, although a PCM coder may be shared among a number of subscribers while a DM coder cannot. With the advent of large-scale integration techniques, however, coder costs have become less important. Current-generation channel banks in fact utilize one coder per subscriber to improve overall reliability and eliminate crosstalk caused by switching the coder from one input to the next.

Filters

Filtering in a PCM coder and decoder requires sharp cutoff to $f_s/2$ Hz in accordance with the Nyquist sampling frequency. This requirement for sharp cutoff filtering is in contrast to a DPCM or DM coder and decoder where the sampling frequency is typically much larger than the message bandwidth so that a gradual rolloff can be used for the filter characteristic. As a consequence filter design for DPCM and DM is simpler than for PCM.

3.5 VOICE CODERS (VOCODERS)

Vocoders are based on the speech model shown in Figure 3.30, in which the source of sound is separated from the vocal tract filter. Speech sound is assumed to be either voiced ("buzz"), corresponding to the periodic flow of

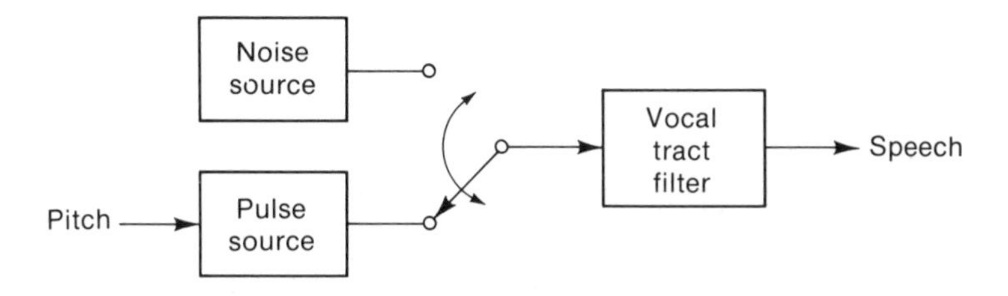

Figure 3.30 Model of Human Speech

air generated by the vocal cords, or unvoiced ("hiss"), corresponding to a turbulent flow of air past a constriction in the vocal tract. Unvoiced sound is an acoustic noise source that can be represented by a random noise generator; voiced sound is an oscillation of the vocal cords that can be represented by a periodic pulse generator. Sound, voiced or unvoiced, passes through the vocal tract, which includes the throat, tongue, mouth, and nasal cavity. The moving parts of the vocal tract, called the articulators, modulate the sound thereby creating speech signals. During voiced sounds the articulators assume different positions causing resonances in the vocal tract that impose spectral peaks called *formants* (Figure 3.31). In Figure 3.31 the speech signal is assumed to be voiced, resulting in a spectrum with equally spaced harmonics at the fundamental pitch of the speaker. The spectral envelope is determined by the vocal tract shape, which is determined by the spoken sound. Should the sound be unvoiced, the periodic structure would disappear leaving only the spectral envelope.

All vocoders work on the same basic principles. Speech is analyzed to determine whether the excitation function (sound) is voiced or unvoiced; if

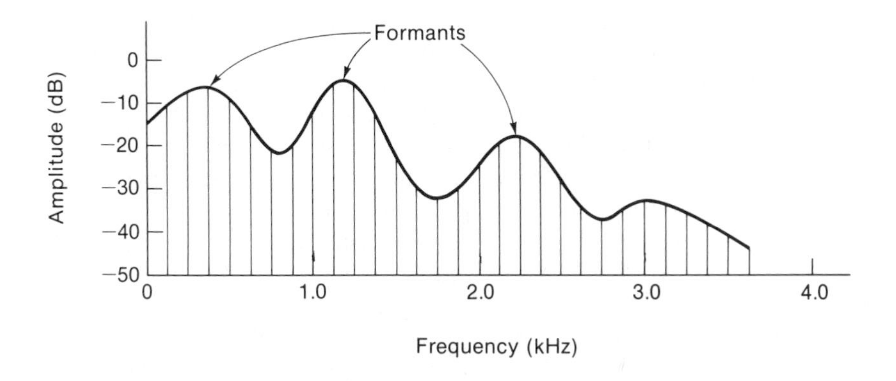

Figure 3.31 Spectral Content of Typical Speech

it is voiced, the vocoder then estimates the fundamental pitch. Simultaneously, the vocal tract filter transfer function (spectral envelope) is also estimated. Since speech is a nonstationary process, this voiced/unvoiced decision, pitch extraction, and spectral estimate must be updated with each successive speech segment. These principles have been implemented in vocoders in a variety of forms: In the **channel vocoder**, values of the short-time spectrum are evaluated by specific frequencies; the **formant vocoder** evaluates the amplitude of only major spectral peaks (formants); in the **LPC vocoder**, linear predictive coefficients describe the spectral envelope. All such vocoders depend on the accuracy of the model in Figure 3.30, which places a fundamental limitation on the quality of vocoders. Typical low-bit-rate vocoders produce an unnatural synthetic speech that although intelligible tends to obscure speaker recognition. The value of vocoders lies in the inherent bandwidth reduction.

3.5.1 Channel Vocoders

The strategy of the channel vocoder is to represent the spectral envelope by samples taken from contiguous frequency bands within the speech signal. In parallel, the voiced/unvoiced decision and pitch extraction are performed; these operations determine the fine-grain structure of the speech spectrum. A block diagram of a channel vocoder analyzer and synthesizer is shown in Figure 3.32. The outputs of a bank of bandpass filters are connected to a rectifier and low-pass filter. The signal $X_i(t)$ represents the smoothed amplitude of the input speech for the ith frequency band. The vocal tract filter can be reasonably well described by 16 channel values taken every 20 ms. Low-pass filtering limits the bandwidth of each spectral channel to about 25 to 30 Hz. The excitation signals require another 50 Hz of bandwidth, so that vocoded speech may be transmitted in a bandwidth of about 500 Hz for a bandwidth savings of 7:1.

At the synthesizer, the speech signal is regenerated by modulating the excitation function (noise or pulse source) with the samples of the vocal tract filter. After bandpass filtering and summation, the result is synthetic speech that reproduces the short-term spectrum of the original speaker.

Despite some 40 years of research in channel vocoders [38], no optimum design exists today and channel vocoders still produce the unnatural machinelike speech that was characteristic of early vocoders [39]. Their chief application was in early attempts at providing digital voice to facilitate encryption in military applications. A digital channel vocoder operating at 2400 b/s could be readily encrypted and transmitted by existing 3-kHz telephone channels. In a 2400-b/s vocoder there are nominally 3 bits per spectral channel, or 48 bits for a 16-channel vocoder, plus 6 bits for pitch information resulting in a total of 54 bits per speech sample. A frame synchronization bit and a voiced/unvoiced decision bit usually occupy a least-weight bit position of one of the spectral channels. A transmission rate of 2400 b/s results if the speech signal is sampled every 22.5 ms. Further bit rate (or bandwidth) reduction was realized with the formant vocoder, which

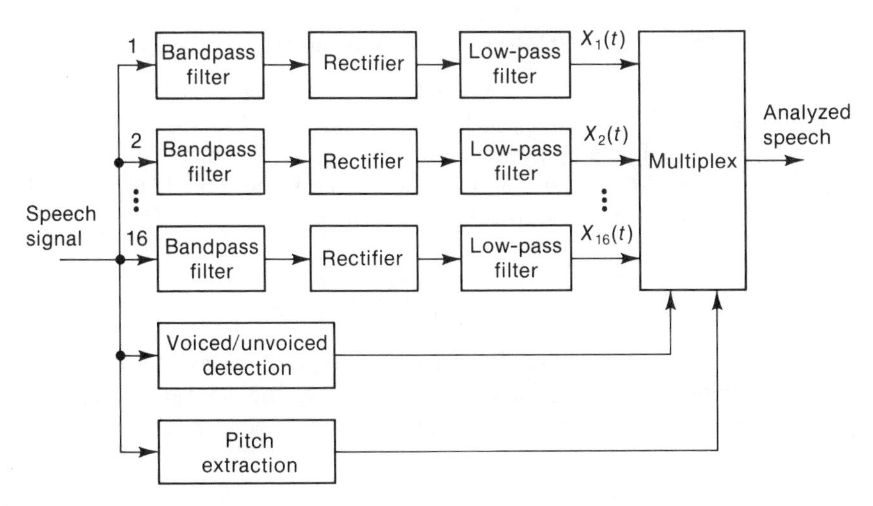

(a) Channel vocoder analyzer

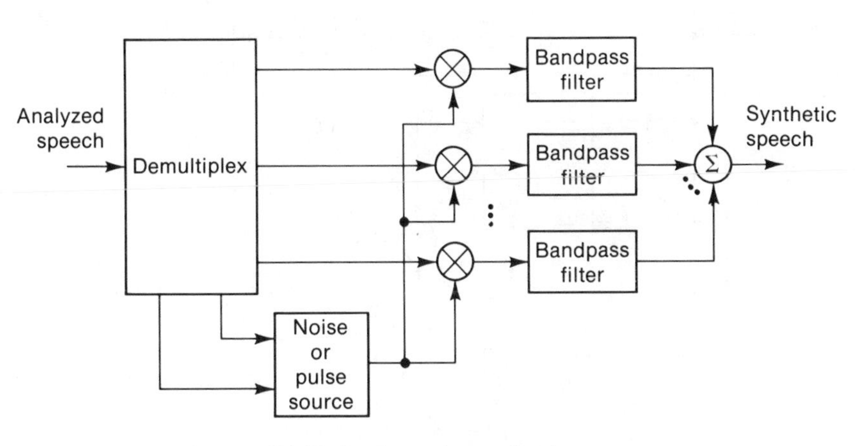

(b) Channel vocoder synthesizer

Figure 3.32 Block Diagram of Channel Vocoder

specified only the frequencies of the spectral peaks (formants) and their amplitudes. However, the added complexity and even poorer voice quality compared to the channel vocoder have limited the application of formant vocoders. Voice-excited vocoders (VEV) [40] provide improved voice quality at the expense of higher data rates (typically 9.6 kb/s). The VEV performs PCM on the frequency band below 900 Hz, which replaces the voiced/unvoiced decision and pitch extraction of the channel vocoder. The spectral information is handled in a similar way to the channel vocoder, except that the synthesizer modulators are excited by the PCM information below 900 Hz.

3.5.2 Linear Predictive Coding (LPC) Vocoders

The linear predictive coding (LPC) vocoder is based on the concept that a reasonable prediction of a speech sample can be obtained as a linear weighted sum of previously measured samples. In the block diagram of Figure 3.33, the LPC analyzer and synthesizer both use a predictor filter of order P that provides an estimate $\hat{S}_n$ expressed as

$$\hat{S}_n = \sum_{m=1}^{P} a_m S_{n-m} \tag{3.84}$$

where the a_m are known as predictor coefficients. The order of the predictor filter is typically 8 to 12, where, for example, the spectral resolution provided by $P = 8$ matches that of the 16-channel vocoder. The predicted

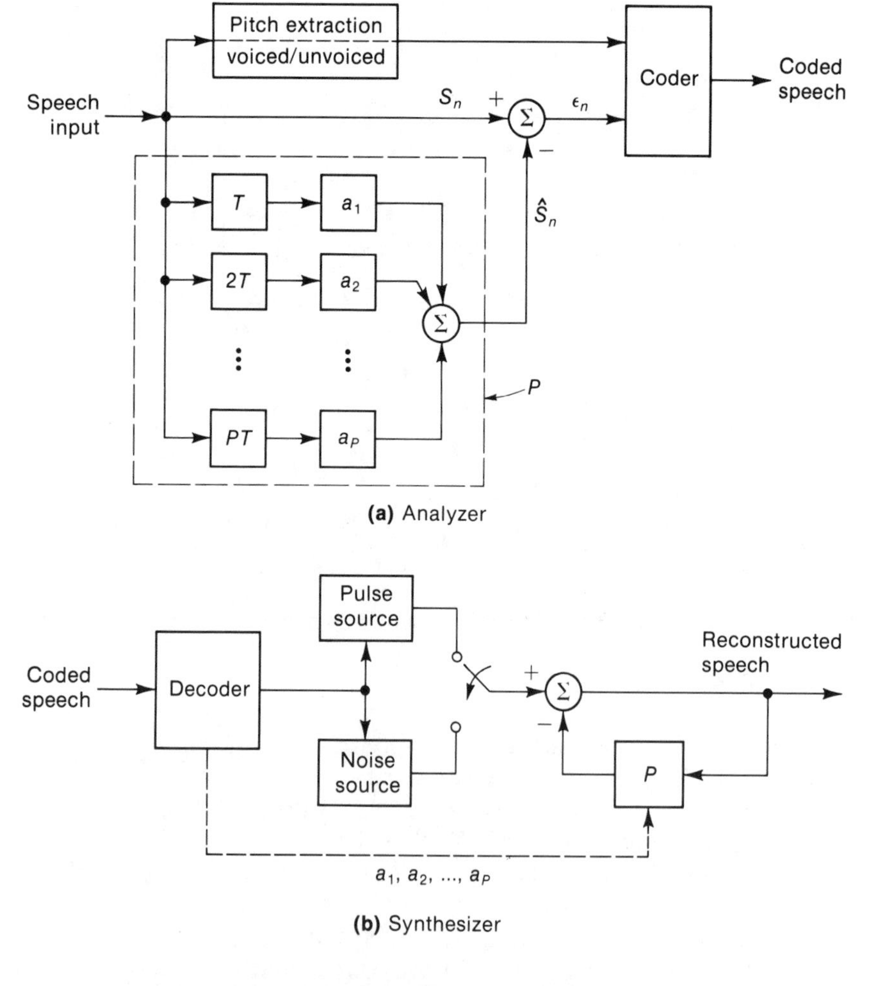

Figure 3.33 Linear Predictive Coding (LPC) Vocoder

value $\hat{S}_n$ of the speech signal is next subtracted from the signal value S_n to form the difference ϵ_n, which is coded and transmitted to the receiver.

Voicing and pitch extraction are estimated as they are in a channel vocoder. At the synthesizer the inverse filter to the analyzer is excited by the pulse (voiced) or noise (unvoiced) source. Predictor coefficients are generally updated every 10 to 25 ms (approximately the syllabic rate). The difference between methods of implementing LPC vocoders lies within the method of determining the predictor coefficients [41–42]. The implementation of LPC has proved simpler than channel vocoders, one of the significant factors in the growing popularity of LPC. An emerging standard for LPC is based on a 2400-b/s data rate, 10 predictor coefficients, and a frame length of 54 bits [43].

3.5.3 Adaptive Predictive Coding (APC) Vocoders

Because of the nonstationary nature of speech signals, a fixed predictor cannot efficiently predict the signal values at all times. Ideally the predictor coefficients must vary with changes in the spectral envelope and changes in the periodicity of voiced speech. Adaptive prediction of speech is a deconvolving process that is done conveniently in two separate stages as shown in Figure 3.34*a*: One prediction (P_1) is with respect to the vocal tract filter and the other (P_2) is with respect to pulse excitation [44]. The short-time spectral envelope predictor can be characterized in Z-transform notation as

$$P_1(Z) = \sum_{K=1}^{P} a_K Z^{-K} \tag{3.85}$$

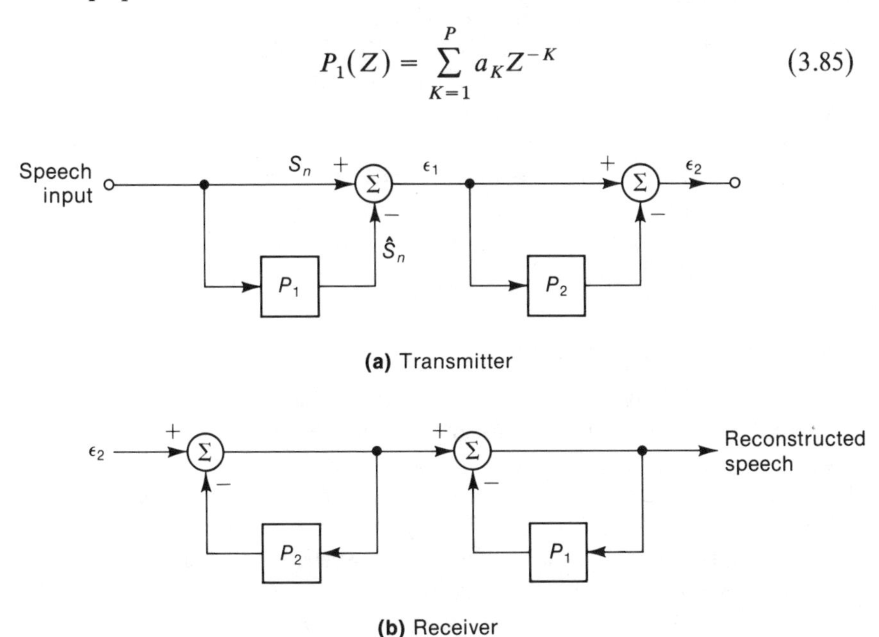

(a) Transmitter

(b) Receiver

Figure 3.34 Adaptive Predictive Coding (APC)

where Z^{-1} represents a delay of one sample interval and $a_1, a_2, \ldots, a_P$ are the P predictor coefficients. The residual ϵ_1 that remains after the first prediction exhibits the periodicity associated with the excitation signal when the speech is voiced. The second predictor is simply a delay of M speech samples and a weighting factor β, so that

$$P_2(Z) = \beta Z^{-M} \tag{3.86}$$

In most cases, the delay M is chosen to be equivalent to the pitch period or an integral number of pitch periods [45, 46]. The receiver structure for the APC vocoder is depicted in Figure 3.34*b* and is the inverse of the transmitter. This receiver filter is determined by the values of β, M and the a_i's, each of which has been quantized, coded, transmitted, and decoded at the receiver. There are two limiting factors in APC vocoder performance: One is the existence and significance in listener perception of small variations in the periodicity of voiced speech; the other is degradation due to incorrect decoding of the vocal tract filter parameters. Although the first limitation is nontrivial, the second problem is readily solved via error correction coding.

As to implementations of APC, Atal and Schroeder [44] claim that 10-kb/s APC compares favorably with logarithmic ($\mu = 100$) PCM at 40 kb/s. Detailed testing of 8-kb/s APC indicated performance equivalent in voice quality to 16-kb/s CVSD.

3.6 SUMMARY

The summary for this chapter is presented in the form of a comparison of digital coders for speech. Comparative evaluations of speech processing algorithms must be tentative because of ongoing developments in speech coders. Nevertheless, some observations can be made based on past experience, current technology, and future trends. The choice of a speech coder for a given application involves a tradeoff of three elements: transmission bit rate, quality, and complexity (cost). Each of these elements is addressed in the following paragraphs.

3.6.1 Transmission Bit Rate

Figure 3.35 compares transmission bit rates for speech coders currently of interest. The lowest practical rates are 1.2 to 2.4 kb/s, used with formant and channel vocoders. Bit rates of 2.4 and 4.8 kb/s are most typical for LPC, while APC rates range from 8 to 16 kb/s. Common bit rates for CVSD and ADPCM are 16 and 32 kb/s. Linear DM or DPCM typically requires 32 to perhaps 64 kb/s. PCM is at the high end of the spectrum, ranging from 48-kb/s (6-bit) logarithmic to 96-kb/s (12-bit) linear PCM. Of course, 64-kb/s logarithmic PCM is the internationally recognized standard for digital speech.

To segregate transmission bit rate requirements for digital speech, two general categories are commonly used. **Narrowband coders** utilize rates up to and including 9.6 kb/s that can be handled by 3-kHz telephone channels

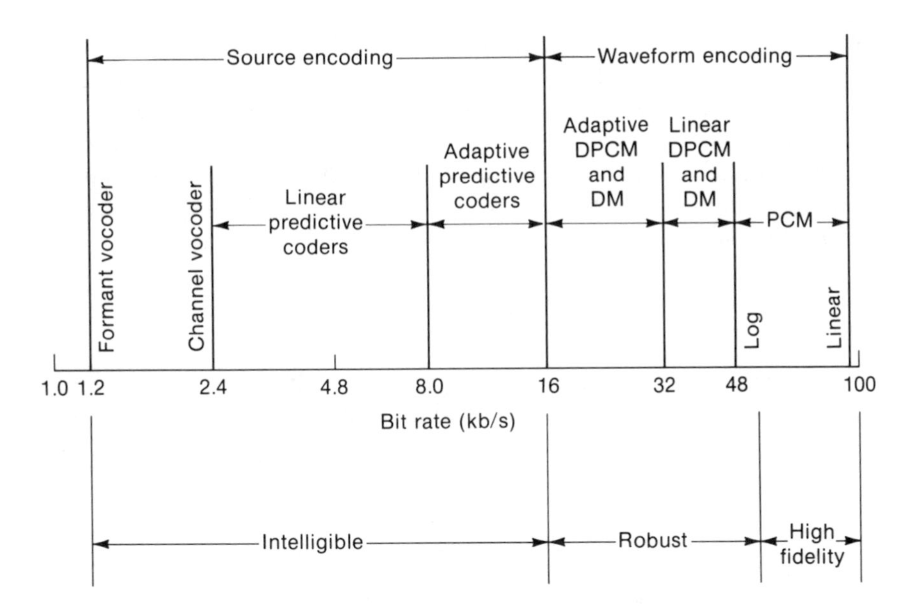

Figure 3.35 Transmission Bit Rates and Quality for Speech Coders

using modems. **Wideband coders** exceed the transmission rate (generally 9.6 kb/s) that can be handled by a nominal telephone channel.

3.6.2 Quality

Figure 3.35 also indicates the quality of speech reproduction that can presently be attained by a prescribed bit rate. The quality indication is roughly divided into three categories: intelligible, robust, and high fidelity. Vocoders, having the lowest bit rates, fall into the intelligible range; they are characterized as highly intelligible but suffering from some loss of naturalness and reduced speaker recognition. At rates above 16 kb/s, robust speech is obtainable by using techniques such as CVSD, ADPCM, or logarithmic PCM. These schemes depend less on individual speaker characteristics and maintain adequate performance with transmission impairments (tandeming or transmission errors). Hence most coders operating in this range provide adequate telephone toll quality. High-fidelity speech can be provided at rates above 64 kb/s, where input signal bandwidth may exceed the nominal telephone channel. This grade of quality is appropriate for radio broadcast material, including music.

Numerous comparisons of speech coders have been made, both objective and subjective, some of which have been referenced here. For further subjective evaluation, the reader is referred to speech recordings included on plastic records with two recent publications [1, 23]. Another subjective comparison [47] studied the application of efficient coders to a telephone network. This study concluded that 64-kb/s, $\mu = 255$ PCM can be introduced into the telephone network without subjective penalty. Other coding techniques such as 48-kb/s, $\mu = 255$ PCM; 32-kb/s ADPCM; and 37.7-kb/s ADM were shown to introduce significant subjective degradation with A/D–D/A tandem encodings. This degradation with tandem encodings can be significantly reduced, however, by use of digital-to-digital tandem encoding, for example, between 32 kb/s ADPCM and 64 kb/s PCM as reported in [48].

3.6.3 Complexity

The speech coders described here range widely in cost, reliability, maintainability, and size. Some indication of relative performance is possible by consideration of coder complexity. Figure 3.36 illustrates coder

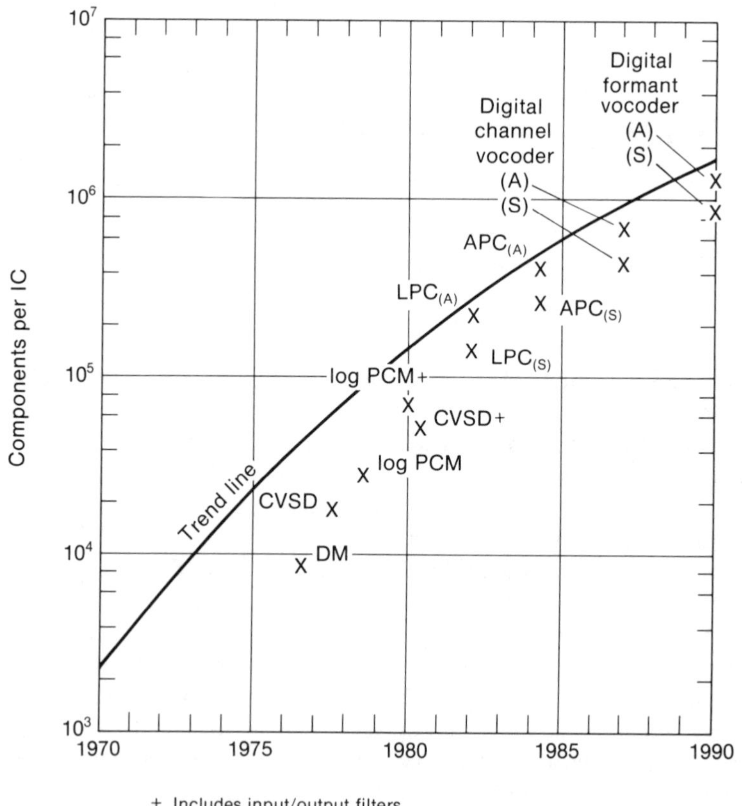

+ Includes input/output filters
(A) Analyzer
(S) Synthesizer
Source: Reprinted by permission of Gnostic Concepts, Inc.

Figure 3.36 Relative Voice Coder Complexity and Trends

complexity by a count of components required for large-scale integration. The trend line represents an upper bound of forecasts made up to 1990. By 1980 coder chips had been developed and utilized for log PCM and CVSD, including all functions (even multiplexing for 24 and 30-channel PCM channel banks) except input/output filters. During the early 1980s, filters were integrated with CVSD and log PCM chips. Projected component counts indicate the relative complexity of source coders such as LPC, APC, channel vocoders, and formant vocoders vis-à-vis log PCM and CVSD.

REFERENCES

1. J. L. Flanagan and others, "Speech Coding," *IEEE Trans. on Comm.*, vol. COM-27, no. 4, April 1979, pp. 710–736.
2. B. P. Lathi, *An Introduction to Random Signals and Communication Theory* (Scranton: International Textbook Company, 1968).
3. Bell Telephone Laboratories, *Transmission Systems for Communication*, 4th ed. rev. (Winston-Salem: Western Electric Company, 1971).
4. B. D. Holbrook and J. T. Dixon, "Load Rating Theory for Multi-Channel Amplifiers," *Bell System Technical Journal* 43(1939):624–644.
5. M. D. Paez and T. H. Glisson, "Minimum Mean-Squared Error Quantization in Speech PCM and DPCM Systems," *IEEE Trans. on Comm.*, vol. COM-20, no. 2, April 1972, pp. 225–230.
6. B. Smith, "Instantaneous Companding of Quantized Signals," *Bell System Technical Journal* 36(1957):653–709.
7. CCITT Yellow Book, vol, III.3, *Digital Networks—Transmission Systems and Multiplexing Equipment* (Geneva: ITU, 1981).
8. H. Mann, H. M. Stranbe, and C. P. Villars, "A Companded Coder for an Experimental PCM Terminal," *Bell System Technical Journal* 41(1962): 1173–1226.
9. S. M. Schreiner and A. R. Vallarino, "48-Channel PCM System," 1957 *IRE Nat. Conv. Rec.*, pt. 8, pp. 141–149.
10. A. Chatelon, "Application of Pulse Code Modulation to an Integrated Telephone Network, pt. 2—Transmission and Encoding," *ITT Elec. Comm.* 38(1963):32–43.
11. H. Kaneko, "A Unified Formulation of Segment Companding Laws and Synthesis of Codecs and Digital Compandors," *Bell System Technical Journal* 49(1970):1555–1588.
12. CCITT Green Book, vol. III. 3, *Line Transmission* (Geneva: ITU, 1973).
13. F. DeJager, "Delta Modulation: A Method of PCM Transmission Using a 1-Unit Code," *Philips Research Report*, December 1952, pp. 442–466.
14. H. Van de Weg, "Quantization Noise of a Single Integration Delta Modulation System with an *N*-digit Code," *Philips Research Report*, October 1953, pp. 367–385.
15. J. B. O'Neal, Jr., "Delta Modulation Quantizing Noise Analytical and Computer Simulation Results for Gaussian and Television Input Signals," *Bell System Technical Journal* 45(1966):app. A, pp. 136–139.

16. F. B. Johnson, "Calculating Delta Modulation Performance," *IEEE Trans. on Audio and Electroacoustics*, vol. AU-16, March 1968, pp. 121–129.
17. J. E. Abate, "Linear and Adaptive Delta Modulation," *Proc. IEEE* 55(3)(1967):298–308.
18. H. R. Schindler, "Delta Modulation," *IEEE Spectrum*, October 1970, pp. 69–78.
19. M. R. Winkler, "High Information Delta Modulation," *IEEE Inter. Conv. Record*, pt. 8, 1963, pp. 260–265.
20. M. R. Winkler, "Pictorial Transmission with HIDM," *IEEE Inter. Conv. Record*, pt. 1, 1965, pp. 285–291.
21. R. H. Bosworth and J. C. Candy, "A Companded One-Bit Coder for Television Transmission," *Bell System Technical Journal* 48(1)(1969):1459–1479.
22. N. S. Jayant, "Adaptive Delta Modulation with a One-Bit Memory," *Bell System Technical Journal* 49(1970):321–342.
23. N. S. Jayant, "Digital Coding of Speech Waveforms: PCM, DPCM, and DM Quantizers," *Proc. IEEE* 62(5)(May 1974):611–632.
24. J. A. Greefkes and F. DeJager, "Continuous Delta Modulation," *Philips Research Report* 23(2)(1968):233–246.
25. N. S. Jayant, "Adaptive Quantization with a One-Word Memory," *Bell System Technical Journal* 52(1973):1119–1144.
26. P. Cummiskey, N. S. Jayant, and J. L. Flanagan, "Adaptive Quantization in Differential PCM Coding of Speech," *Bell System Technical Journal* 52(1973):1105–1118.
27. AT&T Contribution to CCITT Question 7/Study Group XVIII, "32 kb/s ADPCM-DLQ Coding," April 1982.
28. R. Steele, *Delta Modulation Systems* (New York: Wiley, 1975).
29. E. D. Harras and J. W. Presusse, *Communications Performance of CVSD at 16/32 Kilobits/Second*, U.S. Army Electronics Command, Ft. Monmouth, N.J., January 1974.
30. N. S. Jayant and K. Shipley, "Multiple Delta Modulation of a Speech Signal," *Proc. IEEE* 59(September 1971):1382.
31. CCITT Yellow Book, vol. III.1, *General Characteristics of International Telephone Connections and Circuits* (Geneva: ITU, 1981).
32. J. B. O'Neal, "Delta Modulation of Data Signals," *IEEE Trans. on Comm.*, vol. COM-22, no. 3, March 1974, pp. 334–339.
33. P. J. May, C. J. Zarcone, and K. Ozone, "Voice Band Data Modem Performance over Companded Delta Modulation Channels," *Conf. Rec.* 1975 *Inter. Conf. on Comm.*, June 1975, pp. 40.16–40.21.
34. J. B. O'Neal, "Waveform Encoding of Voiceband Data Signals," *Proc. IEEE* 68(2)(February 1980):232–247.
35. E. V. Stansfield, "Limitations on the Use of Delta Modulation Links for Data Transmission," SHAPE Technical Centre Report STC CR-NICS-39, January 1979.
36. J. Evanowsky, "Test and Evaluation of Reduced Rate Multiplexers," Rome Air Development Center, Rome, N.Y., April 1981.

37. J. E. Hamant, O. P. Connell, and H. S. Walczyk, "Evaluation of a CVSD Multiplexer," U.S. Army Communications-Electronics Engineering Installation Agency, Fort Huachuca, Ariz., September 1977.

38. B. Gold, "Digital Speech Networks," *Proc. IEEE* 65(12)(December 1977):1636–1658.

39. H. Dudley, "The Vocoder," *Bell Labs Record* 17(1939):122–126.

40. B. Gold and J. Tierney, "Digitized Voice-Excited Vocoder for Telephone Quality Inputs Using Bandpass Sampling of the Baseband Signal," *J. Acoust. Soc. Amer.* 37(4)(April 1965):753–754.

41. B. Atal and S. L. Hanauer, "Speech Analysis and Synthesis by Linear Prediction of the Speech Wave," *J. Acoust. Soc. Amer.* 50(1971):637–655.

42. J. Marhoul, "Linear Prediction: A Tutorial Review," *Proc. IEEE* 63(1975): 561–580.

43. Proposed U.S. Federal Standard 1015, *Analog to Digital Conversion of Voice by* 2400 *Bit/Second Linear Predictive Coding*, National Communications System, Washington, D.C., July 1983.

44. B. S. Atal and M. R. Schroeder, "Adaptive Predictive Coding of Speech Signals," *Bell System Technical Journal* 49(1970):1973–1986.

45. E. E. David and H. S. McDonald, "Note on Pitch Synchronous Processing of Speech," *J. Acoust. Soc. Amer.*, November 1956, pp. 1261–1266.

46. A. H. Frei, H. R. Schindler, P. Vettiger, and E. Von Felten, "Adaptive Predictive Speech Coding Based on Pitch-Controlled Interruption/Reiteration Techniques," 1973 *Inter. Conf. on Comm.*, pp. 46.12–46.16.

47. W. R. Daumer and J. R. Cavanaugh, "A Subjective Comparison of Selected Digital Codec's for Speech," *Bell System Technical Journal* 57(1978):3119–3165.

48. D. W. Petr, "32 kb/s ADPCM-DLQ Coding for Network Applications," 1982 GLOBECOM, pp. A8.3.1–A8.3.5.

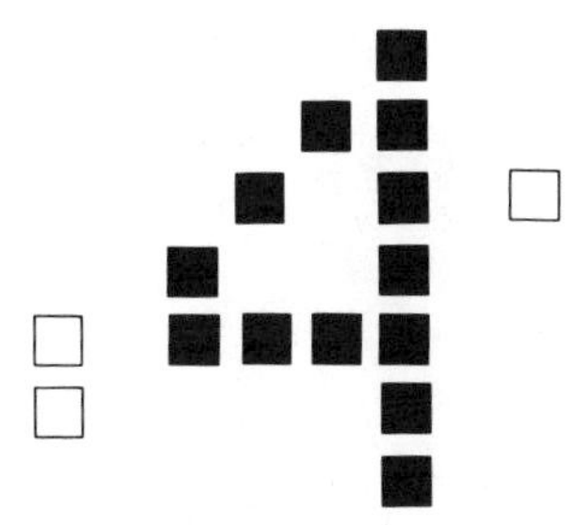

Time-Division Multiplexing

OBJECTIVES

- Explains the basis of multichannel digital communications systems: time-division multiplexing (TDM)
- Describes the techniques and performance of TDM frame synchronization
- Discusses the multiplexing of asynchronous signals by means of pulse stuffing and transitional coding
- Summarizes the CCITT recommendations that apply to the three different digital multiplex hierarchies
- Describes the multiplex equipment specified by these digital hierarchies
- Considers statistical multiplexing for voice and data transmission
- Summarizes the essential TDM design and performance considerations

4.1 INTRODUCTION

The basis of multichannel digital communications systems is **time-division multiplexing (TDM)** in which a number of channels are interleaved in time into a single digital signal. Each channel input is periodically sampled and assigned a certain time slot within the digital signal output. An application of TDM to voice channels is illustrated in Figure 4.1. At the transmit end (multiplex), four voice channels are sequentially sampled by a switch resulting in a train of amplitude samples. The coder then sequentially converts each sample into a binary code using an A/D technique as

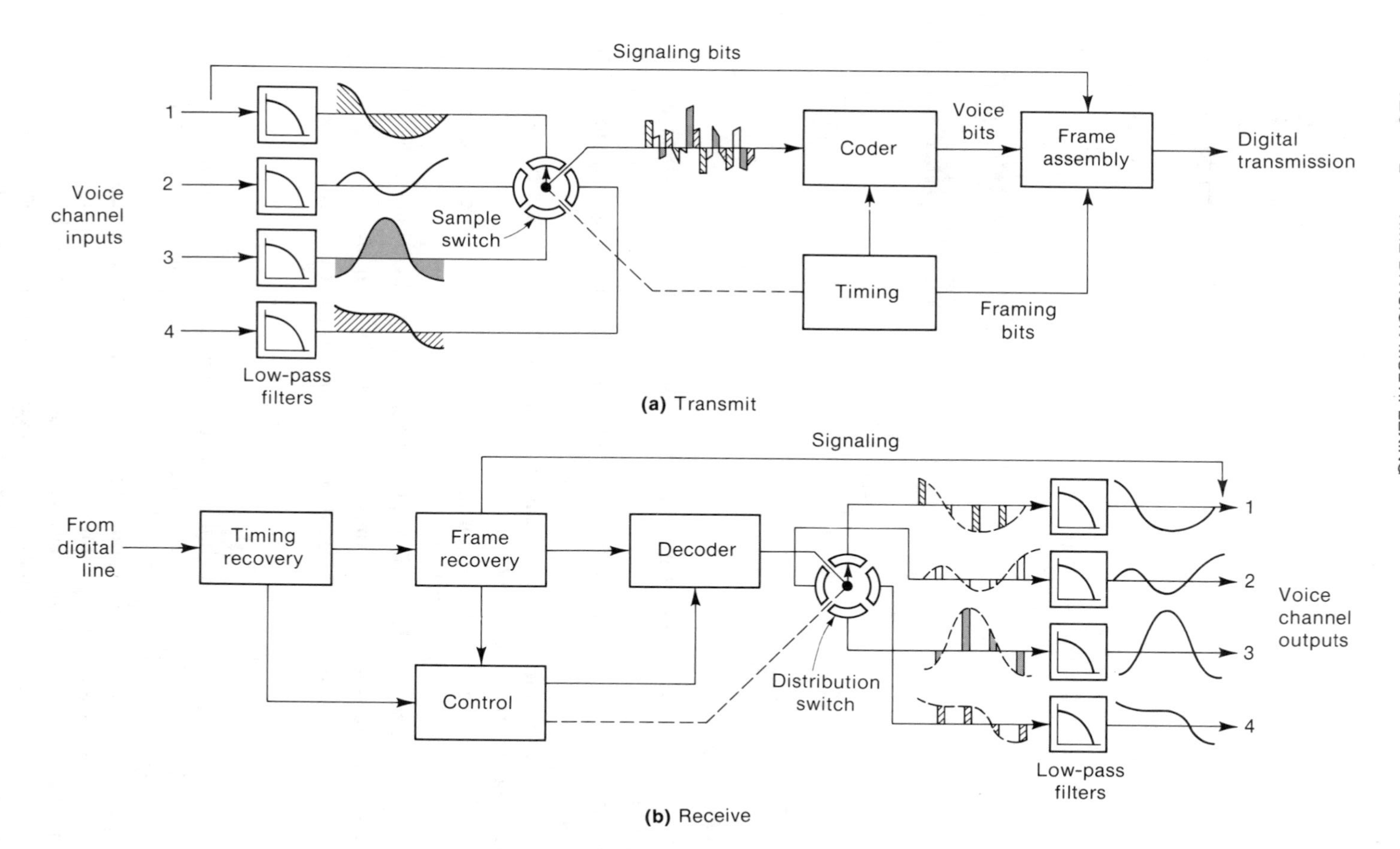

Figure 4.1 Block Diagram of Time-Division Multiplexing

described in Chapter 3. The coder output is a string of binary digits representing channel 1, channel 2, and so on. These voice bits are combined with framing bits for multiplexer/demultiplexer synchronization and signaling bits for telephone network control and are then transmitted over a digital line. The receiver or demultiplexer recovers timing and framing synchronization, which allows the decoder to assign each sample to the proper channel output.

The framing operation of a time-division multiplexer/demultiplexer pair is illustrated in Figure 4.2. The output of the multiplexer is shown to be a string of bits allocated successively to each of K channel inputs and to framing bits. The smallest group of bits containing at least one sample from each channel plus framing bits is known as a **frame**. The framing bits form a repetitive pattern that is combined with channel bits within the multiplexer. The demultiplexer must then successfully recognize the contents of the framing pattern to distribute incoming bits to the proper channels. Techniques and performance of TDM frame synchronization are described in Section 4.2.

If a multiplexer assigns each channel a time slot equal to one bit, the arrangement is known as **bit interleaving**. In this case, the multiplexer can be thought of as a commutator switching sequentially from one channel to the next without the need for buffer storage. A second arrangement is to accept a group of bits, making up a word or character, from each channel in sequence. This commonly used scheme is applicable where the incoming channels are character or word oriented; hence this technique is known as **word interleaving** or **character interleaving**. This scheme introduces the need for buffering to accumulate groups of bits from each channel while waiting for transfer to the multiplexer. The sampling sequence shown in Figure 4.2 can easily be extended to cover those cases where the incoming channels are not all at the same rate but a fixed relationship exists among the channel rates. The frame length is then determined by the lowest common multiple of the incoming channel rates, as shown in the example of Figure 4.3.

Thus far we have assumed that incoming channels are continuous and **synchronous**—that is, with timing provided by a clock which is common to both the channels and the multiplexer. Channel inputs that are not synchronous with multiplexer timing require additional signal processing. When a digital source is **plesiochronous** (nearly synchronous) with the multiplexer, for example, buffering is a commonly used method of compensating for the small frequency differences between each source and multiplexer. This use of buffering is described in more detail in Chapter 10 where we consider independent clock operation for network synchronization. For digital signals with larger frequency offsets, buffering may prove to be unacceptable because of the large buffer lengths required or, conversely, because of high buffer slip (reset) rate. For **asynchronous** signals, whose transitions do not necessarily occur at multiples of a unit interval, another multiplex interface is more appropriate, most commonly pulse stuffing (Section 4.3) or less commonly transitional coding (Section 4.4).

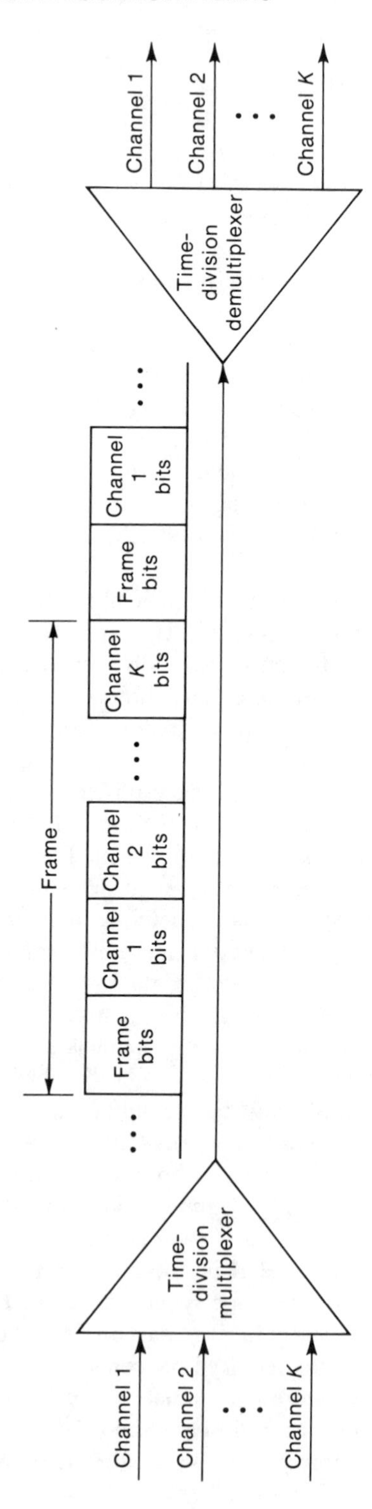

Figure 4.2 TDM Frame Structure

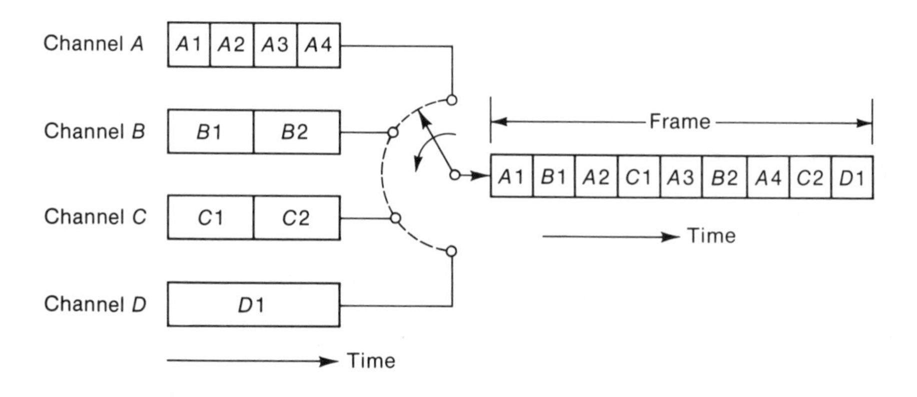

Figure 4.3 Multiplexing of Channels Having Different Bit Rates (Adapted from Bylanski and Ingram [1])

The use of TDM to build bigger and bigger channel capacities has led to different digital hierachies in North America, Japan, and Europe. The different multiplex equipment associated with these digital hierarchies is now standardized, however. Throughout this chapter we will consider examples based on these standards, and Section 4.5 summarizes the applicable CCITT recommendations. In Section 4.6, we consider the case where channel inputs are inactive part of the time. This case suggests a multiplexing scheme where a channel is given access to the multiplexer only during periods of activity, thus allowing more channels to be handled than by conventional TDM techniques. Multiplex designs of this type have been developed for voice transmission, which is called **speech interpolation**, and for data transmission, which is called **statistical multiplexing**.

4.2 FRAME SYNCHRONIZATION

The frame synchronization scheme illustrated in Figure 4.4*a* is characteristic of most time-division multiplexers. A specific code generated by the frame bits is interleaved and transmitted along with channel bits. During frame synchronization the demultiplexer sequentially searches the incoming serial bit stream for a match with the known frame code. Frame search circuitry makes an accept/reject decision on each candidate position based on the number of correct versus incorrect matches. Once frame acceptance has occurred, the selected frame position is continuously monitored for correct agreement. This *forward-acting* frame synchronization scheme thus provides for automatic recognition and recovery from loss of frame synchronization, since the frame monitor initiates a new frame search upon loss of synchronization. There are three inherent drawbacks to this technique: slight reduction in transmission efficiency due to overhead, finite possibility

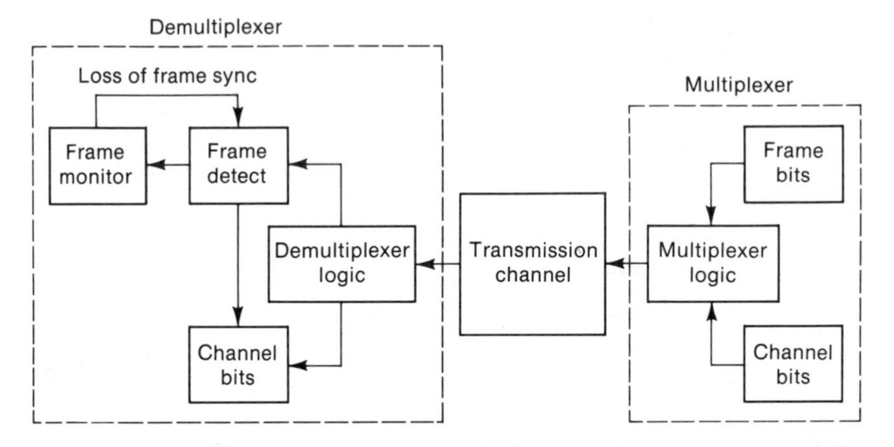

(a) Forward-acting frame synchronization

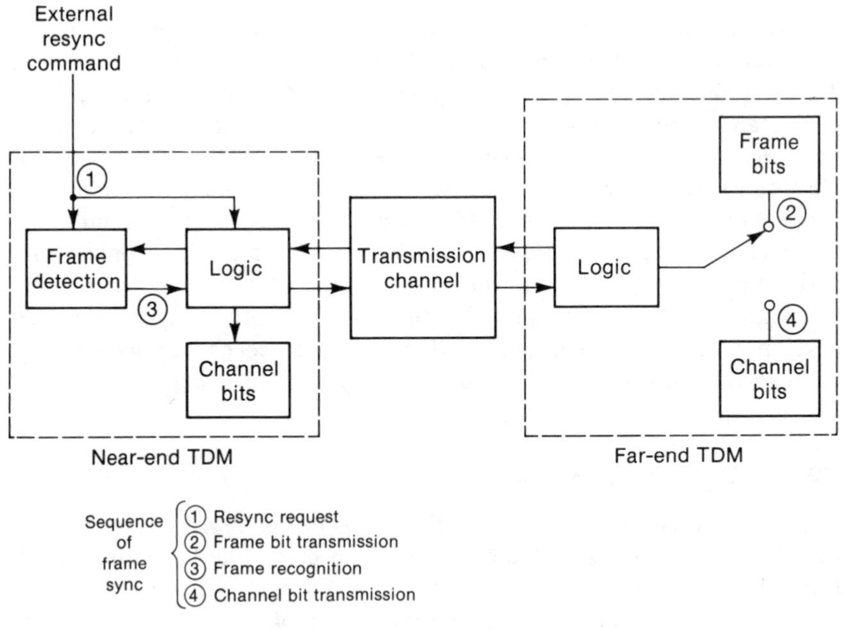

(b) Handshaking frame synchronization

Figure 4.4 Types of Frame Synchronization

of false frame synchronization due to random data matching the frame pattern, and finite possibility of inadvertent loss of frame synchronization due to transmission errors that corrupt the frame pattern.

A second type of frame synchronization involves a sequence of actions known as *handshaking* (Figure 4.4*b*). To initiate frame synchronization, a request for synchronization or resynchronization is transmitted by the near-end TDM to the far-end TDM. A preamble of 100 percent framing bits is then transmitted back to the near-end TDM until a frame recognition signal is activated. The transmission of frame bits is then terminated and transmission of 100 percent channel bits is commenced. As long as frame synchronization is maintained, no further action is required. However, loss of synchronization at other levels in a transmission system will also disrupt frame synchronization. Since the demultiplexer has no frame to monitor continuously, this loss in frame synchronization would go unrecognized unless signaled via external means. The advantage of this frame strategy is that once frame synchronization is established, no additional overhead bits are required to maintain synchronization. Moreover, frame acquisition is accomplished more quickly than forward-acting schemes since there exists no ambiguity between frame bits and channel bits.* Once frame is acquired, no accidental loss of synchronization can be caused by the TDM itself, since no frame bits exist to monitor for loss of synchronization. If loss of synchronization occurs due to any other cause, however, this type of TDM must rely on external equipment to recognize the loss of synchronization and command a resynchronization.

An application of handshaking (also called *cooperative*) frame synchronization is a synchronous TDM in which standard low-speed rates (say 1.2 kb/s) are combined into a standard high-speed rate (say 4×1.2 kb/s $\Rightarrow$ 4.8 kb/s). By not using continuous overhead, the combined rate can be constrained to standard transmission rates (say 4.8 kb/s modem operation over a telephone circuit). Such applications of synchronous multiplexing are an exception, however; the predominant application of TDM warrants the use of forward-acting frame synchronization. The following sections that deal with frame synchronization assume the use of forward-acting schemes, but the techniques of analysis can be applied to the simpler handshaking synchronization scheme as well.

4.2.1 Frame Structure

Frame synchronization bits may be inserted one at a time (**distributed**) or several at a time (**burst** or **bunched**) to form an N-bit frame pattern. Figure 4.5 represents a distributed frame structure where frame bits are inserted one at a time at the beginning or end of the frame. Each frame contains M

*Note that for satellite applications of the handshaking synchronization scheme, propagation delay becomes the dominant contributor to resynchronization times due to the two round trips required through the transmission channel.

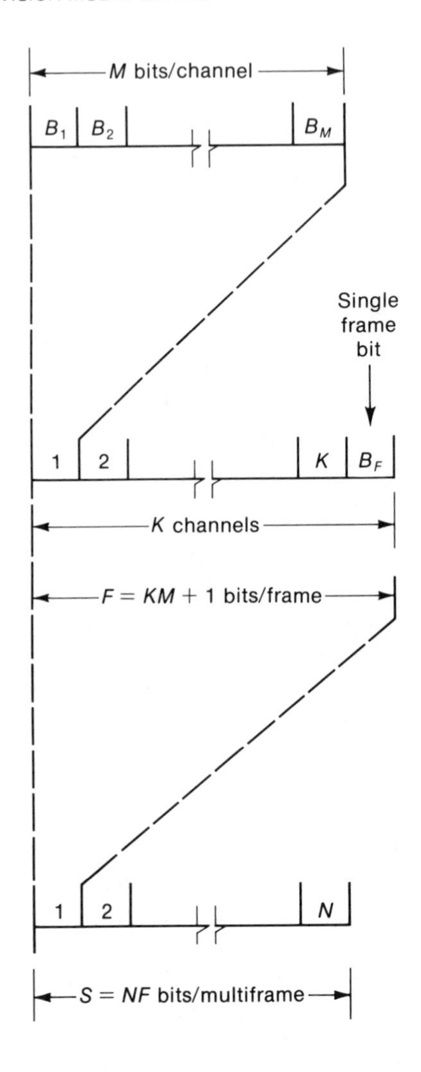

Figure 4.5 Distributed Frame Structure

bits for each of K channels plus one frame bit for a total of $F = KM + 1$ bits. To transmit the complete N-bit frame pattern, a total of NF transmission bits are required, which forms a **multiframe** (sometimes called superframe or major frame). The advantage of distributing frame bits is immunity from the error bursts that might occur, for example, in a radio channel. An error burst of length F bits can affect at most one synchronization bit rather than several or all of the frame bits. The disadvantage of the distributed frame structure is the additional time required to acquire frame synchronization as compared to a bunched frame structure. An example of

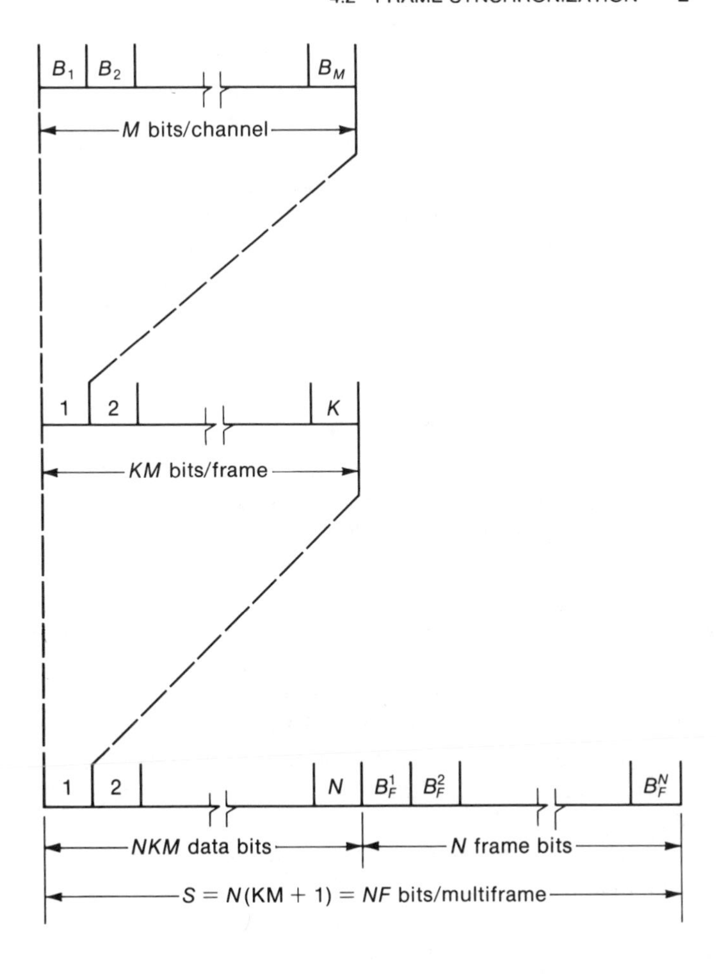

Figure 4.6 Bunched Frame Structure

a multiplexer that employs the distributed frame structure is the 24-channel PCM multiplexer specified by CCITT Rec. G.733 [2].

Figure 4.6 illustrates the bunched frame structure, with the N-bit frame code appearing as contiguous bits at the beginning or end of the multiframe. Note that the total multiframe length is NF bits, so that the ratio of frame bits to total bits per multiframe is identical to that of the distributed frame. The advantage of the bunched frame is that less time is required to acquire frame synchronization when compared to the time required with the distributed frame. One disadvantage lies in the fact that error bursts may effect most or all of a particular N-bit pattern. Another disadvantage is that during the insertion of the frame pattern in N consecutive bit positions, the TDM must store incoming bits from each channel, which results in a design penalty for the extra buffering required in the multiplexer and a performance penalty for residual jitter from demultiplexer smoothing of the reconstructed data. The 2.048-Mb/s PCM multiplexer and the 8.448-Mb/s second-level multiplexer recommended by CCITT Rec. G.732 and G.742, respectively, are examples of multiplexers using bunched frame structure [2].

These two definitions of frame structure represent the two extremes of frame organization. In practice, a TDM may use a combination of the two types in organizing frame structure—as for example the 6.312-Mb/s second-level multiplexer recommended by CCITT Rec. G.743 [2].

4.2.2 Frame Synchronization Modes

TDM frame synchronization circuits commonly employ two modes of operation. The initial mode, called the **frame search** or **frame acquisition mode**, searches through all candidate framing bits until detection occurs. Frame acquisition is declared when a candidate frame alignment position meets the specified acceptance criterion, which is based on correlation with the known frame pattern. With acquisition, the frame synchronization circuit switches to a second mode of operation, called the **frame maintenance mode**, which continuously monitors for correlation between the received framing pattern and the expected framing pattern. If the maintenance mode detects a loss of synchronization as indicated by loss of required correlation, a loss of frame synchronization is declared and the search mode is reentered. A third and sometimes fourth mode of frame synchronization are occasionally used as intermediate modes between search and maintenance. In general, these additional modes only marginally enhance frame synchronization performance [3, 4]. We limit our treatment here to the more typical two-mode operation, which is depicted in a state transition diagram (Figure 4.7). The strategies employed with the search and maintenance modes, described in the following paragraphs, are considered typical of current TDM design. Variations on these basic approaches will also be discussed.

Frame synchronization performance is characterized by specifying:

- The time to acquire initial synchronization, or **frame acquisition time**
- The time that frame synchronization is maintained as a function of BER, or **time to loss of frame alignment**
- The time to reacquire frame synchronization after a loss of synchronization, or **frame reacquisition time**

To be complete, a specification must state the transmission bit error rate and the required probabilities of acquisition, maintenance, or reacquisition within the specified times. Frame synchronization times are usually specified by the statistical mean and sometimes also the variance [4], which will be used here.

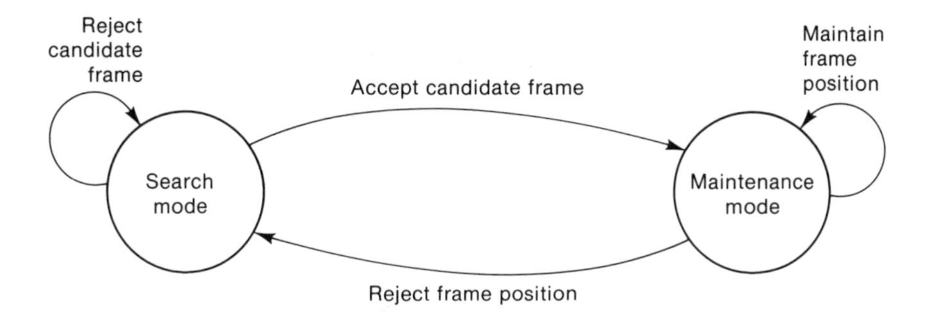

Figure 4.7 State Diagram of Two-Mode Frame Synchronization

Before proceeding with a description and analysis of frame synchronization modes, it is necessary to state the assumptions:

1. Sequential search is assumed in which only a single frame alignment candidate is under consideration at any given time. This is the most commonly used method of frame search. However, the time to acquire frame synchronization can be reduced with parallel search of all candidate positions in a frame.
2. In the demultiplexer, bit timing is locked to timing of the transmit TDM, so that the time to acquire or maintain frame synchronization is independent of other synchronization requirements (such as bit synchronization).
3. Channel bits are assumed equally likely and statistically independent—that is, for binary transmission, $P(0) = P(1) = \frac{1}{2}$. Therefore the probability that a channel bit will match a frame synchronization bit is equal to $\frac{1}{2}$. Bit errors are also assumed statistically independent, thus allowing use of Bernoulli trials formulation where the bit error probability p_e is constant.

Frame Search Mode

The search mode provides a sequential search comparing each candidate frame alignment position with the known frame pattern until a proper match is found. For the distributed frame structure each candidate position is scanned one bit at a time at F-bit intervals for up to N times, where F is the frame length and N is the number of frames in the pattern. For the bunched frame structure each candidate is scanned by examining the last N bits. The criterion for acceptance of a candidate position is that the N-bit comparison must yield ϵ or fewer errors, while rejection occurs for more than ϵ errors. When acceptance occurs, the frame maintenance mode is activated; for rejection the search mode shifts one bit position and tests the next candidate position using the same accept/reject criterion. This op-

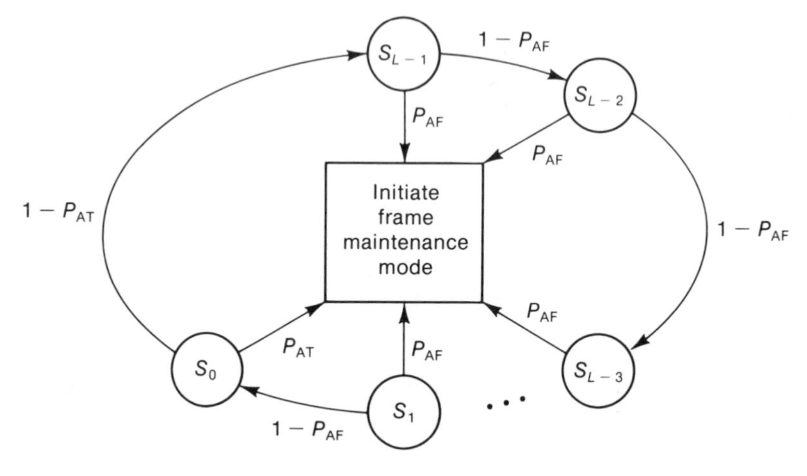

P_{AF} = Prob [accept false candidate]
P_{AT} = Prob [accept true candidate]
S_i = search mode state, i bit positions away from true frame alignment ($i = 0 \to$ true frame alignment; $i \neq 0 \to$ false frame alignment)
L = number of candidate alignment positions

Figure 4.8 State Diagram of Frame Search Mode

eration of the search mode can be conveniently described in terms of a set of states (candidate positions) and probabilities of transitions between these states (accept/reject probabilities), as shown in Figure 4.8. Thus there are four transition probabilities [5]:

1. Probability of accepting a false position (P_{AF}):

$$P_{AF} = \sum_{i=0}^{\epsilon} \binom{N}{i} (0.5)^N \tag{4.1}$$

2. Probability of rejecting a false position (P_{RF}):

$$\begin{aligned} P_{RF} &= 1 - P_{AF} \\ &= \sum_{i=\epsilon+1}^{N} \binom{N}{i} (0.5)^N \end{aligned} \tag{4.2}$$

3. Probability of accepting the true position (P_{AT}):

$$P_{AT} = \sum_{i=0}^{\epsilon} \binom{N}{i} (1 - p_e)^{N-i} p_e^i \tag{4.3}$$

4. Probability of rejecting the true position (P_{RT}):

$$\begin{aligned} P_{RT} &= 1 - P_{AT} \\ &= \sum_{i=\epsilon+1}^{N} \binom{N}{i} (1 - p_e)^{N-i} p_e^i \end{aligned} \tag{4.4}$$

The probability of acquiring true alignment during the first search of all possible candidates is the probability of rejecting all random bit positions and accepting the true frame position. Should no match be found during the first search, even at the true frame position, additional searches are made until a match is found. Assuming that search begins with the first (channel) bit of the frame, the probability that the synchronization position selected by the search mode is the true position is given by

$$P_T = P_{RF}^{L-1}P_{AT} + \left(P_{RF}^{L-1}\right)^2 P_{AT}(1 - P_{AT}) + \left(P_{RF}^{L-1}\right)^3 P_{AT}(1 - P_{AT})^2 + \cdots \tag{4.5}$$

where L is the number of candidate positions. For the distributed frame structure the number of candidates is $L = F$; for the bunched frame structure the number of candidates is $L = NF$. The geometric series in (4.5) reduces to

$$\begin{aligned} P_T &= P_{RF}^{L-1}P_{AT}(1 + c + c^2 + \cdots) \\ &= \frac{P_{RF}^{L-1}P_{AT}}{1 - c} \end{aligned} \tag{4.6}$$

where

$$c = P_{RF}^{L-1}(1 - P_{AT}) \tag{4.7}$$

Similarly, the probability of the search mode inadvertently acquiring a false frame position is found to be

$$P_F = \frac{1 - P_{RF}^{L-1}}{1 - c} \tag{4.8}$$

From (4.6) and (4.8), we find that the total probability of acquisition in the search mode is, as expected,

$$P_F + P_T = 1 \tag{4.9}$$

Frame Maintenance Mode

Two commonly used schemes for the frame maintenance mode are examined here. In both schemes, each frame pattern is first tested against a pass/fail criterion, similar or identical to the accept/reject criterion used in the search mode. Either scheme must be able to recognize actual loss of frame alignment and initiate frame reacquisition. If a maintenance mode scheme were employed that rejected the frame position based on a single failed test, however, the time to hold true frame synchronization would probably be unacceptably small in the presence of transmission bit errors. For this reason, the maintenance mode reject criterion is based on more than one frame pattern comparison, which results in lower probability of inadvertent loss of frame synchronization. This protection against false loss of synchronization is known as the *hysteresis* or *flywheel* provided by the maintenance mode. The first scheme to be considered requires that r successive tests fail before the search mode is initiated. The second scheme is based on an up-down counter, which is initially set at its highest count, $M + 1$. For each test of the received frame pattern, agreement results in an increment of 1 and disagreement results in a decrement of D. A loss of frame is declared and the search mode initiated if the counter reaches its minimum (zero) state.

For the first scheme, we have to calculate the number of tests required before r successive misalignments are identified. This calculation can be done using results from Feller, who derived expressions for the mean μ_n and variance σ_n^2 of the number of tests n required to observe runs of length r [6]:

$$\mu_n = \frac{1 - p^r}{qp^r} \tag{4.10}$$

$$\sigma_n^2 = \frac{1}{(qp^r)^2} - \frac{2r + 1}{qp^r} - \frac{p}{q^2} \tag{4.11}$$

Here p is the probability of misalignment detected in any single test and $q = 1 - p$.

For the second scheme employing an up-down counter, the calculation of number of tests to detect misalignment is facilitated with the state diagram of Figure 4.9. This figure shows that if the counter value is within D counts of zero, then the counter moves to state zero when frame disagreement occurs. Similarly, if the counter is already at its maximum value, succeeding frame agreements leave the counter at its maximum value. Once the counter arrives at its zero state, the search mode is activated. In general, the calculation of the number of tests n required before the zero state is reached is difficult and requires numerical techniques. For $D = 1$, however, the problem reduces to the classic simple random walk, which has been analyzed by several authors [6, 7, 8]. These results may be used to find an analytic expression for the expected value of n:

$$\mu_n = \frac{M}{p - q} + \frac{q^{M+1}}{p^M(q - p)^2}\left[1 - (p/q)^M\right] \qquad (p \neq q) \tag{4.12}$$

$$= M(M + 1) \qquad (p = q = 0.5) \tag{4.13}$$

where q is the probability of an up count (frame agreement) and $p = 1 - q$ is the probability of a down count (frame disagreement). For a general D, recursive methods exist that are suitable for numerical solution by computer. One such approach is to note that the up-down counter as shown in Figure 4.9 is a finite Markov chain and take advantage of the well-known properties of finite Markov chains to obtain the mean, variance, and even probability density function of n for general choice of p, M, and D [9].

We are also interested in specifying the probability of detection. Since the number of tests is a sum of independent random variables, the central limit theorem may be invoked, which asserts that the number of tests is

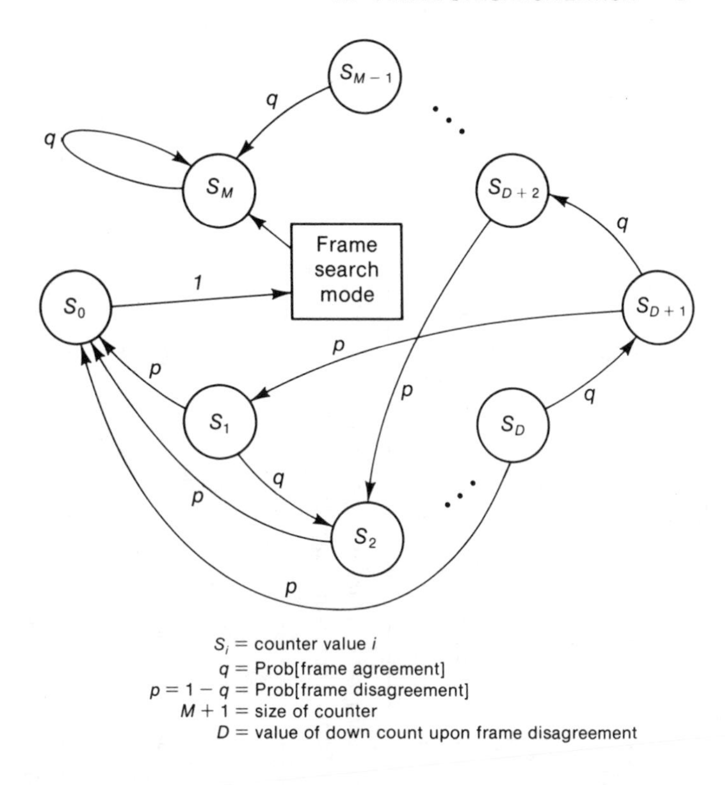

Figure 4.9 State Diagram of Maintenance Mode Using Up-Down Counter

gaussian-distributed with mean μ_n and variance σ_n^2. Hence the number of tests to detect frame misalignment with 90 percent probability, for example, is given by

$$n_{0.9} = \mu_n + 1.28\sigma_n \tag{4.14}$$

and for 99 percent probability

$$n_{0.99} = \mu_n + 2.33\sigma_n \tag{4.15}$$

To specify the number of bit times to detection of misalignment, we note that each maintenance mode test of a frame pattern occurs at regularly spaced intervals. For the distributed and bunched frame structures of Figures 4.5 and 4.6 respectively, each test occurs every NF bits, since there are exactly NF bits between successive frame words for either the distributed or bunched frame structure. Thus Expressions (4.10) through (4.15) given here for number of tests to misalignment detection may be converted to bit times by using the multiplying factor of NF bit times per test.

Table 4.1 Performance of Frame Misalignment Detection

(a) Scheme 1: r successive frame word errors result in reframe.

Detection Probability in Each Test (P_{RF})	Successive Tests Resulting in Frame Rejection (r)	Number of Tests to Detect Misalignment: Mean (μ_n)	Variance (σ_n^2)	90% Probability	99% Probability
0.5	2	6	22	12.0	16.9
0.5	4	30	734	64.7	93.1
0.5	6	126	14,718	281.3	408.7
0.5	8	510	257,790	1160	1693

(b) Scheme 2: Count up 1 for correct frame word, down D for incorrect frame word. Counter size M = 15. Reframe occurs at state zero.

Detection Probability in Each Test (P_{RF})	Value of Down Count upon Frame Disagreement (D)	Number of Tests to Detect Misalignment: Mean (μ_n)	Variance (σ_n^2)	90% Probability	99% Probability
0.5	2	27.5	191	45.2	59.7
0.5	4	10.1	23.8	16.4	21.5
0.5	6	6.4	10.2	10.5	13.8
0.5	8	4.5	7.4	8.0	10.8

Detection of Misalignment

Suppose a true loss of frame synchronization has occurred. The maintenance mode will then be examining random data rather than the true frame pattern. The probability of misalignment detection in a particular frame pattern test depends on the length of the frame pattern and the accept/reject criterion and is specified in Equation (4.2). The number of such tests required before the maintenance mode declares loss of frame is then dependent on the scheme employed. The necessary calculations are made by using appropriate expressions selected from (4.10) through (4.15), where p is equivalent to P_{RF} and n is the number of tests to detect misalignment. Table 4.1 presents selected examples of performance for the two maintenance mode schemes considered here. To convert these results from number of tests to number of bit times, recall that there are NF bit times per test in both the bunched and distributed frame structure.

Frame Alignment Time

From Figure 4.7 we see that the time to true frame synchronization involves time spent in both the search and maintenance modes. Each time the search mode accepts a candidate position, the maintenance mode is entered. If false alignment is acquired by the search mode, the maintenance mode will eventually reject that false alignment and return to the search mode. The search mode then accepts another candidate position and reenters the maintenance mode. This cycle continues until true frame alignment is found in the search mode.

As indicated earlier, the time to true alignment is usually specified by its mean value, which we will derive here for both the bunched and distributed frame structures. First suppose that true frame alignment has been lost and that detection of loss of frame alignment has taken place in the maintenance mode. We are now interested in the time for the combined action of the search and maintenance modes to *acquire* true frame alignment. The time to *reacquire* frame alignment can then be obtained by summing the time to detect misalignment with the time to acquire true alignment.

Bunched Frame Structure. Assuming a random starting position, the search mode examines an average of $(NF - 1)/2$ false positions, each with a probability P_{AF} of being accepted, before reaching the true alignment position. For those positions that are rejected, the search mode shifts one bit position and examines the next candidate position, so that eventually all $(NF - 1)/2$ positions are examined. On average, however, there will be $(P_{AF})(NF - 1)/2$ false candidates accepted. For each of these there will be a mean number of tests to reject false synchronization, μ_n, in the maintenance mode, with NF bits per maintenance mode test for the bunched frame structure. The mean number of bits to reach the true frame alignment position, μ_T, is then the sum of the time to shift through $(NF - 1)/2$

positions and the time for the maintenance mode to reject all false frame alignment positions:

$$\mu_T = \left(\frac{NF - 1}{2}\right)[1 + (P_{AF})(\mu_n)(NF)] \text{ bit times} \tag{4.16}$$

The *maximum average frame acquisition time* is also sometimes used as a specification, in which case the search mode is assumed to start at maximum distance from the true frame position. In this case

$$(\mu_T)_{max} = (NF - 1)[1 + (P_{AF})(\mu_n)(NF)] \text{ bit times} \tag{4.17}$$

Distributed Frame Structure. Assuming a random starting position, the search mode examines an average of $(F - 1)/2$ false positions before reaching the true alignment position. The time to test each candidate is NF bit times, so that $NF(F - 1)/2$ bit times are required to test $(F - 1)/2$ candidates. For the distributed frame structure, we assume that any cyclic version of the N-bit frame pattern will be accepted by the search mode. Therefore each false alignment position has a probability NP_{AF} of being accepted. Then, on the average, there will be $(NP_{AF})(F - 1)/2$ false candidates accepted. For each of these there will be a mean number of tests in the maintenance mode to reject false positions, μ_n, with NF bits per maintenance mode test. Thus the mean number of bits to reach true frame alignment is

$$\mu_T = NF\left(\frac{F - 1}{2}\right)(NP_{AF}\mu_n + 1) \tag{4.18}$$

The maximum average frame acquisition time is then

$$(\mu_T)_{max} = NF(F - 1)(NP_{AF}\mu_n + 1) \tag{4.19}$$

Example 4.1

The 2.048-Mb/s PCM multiplex equipment specified by CCITT Rec. G.732 contains a 7-bit bunched frame alignment word that is repeated every 512 bits (see Figure 4.30). Frame alignment requires the presence of the correct frame alignment word without error. Frame alignment is assumed lost when three consecutive frame words are received in error. Assuming a random starting position, find the mean time for the demultiplexer to acquire frame alignment. Assume the transmission error rate to be zero.

Solution

For each test of a false frame alignment word, the probability of acceptance is given by (4.1):

$$P_{AF} = \left(\frac{1}{2}\right)^7 = 7.8125 \times 10^{-3}$$

The number of tests for rejection of false alignment by the maintenance mode is found from (4.10) with $p = P_{RF} = 1 - P_{AF}$ and $q = P_{AF}$:

$$\mu_{RF} = \frac{1 - (0.9921875)^3}{(7.8125 \times 10^{-3})(0.9921875)^3} = 3.05 \text{ tests}$$

Then from (4.16),

$$\mu_T = \frac{511}{2}\left[1 + (7.8125 \times 10^{-3})(3.05)(512)\right]$$
$$= 3373 \text{ bit times}$$

or

$$\mu_T = \frac{3373 \text{ bit times}}{2.048 \times 10^6 \text{ b/s}} = 1.65 \text{ ms}$$

Note that from (4.17) the maximum average frame acquisition time would then be 3.3 ms.

False Rejection of Frame Alignment

In the presence of transmission errors, frame bits may be corrupted, causing the maintenance mode to declare loss of frame and to initiate realignment in the search mode. Calculation of the time between such false rejections begins with the probability of false rejection in a particular test, given by (4.4). For the maintenance mode scheme that requires r consecutive test rejections, the mean and variance of the number of tests to false rejection are given by (4.10) and (4.11), with $p = P_{RT}$. For the maintenance mode scheme employing an up-down counter, the statistics of the number of tests to false rejection can be obtained by recursive techniques using the Markov chain model as previously referenced. Once again invoking the central limit theorem, we note that the number of tests is gaussian-distributed so that (4.14) and (4.15) apply for the 90 percent and 99 percent probabilities. Because false resynchronization occurs infrequently, however, the mean is commonly the only statistic used as a specification. Using the statistical mean, false resynchronization performance is illustrated in Figures 4.10 and 4.11 for the two maintenance mode schemes.

Example 4.2

For the 2.048-Mb/s multiplexer described in Example 4.1, find the mean time between loss of frame alignment due to a transmission error rate of 10^{-3}.

Solution

During operation of the maintenance mode, the probability of rejecting the true alignment word is found from (4.4):

$$P_{RT} = 1 - P_{AT}$$
$$= 1 - (1 - p_e)^7 = 7 \times 10^{-3}$$

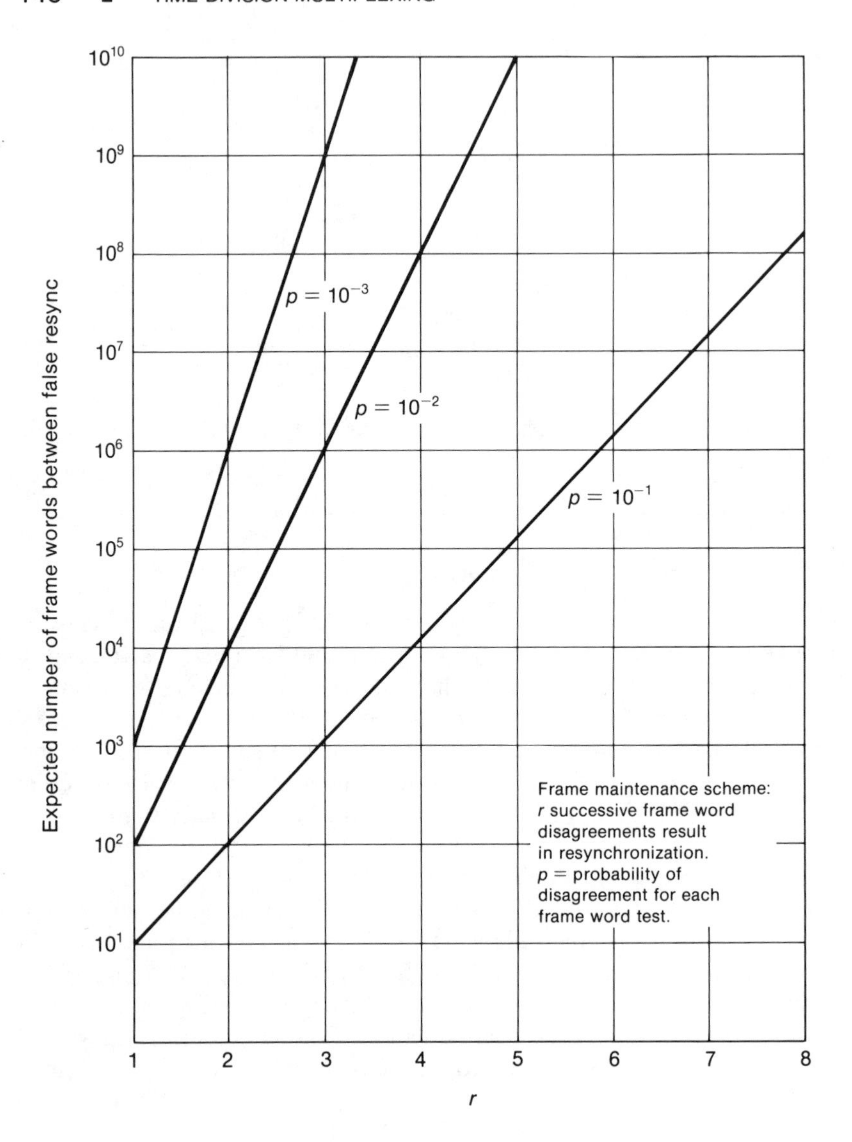

Figure 4.10 False Resynchronization in Maintenance Mode (Scheme 1)

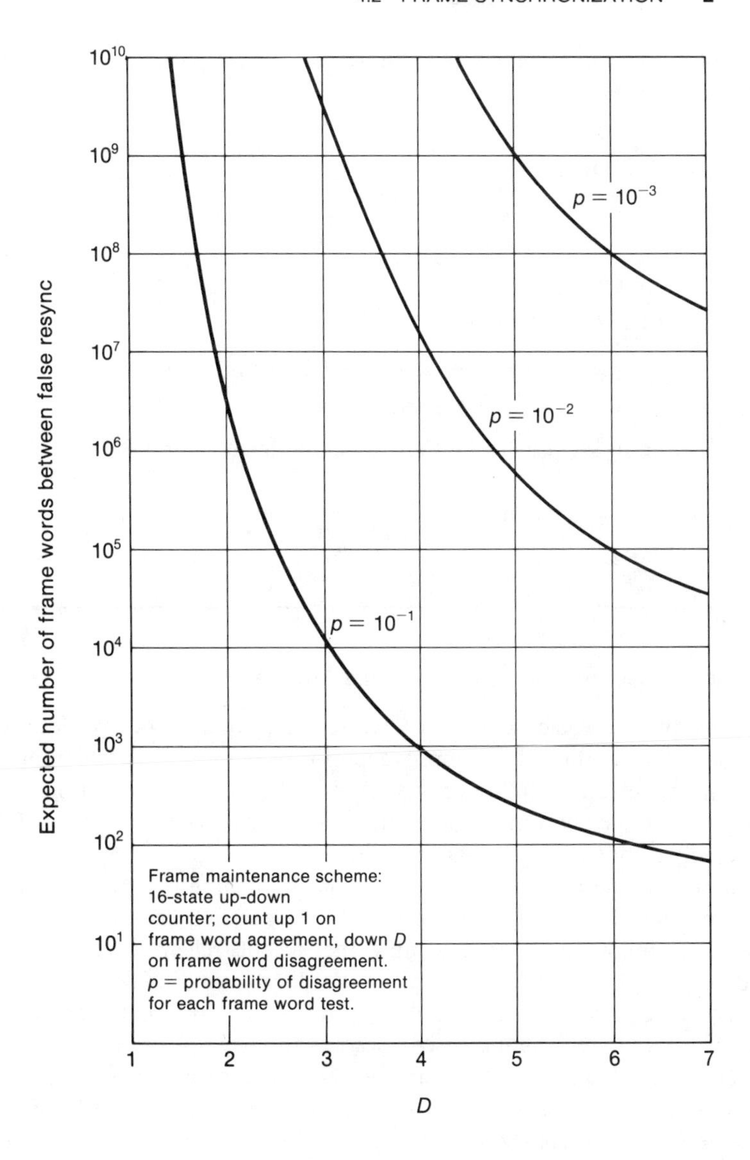

Figure 4.11 False Resynchronization in Maintenance Mode (Scheme 2)

The mean number of tests to reject the true alignment pattern is found from (4.10) with $p = P_{RT}$:

$$\mu_{RT} = \frac{1 - (P_{RT})^3}{(1 - P_{RT})(P_{RT})^3}$$
$$\approx \frac{1}{(7 \times 10^{-3})^3} = 2.9 \times 10^6 \text{ tests}$$

Since there are 512 bits between frame words,

$$\mu_{RT} \text{ (bits)} = (2.9 \times 10^6 \text{ tests})(512 \text{ bits/test})$$
$$= 1.5 \times 10^9 \text{ bits}$$

The mean time in seconds to loss of frame alignment is then

$$\mu_{RT} \text{ (seconds)} = \frac{1.5 \times 10^9 \text{ bits}}{2.048 \times 10^6 \text{ b/s}} = 732 \text{ s}$$

Variations of Search and Maintenance Modes

Variations of the basic search or maintenance mode strategy described above have been utilized to improve performance—that is, to minimize frame synchronization times or occurrences of spurious rejections. The choice of frame synchronization strategy is determined by a tradeoff between complexity and performance.

One technique that reduces frame acquisition time is a parallel search of all candidate frame alignment positions. In parallel, each of the candidates is continuously checked against the accept/reject criterion; failed candidates are eliminated until only one candidate remains. Parallel search is clearly faster than serial search but requires additional storage and comparator circuitry. The increased speed in frame acquisition depends on how quickly all false candidates are rejected. The probability of any one false candidate being rejected in n or less comparisons is $1 - P_{AF}^n$. For an F-bit frame, the probability of rejecting all $(F - 1)$ false candidates in n or less comparisons is

$$\text{Prob[rejection time} < n] = (1 - P_{AF}^n)^{F-1} \tag{4.20}$$

Of course, after all false candidates have been eliminated, the remaining position is accepted as the true frame alignment, so that (4.20) is also the probability of true frame acquisition in n or less comparisons.

To minimize false (spurious) losses of synchronization, a maintenance mode can lock to the old frame position while the search mode looks for the actual frame position. Upon declaration of loss of frame, frame is held at the old frame position while in parallel a search mode is initiated. The old frame position is held until the search mode locates the correct frame position. By this process, if a spurious loss of frame occurs the search mode will return to the same position that was held when loss of frame was declared. No loss of synchronization occurs for this process, since the true frame position is held during the search mode. Moreover the time to reacquire true frame alignment is reduced if the search mode starts with examination of bit positions immediately adjacent to the old frame position, since these are the most likely true frame positions.

Finally, the choice of type of search and maintenance mode and parametric values for each mode involves a tradeoff between quick synchronization times and immunity to transmission errors and false synchronization patterns. In the search mode, performance is determined by the frame structure (distributed versus burst), by the frame pattern length (and code itself as shown in Section 4.2.3.), and by the number of errors allowed in each comparison. Expressions (4.1) to (4.9) can be used to evaluate these tradeoffs for the search mode. Maintenance mode performance is usually determined by the same criteria as the search mode plus the number of frame miscompares required before loss of frame is declared. Table 4.1 and Figures 4.10 and 4.11 can be used to trade off misalignment detection with false resynchronization in the maintenance mode.

4.2.3 Choice of Frame Pattern

For the distributed frame structure of Figure 4.5, each test of a candidate frame position is independent of previous tests. Further, each candidate position contains either all channel bits or all frame bits. In this case, the probability of random bits exactly matching an N-bit framing pattern is 2^{-N}. This probability can be made arbitrarily small simply by lengthening the pattern length, but at the penalty of reduced transmission efficiency and increased synchronization times. If the channel bits are assumed to be random with equally likely levels—that is, $P(0) = P(1) = \frac{1}{2}$—then any choice of frame pattern is equally good for a selected length N.

In contrast, for the bunched frame structure of Figure 4.6, successive tests for the frame pattern are not independent and certain tests will contain a mix of channel bits and framing bits. If the overlapping positions are in agreement, the probability of accepting a false position is increased, thus degrading overall performance. Consider Figure 4.12, for example, which shows a TDM that is out of frame alignment by exactly one bit time. For the candidate 7-bit pattern being examined, six of the positions overlap with the true alignment pattern while one does not. Since the code selected for this example consists of all 1's, the six overlapping positions all match with the true frame pattern, and the seventh (channel bit) position will also match if it is a 1. Thus the probability of frame pattern imitation has been reduced to $\frac{1}{2}$ rather than $(\frac{1}{2})^N$ as would be the case for purely random bits. This argument can easily be extended to other degrees of overlap and for other poor choices of frame alignment pattern.

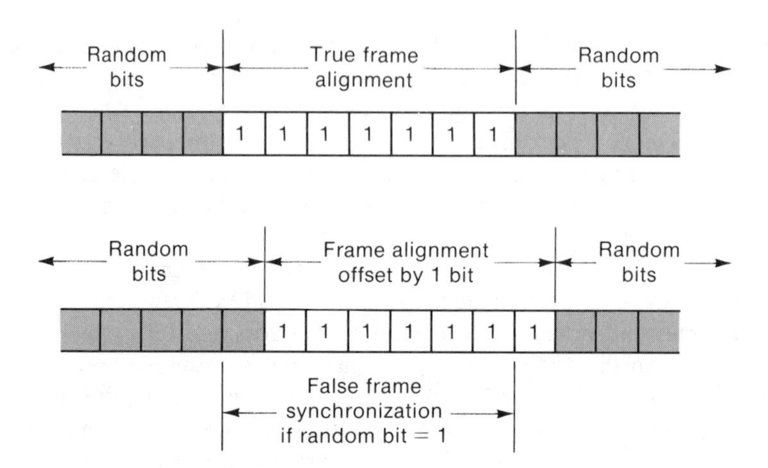

Figure 4.12 False Synchronization for Bunched Frame

Certain framing patterns have been discovered that minimize the degree of agreement for shifted versions of the frame pattern. The autocorrelation function $R_x(k)$ measures this property of a frame pattern. If the pattern is represented by $x_1, x_2, \ldots, x_m$, then it is desired that the autocorrelation

$$R_x(k) = \sum_{i=1}^{m-k} x_i x_{i+k} \qquad \text{for } x_i = \pm 1 \tag{4.21}$$

be small for $k \neq 0$. R. H. Barker investigated codes that have the property

$$|R_x(k)| \leq 1 \qquad k \neq 0 \tag{4.22}$$

and found three such codes of length 3, 7, and 11. When represented with 1's and 0's, these are [5]:

110
1110010
11100010010

For each Barker code, the complement, reflected (mirror image), and reflected complement versions also obey Expression (4.22) and thus are also good synchronization codes. M. W. Williard has also derived a set of codes that exhibit similar good synchronization properties, although they do not meet the Barker property of (4.22); however, Williard has shown that the probability of false synchronization in any overlap condition is less than the probability of false synchronization due to random bits [10].

Example 4.3

Show that 7-bit code 1110010 meets the autocorrelation property of (4.22) necessary for Barker codes.

Solution

To determine its autocorrelation, this 7-bit binary code must first be rewritten as the sequence $x_1, x_2, \ldots, x_7$:

$$1110010 \rightarrow +++--+-$$

Now applying (4.21) we find the autocorrelation as a function of k:

$$\begin{array}{ll} k = 0 & R(0) = 7 \\ k = 1 & R(1) = 0 \\ k = 2 & R(2) = -1 \\ k = 3 & R(3) = 0 \\ k = 4 & R(4) = -1 \\ k = 5 & R(5) = 0 \\ k = 6 & R(6) = -1 \end{array}$$

which is sketched in Figure 4.13. Since the autocorrelation $R(k)$ is an even function, this code is in fact a Barker code.

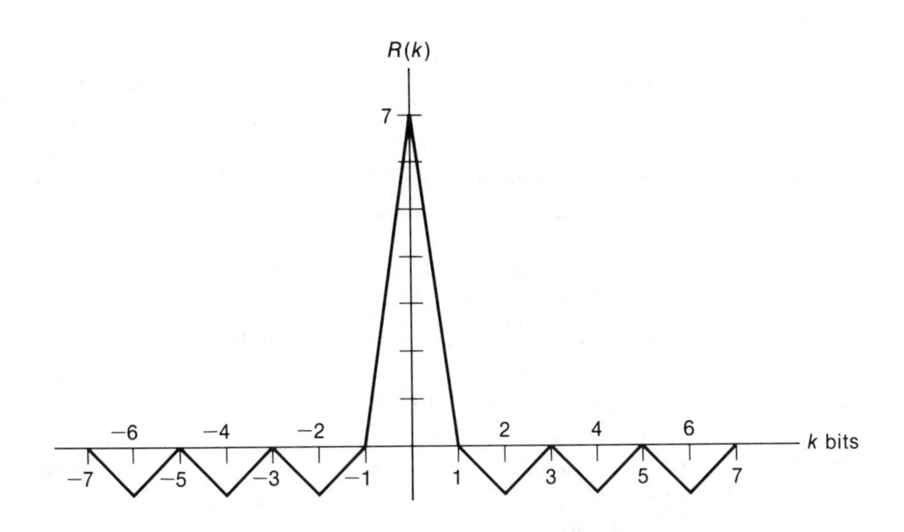

Figure 4.13 Autocorrelation of 7-bit Barker Code

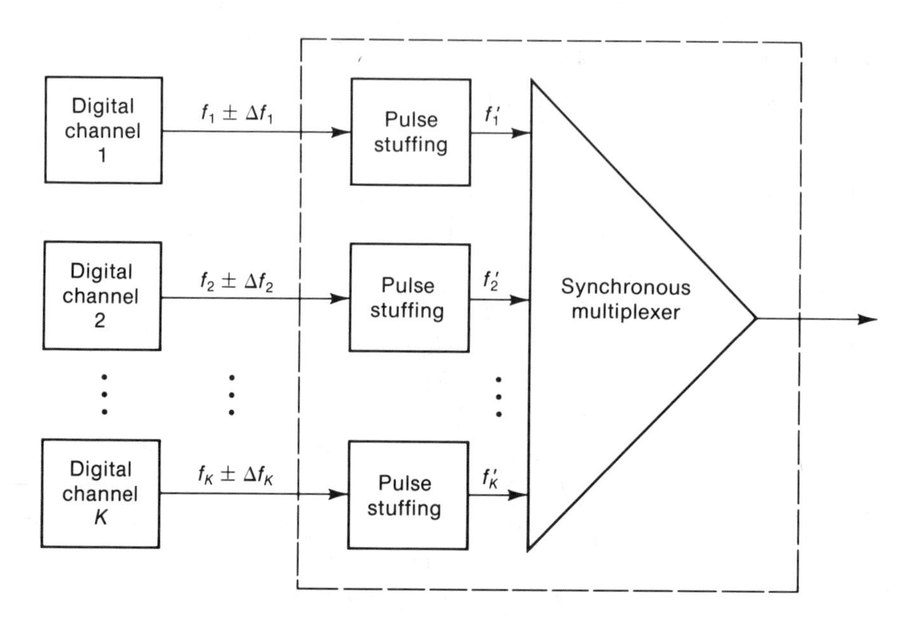

Figure 4.14 Pulse Stuffing Multiplexer

4.3 PULSE STUFFING

A time-division multiplexer may be categorized as providing either synchronous or asynchronous interface with each information channel. A synchronous interface implies that each source which interfaces the TDM is timed by the same clock that provides timing within the TDM. For this case, there is no ambiguity between the arrival time of each channel bit and the time to multiplex each channel bit. If each digital source is timed by its own internal clock, however, the interface is said to be asynchronous. For this case, slight differences between the source clock and TDM clock arise due to clock inaccuracies and instabilities; the two clocks then wander apart and eventually cause a bit to be added or deleted in the TDM, resulting in loss of synchronization. To compensate for this lack of synchronization between each source and the TDM, the technique known as **pulse stuffing*** can be used to convert each asynchronous source to a rate that is synchronous with the TDM clock frequency. Each digital source is clocked at some nominal rate f_i that ranges to $f_i \pm \Delta f_i$, where the tolerance $\pm \Delta f_i$ is dependent on clock performance parameters but is always much smaller than f_i. A typical tolerance value is ± 50 parts per million (ppm) for rates of 1.544 or 2.048 Mb/s. The TDM must be capable of accepting each nominal input rate

*Other terms also used are **justification** (by the CCITT) and **bit stuffing**.

over its full range and must provide a conversion of each channel input to a rate that is synchronous with the internal clock of the TDM. As illustrated in Figure 4.14, pulse stuffing provides a conversion from the nominal source frequency f_i to a frequency f_i' for synchronous multiplexing in the TDM. Two commonly applied techniques for accomplishing this synchronous conversion are **positive pulse stuffing** and **positive-negative pulse stuffing**, which are described in the following sections.

4.3.1 Positive Pulse Stuffing

With positive pulse stuffing each channel input rate f_i is synchronized to a TDM channel rate f_i' that is higher then the maximum possible frequency of the channel input, $f_i + \Delta f_i$. Figures 4.15 and 4.16 show typical block diagrams of the transmit and receive sections, respectively, of a positive pulse stuffing TDM. Channel input bits are written into a digital buffer at a rate of f_i and read out of the buffer at a rate f_i'. This buffer is often referred to as an **elastic buffer**, since the length of its contents varies with differences between the write and read clocks. Since f_i' is at a rate slightly higher than f_i, there is a tendency to deplete the buffer contents. To avoid emptying the buffer, the buffer fill is monitored and compared to a preset threshold. When this threshold is reached, a request is made to stuff a time slot with a stuff (dummy) bit. At the next available stuff opportunity, the read clock is inhibited for a single clock pulse, allowing a stuff bit to be inserted into a designated time slot within the synchronous channel while the asynchronous channel input continues to fill the buffer. Prior to the actual stuffing operation, the precise location of the stuffed time slot is coded into the overhead channel and transmitted to the receive end (demultiplexer) of the TDM. Here the received channel bits are written into an elastic buffer, but stuff bits are prevented from entering the buffer by the overhead channel, which drives a clock inhibit circuit. Bits are read out of the buffer by a smoothed clock. This smoothed clock is derived by using the buffer fill, which is a measure of the difference between phases of the write and read clocks and thus can be used as a phase detector in a clock recovery circuit. The smoothing loop is achieved by using the buffer fill to drive a phase-locked loop, which consists of a filter plus voltage-controlled oscillator.

The timing relationship between the asynchronous channel input and synchronous TDM channel is illustrated in Figure 4.17. In this figure, which exaggerates the typical frequency difference for the sake of illustration, each synchronous output bit has a period which is four-fifths the period of the asynchronous input bit so that the instantaneous output rate is 1.25 times the input rate. Every fifth bit of the synchronous output is converted to a stuff (dummy) bit. Because of the rate difference, the phase of the output signal increases linearly with respect to the input signal. When this phase difference accumulates to 1 bit, the phase is reset by inserting a dummy bit within the synchronous channel output.

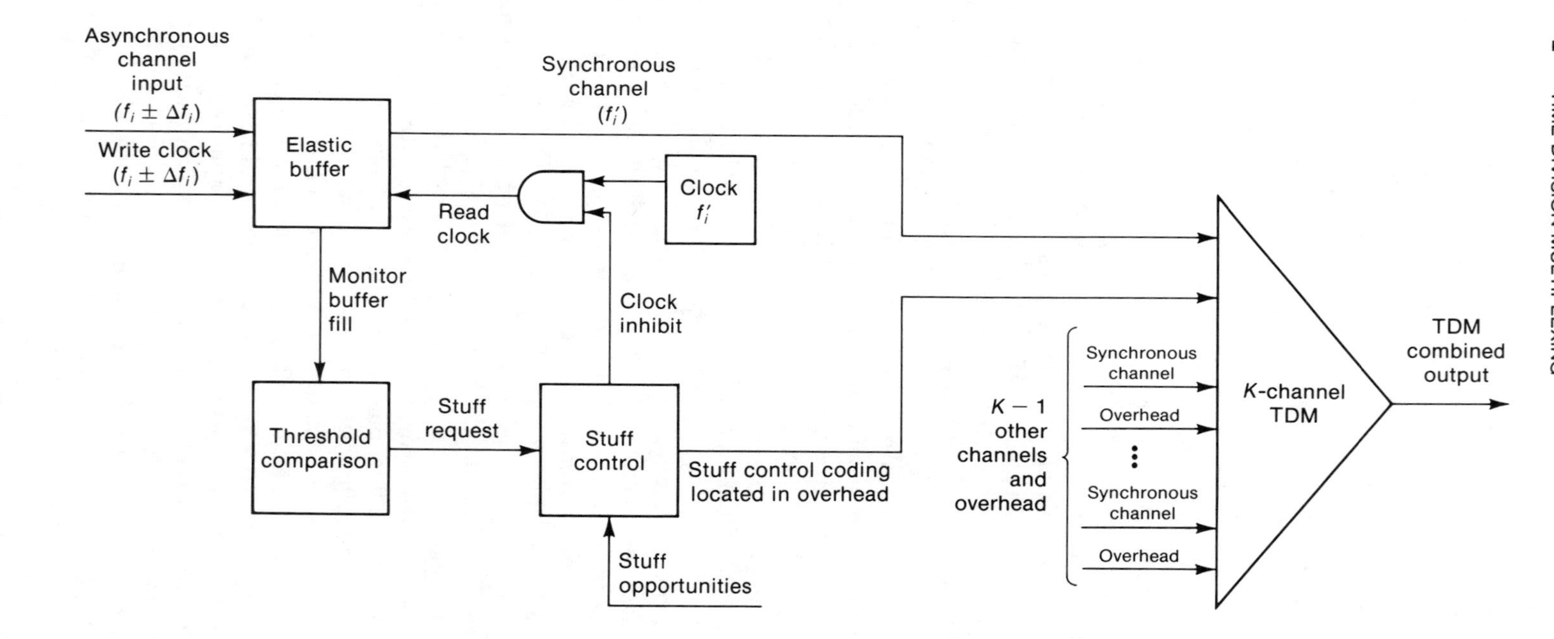

Figure 4.15 Block Diagram of Positive Pulse Stuffing Multiplexer

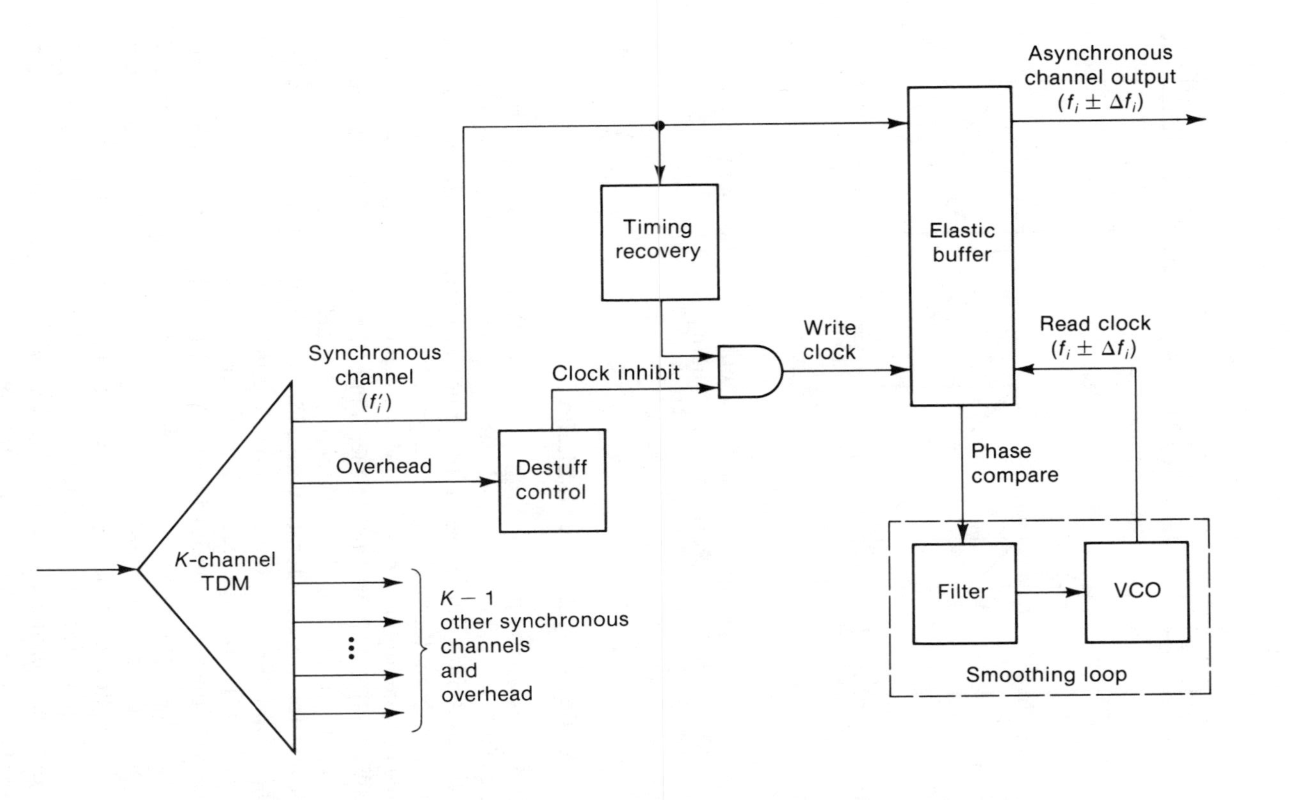

Figure 4.16 Block Diagram of Positive Pulse Stuffing Demultiplexer

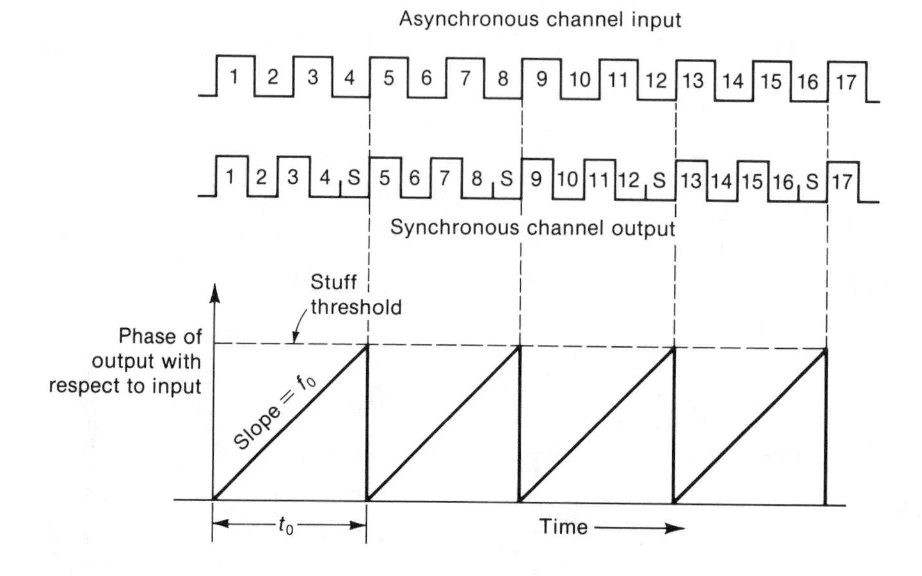

Figure 4.17 Asynchronous / Synchronous Timing Relationships for Positive Pulse Stuffing

4.3.2 Positive-Negative Pulse Stuffing

In a positive-negative pulse stuffing technique, each channel input rate f_i is synchronized to a TDM channel rate that is identical to the nominal channel input rate. Buffering of the input is provided just as shown in Figure 4.15. Now the buffer contents may deplete or spill, however, since both the channel input rate and TDM channel rate may have either negative or positive offset and drift about the nominal frequency f_i. If the channel input rate is lower than the multiplexer channel rate, the buffer tends to deplete and for this case stuff bits are added (positive stuff); if the input rate is higher, the buffer tends to spill and here information bits are subtracted (negative stuff) from the channel input and transmitted as part of the overhead channel. For positive-negative pulse stuffing, two actions may occur with each stuff opportunity: positive stuff or negative stuff (or spill). In a similar scheme called **positive-zero-negative pulse stuffing**, three actions may occur with each stuff opportunity: no stuff, positive stuff, or negative stuff. Stuff/spill locations are coded and spilled information bits are added to an overhead channel to allow the receiving end to properly remove positive stuff bits and add negative stuff information bits.

Because the multiplexer channel rate is equal to the nominal asynchronous channel rate, synchronous operation is easily facilitated by providing common clock to the TDM and digital source and ignoring the stuff/spill code since it will be inactive. Moreover, since stuffing circuitry is independent for each channel, synchronous channels can be mixed with asynchronous channels. These advantages are not found with positive pulse stuffing because of the frequency offset employed with the multiplexer channel rate. Although positive pulse stuffing is prevalent among higher-order multiplexers, there are positive-zero-negative stuffing schemes under consideration by the CCITT [2].

4.3.3 Stuff Coding

Pulse stuffing multiplexers provide each channel with one time slot for stuffing in each frame or multiframe. This time slot contains either a dummy bit when stuffing has occurred or an information bit when no stuffing is required. Each time slot that is available for stuffing has an associated stuff control bit or code that allows the demultiplexer to interpret each stuffing time slot properly. Incorrect decoding of a stuffing time slot due to transmission errors causes loss of bit count integrity in the associated channel since the demultiplexer will have incorrectly added or deleted a bit in the derived asynchronous channel. To protect against potential loss of synchronization, each stuffing time slot is redundantly signaled with a code, where two actions (stuff/no stuff) are signaled for a positive pulse stuffing TDM and three actions (no stuff/positive stuff/negative stuff) are signaled for a positive-zero-negative pulse stuffing TDM. At the receiver, majority logic applied to each code word determines the proper action. As an example, a 3-bit code is commonly used for positive pulse stuffing in which the code 111 signals stuffing and 000 signals no stuffing. In this case, if two or more 1's are received for a particular code, the associated stuffing time slot is detected as a stuffing bit whereas two or more 0's received results in detection of an information bit in the stuffing time slot.

Performance is specified by the probability of decoding error per stuff code, and from this probability the time to loss of synchronization due to incorrect decoding of a stuff code may be obtained. If we assume independent probability of error from code bit to code bit, then the probability of incorrect decoding for a particular stuff code is given by the binominal probability distribution:

$$\text{Prob[stuff code error]} = P(\text{SCE}) = \sum_{i=(N+1)/2}^{N} \binom{N}{i} p_e^i (1 - p_e)^{N-i} \quad (4.23)$$

where N = code length

p_e = average probability of bit error over transmission channel

For small probability of bit error, this expression can be approximated by

$$P(\text{SCE}) \approx \binom{N}{x} p_e^x \quad (4.24)$$

where $x = (N + 1)/2$. For the case of a 3-bit code,

$$P(\text{SCE}) \approx 3p_e^2 \tag{4.25}$$

Knowing the rate of stuffing opportunities, f_s, given in stuffs per unit time, the mean time to loss of BCI due to stuff code error can be calculated as

$$\text{Mean time to loss of BCI} = \frac{1}{f_s P(\text{SCE})} \tag{4.26}$$

Stuff code bits are also distributed within the frame to protect against error bursts and to minimize the number of overhead bits inserted in the information channel at any one time. By distributing the code bits, the probability of an error burst affecting all bits of a particular code is reduced, which therefore increases the effectiveness of the redundancy coding. Distributed code bits also result in smaller buffer sizes required to store information bits during multiplexing and demultiplexing and in improved jitter performance. In the demultiplexer, the clock recovery circuit for each channel must smooth the gaps created by deletion of overhead bits. The jitter resulting from this smoothing process is minimized when overhead bits are distributed rather than bunched.

4.3.4 Pulse Stuffing Jitter

The removal of the stuffed bits at the demultiplexer causes timing jitter at the individual channel output, which has been called **pulse stuffing jitter**. As shown in Figure 4.17, pulse stuffing jitter is a sawtooth waveform with a peak-to-peak value of 1 bit interval and a period t_0 seconds. The *actual stuffing rate* occurs at average frequency $f_0 (= 1/t_0)$ stuffs per second and is equal to the difference between the multiplexer synchronous rate and the channel input rate. The slope of the sawtooth in Figure 4.17 thus equals the actual stuffing rate. The *maximum stuffing rate* f_s is a design parameter whose value is selected to minimize pulse stuffing jitter. Moreover, the maximum stuffing rate sets a limit on allowable frequency difference between the channel input and multiplexer clocks. This relationship is described by the stuffing ratio ρ, defined as the ratio of the actual stuffing rate to the maximum stuffing rate:

$$\rho = \frac{f_0}{f_s} \leq 1 \tag{4.27}$$

For pulse stuffing jitter to be the periodic sawtooth waveform shown in Figure 4.18*a*, stuff bits need to be inserted at the specific instant that the phase error crosses threshold. Only discrete time slots are available for stuff bits at the maximum stuffing rate, however, and phase threshold crossings do not always occur at these times. The frequency offset f_0 tends to change slowly in time due to small differences between the multiplexer and channel input clocks. With these changes in f_0, the interval between threshold crossings also varies. In the case of Figure 4.18*b*, stuffing usually occurs once in every five opportunities, but occasionally at every four opportunities. The corresponding stuffing ratio ρ is slightly greater than 1:5 rather than exactly 1:5 as in Figure 4.18*a*. This effect is called **waiting time jitter** because of the delay or waiting time between the initiation of a stuff request and the time at which the stuff is actually accomplished. The result of waiting time jitter is a low-frequency sawtooth-shaped jitter superimposed on high-frequency pulse stuffing jitter.

The peak waiting time jitter occurs when the phase error crosses the threshold immediately after an available stuff opportunity; then the error accumulates until the next possible stuff opportunity. The peak-to-peak phase error that can accumulate during this interval equals the frequency offset times the interval between stuff opportunities:

$$\phi_\varepsilon = f_0 t_s \text{ bits} \tag{4.28}$$

where $t_s = 1/f_s$ seconds. From the definition of stuffing ratio given in (4.27), the peak-to-peak waiting time jitter can also be expressed as

$$\phi_\epsilon = \rho \text{ bits} \tag{4.29}$$

At $\rho = 1$, peak-to-peak waiting time jitter is 1 bit. At $\rho = \frac{1}{2}$, the peak-to-peak jitter equals $\frac{1}{2}$ bit; likewise, at $\rho = \frac{1}{3}, \frac{1}{4}$, and $\frac{1}{5}$, the peak-to-peak jitter equals $\frac{1}{3}$, $\frac{1}{4}$, and $\frac{1}{5}$ bits. In general, for $\rho = 1/n$ the peak-to-peak

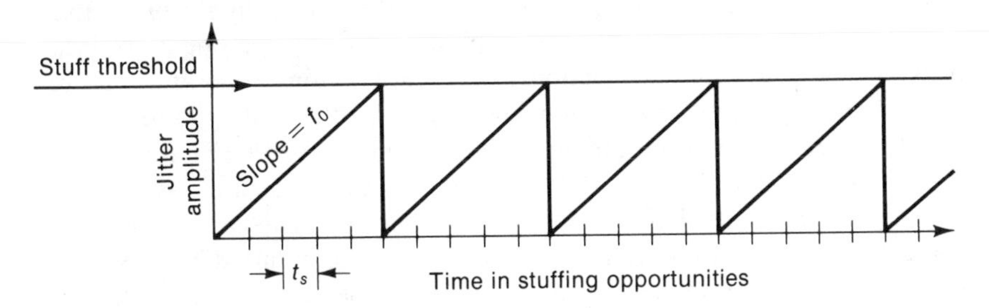

(a) Pulse stuffing jitter without waiting time jitter ($\rho = \frac{1}{5}$)

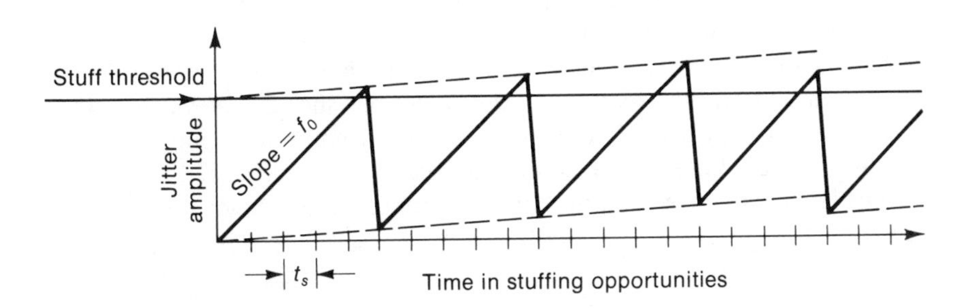

(b) Pulse stuffing jitter with superimposed waiting time jitter as indicated by dashed envelope ($\rho = \frac{1}{5}+$)

Figure 4.18 Waiting Time Jitter

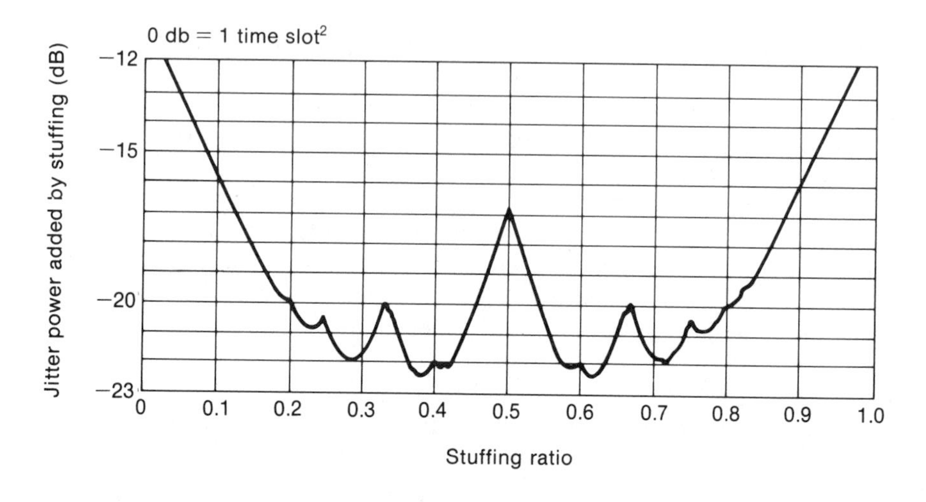

Figure 4.19 Waiting Time Jitter due to Pulse Stuffing (Courtesy CCITT [12]) (Smoothing Loop Bandwidth = 0.15f_s)

jitter equals $1/n$. Figure 4.19 illustrates this relationship in a plot of jitter amplitude versus stuffing ratio. This plot shows that waiting time jitter peaks when ρ is a simple fraction $1/n$. Hence waiting time jitter can be minimized by selecting a nonrational number n for the stuffing ratio.

In practice, the pulse stuffing recommended by the CCITT for second, third, and fourth-order multiplex has stuffing ratios that are less than 0.5 and not equal to simple fractions in order to avoid significant jitter peaks. As seen in Figure 4.19, waiting time jitter would be minimized by selecting a value of ρ near but not exactly equal to $\frac{1}{3}$ or $\frac{2}{5}$. Because the actual stuffing ratio varies due to drift in system clocks, the system designer is also interested in the range of values that the stuffing ratio may experience. By judiciously selecting the maximum stuffing rate based on expected clock frequency variations and known jitter characteristics (Figure 4.19), the stuffing ratio can be contained within a range that avoids large peaks of waiting time jitter. However, D. L. Duttweiler has shown that much of this advantage is lost if jitter is already present on a multiplexer input. As the amount of input jitter is increased, the peaks and valleys in Figure 4.19 become less pronounced and the choice of stuffing ratio becomes less critical [11].

Example 4.4

The 6.312-Mb/s second-order multiplexer described in CCITT Rec. G.743 combines four 1.544-Mb/s channels using positive pulse stuffing with the following characteristics (also see Table 4.5):

Frame length: 294 bits
Information bits per frame: 288 bits
Maximum stuffing rate per channel: $f_s = 5367$ stuffs per second
Stuff code length: 3 bits
Tolerance on 1.544-Mb/s channels: $\Delta f = 50$ ppm

(a) Determine the nominal, maximum, and minimum stuffing ratios. **(b)** Find the mean time to loss of BCI per channel due to stuff decoding errors, with an average bit error rate of 10^{-4}.

Solution

(a) The multiplexer synchronous rate per channel is given by

$$f_i' = \left(\frac{6.312 \times 10^6}{4}\right)\left(\frac{288}{294}\right) = 1{,}545{,}796 \text{ b/s}$$

The nominal, maximum, and minimum channel input rates are

$$\text{Nominal:} \quad f_i = 1{,}544{,}000 \text{ b/s}$$

$$\text{Maximum:} \quad f_i + \Delta f = 1{,}544{,}000 + \left(\frac{50}{10^6} \times 1.544 \times 10^6\right) = 1{,}544{,}077 \text{ b/s}$$

$$\text{Minimum:} \quad f_i - \Delta f = 1{,}544{,}000 - \left(\frac{50}{10^6} \times 1.544 \times 10^6\right) = 1{,}543{,}923 \text{ b/s}$$

The corresponding stuffing ratios, defined by (4.27), are then

$$\text{Nominal:} \quad \rho = \frac{f_i' - f_i}{f_s} = \frac{1796}{5367} = 0.3346$$

$$\text{Maximum:} \quad \rho = \frac{f_i' - (f_i - \Delta f)}{f_s} = \frac{1873}{5367} = 0.3490$$

$$\text{Minimum:} \quad \rho = \frac{f_i' - (f_i + \Delta f)}{f_s} = \frac{1719}{5367} = 0.3203$$

(b) From (4.25), the probability of decoding error per stuffing time slot is

$$3(10^{-4})^2 = 3 \times 10^{-8}$$

From (4.26), the mean time to loss of BCI per 1.544-Mb/s channel is

$$\frac{1}{(5367)(3 \times 10^{-8})} = 6.2 \times 10^3 \text{ s} = 1.7 \text{ hr}$$

4.3.5 Smoothing Loop

The purpose of the phase-locked loop (PLL) shown in the demultiplexer of Figure 4.16 is to derive the original channel input clock. The timing signal at

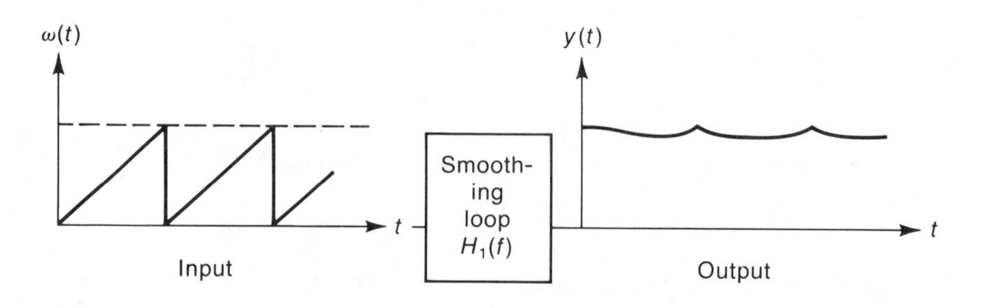

Figure 4.20 Model of Smoothing Loop Effect on Jitter

the input to the PLL will have an instantaneous frequency of f_i', the multiplexer channel rate, but with occasional timing cycles deleted so that the long-term average frequency is f_i. This gapped version of the original channel timing signal is then smoothed by the PLL to provide an acceptable recovered timing signal.

The effect of a smoothing loop on stuffing jitter is illustrated in Figure 4.20. First-order phase-locked loops have been found adequate to attenuate stuffing jitter properly, as will be verified here. This loop has a cutoff

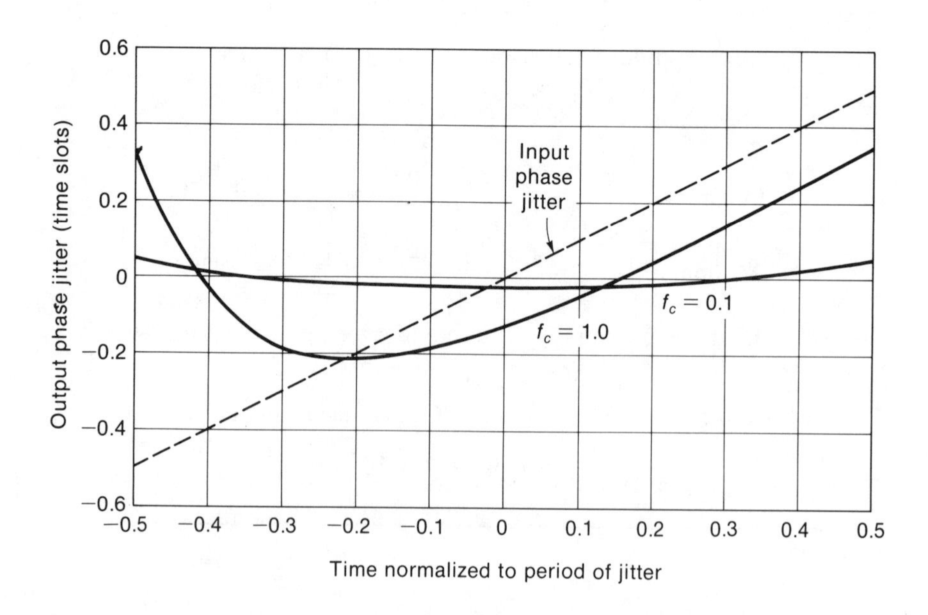

Figure 4.21 Stuffing Jitter Waveform at Output of First-Order Smoothing Loop

frequency that is low compared to the stuffing rate. In this way, jitter on the output signal due to removal of stuffing bits can be controlled to acceptable levels. The closed-loop transfer function of a first-order loop is given by

$$H_1(f) = \frac{f_c}{f_c + jf} \tag{4.30}$$

The corresponding impulse response is then

$$h_1(t) = 2\pi f_c e^{-2\pi f_c t} \tag{4.31}$$

where f_c is the 3-dB cutoff frequency of the smoothing filter normalized to the actual stuffing rate f_0. The periodic sawtooth waveform representing unsmoothed pulse stuffing jitter can be expressed as

$$\omega(t) = t - n \qquad n - 0.5 < t < n + 0.5;\ n = 0, \pm 1, \pm 2, \ldots \tag{4.32}$$

where the period has been normalized by the reciprocal of the stuffing rate, f_0, and the sawtooth peak-to-peak amplitude has been normalized by 1 bit. Note that the waveform given in (4.32) neglects the presence of waiting time jitter. This turns out to be a reasonable assumption, since waiting time jitter is a low-frequency jitter and is not significantly attenuated by the smoothing loop, whereas pulse stuffing jitter is at a higher frequency and can be significantly attenuated by the loop. The output of the first-order smoothing loop with the periodic sawtooth input is obtained by applying the convolution theorem, which results in

$$y(t) = \int_{-\infty}^{t} \omega(\alpha) 2\pi f_c e^{-2\pi f_c (t-\alpha)}\, d\alpha \tag{4.33}$$

Upon carrying out the integration, the output jitter waveform over the period $-0.5 < t < 0.5$ is

$$y(t) = t - \frac{1}{2\pi f_c} + e^{-2\pi f_c t}\left(\frac{e^{-\pi f_c}}{1 - e^{-2\pi f_c}}\right) \tag{4.34}$$

The phase jitter $y(t)$ given by (4.34) is plotted in Figure 4.21. Note that the smoothing loop with narrower bandwidth provides greater jitter attenuation. In Figure 4.21, as before, f_c is the 3-dB cutoff frequency of the smoothing loop normalized to the stuffing frequency.

To determine the peak-to-peak output jitter from Figure 4.21 note that the waveform of (4.34) has a maximum value at the two extremes ($t = \pm 0.5$) and minimum value in between. The minimum point is found by differentiating $y(t)$, setting this equal to zero, and solving for t. From the minimum and maximum values, the peak-to-peak output jitter can be expressed as

$$\Delta y_{p-p} = \frac{1}{1 - e^{-2\pi f_c}} - \frac{1}{2\pi f_c}\left[1 + \ln\left(\frac{2\pi f_c}{1 - e^{-2\pi f_c}}\right)\right] \tag{4.35}$$

A plot of this peak-to-peak stuffing jitter is shown in Figure 4.22 as a function of f_c, the smoothing loop bandwidth normalized to stuffing rate. As expected, stuffing jitter is significantly attenuated with narrow loop bandwidths.

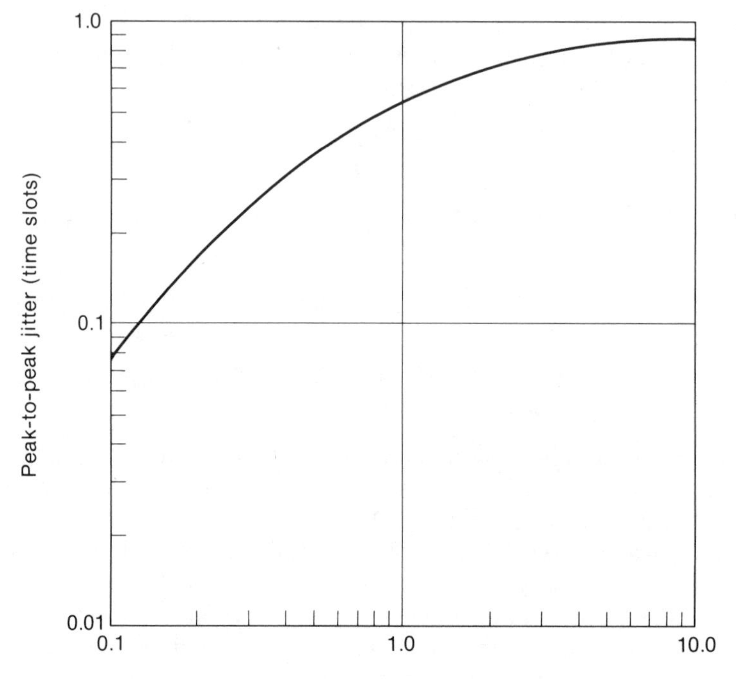

Figure 4.22 Peak-to-Peak Jitter at Output of First-Order Smoothing Loop

Example 4.5

For the 6.312-Mb/s multiplexer described by CCITT Rec. G.743, the two specifications pertaining to pulse stuffing jitter are:

1. With no jitter at the input to the multiplexer and demultiplexer, the jitter at the demultiplexer output should not exceed one-third of a time slot peak-to-peak.
2. The gain of the jitter transfer function should not exceed the limits given in Figure 4.23.

Find the maximum cutoff frequency f_c for the smoothing loop that will satisfy these two requirements.

Solution

From Figure 4.22 we see that the first requirement is met with a cutoff frequency

$$f_c = 0.45 \text{ Hz/stuffing rate}$$

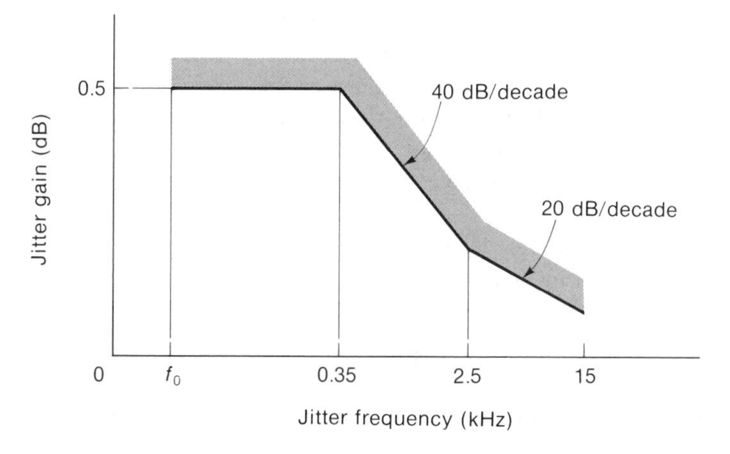

Note: The frequency f_0 should be as low as possible taking into account the limitations of measuring equipment.

Figure 4.23 Required Jitter Attenuation for 6.312-Mb/s Pulse Stuffing Multiplexer (Courtesy CCITT [2])

In terms of hertz and for a nominal stuffing rate of 1796 stuffs per second (from Example 4.4),

$$f_c\ (\text{Hz}) = (0.45)(1796) = 808\ \text{Hz}$$

while for the minimum stuffing rate of 1719 stuffs per second (also from Example 4.4),

$$f_c\ (\text{Hz}) = (0.45)(1719) = 773\ \text{Hz}$$

For the second requirement on pulse stuffing jitter, examination of Figure 4.23 shows that the 3-dB cutoff frequency must be at about 625 Hz or less. Further, the smoothing loop must exhibit a rolloff characteristic as shown in Figure 4.23 and limit jitter gain to 0.5 dB below 350 Hz.

A potential problem in introducing a pulse stuffing multiplexer into a digital transmission system is the effect that pulse stuffing jitter has on bit synchronizers in the system. Excessive jitter can cause a phase-locked loop used for bit synchronization to lose lock or slip bits. Since timing derived by the PLL is used to make bit decisions, it is not enough that the loop merely maintain lock. It must also track instantaneous bit timing, or bit decisions will be made with a timing error. The ability of the bit synchronizer to track input jitter is improved by increasing the loop bandwidth but at the expense of added noise bandwidth, so that bit synchronizer bandwidths are generally no greater than necessary for proper timing recovery.

To analyze the effect of pulse stuffing jitter on bit synchronizers, a model of a first-order smoothing loop cascaded with a first-order bit synchronizer is needed as shown in Figure 4.24. This model represents a first-order loop as a phase detector, a multiplier of constant gain (α or β), and an integrator. The transfer function of this cascade from input signal to bit synchronizer phase detector output is

$$H(f) = H_1(f) \cdot H_2(f) = \frac{f_c(jf)}{(f_c + jf)(kf_c + jf)} \tag{4.36}$$

By partial fractions this expression can be expanded to

$$H(f) = \frac{1}{(1-k)}\left(\frac{f_c}{f_c + jf} - \frac{kf_c}{kf_c + jf}\right) \tag{4.37}$$

We can convolve the sawtooth waveform representing the jitter input as given in (4.32) with $H(f)$, similar to that done earlier in (4.33), to arrive at

$$\theta_e(t) = \frac{1}{2\pi kf_c} + \frac{1}{1-k}\left(e^{-2\pi f_c t} \cdot \frac{e^{-\pi f_c}}{1 - e^{-2\pi f_c}} - e^{-2\pi kf_c t} \cdot \frac{e^{-\pi kf_c}}{1 - e^{-2\pi kf_c}}\right) \tag{4.38}$$

This is an expression for phase error over one stuffing jitter period, $-0.5 < t < 0.5$. This waveform has a maximum value at the extremes ($t = \pm 0.5$)

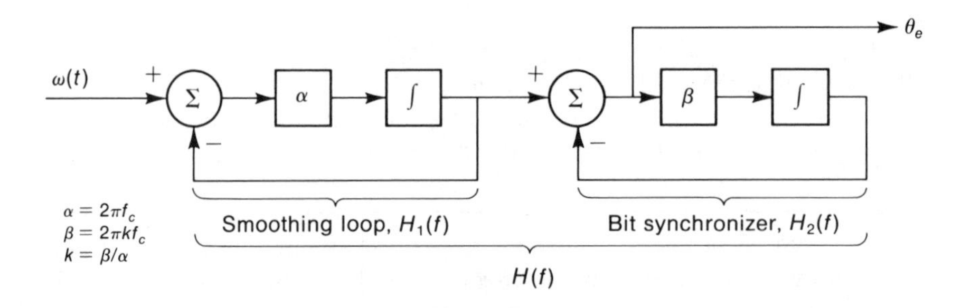

Figure 4.24 Block Diagram of First-Order Smoothing Loop Cascaded with First-Order Bit Synchronizer

and a minimum point inbetween. The minimum point is found by differentiating $\theta_e(t)$, setting this equal to zero, and solving for t.

From the maximum and minimum values of (4.38), the peak-to-peak phase error is found to be

$$\Delta\theta_{e_{p-p}} = \frac{1}{1-k} \cdot \frac{1}{1-e^{-2\pi f_c}} \left[1 - \frac{1-e^{-2\pi f_c}}{1-e^{-2\pi k f_c}} + \left(\frac{1-k}{k} \right) \cdot \left(k \cdot \frac{1-e^{-2\pi f_c}}{1-e^{-2\pi k f_c}} \right)^{1/(1-k)} \right] \tag{4.39}$$

Using (4.39), curves are shown in Figure 4.25 for peak timing error when a first-order smoothing loop is interfaced with a first-order bit synchronizer. This figure presents the results for a constant ratio of bit synchronizer loop bandwidth to smoothing loop filter bandwidth. The curves only consider the effects of stuffing jitter since Expression (4.32) does not include waiting time jitter, which is so low in frequency that its effects are negligible for bit

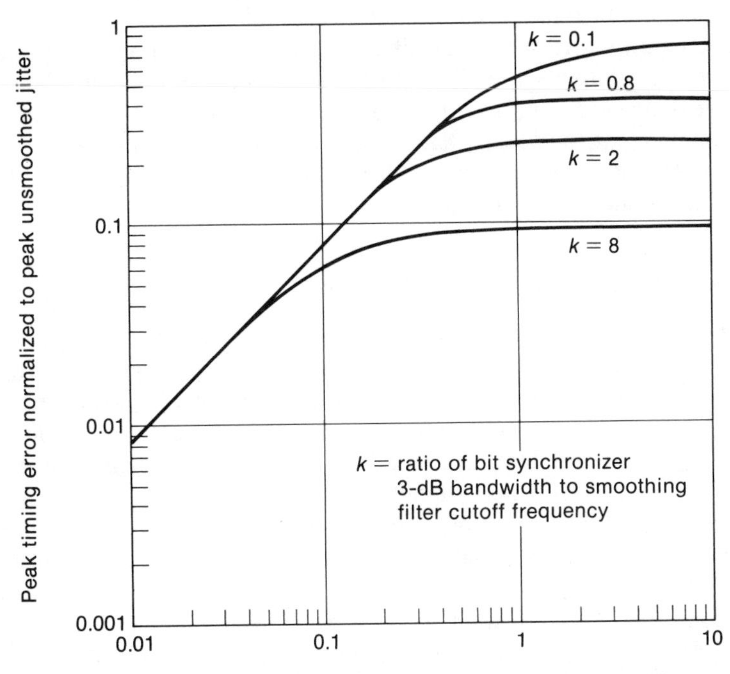

Figure 4.25 Effect of Pulse Stuffing Jitter on Timing Recovery for a First-Order Bit Synchronizer

synchronizers. To illustrate the significance of these curves, first note that when the smoothing loop filter has a bandwidth equal to the stuffing rate ($f_c = 1$), the timing error is significantly decreased only when the bit synchronizer has a bandwidth approximately 10 times the stuffing rate. When the bit synchronizer bandwidth and the smoothing filter bandwidth are both equal to the stuffing rate ($k = 1, f_c = 1$), the timing error is approximately 30 percent of a bit, which corresponds to an 8.0-dB loss in signal-to-noise ratio. Figure 4.25 points out the importance of minimizing pulse stuffing jitter by proper smoothing loop design to avoid excessive degradation in downstream bit synchronizers.

Another concern in the application of pulse stuffing multiplexers is the degree to which jitter accumulates along a digital path. Waiting time jitter has been shown to accumulate according to $\sqrt{N}$, where N is the number of multiplex/demultiplex pairs connected in tandem [11]. Proper choice of the stuffing ratio in each multiplexer/demultiplexer pair limits the accumulation of waiting time jitter. As observed earlier, waiting time jitter has little effect on bit synchronizers, although it does produce distortion in PCM systems due to the irregular spacing of reconstructed analog samples. A CCITT study of jitter accumulation along a digital path indicates that PCM multiplexers and pulse stuffing multiplexers tolerate high input jitter and cause low output jitter [13]. This jitter-reducing effect of digital multiplexers means that limitations in jitter amplitude such as those shown in Figure 4.23 will not be exceeded in a path of tandemed multiplexers. The same study also pointed out, however, that digital repeatered lines tolerate only low input jitter and generate relatively high output jitter. This question of jitter and its accumulation in repeatered lines is considered in Chapter 8.

4.4 MULTIPLEXING OF ASYNCHRONOUS DATA

Many low-speed terminals transmit characters asynchronously, where characters have a uniform length but do not necessarily occur at a uniform rate. Teletypewriters and other keyboard terminals produce alphanumeric symbols, one at a time, in character format. Each character is made up of several information bits plus 2 or 3 bits for character synchronization; for example, the CCITT International Alphabet No. 2 uses five information bits. In the United States, most devices use the 7-bit ASCII code (American Standard Code for Information Interchange), which is compatible with CCITT Alphabet No. 5. As shown in the example of Figure 4.26, character synchronization for the ASCII code is provided by a start bit of value 0 (space) and two stop bits of value 1 (mark). The ASCII code also provides a parity bit for error detection, thus resulting in an 11-bit character.

Several techniques are used in the digital multiplexing of asynchronous data. A form of synchronous multiplexing is possible by using buffers and some means of character detection to assemble characters in blocks of data. Each block of data is delimited or framed by a synchronization code. Moreover, each character's start and stop pulses are dropped at the multi-

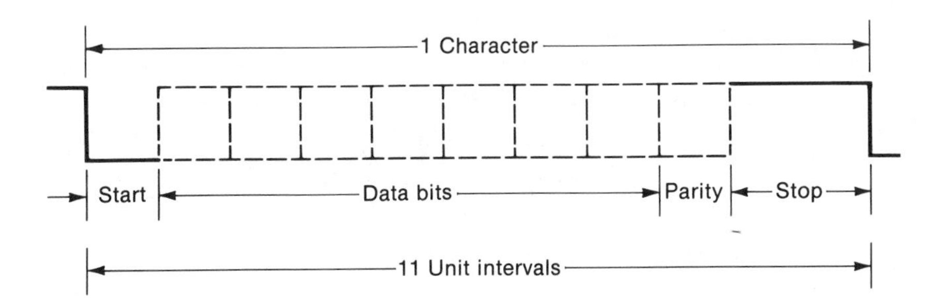

Figure 4.26 Format of Character from CCITT Alphabet No. 5 or ASCII Code

plexer and reinserted by the demultiplexer before interface with the receiving data terminal. This technique results in high transmission efficiency, since only the character's data bits are transmitted, and avoids the problem of handling stop pulses that are not an integer number of bit intervals. A form of pulse stuffing can also be used to compensate for variation in arrival time of characters and to absorb any fraction of a bit used as part of the stop pulse. Pulse stuffing is also a highly efficient transmission technique, but the system designer must be aware of the effects of pulse stuffing jitter and stuff code errors.

A low-speed asynchronous data signal can also be converted to a synchronous signal by simply sampling the data at a very high rate, a technique that is inexpensive to implement. **Transitional coding** is a similar technique that uses extra bits to code the data transition times. Decoding in the receiver provides reconstruction of the transition time and binary value for each data bit. The disadvantage of both techniques is the high transmission bit rate required by the sampling or encoding process and the jitter resulting from reconstruction of the data transitions. Of the two schemes, transitional coding is the more commonly used, since it results in more efficient transmission.

The simplest form of transitional coding is the 2-bit version illustrated in Figure 4.27*a*. Shown are the data signal and corresponding 2-bit code. When a data transition occurs, the two coded bits indicate the transition time and data bit value. The first bit following the transition (indicated as X) is a timing bit whose value indicates whether the data transition occurred in the first or second half of the sampling clock period. After this first half/second half bit has been transmitted, the coded bits are set equal to the data bit until another transition occurs. Since 2 bits of code are required for each data transition, the code rate is twice that of the data—one of two disadvantages of transitional coding. The second disadvantage is the jitter introduced when reconstructing the data transitions at the receiver, as

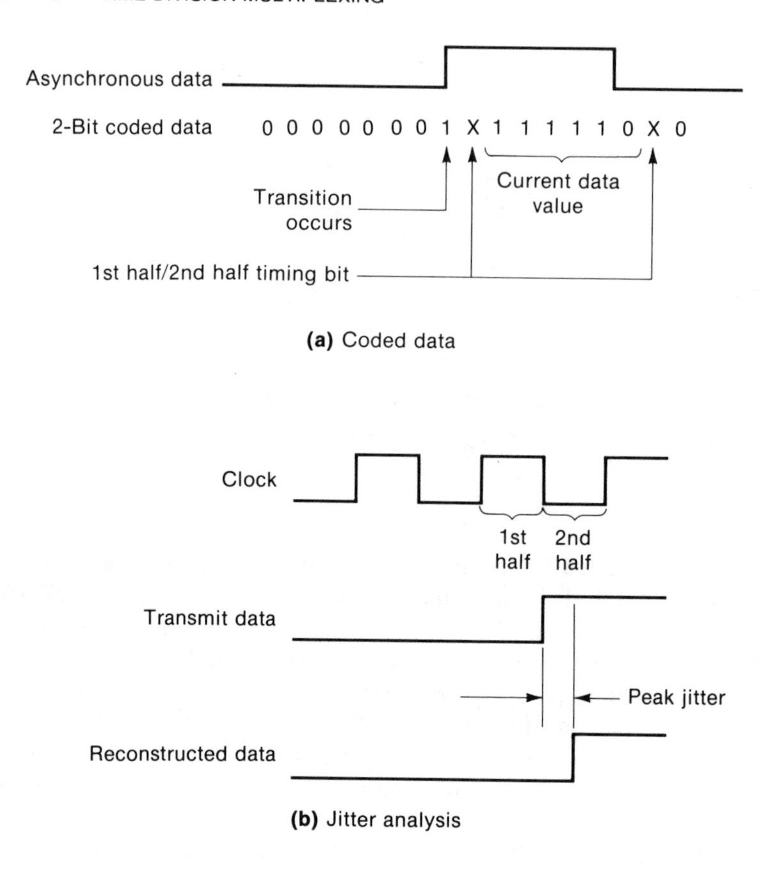

Figure 4.27 Two-Bit Transitional Coding

illustrated in Figure 4.27*b*. The clock is shown partitioned into the first and second half of a cycle. In the receiver, the transition is reconstructed in the center of the first or second half-clock cycle. Peak jitter occurs when the data transition is precisely at the boundary between the two half-clock cycles and amounts to one-quarter of the clock cycle. Since the data rate is one-half that of the sampling clock, the peak jitter is therefore one-eighth of the data bit interval, which is tolerable in most data terminals.

Jitter performance can be improved by increasing the number of code bits to partition the clock cycle further for data reconstruction. For example, the 3-bit coder recommended by CCITT Rec. R.111 [14] uses 2 bits to identify the quarter cycle in which each data transition occurs. Table 4.2 shows the assignment of the 3 bits—B_1, B_2, and B_3—according to the data bit value and position of the transition. This scheme clearly reduces the jitter compared to 2-bit coding but increases the transmission rate by 50 percent since 3-bit coding requires a code rate three times that of the data

Table 4.2 CCITT Rec. R.111 for Transitional Coding of Asynchronous Data

Position of Transition in a Group of Four Sampling Pulses	*Code Character for Transition from 1 to 0 in Data Signal*			*Code Character for Transition from 0 to 1 in Data Signal*		
	B_1	B_2	B_3	B_1	B_2	B_3
First quarter	0	0	0	1	1	1
Second quarter	0	0	1	1	1	0
Third quarter	0	1	0	1	0	1
Fourth quarter	0	1	1	1	0	0

rate. The peak jitter J_p due to transitional coding or simple sampling can be expressed in percentage by

$$J_p = \frac{1}{2^N}\left(\frac{f_D}{f_C}\right) \times 100 \tag{4.40}$$

where N is the number of code bits ($N = 1$ for simple sampling), f_D is the data rate, and f_C is the clock rate. A typical limit imposed on transitional coders is a maximum of 10 percent peak jitter, which is acceptable to most asynchronous data terminals.

4.5 DIGITAL MULTIPLEX HIERARCHIES

Three different digital multiplex hierarchies have emerged and been adopted by the CCITT. The three hierarchies reflect choices made by AT&T and followed by the United States and Canada, by Nippon Telegraph and Telephone (NTT) and followed by Japan, and by the Conférence Européene des Administrations des Postes et Télécommunications (CEPT) and followed by most countries outside North America and Japan. The bit rate structure and channel capacity of these hierarchies are shown in Figure 4.28. This section describes the multiplex equipment specified by these digital hierarchies.

4.5.1 PCM Multiplex

The first-level multiplexer of each CCITT digital hierarchy is a PCM multiplexer operating at a bit rate of 1.544 Mb/s in the North American and Japanese hierarchy and 2.048 Mb/s in the CEPT hierarchy. Table 4.3

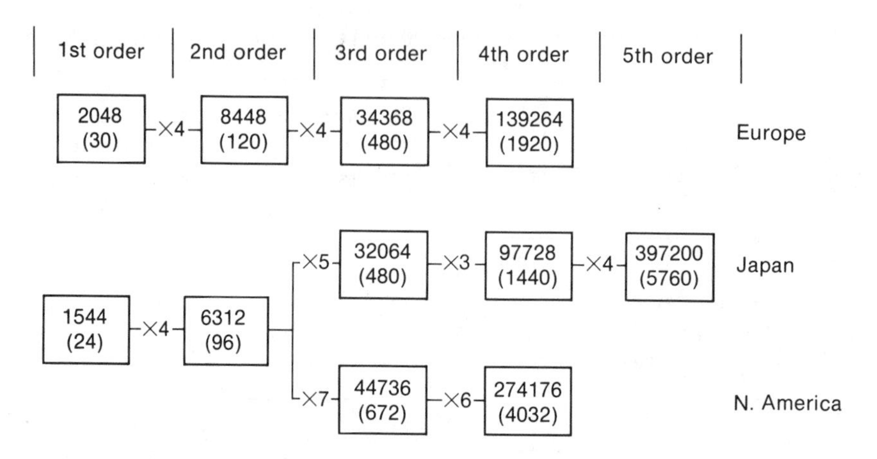

Notes: 1. The top figures in each box indicate bit rates in kb/s.
2. The figures in parentheses indicate capacity in number of 64 kb/s time slots.
3. The figures ×*N* indicate the number of multiplexed digital channels.

Figure 4.28 Multiplex Hierarchies Recommended by CCITT

lists characteristics of these two PCM standards as prescribed by the CCITT.

Figure 4.29 illustrates the 1.544-Mb/s frame structure, which provides 24 channels of 64-kb/s PCM speech signals. This grouping of 24 voice channels with a transmission rate of 1.544 Mb/s is known as a **digroup**. Each frame consists of 24 eight-bit words and one synchronization bit for a total of 193 bits. Each channel is sampled 8000 times per second; thus there are 8000 frames per second and a resulting aggregate rate of (8000 frames) × (193 bits/frame) = 1,544,000 b/s. Each multiframe consists of 12 frames. The 12 synchronizing bits per multiframe are used for frame and multiframe synchronization. This 12-bit sequence is subdivided into two sequences. The frame alignment pattern is 101010 and is located in odd-numbered frames; the multiframe alignment pattern is 001110 and is located in even-numbered frames. The multiframe structure also contains signaling bits that are used to carry supervisory information (on-hook, off-hook) and dial pulses for each voice channel. This signaling function is accomplished by time-sharing the least significant bit (B8) between speech and signaling. The B8 bits carry speech sample information for five frames, followed by signaling information in every sixth frame. Two separate signaling channels are provided, which alternate between the sixth and twelfth frame of each multiframe. Signaling bits carried by frame 6 are called A and those carried by frame 12 are called B signaling bits. This form of signaling is known as *robbed bit* signaling.

Table 4.3 CCITT Characteristics of PCM Multiplex Equipment [2]

Characteristic	*1.544 Mb/s PCM Standard (CCITT Rec. G.733)*	*2.048 Mb/s PCM Standard (CCITT Rec. G.732)*
Encoder		
Compander law	μ-law	*A*-law
No. of segments	15	13
Zero code suppression	Yes	No
Bit Rate	1.544 Mb/s ± 50 ppm	2.048 Mb/s ± 50 ppm
Timing signal	Internal source or incoming digital signal or external source	Internal source or incoming digital signal or external source
Frame structure	See Figure 4.29	See Figure 4.30
No. of bits per channel	8	8
No. of channels per frame	24	32
No. of bits per frame	193	256
Frame repetition rate	8000	8000
Signaling	Least significant bit of each channel every sixth frame	Time slot 16 used; rate $\leq$ 64 kb/s

The CEPT standard for PCM multiplexing is illustrated in Figure 4.30. Each frame consists of 30 eight-bit PCM words (time slots 1–15 and 17–31), an eight-bit frame synchronization word (time slot 0), and an eight-bit signaling word (time slot 16)—a total of 32 eight-bit words or 256 bits. Since there are 8000 frames transmitted per second, the resulting aggregate rate is 2,048,000 b/s. The multiframe structure shown in Figure 4.30*b* comprises 16 frames, with signaling information inserted in time slot 16 of each frame. The first eight-bit signaling time slot is actually used for multiframe alignment. The next 15 time slots for signaling provide four signaling bits per channel for all 30 speech channels.

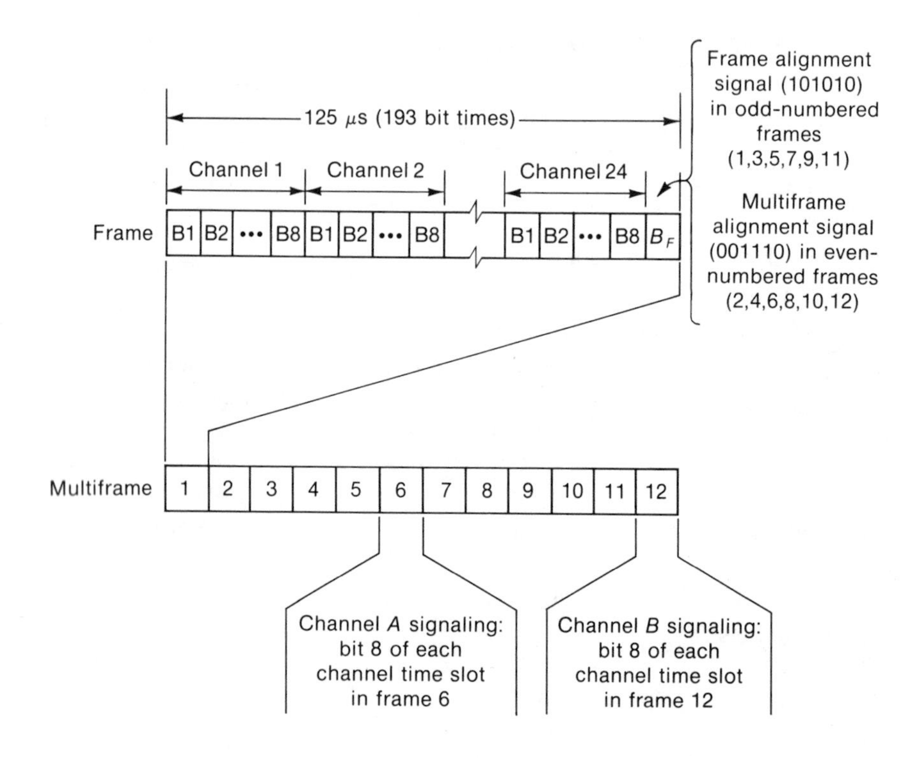

Figure 4.29 Frame and Multiframe Format for 24-Channel PCM Multiplex

4.5.2 Higher-Order Digital Multiplex

Higher-order digital multiplex standards have also been developed and codified by the CCITT. These standards have evolved from practices in North America, Japan, and Europe. Characteristics of higher-order multiplex hierarchies are shown in Table 4.4. Two standards exist for second-order multiplexers, corresponding to the North America/Japan and CEPT hierarchies and identified by the CCITT as Rec. G.743 and G.742, respectively [2]. Tables 4.5 and 4.6 show the frame structure for the two standard second-order multiplexers, which operate at rates of 6.312 Mb/s and 8.448 Mb/s, respectively.

Three standards have emerged for third and fourth-order multiplexers, corresponding to North America, Japan, and CEPT hierarchies. For details on the frame structure of these multiplexers, refer to the appropriate CCITT recommendations listed in Table 4.4. As observed in Table 4.4, some common features exist in the various standards, such as the multiplexing method and pulse stuffing technique. However, the divergence of essential features such as channel and aggregate bit rates has precluded interface

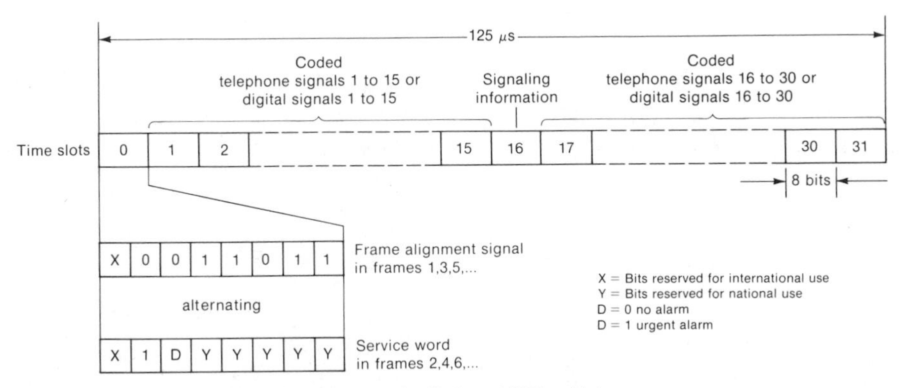

(a) Framing format for 30-channel PCM multiplex

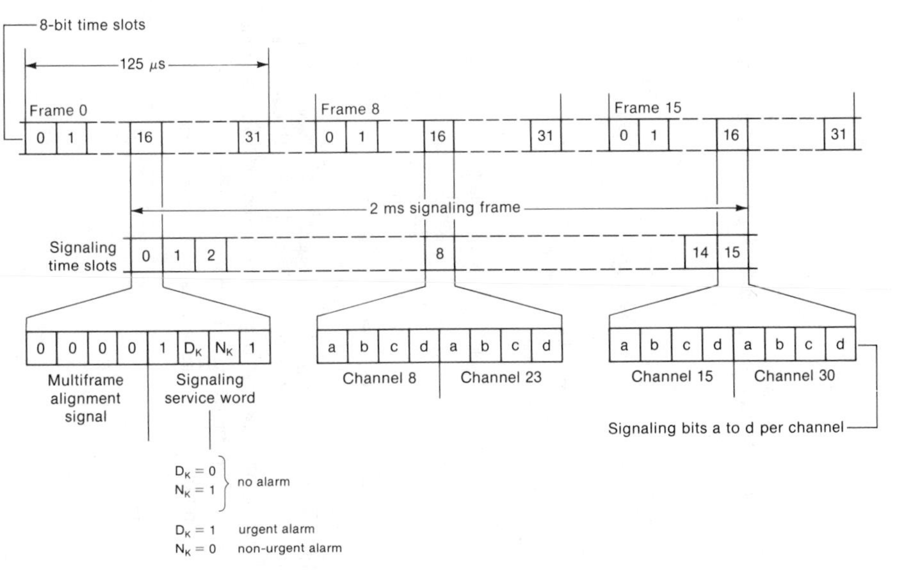

(b) Multiframe format for 30-channel PCM multiplex

Figure 4.30 Frame Structure for 30-Channel PCM (Reprinted by Permission of Siemens AG)

between the various hierarchies except at the lowest level (zero order: voice channel).

4.5.3 Multiplex Equipment

As illustrated in Figure 4.31, AT&T utilizes multiplexers at four levels, where the resulting digital signals (DS) at rates of 1.544, 6.312, 44.736, and

Table 4.4 Characteristics of Digital Multiplex Equipment [2]

	Second Order		*Third Order*			*Fourth Order*		
Multiplex Characteristic	*CCITT Rec. G.743*	*CCITT Rec. G.742*	*CCITT Rec. G.751*	*CCITT Rec. G.752*	*CCITT Rec. G.752*	*CCITT Rec. G.751*	*North America*	*Japan NTT*
Aggregate bit rate (kb/s)	6312	8448	34,368	32,064	44,736	139,264	274,176	97,728
Tolerance on aggregate rate (ppm)	±30	±30	±20	±10	±20	±15	±10	
Aggregate digital interface[a]	B6ZS	HDB3	HDB3	Scrambled AMI	B3ZS	CMI		
Channel bit rate (kb/s)	1544	2048	8448	6312	6312	34,368	44,736	32,064
No. of channels	4	4	4	5	7	4	6	4
Channel digital interface[a]	Bipolar (AMI)	HDB3	HDB3	B6ZS	B6ZS	HDB3	B3ZS	Scrambled AMI
Frame structure	See Table 4.5	See Table 4.6	See CCITT Rec. G.751	See CCITT Rec. G.752	See CCITT Rec. G. 752	See CCITT Rec. G.751		
Pulse stuffing technique		Positive pulse stuffing with 3-bit stuff control word and majority vote decoding						
Multiplexing method		Cyclic bit interleaving in channel numbering order						
Timing signal		Derive from external as well as from internal source						

[a]See Chapter 5 for a description of baseband transmission codes.

Table 4.5 Frame Structure of 6.312-kb/s Second-Order Multiplex

Frame Structure

M_0	I_{1-48}	C_{11}	I_{1-48}	F_0	I_{1-48}	C_{12}	I_{1-48}	C_{13}	I_{1-48}	F_1	S_1	I_{2-48}			
M_1	I_{1-48}	C_{21}	I_{1-48}	F_0	I_{1-48}	C_{22}	I_{1-48}	C_{23}	I_{1-48}	F_1	I_1	S_2	I_{3-48}		
M_1	I_{1-48}	C_{31}	I_{1-48}	F_0	I_{1-48}	C_{32}	I_{1-48}	C_{33}	I_{1-48}	F_1	I_1	I_2	S_3	I_{4-48}	
X	I_{1-48}	C_{41}	I_{1-48}	F_0	I_{1-48}	C_{42}	I_{1-48}	C_{43}	I_{1-48}	F_1	I_1	I_2	I_3	S_4	I_{5-48}

Frame length: 294 bits
Multiframe length: 1176 bits
Bits per channel per multiframe: 288 bits
Maximum stuffing rate per channel: 5367 b/s
Nominal stuffing ratio: 0.334

F_i = frame synchronization bit with logic value i
M_i = multiframe synchronization bit with logic value i
I_i = information bits from channels
C_{ij} = jth stuff code bit of ith channel
S_i = bit from ith channel available for stuffing
X = alarm service bit

Table 4.6 Frame Structure of 8.448-Mb / s Second-Order Multiplex

Frame Structure

F_1	F_1	F_1	F_1	F_0	F_1	F_0	F_0	F_0	F_0	X	Y	I_{13}	...	I_{212}
C_{11}	C_{21}	C_{31}	C_{41}	I_5	·	·	·							I_{212}
C_{12}	C_{22}	C_{32}	C_{42}	I_5	·	·	·							I_{212}
C_{13}	C_{23}	C_{33}	C_{43}	S_1	S_2	S_3	S_4	I_9	· · ·					I_{212}

Frame Length: 212 bits
Multiframe length: 848 bits
Bits per channel in a frame: 206 bits
Maximum stuffing rate per channel: 9962 b/s
Nominal stuffing ratio: 0.424

F_i = frame synchronization bit with logic value i
I_i = information bit from channels
X = alarm indication to remote multiplexer
Y = reserved for national use
C_{ij} = jth stuff code bit of ith channel
S_i = bit from ith channel available for stuffing

274.176 Mb/s are denoted DS-1, DS-2, DS-3, and DS-4. Transmission (T) lines for these rates are denoted T1, T2, T3, and T4; multiplex (M) equipment is denoted by the channel and aggregate level—for example, an M12 combines four DS-1 signals into a DS-2 signal.

AT&T uses the term **digital channel bank** to denote a PCM multiplexer. The D1 channel bank (D for digital, 1 for first generation) was placed in service in 1962. The D1 used 7-bit logarithmic (μ-law) PCM coding and combined 24 channels into the DS-1 rate [15]. In 1969, the D2 channel bank was introduced, which provided the 8-bit coding scheme used in all later channel banks [16]. The D2 combines up to 96 channels into four separate DS-1 signals. The third-generation channel bank, the D3, was introduced in 1972 for 24-channel (1.544 Mb/s) application and to replace the D1 [17]. Because of differences in signaling format and companding characteristic, the D1 is not end-to-end compatible with the later-generation channel banks. However, the D1D, a modified D1 that is compatible with a D3, was also introduced to upgrade existing D1 channel banks [18]. In 1976 the

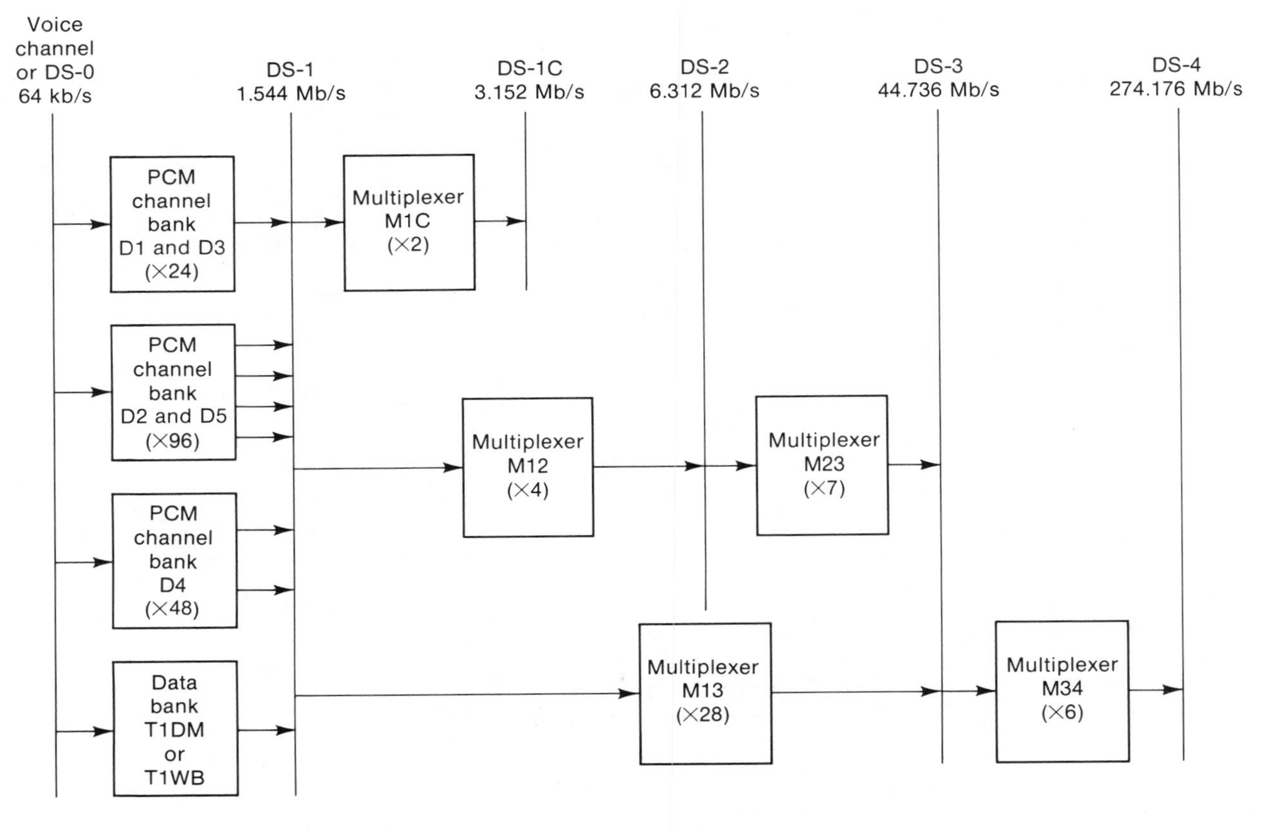

Figure 4.31 North American Digital Hierarchy

Table 4.7 Comparison of PCM Channel Banks

Characteristic	*D1*	*D2*	*D1D*	*D3*	*D4*	*D4E*	*D5*
No. of voice channels	24	24, 48, 72, or 96	24	24	24 or 48	30	24, 48, 72, or 96
Sampling sequence	1, 13, 2, 14, 3, 15, 4, 16, 5, 17, 6, 18, 7, 19, 8, 20, 9, 21, 10, 22, 11, 23, 12, 24	12, 13, 1, 17, 5, 21, 9, 15, 3, 19, 7, 23, 11, 14, 2, 18, 6, 22, 10, 16, 4, 20, 8, 24	Same as D1	1, 2, 3, 4, 5, 6, 7, 8, 9, 10, 11, 12, 13, 14, 15, 16, 17, 18, 19, 20, 21, 22, 23, 24,	Options for three sampling sequences corresponding to D1, D2, D3	1, 2, 3, 4, . . . , 30	Options for D1, D2, D3, D4
No. of coding bits per channel	7	8	8	8	8	8	8
Signaling bit location	Eighth bit of every channel	Eighth bit of every channel of every sixth frame	Same as D2	Same as D2	Same as D2	Time slot 16	Same as D2
Companding characteristic	$\mu = 100$ nonlinear diode characteristic	$\mu = 255$ 15-segment piecewise linear approximation	Same as D2	Same as D2	Same as D2	$A = 87.6$ 13-segment piecewise linear approximation	Same as D2
Transmission line	T1	Up to four independent T1's	T1	T1	T1, T1C, or T2	T1E (2.048 Mb/s)	T1, T1C, or T2

versatile D4 channel bank was introduced; it provides up to 48-channel operation [19]. The D4 has the following options for transmission line interface: two independent T1 lines, a single T1C line (3.152 Mb/s),* and two D4's ganged and interfaced with a T2 line. It should also be noted that AT&T has developed a D4E that is compatible with the 2.048-Mb/s multiplexer of CCITT Rec. G.732 and operates over a T1E (2.048 Mb/s) transmission line. Finally, the D5 channel bank was announced in 1982 and provides up to 96 channel operation. A comparison of these PCM channel banks is presented in Table 4.7.

Digital multiplexers operating at the DS-1 level are used to provide data channels at 64 kb/s (designated the DS-0 level) and 2.4, 4.8, or 9.6 kb/s (designated subrate data channels). One is designated the T1 data multiplexer (T1DM) and provides up to 23 data channels; a second, designated the T1WB4, provides up to 12 data channels while also providing PCM voice in the remaining channels [20].

The DS-2 level is formed by multiplexing four DS-1's using the M12 or by direct interface with the output of a video coder [21]. The DS-3 level at 44.736 Mb/s is formed by the M23, which multiplexes seven DS-2 signals, or by the M13, which directly multiplexes 28 DS-1 signals. A mastergroup coder (600 FDM channels) and commercial color TV systems are also available at the DS-3 rate. The fourth level is formed by the M34 by multiplexing six DS-3 signals into a 274.176-Mb/s DS-4 signal. The M12, M23, and M34 follow CCITT recommendations for second, third, and fourth-order multiplexers.

4.6 STATISTICAL MULTIPLEXING AND SPEECH INTERPOLATION

With conventional time-division multiplexing, each channel is assigned a fixed time slot. Fixed frames are transmitted continuously between the multiplexer and demultiplexer. This means that each of the time slots belonging to an inactive channel must be filled with an "idle" character in order to maintain proper framing. Since most data terminals do not operate continuously, the average activity tends to be low compared to the available transmission bandwidth. Figure 4.32 illustrates the improved transmission efficiency that can be provided with statistical time-division multiplexing (STDM) by dynamically allocating transmission bandwidth to whichever terminals are active at any one time. Statistical multiplexing results in transmission efficiency improvements by a factor of 2 to 1 or more, as in the example of Figure 4.32.

As data arrive at the statistical multiplexer, they are placed in a buffer that has been dynamically assigned for each active channel. Frames of data are then assembled as the data are systematically removed from these

*T1C and M1C are designations for the transmission line and multiplexer for DS-1C (3.152 Mb/s) operation.

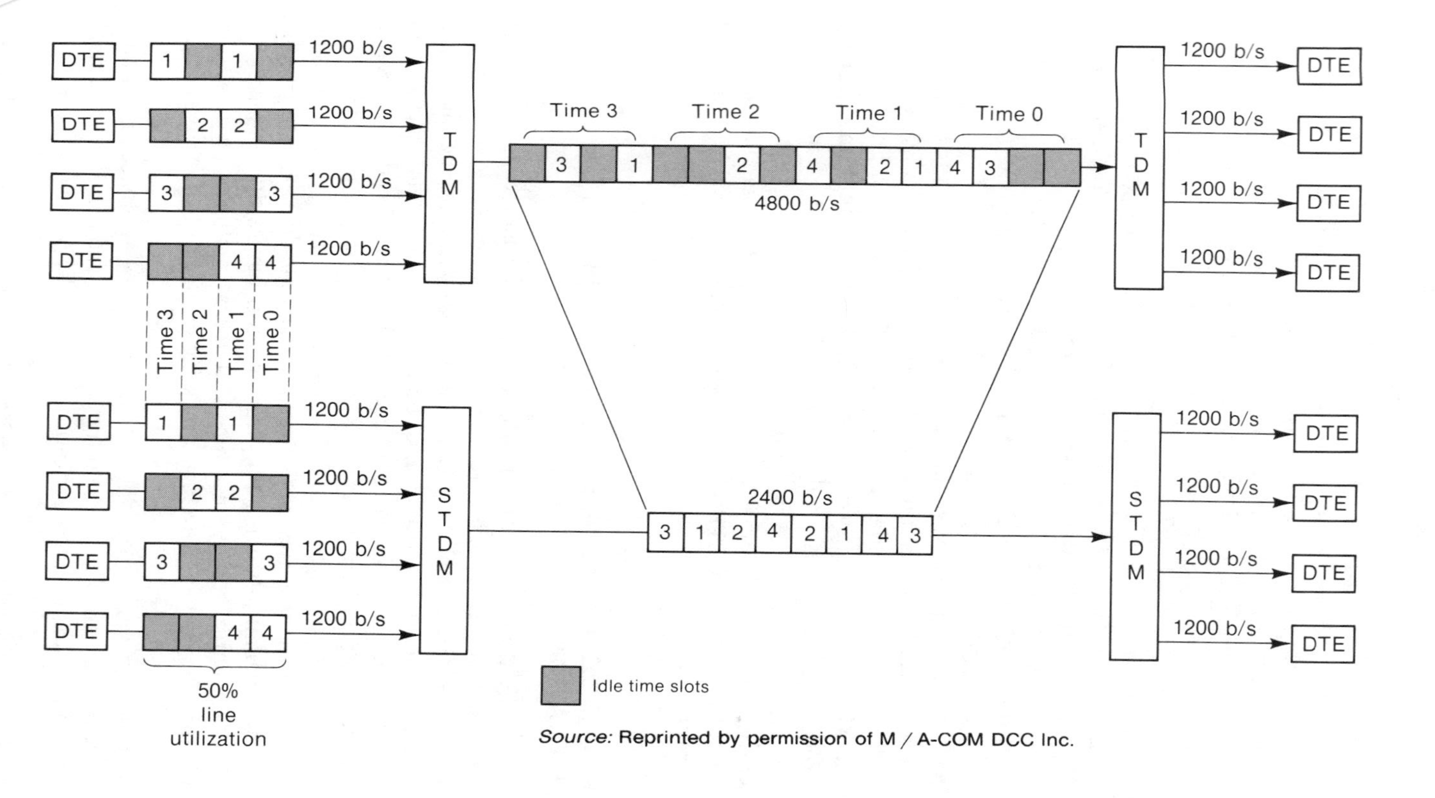

Source: Reprinted by permission of M / A-COM DCC Inc.

Figure 4.32 Comparison of Conventional TDM with Statistical TDM

buffers. However, buffer overflow can occur during prolonged peak traffic activity or excessive retransmission of data unless the terminal equipment is signaled to cease transmission of data. Transmission flow control on incoming channels can be invoked by using either in-band or out-of-band traffic control signals. In-band signals are generally nonprinting control characters, which are transmitted to the terminal by the multiplex in-band as if they were data characters. Out-of-band traffic control is achieved by placing control signals on the separate control lines available in data terminals.

Should the terminal equipment continue sending data during overload, buffer overflow will occur. The buffer recovers from overflow by dumping data of one or more terminals, transmitting a "data lost" message to the affected terminal, and possibly disconnecting the terminal. When total buffer utilization drops below the preset overload condition, flow control will be released by either in-band or out-of-band signals to allow transmission to be resumed.

Most statistical multiplexers also utilize the CCITT Rec. X.25 level 2 link protocol procedure. This procedure calls for transmission of information in variable-length frames. The fields within the frame are:

- *Flag* synchronizes the demultiplexer with the rest of the information in the frame.
- *Address* indicates the source or destination of a particular message.
- *Control* identifies the type of message.
- *Data* is a variable-length field into which data from input channels are packed.
- *Cyclic redundancy check* (CRC) performs error detection.

The error checking function of the link protocol involves a retransmission of the frame in error. This is made possible by retaining each frame in buffer storage until an acknowledgment is received from the far end of the link.

Additional features beyond those described earlier for statistical multiplexers are found in *intelligent* multiplexers enabled by the use of microprocessors. The selection of individual channel interface by data rate, clock type, code type, and so forth is typically done by front panel switches or even alphanumeric key pad to facilitate reconfiguration. Often the far-end multiplexer can be configured remotely by sending configuration settings down the transmission line, thus making it unnecessary ever to physically reconfigure the remote unit. Another typical feature of intelligent multiplexers is the ability to store a backup configuration and electronically switch from primary to backup upon demand. Built-in diagnostics are generally provided for automatic self-test after reconfiguration and periodically during normal operation. Not only the multiplexer itself is checked but also the modem, transmission line, and other devices connected to it. Tests are conducted via a series of loopbacks wherein test characters or signals are transmitted, looped, and received without interfering with the terminals or modems connected to the multiplexer.

A form of statistical multiplexing can also be applied to voice channels in order to realize concentrations of 2:1 or more. The technique is known as **time assignment speech interpolation (TASI)** for analog transmission and **digital speech interpolation (DSI)** for digital transmission. The principle involved is the filling of the silent gaps and pauses occurring in one voice channel with speech from other channels. The receiver reorders the bursts of speech into the required channel time slots. The technology consists of digital signal processing (such as PCM), buffer memory for speech storage, and microprocessors for switching and diagnostics. Speech interpolation devices have been employed by the U.S. Bell System [22], COMSAT [23], and European telephone companies [24]. Common configurations include 48 voice channels concentrated onto 24 transmission channels (abbreviated 48:24) for North American systems and 60:30 for European systems.

The operation of speech interpolation devices is similar to that of statistical multiplexers for data. As shown in Figure 4.33, incoming speech is recognized by speech activity detectors and simultaneously is stored in a buffer dynamically assigned to that voice channel. Data modem signals are also recognized and provided a dedicated connection, which then reduces the number of channels available for effective speech interpolation. Stored speech bursts are transmitted as soon as transmission bandwidth becomes available. The buffer memory allocated to a channel is of variable length to allow storage of speech until transmission is possible. After transmission of a speech burst the buffer is returned to a pool for future use. Periodically a control signal is transmitted that informs the receiver where to place each incoming speech burst. With sufficiently high speech activity on incoming channels or a large number of data signals, buffer overflow can result in loss of speech segments. Overload conditions can be controlled by blocking any unseized channels from being seized, by dropping segments on a controlled basis [25], or by reducing the code rate per voice channel for DSI systems [26]. The loss of speech segments, due to overload or delay in detecting a speech burst, is known as **clipping** and is the limiting factor in the performance of speech interpolation systems.

4.7 SUMMARY

Here we will summarize essential TDM design and performance considerations. These criteria can be used by the system designer in specifying and selecting TDM equipment.

Organization of a TDM frame depends on the number of channels to be multiplexed, the channel bit rates, the relationship between channel input timing and multiplexer timing, and the number and location of overhead bits. These overhead bits are required for frame alignment and possibly other functions. Their location within the frame can be bunched to facilitate frame alignment in the demultiplexer or distributed to protect against bursts of transmission errors. A fixed timing relationship between each channel and the multiplexer permits synchronous multiplexing, the simplest form of

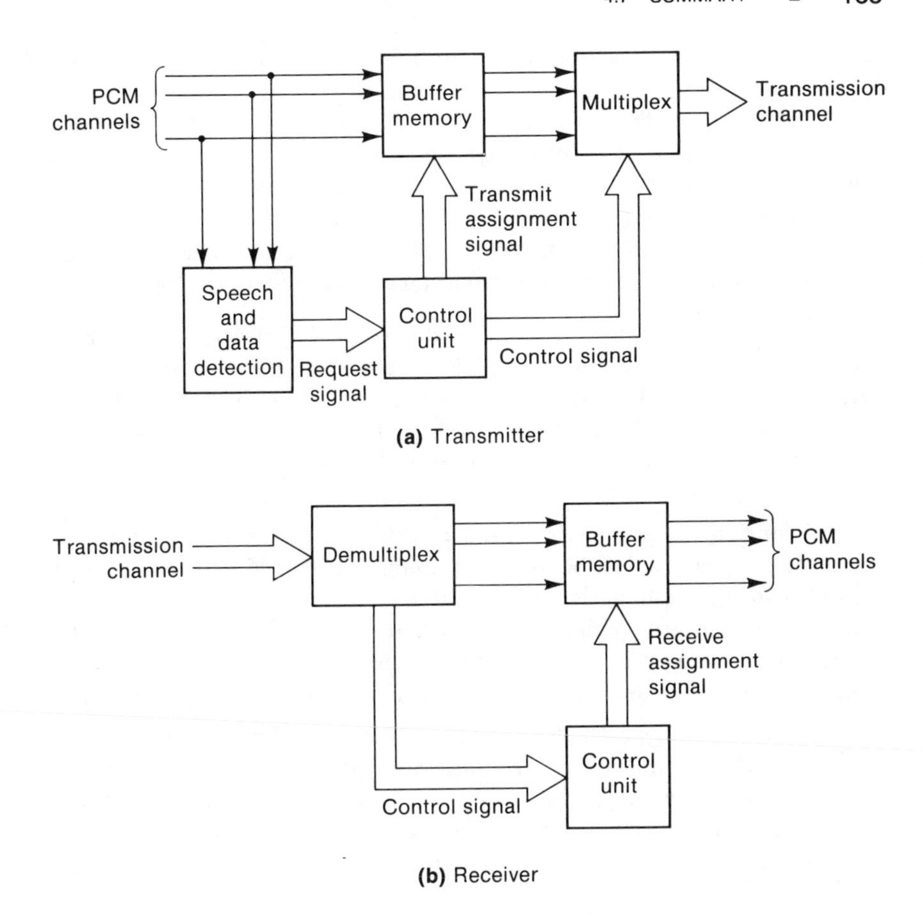

Figure 4.33 Block Diagram of a Speech Interpolation Device

TDM. For asynchronous channels, more complex TDM schemes such as pulse stuffing or transitional encoding are required to synchronize each channel to multiplexer timing. Asynchronous forms of multiplexing require additional overhead bits that must be transmitted within the frame to allow reconstruction of the original asynchronous channels by the demultiplexer. The choice of frame alignment pattern, both its length and content, and the frame synchronization strategy, which typically includes a search mode, maintenance mode, and some accept/reject criteria for each mode, are selected in accord with the following performance characteristics:

- Time to acquire frame alignment
- Time to detect misalignment and reacquire frame alignment
- Time between false resynchronizations

Calculation of these synchronization times is possible by use of the equations given in Section 4.2.

Pulse stuffing is the most efficient and commonly used synchronization scheme for the multiplexing of asynchronous channels. Two basic forms of pulse stuffing are available: positive only or positive-negative. Positive stuffing is less complex, but positive-negative stuffing facilitates the multiplexing of synchronous channels along with asynchronous channels. Four design parameters are of primary importance to the system designer:

- *Coding of stuff decisions*: Errors in the decoding of stuff decisions result in loss of synchronization in the affected channel. Therefore simple redundancy coding of stuff decisions is usually employed to protect each channel against transmission bit errors. The choice of code length determines a key performance characteristic—the time to loss of synchronization due to stuff decoding error, as given by Equation (4.26).
- *Maximum stuffing rate*: This parameter determines the upper limit on the allowed frequency offset between the timing signal used with the channel input and the timing signal internal to the multiplexer. Setting this parameter arbitrarily high is undesirable, however, since an increase in the maximum stuffing rate raises the percentage of overhead bits, lowers the time between stuff decoding errors, and may increase waiting time jitter.
- *Stuffing ratio*: This parameter is given by the ratio of actual stuffing rate —which is determined by the difference in frequency between the asynchronous channel input and the synchronous multiplexer—to the maximum stuffing rate, which is a design parameter. By careful choice of the stuffing ratio, waiting time jitter can be controlled to within specified levels.
- *Smoothing loop*: In the demultiplexer, the timing associated with each asynchronous channel is recovered by using a clock smoothing loop. Design of the loop involves a choice of loop type (a first-order phase-locked loop is typical) and loop bandwidth. This loop also must attenuate the jitter caused by the insertion and deletion of stuff bits. Residual jitter in the recovered clock must be within the range of input jitter tolerated by the interfacing equipment.

Transitional encoding is commonly used with asynchronous data channels because of its simplicity and transparency to data format. The transmission overhead is high, however, typically equal to or greater than the data rate itself. The system designer must also consider the input jitter requirements for the data terminal equipment, since transitional encoding introduces jitter in the process of reconstructing data transitions at the receiver. The magnitude of this jitter can be reduced by increasing the number of coding bits, as indicated in Equation (4.40), but at the expense of increased overhead.

Statistical multiplexing and speech interpolation are two multiplexing techniques that apply to data and voice channels, respectively, using the same principle of filling inactive time of one channel with active time from another channel. Three parameters are of interest to the system designer:

- The *gain* or concentration ratio—that is, the ratio of the number of channel inputs to the number of channel time slots available for transmission
- The allowed blocking of incoming channels or clipping of existing channels due to an overload of channel activity
- The allowed number of channel inputs that require a dedicated, full-period channel for transmission

REFERENCES

1. P. Bylanski and D. G. W. Ingram, *Digital Transmission Systems* (London: Peter Peregrinus Ltd., 1980).
2. CCITT Yellow Book, vol. III.3, *Digital Networks—Transmission Systems and Multiplexing Equipment* (Geneva: ITU, 1981).
3. V. L. Taylor, "Optimum PCM Synchronizing," *Proceedings of the National Telemetering Conference*, 1965, pp. 46–49.
4. M. W. Williard, "Mean Time to Acquire PCM Synchronization," *Proceedings of the National Symposium on Space Electronics and Telemetry*, Miami Beach, October 1972.
5. R. H. Barker, "Group Synchronizing of Binary Digital Systems." In *Communications Theory* (New York: Academy Press, 1953).
6. W. Feller, *An Introduction to Probability Theory and Its Applications* (New York: Wiley, 1957).
7. D. R. Cox and H. D. Miller, *The Theory of Stochastic Processes* (New York: Wiley, 1965).
8. B. Weesakul, "The Random Walk Between a Reflecting and Absorbing Barrier," *Ann. Math. Statistics* 32(1971):765–769.
9. J. G. Kemeny and J. L. Snell, *Finite Markov Chains* (New York: Van Nostrand, 1960).
10. M. W. Williard, "Optimum Code Patterns for PCM Synchronization," *Proceedings of the National Telemetering Conference*, Washington D.C., May 1962.
11. D. L. Duttweiler, "Waiting Time Jitter," *Bell System Technical Journal* 51(1)(January 1972):165–207.
12. CCITT Green Book, vol. III.3, *Line Transmission* (Geneva: ITU, 1973).
13. Federal Republic of Germany, *Jitter Accumulation on Digital Paths and Jitter Performance of the Components of Digital Paths*, vol. COM XVIII, no. 1-E (Geneva: CCITT, 1981), pp. 77–83.
14. CCITT Orange Book, vol. VII, *Telegraph Technique* (Geneva: ITU, 1977).
15. C. G. Davis, "An Experimental Pulse Code Modulation System for Short-Haul Trunks," *Bell System Technical Journal* 41(1)(January 1962):1–24.
16. H. H. Henning and J. W. Pan, "The D2 Channel Bank-System Aspects," *Bell System Technical Journal* 51(8)(October 1972):1641–1657.

17. J. B. Evans and W. B. Gaunt, "The D3 PCM Channel Bank," *1973 International Conference on Communications*, June 1973, pp. 70-1 to 70-5.
18. H. A. Mildonian, Jr., and D. A. Spires, "The DID PCM Channel Bank," *1974 International Conference on Communications*, June 1974, pp. 7E-1 to 7E-4.
19. C. R. Crue and others, "D4 Digital Channel Bank Family: The Channel Bank," *Bell System Technical Journal* 61(9)(November 1982):2611–2664.
20. P. Benowitz and others, "Digital Data System: Digital Multiplexers," *Bell System Technical Journal* 54(5)(May–June 1975):893–918.
21. J. B. Millard and H. I. Maunsell, "The Picturephone System: Digital Encoding of the Video Signal," *Bell System Technical Journal* 50(2)(February 1971):459–479.
22. R. L. Easton and others, "TASI-E Communications System," *IEEE Trans. on Comm.*, vol. COM-30, no. 4, April 1982, pp. 803–807.
23. J. H. Rieser, H. G. Snyderhoud, and Y. Yatsueuka, "Design Considerations for Digital Speech Interpolation," *1981 International Conference on Communications*, June 1981, pp. 49.4.1–49.4.7.
24. D. Lombard and H. L. Marchese, "CELTIC Field Trial Results," *IEEE Trans. on Comm.*, vol. COM-30, no. 4, April 1982, pp. 808–814.
25. P. A. Vachon and P. G. Ruether, "Evolution of COM 2, a TASI Based Concentrator," *1981 International Conference on Communications*, June 1981, pp. 49.6.1–49.6.5.
26. R. P. Gooch, "DCEM: A DSI System for North America and Europe," *1982 International Conference on Communications*, June 1982, pp. 4G.5.1–4G.5.5.

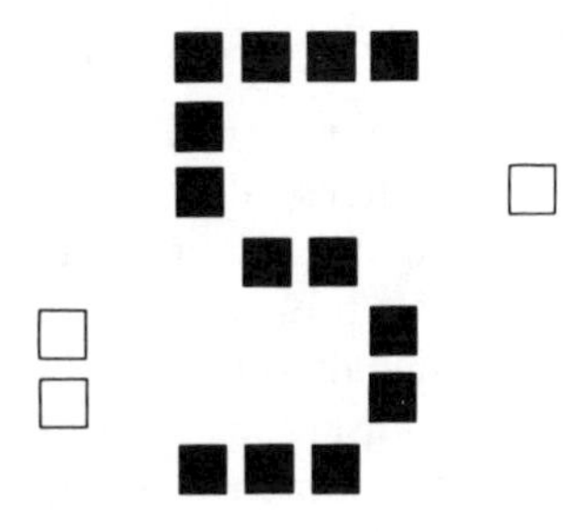

Baseband Transmission

OBJECTIVES

- Describes techniques for transmitting signals without frequency translation
- Discusses the commonly used codes for binary transmission, including presentation of waveforms, properties, and block diagrams for the coder and decoder
- Compares bandwidth and spectral shape for commonly used binary codes
- Compares error performance for binary codes in the presence of noise and other sources of degradation
- Explains how pulse shaping can be used to control the form of distortion called intersymbol interference
- Describes how multilevel transmission increases the data rate packing or spectral efficiency over binary transmission
- Discusses several classes of partial response codes that permit performance superior to that for binary and multilevel transmission
- Explains how the eye pattern shows the effects of channel perturbations and thus indicates overall system "health"
- Discusses the techniques of automatic adaptive equalization of signals distorted by a transmission channel
- Describes data scrambling techniques that minimize interference and improve performance in timing and equalization

5.1 INTRODUCTION

The last two chapters have described techniques for converting analog sources into a digital signal and the operations involved in combining these signals into a composite signal for transmission. We now turn our attention to transmission channels, starting in this chapter with baseband signals—signals used for direct transmission without further frequency translation. Transmission channels in general impose certain constraints on the format of digital signals. Baseband transmission often requires a reformatting of digital signals at the transmitter, which is removed at the receiver to recover the original signal. This signal formatting may be accomplished by shaping or coding, which maintains the baseband characteristic of the digital signal. The techniques described here are applicable to cable systems, both metallic and fiber optics, since most cable systems use baseband transmission. Later in our treatment of digital modulation techniques, we will see that many of the techniques applicable to baseband transmission are also used in radio frequency (RF) transmission.

There are a number of desirable attributes that can be achieved for baseband transmission through the use of shaping or coding. These attributes are:

- *Adequate timing information*: Certain baseband coding techniques increase the data transition density, which enhances the performance of timing recovery circuits—that is, bit and symbol synchronization.
- *Error detection/correction*: Many of the codes considered here have an inherent error detection capability due to constraints in allowed transitions among the signal levels. These constraints in level transitions can be monitored to provide a means of performance monitoring, although error correction is not possible through this property of baseband codes.
- *Reduced bandwidth*: The bandwidth of digital signals may be reduced by use of certain filtering and multilevel transmission schemes. The penalty associated with such techniques is a decrease in signal-to-noise ratio or increase in the amount of intersymbol interference.
- *Spectral shaping*: The shape of the data spectrum can be altered by scrambling or filtering schemes. These schemes may be selected to match the signal to the characteristics of the transmission channel or to control interference between different channels.

5.2 TYPES OF BINARY CODING

Binary codes simply condition binary signals for transmission. This signal conditioning provides a square wave characteristic suitable for direct transmission over cable. Here we discuss commonly used codes for binary transmission, including presentation of waveforms, properties, and block diagrams of the coder and decoder. Later we will compare bandwidth and spectral shape (Section 5.3) and error performance (Section 5.4) for these same binary codes.

5.2.1 Nonreturn-to-Zero (NRZ)

With **nonreturn-to-zero (NRZ)**, the signal level is held constant at one of two voltages for the duration of the bit interval T. If the two allowed voltages are 0 and V, the NRZ waveform is said to be **unipolar**, because it has only one polarity. This signal has a nonzero dc component at one-half the positive voltage, assuming equally likely 1's and 0's. A **polar** NRZ signal uses two polarities, $\pm V$, and thus provides a zero dc component.

Various versions of NRZ are illustrated in Figure 5.1. With NRZ(L), the voltage level of the signal indicates the value of the bit. The assignment of bit values 0 and 1 to voltage levels can be arbitrary for NRZ(L), but the usual convention is to assign 1 to the higher voltage level and 0 to the lower voltage level. NRZ(L) coding is the most common mode of NRZ transmission, due to the simplicity of the transmitter and receiver circuitry. The coder/decoder consists of a simple line driver and receiver, whose characteristics have been standardized by various level 1 (physical) interface standards as discussed in Chapter 2.

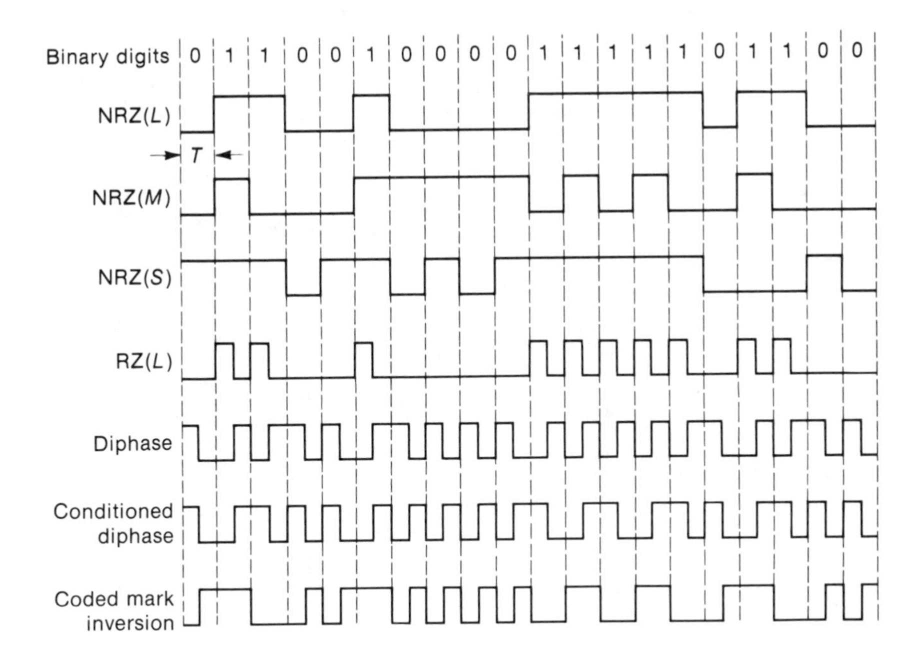

Figure 5.1 Binary Coding Waveforms

In the NRZ(M) format, a level change is used to indicate a mark (that is, a 1) and no level change for a space (that is, a 0); NRZ(S) is similar except that the level change is used to indicate a space or zero. Both of these formats are examples of the general class NRZ(I), also called **conditioned NRZ**, in which level inversion is used to indicate one kind of binary digit. The logic required to generate and receive NRZ(I) is shown in the coder and decoder block diagrams of Figure 5.2. The chief advantage of NRZ(I) over NRZ(L) is its immunity to polarity reversals, since the data are coded by the presence or absence of a transition rather than the presence or absence of a pulse, as indicated in the waveform of Figure 5.2*c*.

As indicated in Figure 5.1, the class of NRZ signals contains no transitions for strings of 1's with NRZ(S), strings of 0's with NRZ(M), and strings of 1's or 0's with NRZ(L). Since lack of data transitions would result in poor clock recovery performance, either the binary signal must be

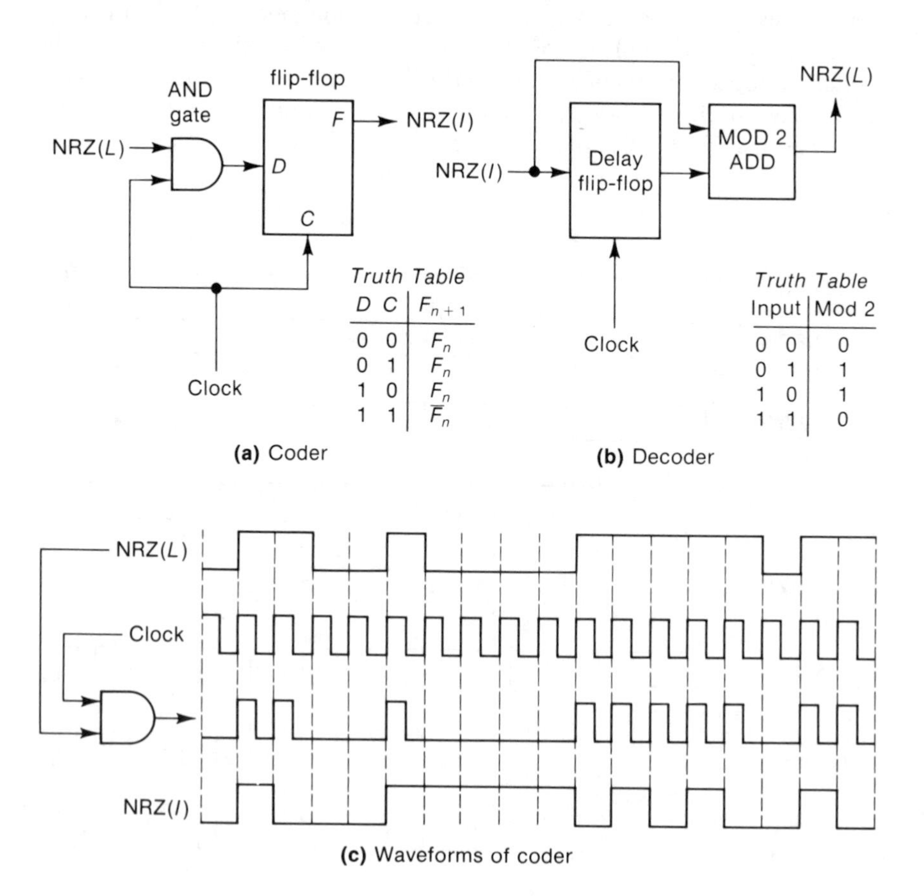

Figure 5.2 Characteristics of NRZ(I) Coding

precoded to eliminate such strings of 1's and 0's or a separate timing line must be transmitted with the NRZ signal. This characteristic has limited the application of NRZ coding to short-haul transmission and intrastation connections.

5.2.2 Return-to-Zero (RZ)

With **return-to-zero (RZ)**, the signal level representing bit value 1 lasts for the first half of the bit interval, after which the signal returns to the reference level (zero) for the remaining half of the bit interval. A zero is indicated by no change, with the signal remaining at the reference level. Its chief advantage lies in the increased transitions vis-à-vis NRZ and the resulting improvement in timing (clock) recovery. The RZ waveform for an arbitrary string of 1's and 0's is shown in Figure 5.1 for comparison with other coding formats. Note that a string of 0's results in no signal transitions, a potential problem for timing recovery circuits unless these signals are eliminated by precoding.

The RZ coder, waveforms, and decoder are shown in Figure 5.3. The RZ code is generated by ANDing NRZ(L) with a clock operating at the system bit rate. Decoding is accomplished by delaying the RZ code $\frac{1}{2}$ bit

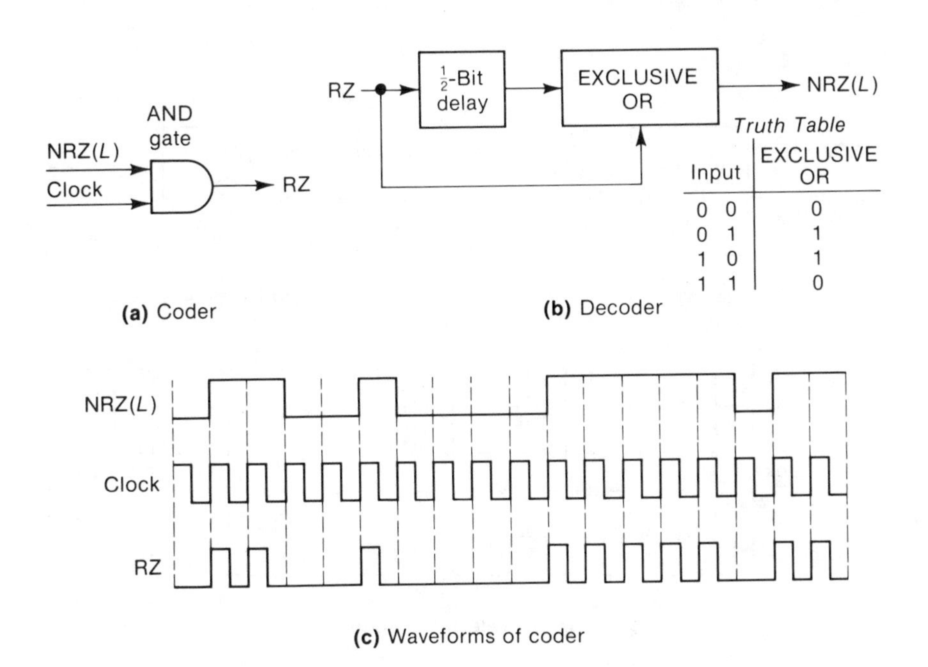

Figure 5.3 Characteristics of Return-to-Zero Coding

and applying the delayed version with the original version to an EXCLUSIVE OR (also known as MOD 2 ADD).

5.2.3 Diphase

Diphase (also called biphase, split-phase, and Manchester) is a method of two-level coding where

$$\begin{aligned} f_1(t) &= \begin{cases} V & 0 \le t \le T/2 \\ -V & T/2 < t \le T \end{cases} \\ f_2(t) &= -f_1(t) \end{aligned} \tag{5.1}$$

This code can be generated from NRZ(L) by EXCLUSIVE OR or MOD 2 ADD logic, as shown in Figure 5.4, if we assume 1's are transmitted as $+V$ and 0's as $-V$. From the diphase waveforms shown in Figure 5.4, it is readily apparent that the transition density is increased over NRZ(L), thus providing improved timing recovery at the receiver—a significant advantage of diphase. Data recovery is accomplished by the same logic employed by the coder.

Diphase applied to an NRZ(I) signal generates a code known as **conditioned diphase**. This code has the properties of both NRZ(I) and diphase signals—namely, immunity to polarity reversals and increased transition density, as shown in Figure 5.1.

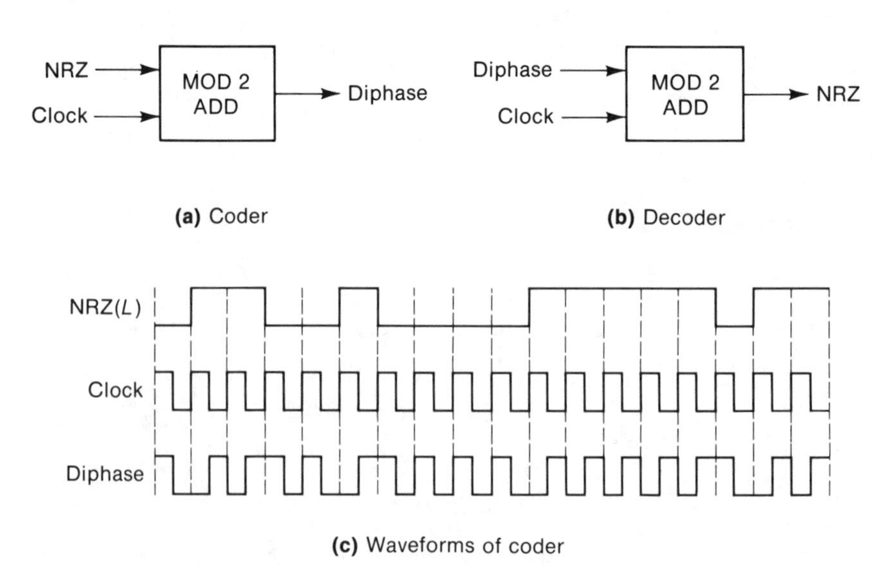

Figure 5.4 Characteristics of Diphase Coding

5.2.4 Bipolar or Alternate Mark Inversion

In **bipolar** or alternate mark inversion (AMI), binary data are coded with three amplitude levels, 0 and $\pm V$. Binary 0's are always coded as level 0; binary 1's are coded as $+V$ or $-V$, where the polarity alternates with every occurrence of a 1. Bipolar coding results in a zero dc component, a desirable condition for baseband transmission. As shown in Figure 5.5, bipolar representations may be NRZ (100 percent duty cycle) or RZ (50 percent duty cycle). Figure 5.6 indicates a coder/decoder block diagram and waveforms for bipolar signals. The bipolar signal is generated from NRZ by use of a 1-bit counter that controls the AND gates to enforce the alternate polarity rule. Recovery of NRZ(L) from bipolar is accomplished by simple full-wave rectification [1].

The numerous advantages of bipolar transmission have made it a popular choice, for example, by AT&T for T1 carrier systems that use 50 percent duty cycle bipolar. Since a data transition is guaranteed with each binary 1, the clock recovery performance of bipolar is improved over that of

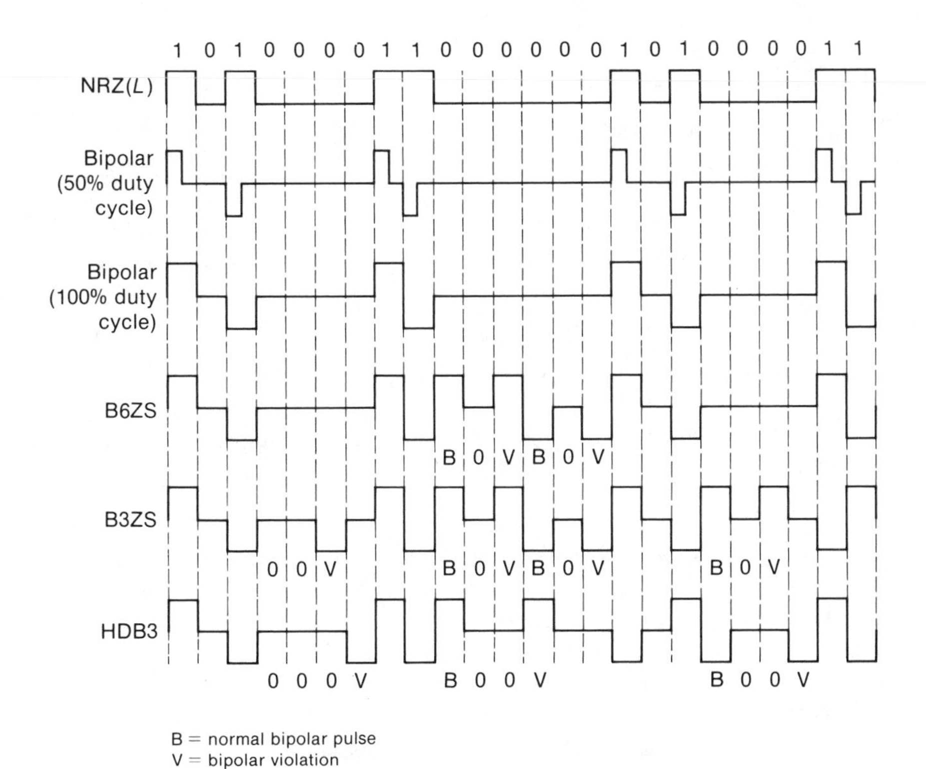

Figure 5.5 Bipolar Coding Waveforms

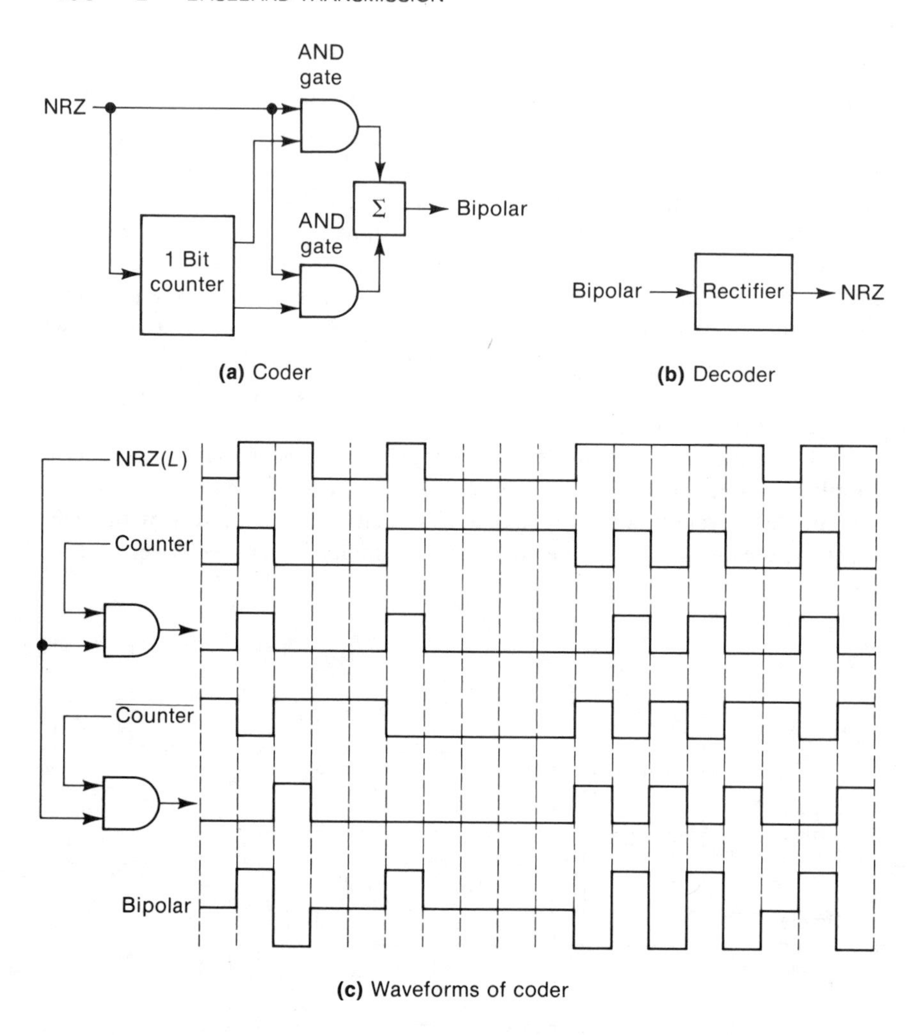

Figure 5.6 Characteristics of Bipolar Coding

NRZ. An error detection capability results from the property of alternate mark inversion. Consecutive positive amplitudes without an intervening negative amplitude (and vice versa) are a **bipolar violation** and indicate that a transmission error has occurred. This property allows on-line performance monitoring at a repeater or receiver without disturbing the data.

Although clock recovery performance of bipolar is improved over that of NRZ, a long string of zeros produces no transitions in the bipolar signal, which can cause difficulty in clock recovery. For T1 carrier repeaters or receivers, the maximum allowed string of consecutive zeros is 14. The common practice in PCM multiplex is to design the encoder in such a way that it satisfies this code restriction. In data multiplex equipment, however, a restriction of the binary data is not practical. The answer to this problem is to replace the string of zeros with a special sequence that contains intentional bipolar violations, which create additional data transitions and improve timing recovery. This "filling" sequence must be recognized and replaced by the original string of zeros at the receiver. Two commonly used bipolar coding schemes for eliminating strings of zeros are described below and illustrated in Figure 5.5 [2, 5]:

Bipolar *N*-Zero Substitution (BNZS)

Bipolar *N*-zero substitution (BNZS) replaces all strings of N zeros with a special N-bit sequence containing bipolar violations. For example, the B6ZS code replaces six consecutive zeros with the sequence B0VB0V (the sequence 0VB0VB can also be used), where B represents a normal bipolar pulse that conforms to the alternating polarity rule, V represents a bipolar violation, and 0 represents no pulse. The B6ZS code is the interface standard for the AT&T T2 system [3] and has been adopted by CCITT for 6.312-Mb/s second-order multiplex interface [4]. The B3ZS code is specified for the 44.736-Mb/s North American standard rate, and B8ZS has been proposed for future 1.544-Mb/s systems (see Chapter 12).

High-Density Bipolar *N* (HDBN)

High-density bipolar *N* (HDBN) limits the number of allowed consecutive zeros to N by replacing the $(N + 1)$th zero with a bipolar violation. Furthermore, in order to eliminate any possible dc offset due to the filling sequence, the coder forces the number of B pulses between two consecutive V pulses to be always odd. Then the polarity of V pulses will always alternate and a dc component is avoided. There are thus two possible $N + 1$ sequences—B00 ··· V or 000 ··· V—where the first bit location is used to ensure an odd number of B pulses between V pulses, the last bit position is always a violation, and all other bit positions are zeros. Two commonly used HDB codes are HDB2, which is identical to B3ZS, and HDB3, which is used for coding of 2.048-Mb/s, 8.448-Mb/s, and 34.368-Mb/s multiplex within the European digital hierarchy [4].

5.2.5 Coded Mark Inversion (CMI)

Coded mark inversion (CMI) is another two-level coding scheme. In this case 1's are coded as either level for the full bit duration T, where the levels alternate with each occurrence of a 1, as in bipolar coding. Zeros are coded so that both levels are attained within the bit duration, each level for $T/2$ using the same phase relationship for each zero. An example of CMI is illustrated in Figure 5.1. This example indicates that CMI significantly improves the transition density over NRZ. Coded mark inversion is specified for the coding of the 139.264-Mb/s multiplex within the European digital hierarchy [4].

5.3 POWER SPECTRAL DENSITY OF BINARY CODES

The power spectral density of a baseband code describes two important transmission characteristics: required bandwidth and spectrum shaping. The bandwidth available in a transmission channel is described by its frequency response, which typically indicates limits at the high or low end. Further bandwidth restriction may be imposed by a need to stack additional signals into a given channel. Spectrum shaping can help minimize interference from other signals or noise. Conversely, shaping of the signal spectrum can allow other signals to be added above or below the signal bandwidth.

Power Spectral Density

Derivation of the power spectral density is facilitated by starting with the autocorrelation function, which is defined for a random process $x(t)$ as

$$R_x(\tau) = E[x(t)x(t+\tau)] \tag{5.2}$$

where $E[\cdot]$ represents the expected value or mean. The power spectral density describes the distribution of power versus frequency and is given by the Fourier transform of the autocorrelation function

$$S_x(f) = \mathscr{F}[R_x(\tau)] = \int_{-\infty}^{\infty} R_x(\tau)e^{-j2\pi f\tau}\, d\tau \tag{5.3}$$

In some cases $S_x(f)$ may be more conveniently derived directly from the Fourier transform of the signal. Assuming the signal $x(t)$ to be zero outside the range $-T/2$ to $T/2$, then the corresponding frequency domain signal is given by

$$\begin{aligned} X(f) &= \int_{-\infty}^{\infty} x(t)e^{-j2\pi ft}\, dt \\ &= \int_{-T/2}^{T/2} x(t)e^{-j2\pi ft}\, dt \end{aligned} \tag{5.4}$$

By Parseval's theorem the average power across a 1-ohm load over the time interval $-T/2$ to $T/2$ is given by

$$\begin{aligned} P &= \frac{1}{T}\int_{-T/2}^{T/2} x^2(t)\, dt \\ &= \frac{1}{T}\int_{-\infty}^{\infty} |X(f)|^2\, df \end{aligned} \tag{5.5}$$

But from the definition of power spectral density, the average power can also be expressed as

$$P = \int_{-\infty}^{\infty} S_x(f)\, df \tag{5.6}$$

where $S_x(f)$ is in units of watts per hertz and P is in units of watts. From (5.5) and (5.6) we obtain

$$S_x(f) = \frac{|X(f)|^2}{T} \tag{5.7}$$

In the following derivations we assume that the digital source produces 0's and 1's with equal probability. Each bit interval is $(0, T)$ and each corresponding waveform is limited to an interval of time equal to T.

5.3.1. Nonreturn-to-Zero (NRZ)

With polar NRZ transmission, we assume that the two possible signal levels are V and $-V$ and that the presence of V or $-V$ in any 1-bit interval is statistically independent of that in any other bit interval. The autocorrelation function for NRZ is well known and is given by

$$R_Q(\tau) = \begin{cases} V^2(1 - |\tau|/T) & |\tau| < T \\ 0 & |\tau| > T \end{cases} \tag{5.8}$$

Using the Fourier transform, we find the corresponding power spectral density to be

$$S_Q(f) = V^2 T \left(\frac{\sin \pi f T}{\pi f T} \right)^2 \tag{5.9}$$

This is shown in Figure 5.7*a*. With most of its energy in the lower frequencies, NRZ is often a poor choice for baseband transmission due to the presence of interfering signals around dc. Moreover, clock recovery at the receiver is complicated by the fact that random NRZ data do not exhibit discrete spectral components.

5.3.2 Return-to-Zero (RZ)

In the RZ coding format, the signal values are

$$f_1(t) = \begin{cases} V & 0 < t \le T/2 \\ 0 & T/2 < t \le T \end{cases} \tag{5.10}$$

$$f_2(t) = 0$$

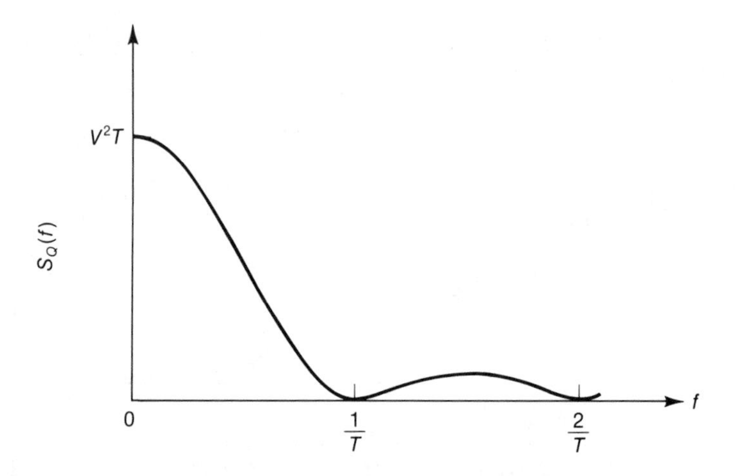

(a) Power spectral density of NRZ

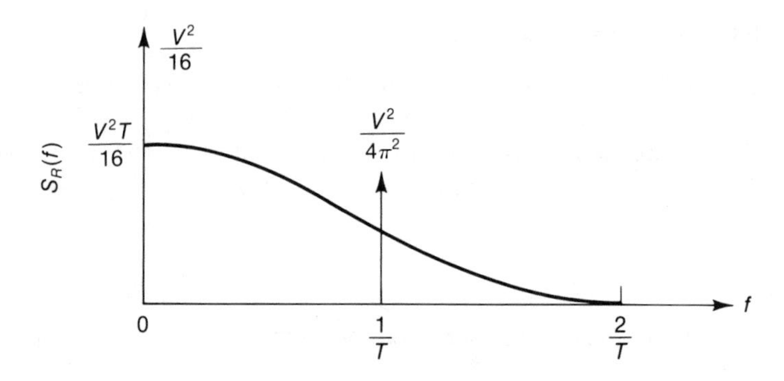

(b) Power spectral density of RZ

Figure 5.7 Power Spectral Density of NRZ and RZ

The power spectral density can be found by first dividing the RZ signal into two components. The first component is a periodic pulse train of amplitude $V/2$ and duration $T/2$, defined by

$$p_1(t) = p_2(t) = \begin{cases} V/2 & 0 < t \le T/2 \\ 0 & T/2 < t \le T \end{cases} \tag{5.11}$$

The second component is a corresponding random pulse train defined by

$$\begin{aligned} r_1(t) &= \begin{cases} V/2 & 0 < t \le T/2 \\ 0 & T/2 < t \le T \end{cases} \\ r_2(t) &= \begin{cases} -V/2 & 0 < t \le T/2 \\ 0 & T/2 < t \le T \end{cases} \end{aligned} \tag{5.12}$$

where now

$$
\begin{aligned}
f_1(t) &= p_1(t) + r_1(t) \\
f_2(t) &= p_2(t) + r_2(t)
\end{aligned}
\tag{5.13}
$$

The random component may be handled in a manner similar to the NRZ case, yielding a continuous power spectral density:

$$S_r(f) = \frac{V^2T}{16}\left[\frac{\sin(\pi fT/2)}{\pi fT/2}\right]^2 \tag{5.14}$$

The periodic component may be evaluated by standard Fourier series analysis as

$$S_p(f) = \sum_{n=-\infty}^{\infty} \frac{V^2}{16}\left[\frac{\sin(n\pi/2)}{n\pi/2}\right]^2 \delta\left(f - \frac{n}{T}\right) \tag{5.15}$$

The power spectral density for RZ is simply the sum of (5.14) and (5.15). This is shown in Figure 5.7*b*. Note that the first zero crossing of RZ is at twice the frequency as that for NRZ, indicating that RZ requires twice the bandwidth of NRZ. However, the discrete spectral lines of RZ permit simpler bit synchronization.

5.3.3 Diphase

In the diphase format, the signal values are given by (5.1). The power spectral density may be found from (5.7) by first determining the Fourier transform of the two signal pulses as follows:

$$
\begin{aligned}
F_1(f) &= j\frac{2V}{\pi f}\sin^2\left(\frac{\pi fT}{2}\right) \\
F_2(f) &= -F_1(f)
\end{aligned}
\tag{5.16}
$$

As determined by W. R. Bennett [6], the power spectral density depends on the difference between the Fourier transforms of the two pulses:

$$S_D(f) = \frac{1}{T}|F_1(f) - F_2(f)|^2 \tag{5.17}$$

or from (5.16)

$$S_D(f) = V^2T\frac{\sin^4(\pi fT/2)}{(\pi fT/2)^2} \tag{5.18}$$

As shown in Figure 5.8, the power spectral density of diphase has relatively low power at low frequencies and zero power at dc. The power spectral density is maximum at $0.743/T$ and has its first null at $2/T$. Hence the bandwidth occupancy is similar to that for RZ, although there are no discrete lines in the diphase power spectral density.

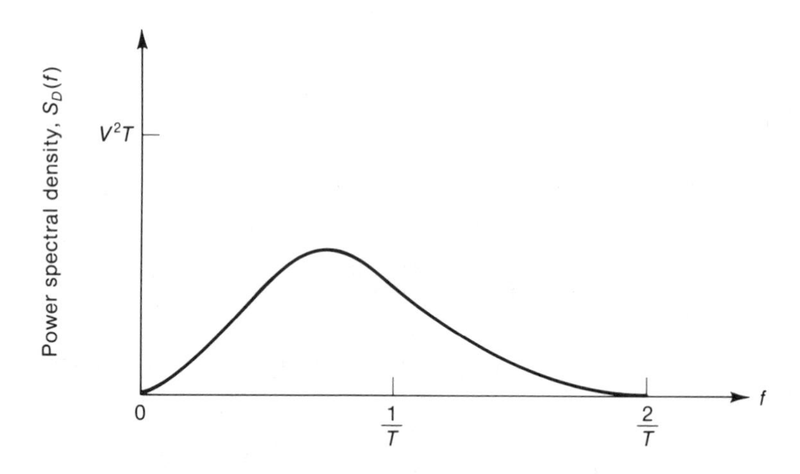

Figure 5.8 Power Spectral Density of Diphase (Manchester) Coding

5.3.4 Bipolar

The bipolar coding format uses three amplitudes: $V, 0, -V$. The bipolar waveforms are given by

$$f_1(t) = 0$$
$$f_2(t) = \begin{cases} \left.\begin{array}{ll} V & 0 < t \leq \alpha T \\ 0 & \alpha T < t \leq T \end{array}\right\} & \text{first 1} \\ \left.\begin{array}{ll} -V & 0 < t \leq \alpha T \\ 0 & \alpha T < t \leq T \end{array}\right\} & \text{next 1} \end{cases} \tag{5.19}$$

where $\alpha = 1$ represents nonreturn-to-zero (100 percent duty cycle) and $\alpha = \frac{1}{2}$ represents return-to-zero (50 percent duty cycle). The power spectral density of bipolar signaling is identical to that of *twinned binary* [1], which can be generated from NRZ by delaying it αT, subtracting it from the original, and dividing by 2 to provide proper amplitude scaling—that is,

$$B(t) = \frac{Q(t) - Q(t - \alpha T)}{2} \tag{5.20}$$

Taking the Fourier transform of both sides of (5.20), we have

$$B(f) = \frac{Q(f)}{2}(1 - e^{-j2\pi f\alpha T}) \tag{5.21}$$

From the definition of power spectral density (5.7),

$$\begin{aligned} S_B(f) &= \frac{|Q(f)|^2}{4T}|1 - e^{-j2\pi f\alpha T}|^2 \\ &= S_Q(f)\sin^2\pi f\alpha T \end{aligned} \tag{5.22}$$

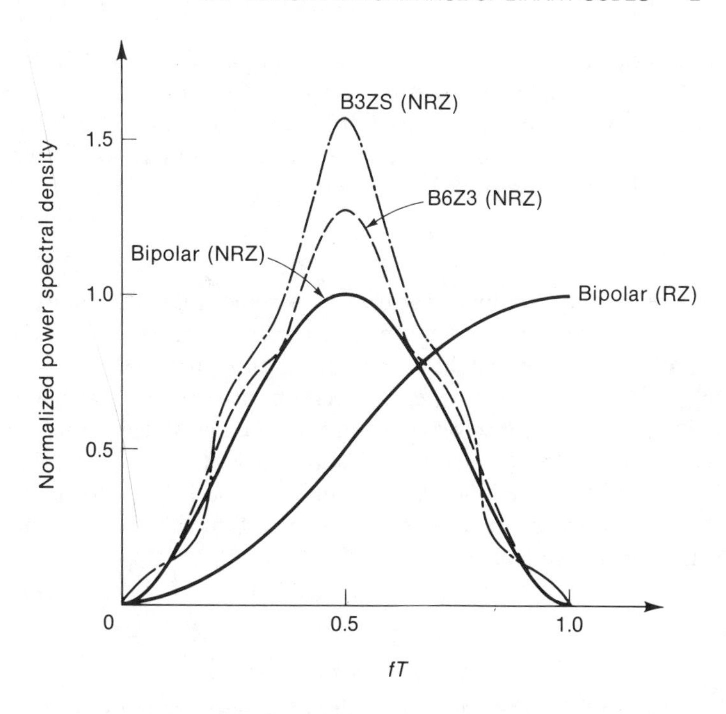

Figure 5.9 Power Spectral Density for Bipolar Codes

The resulting power spectral density is shown in Figure 5.9 for both bipolar return-to-zero and bipolar nonreturn-to-zero. The class of bipolar codes that use filling sequences to substitute for strings of zeros have power spectral densities that are similar to simple bipolar, but each member of this class differs slightly as shown in Figure 5.9. Note the lack of discrete lines in any of the bipolar power spectral densities. The bandwidth occupancy of bipolar NRZ and bipolar RZ is similar to NRZ and RZ, respectively, although bipolar has relatively low power at low frequencies and zero power at dc—another advantage of bipolar over NRZ coding.

5.4 ERROR PERFORMANCE OF BINARY CODES

Another important characteristic of binary codes is their error performance in the presence of noise and other sources of degradation. This performance characteristic is dependent on the signaling waveform and the nature of the noise. Here we will assume the noise to be gaussian. This is a reasonable assumption for linear baseband channels, where gaussian noise is the predominant (and sometimes only) source of signal corruption. The following analysis provides probability of error expressions for the binary codes already considered. This analysis also builds the foundation for subsequent analysis of more complex modulation schemes and transmission channels.

We begin with the simplest case by considering a polar NRZ signal having amplitudes $\pm V$ representing binary digits 0 and 1. At the receiver the two signals are observed to be

$$\begin{aligned} y_1 &= V + n \\ y_0 &= -V + n \end{aligned} \tag{5.23}$$

where n represents additive noise. We assume that this noise n is gaussian-distributed with zero mean and variance σ^2. Since the noise is random and has some probability of exceeding the signal level, there exists a probability of error in the receiver decision process. To calculate this probability of error, the receiver decision rule must be stated. Intuitively, if the noise is symmetrically distributed about $\pm V$, the decision threshold should be set at zero. Then the receiver would choose 1 if $y > 0$ and choose 0 if $y < 0$. For a transmitted 1, an error occurs if at the decision time the noise is more negative than $-V$, with probability

$$P(e|1) = P(n < -V) \tag{5.24a}$$

and similarly

$$P(e|0) = P(n > V) \tag{5.24b}$$

Now let $P(0)$ and $P(1)$ be the probability of 0 and 1 at the source, where $P(0) + P(1) = 1$. The total probability of error may then be expressed as

$$P(e) = P(e|0)P(0) + P(e|1)P(1) \tag{5.25}$$

Since n is gaussian with probability density function

$$P(n) = \frac{1}{\sigma\sqrt{2\pi}} e^{-n^2/2\sigma^2} \tag{5.26}$$

it follows that y_1 and y_0 are also gaussian with variance σ^2 but with mean $\overline{y_1} = V$ and $\overline{y_0} = -V$. The probabilities of error $P(e|1)$ and $P(e|0)$ are simply the shaded areas under the curves $p(y_1)$ and $p(y_0)$ in Figure 5.10.

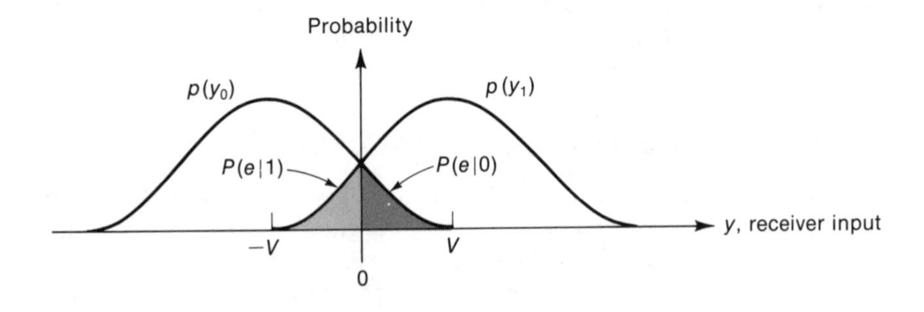

Figure 5.10 Probability Density Functions for Binary Transmission (Polar NRZ) over Additive Gaussian Noise Channel

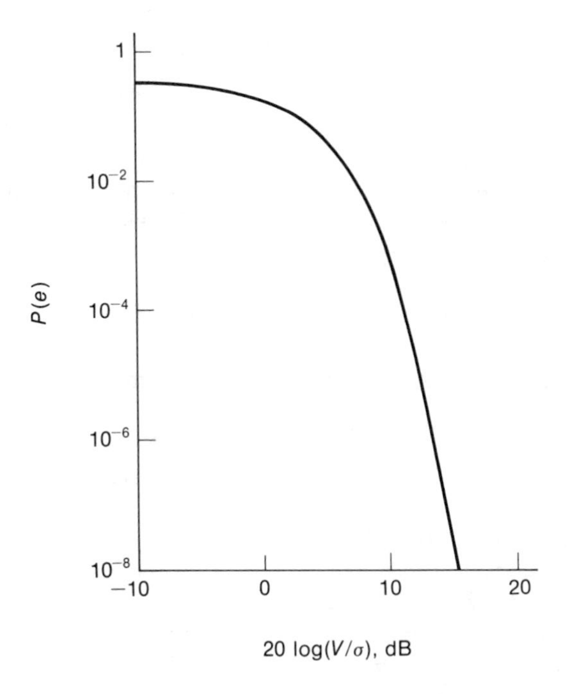

Figure 5.11 Probability of Error versus V/σ for Polar NRZ

In equation form these probabilities are

$$P(e|1) = \int_{-\infty}^{0} p(y_1)\, dy_1 = \frac{1}{\sigma\sqrt{2\pi}} \int_{-\infty}^{0} e^{-(y_1 - V)^2/2\sigma^2}\, dy_1$$
$$P(e|0) = \int_{0}^{\infty} p(y_0)\, dy_0 = \frac{1}{\sigma\sqrt{2\pi}} \int_{0}^{\infty} e^{-(y_0 + V)^2/2\sigma^2}\, dy_0 \tag{5.27}$$

By setting the decision threshold equal to zero, we observe from Figure 5.10 that the shaded areas are equal and further that the sum of the two areas is at a minimum. If we further assume that the transmitted binary digits are equiprobable (that is, $P(0) = P(1) = \frac{1}{2}$), then placing the decision threshold at zero also results in a minimum total probability of error, as given by (5.25). Thus $P(e)$ can be obtained directly from (5.27), where a change of variable $Z = (y_1 - V)/\sigma = (y_0 + V)/\sigma$ results in

$$P(e) = \frac{1}{\sqrt{2\pi}} \int_{V/\sigma}^{\infty} e^{-Z^2/2}\, dZ \tag{5.28}$$

We recognize this as the complementary error function* so that (5.28) can

*There exist several definitions of $\mathrm{erfc}(x)$ in the literature. These are all essentially equivalent except for minor differences in the choice of constants.

be rewritten as

$$P(e) = \operatorname{erfc}\left(\frac{V}{\sigma}\right) \tag{5.29}$$

which is plotted in Figure 5.11.

These results can be related to signal-to-noise ratio by noting that the average signal power S is equal to V^2 and that the average noise power N is equal to σ^2. Hence (5.29) can be rewritten

$$P(e) = \operatorname{erfc}\sqrt{\frac{S}{N}} \qquad \text{(polar NRZ)} \tag{5.30}$$

The unipolar $(0, V)$ NRZ case follows in the same manner except that $S = V^2/2$ so that

$$P(e) = \operatorname{erfc}\sqrt{\frac{S}{2N}} \qquad \text{(unipolar NRZ)} \tag{5.31}$$

This derivation of probability of error applies only to those cases where the binary decision for each digit is made independently of all other bit decisions. Hence NRZ(L), RZ(L), and diphase are all characterized by the expressions derived to this point. However, certain binary codes make use of both the present bit value and previous bit values in forming the coded signal. The receiver must first make each bit decision followed by decoding using the present and previous bit values. As an example consider NRZ(I), which falls into this class of binary codes. The probability of error of the decoded signal depends on the binary decision made for both the present and previous bit; that is,

$$\begin{aligned} P(e) &= P[\text{present bit correct and previous bit incorrect}] \\ &\quad + P[\text{present bit incorrect and previous bit correct}] \\ &= (1-p)p + p(1-p) \\ &\approx 2p \qquad (\text{for small } p) \end{aligned} \tag{5.32}$$

where p is the probability of error in the binary decision as given in (5.29). In addition to NRZ(I), conditioned diphase has a probability of error given by (5.32).

For bipolar coding with levels $\pm V$ and 0, the probability of error in the receiver can be written

$$P(e) = P(e|V)P(V) + P(e|-V)P(-V) + P(e|0)P(0) \tag{5.33}$$

As shown in Figure 5.12, the two decision thresholds are set midway between the three signal levels. The receiver decision rule is identical to the two-level case previously described: Select the signal level that is closest to the received signal. Expression (5.33) can then be rewritten

$$\begin{aligned} P(e) &= P\left(n < \frac{-V}{2}\right)P(V) + P\left(n > \frac{V}{2}\right)P(-V) \\ &\quad + \left[P\left(n < \frac{-V}{2}\right) + P\left(n > \frac{V}{2}\right)\right]P(0) \end{aligned} \tag{5.34}$$

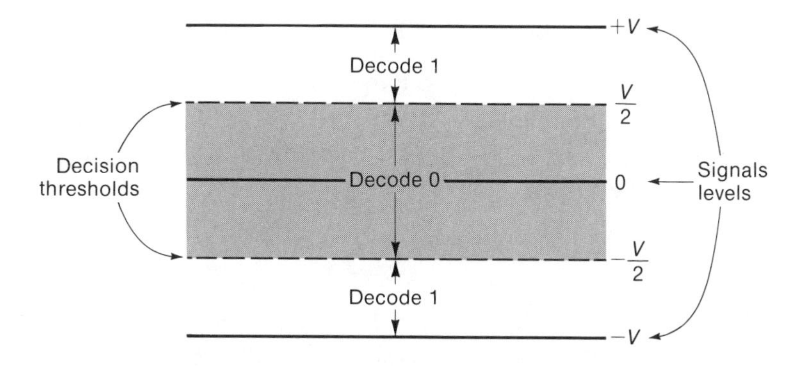

Figure 5.12 Bipolar Decoding Rules

Because of the symmetry of gaussian-distributed noise, all the noise terms in (5.34) are equal. If we assume equiprobable 1's and 0's at the transmitter, then (5.34) reduces to

$$P(e) = \frac{3}{2}\operatorname{erfc}\left(\frac{V}{2\sigma}\right) \qquad \text{(bipolar)} \tag{5.35}$$

which simply indicates that bipolar coding has a probability of error that is 3/2 times that of unipolar NRZ.

5.5 PULSE SHAPING AND INTERSYMBOL INTERFERENCE

In discussing binary codes thus far, we have assumed the transmission channel to be linear and distortionless. In practice, however, channels have a limited bandwidth, and hence transmitted pulses tend to be spread during transmission. This pulse spreading or **dispersion** causes overlap of adjacent pulses, giving rise to a form of distortion known as **intersymbol interference** (ISI). Unless this interference is compensated, the effect at the receiver may be errored decisions.

One method of controlling ISI is to shape the transmitted pulses properly. To understand this problem, consider a binary sequence $a_1 a_2 \cdots a_n$ transmitted at intervals of T seconds. These digits are shaped by the channel whose impulse response is $x(t)$. The received signal can then be written

$$y(t) = \sum_{n=-\infty}^{\infty} a_n x(t - nT) \tag{5.36}$$

One obvious means of restricting ISI is to force the shaping filter $x(t)$ to be zero at all sampling instants nT except $n = 0$. This restriction can be seen mathematically by rearranging (5.36) for a given time $t = nT$:

$$y_n = \sum_k a_k x_{n-k} \tag{5.37}$$

where $y_n = y(nT)$ and $x_{n-k} = x[(n-k)T]$. Then, for time $t = 0$, Equation (5.37) can be written

$$y_0 = a_0 x_0 + \sum_{k \neq 0} a_k x_{-k} \tag{5.38}$$

The first term of (5.38) represents the desired signal sampled at time $t = 0$; the second term arises from the overlap of other pulses contributing to the desired pulse at the zeroth sampling time. Clearly if the impulse function is zero for all sampling instants nT except $n = 0$, the second term of (5.38) disappears, thus eliminating intersymbol interference.

Example 5.1

In the following channel impulse response, let $x_{-1} = \frac{1}{4}$, $x_0 = 1$, $x_1 = -\frac{1}{2}$, and $x_i = 0$ for $i \neq -1, 0, 1$:

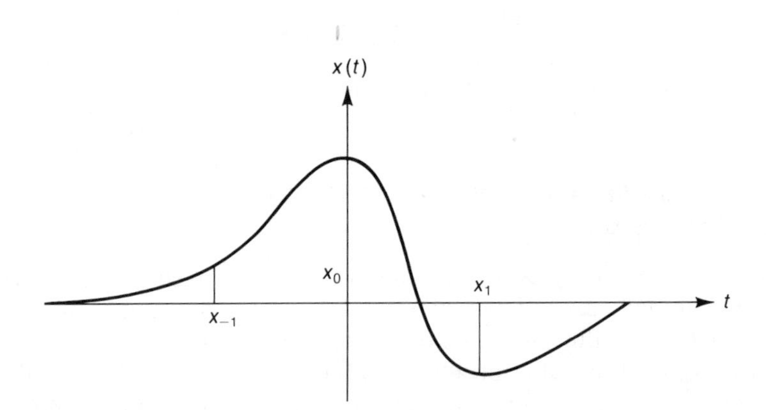

The sample x_{-1} interferes with the previous pulse, while the sample x_1 interferes with the succeeding pulse. The interference with past as well as future symbols is possible because of the time delay through a channel. The received pulse at time zero is then the accumulation of the precursor and tail of two other transmitted pulses plus the value of the pulse transmitted at time zero:

$$\begin{aligned} y_0 &= a_1 x_{-1} + a_0 x_0 + a_{-1} x_1 \\ &= \tfrac{1}{4} a_1 + a_0 - \tfrac{1}{2} a_{-1} \end{aligned}$$

One pulse shape that produces zero ISI is the function

$$x(t) = \frac{\sin \pi t/T}{\pi t/T} \tag{5.39}$$

This is the impulse response of an ideal low-pass filter as shown in Figure 5.13. Note that $x(t)$ goes through zero at equally spaced intervals that are multiples of T, the sampling interval. Moreover, the bandwidth W is observed to be π/T rad/s or $1/2T$ hertz. If $1/2W$ is selected as the sampling interval T, pulses will not interfere with each other. This rate of $2W$ pulses per second transmitted over a channel with bandwidth W hertz is called the **Nyquist rate** [7]. In practice, however, there are difficulties with this filter shape. First, an ideal low-pass filter as pictured in Figure 5.13 is not physically realizable. Second, this waveform depends critically on timing precision. Variation of timing at the receiver would result in excessive ISI because of the slowly decreasing tails of the $\sin x/x$ pulse.

This ideal low-pass filter can be modified to provide a class of waveforms described by Nyquist that meet the zero ISI requirement but are simpler to attain in practice. Nyquist suggested a more gradual frequency cutoff with a filter characteristic designed to have odd symmetry about the ideal low-pass cutoff point. Satisfaction of this characteristic gives rise to a set of Nyquist pulse shapes that obey the property of having zeros at uniformly spaced intervals. This set of pulse shapes is said to be *equivalent* since their filter characteristics all lead to the same sample sequence $\{x_n\}$ [8]. To maintain a rate of $2W$ pulses per second, this set of filter characteristics requires additional bandwidth over the so-called **Nyquist bandwidth**, which is defined by the ideal low-pass filter of Figure 5.13.

An example of a commonly used filter that meets this Nyquist criterion is the **raised-cosine** characteristic. This spectrum consists of a flat magnitude at low frequencies and a rolloff portion that has a sinusoidal form. The

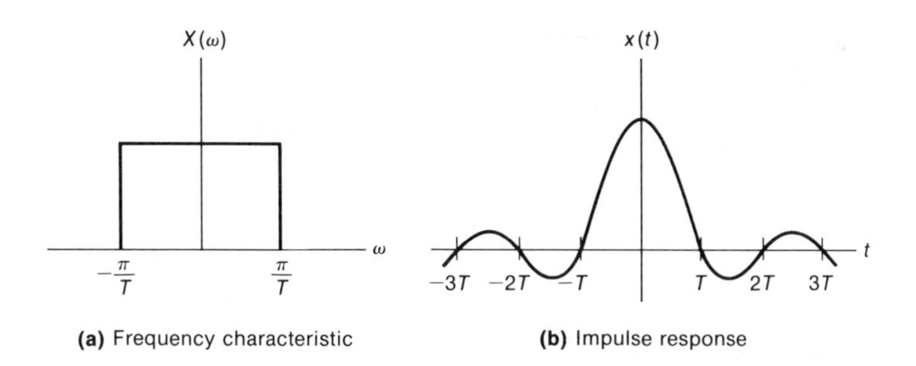

Figure 5.13 sin x / x Filter Characteristic and Pulse Shape

raised-cosine characteristic is given by

$$X(\omega) = \begin{cases} T & 0 \le |\omega| \le \frac{\pi}{T}(1-\alpha) \\ \frac{T}{2}\left\{1 - \sin\left[\frac{T}{2\alpha}\left(|\omega| - \frac{\pi}{T}\right)\right]\right\} & \frac{\pi}{T}(1-\alpha) \le |\omega| \le \frac{\pi}{T}(1+\alpha) \\ 0 & \omega > \frac{\pi}{T}(1+\alpha) \end{cases} \tag{5.40}$$

The parameter α is defined as the amount of bandwidth used in excess of the Nyquist bandwidth divided by the Nyquist bandwidth itself. The corresponding impulse response is

$$x(t) = \left(\frac{\sin \pi t/T}{\pi t/T}\right)\left[\frac{\cos \alpha\pi t/T}{1 - (4\alpha^2 t^2/T^2)}\right] \tag{5.41}$$

Plots of $x(t)$ and $X(\omega)$ are shown in Figure 5.14 for three values of α. Note that the case for $\alpha = 0$ coincides with the ideal low-pass filter depicted earlier in Figure 5.13. The case for $\alpha = 1$ is referred to as full (100 percent) raised cosine and doubles the bandwidth required over the Nyquist bandwidth. Larger values of α lead to faster decay of the leading and trailing oscillations, however, indicating that synchronization is less critical than with $\sin x/x$ ($\alpha = 0$) pulses.

5.6 MULTILEVEL BASEBAND TRANSMISSION

With binary transmission, each symbol contains 1 bit of information that is transmitted every T seconds. The symbol rate or **baud** is given by $1/T$ and is equivalent to the bit rate for binary transmission. According to Nyquist theory, a data rate of $2W$ bits per second (b/s) can be achieved for the ideal

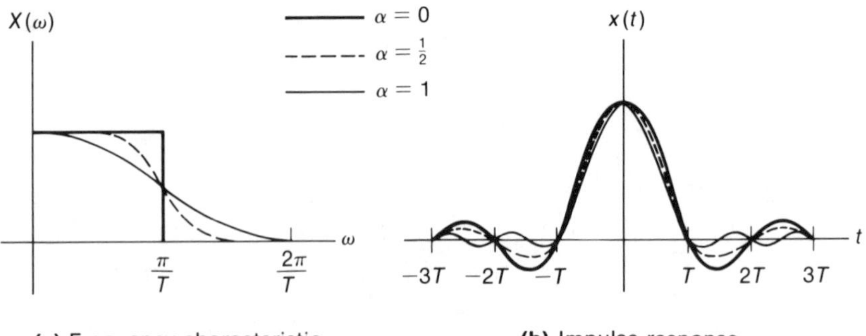

Figure 5.14 Raised-Cosine Filter Characteristic and Pulse Shape

channel of bandwidth W hertz. The achievable data rate may be increased by using additional levels with the input symbols a_k. Usually the number of levels M is a power of 2 so that the number of bits transmitted per symbol can be expressed as $m = \log_2 M$. The data rate R can then be stated as

$$R = 2W \log_2 M = \frac{1}{T} \log_2 M \qquad \text{b/s} \tag{5.42}$$

Multilevel transmission thus increases the data rate packing or spectral efficiency by the factor m. To characterize the spectral efficiency, we can normalize the data rate to obtain

$$\frac{R}{W} = 2 \log_2 M \qquad \text{bps/Hz} \tag{5.43}$$

This is a commonly used characterization of digital baseband and RF systems. As an example, for $M = 8$ level transmission through the theoretical Nyquist channel, the spectral efficiency is 6 bps/Hz. In practice, the signal-to-noise ratio required for acceptable error performance forces a limit on the number of levels that can be used. Moreover, since the theoretical Nyquist channel is unattainable the spectral efficiency given by (5.43) is an upper bound. As a more practical example, consider a raised-cosine system with $\alpha = 1$ and $M = 8$; here the spectral efficiency drops to 3 bps/Hz.

The error performance of M-ary baseband transmission can be found by extending the case for binary transmission, given by (5.29). Assume that the M transmission levels are uniformly spaced with values of $\pm d, \pm 3d, \ldots, \pm(M-1)d$, that the M levels are equally likely, and that the transmitted symbols form an independent sequence $\{a_k\}$. Let the pulse-shaping filter be $\sin x/x$ and assume that no ISI is contributed by the channel. At the receiver, slicers are placed at the thresholds $0, \pm 2d, \ldots, \pm(M-2)d$. Errors occur when the noise level at a sampling instant exceeds d, the distance to the nearest slicer. Note, however, that the two outside signals, $\pm(M-1)d$, are in error only if the noise component exceeds d and has polarity opposite that of the signal. The probability of error is then given by

$$\begin{aligned} P(e) &= P[e \mid \pm(M-1)d] P[\pm(M-1)d] \\ &\quad + P[e \mid \pm d, \pm 3d, \ldots, \pm(M-3)d] \\ &\quad\quad P[\pm d, \pm 3d, \ldots, \pm(M-3)d] \\ &= 2\left(1 - \frac{1}{M}\right) P(|n| > d) \end{aligned} \tag{5.44}$$

For zero mean gaussian noise with variance σ^2, this probability of error can be expressed by

$$P(e) = 2\left(1 - \frac{1}{M}\right) \operatorname{erfc}\left(\frac{d}{\sigma}\right) \tag{5.45}$$

This result can be expressed in terms of signal-to-noise (S/N) ratio by first noting that the average transmitted symbol power is given by

$$S = \overline{a^2} = \frac{d^2(M^2 - 1)}{3} \tag{5.46}$$

Assuming additive gaussian noise to have average power $N = \sigma^2$, the probability of error can now be written as

$$P(e) = 2\left(1 - \frac{1}{M}\right)\operatorname{erfc}\left[\left(\frac{3}{M^2 - 1}\frac{S}{N}\right)^{1/2}\right] \tag{5.47}$$

Some important conclusions and comparisons can be made from the probability of error expression in (5.47). A comparison of the binary $P(e)$ expression derived earlier (Equation 5.30) with (5.47) for $M = 2$ reveals that the two expressions are identical, as expected. The effect of increased M, however, is that the S/N ratio is modified by $3/(M^2 - 1)$. Thus an $M = 4$ transmission system requires about 7 dB more power than the $M = 2$ case.

Another important comparison is that of bandwidth and bit rate. This comparison is facilitated by use of the normalized data rate given in (5.43) and by writing the argument within the erfc of (5.47) as

$$\Gamma = \frac{3}{M^2 - 1}\frac{S}{N} \tag{5.48}$$

After solving for M^2 in terms of Γ and S/N, the normalized data rate can be expressed as

$$\frac{R}{W} = \log_2\left[\left(\frac{3}{\Gamma}\right)\left(\frac{S}{N}\right) + 1\right] \text{ bps/Hz} \tag{5.49}$$

This expression can now be compared with Shannon's theoretical limit on channel capacity C over a channel limited in bandwidth to W hertz with average noise power N and average signal power S [9, 10]:

$$\frac{C}{W} = \log_2\left(\frac{S}{N} + 1\right) \text{ bps/Hz} \tag{5.50}$$

which represents the minimum S/N required for nearly error-free performance. A comparison between (5.49) and (5.50) reveals the difference to be the factor $3/\Gamma$. For a probability of error of 10^{-5}, $\Gamma \approx 20$, where the $2(1 - 1/M)$ factor has been ignored. Thus an additional 8 dB of S/N ratio is required to attain Shannon's theoretical channel capacity. The conclusion that can be made is that M-ary transmission may be spectrally efficient but suffers from effective loss of S/N ratio relative to theoretical channel capacity.

5.7 PARTIAL RESPONSE CODING

To this point we have maintained that the Nyquist rate of $2W$ symbols per second for a channel of bandwidth W hertz can be attained only with a nonrealizable, ideal low-pass filter. Even raised-cosine systems require excessive bandwidth, so that practical symbol rate packing appears to be on the order of 1 symbol/Hz even though the theoretical limit is 2 symbols/Hz. This limit applies only to zero-memory systems, however—that is, where transmitted pulse amplitudes are selected independently.

In the following paragraphs we will discuss a class of techniques that introduce correlation between the amplitudes to permit practical attainment of the Nyquist rate. A. Lender first described this technique in a scheme termed **duobinary** [11, 12], which combines two successive pulses together to form a multilevel signal. This combining process introduces prescribed amounts of intersymbol interference, resulting in a signal that has three levels at the sampling instants. Kretzmer has classified and tabulated these schemes, which are more generally called **partial response** [13] or **correlative coding** [14]. Partial response coding provides the capability of increasing the transmission rate above W symbols per second but at the expense of additional transmitted power. Appropriate choice of precoder or filter makes available a variety of multilevel formats with different spectral properties. Furthermore, certain constraints on level transitions in the received signal make possible some error detection.

The basic ideas behind partial response schemes can be illustrated by considering the example of duobinary (or *class 1 partial response*). Consider an input sequence $\{a_k\}$ of binary symbols spaced T seconds apart. The transmission rate is $2W$ symbols per second over an ideal rectangular low-pass channel of bandwidth W hertz and magnitude T. These symbols are passed through the digital filter shown in Figure 5.15a, which simply sums two successive symbols to yield at the transmitter output

$$y_k = a_k + a_{k-1} \tag{5.51}$$

The cascade of the digital filter $H_1(\omega)$ and the ideal rectangular filter $H_2(\omega)$ is equivalent to

$$|H(\omega)| = |H_1(\omega)||H_2(\omega)| = 2T\cos\frac{\omega T}{2} \qquad |\omega| \le \frac{\pi}{T} \tag{5.52}$$

since

$$\begin{aligned} H_1(\omega) &= 1 + e^{-j\omega T} \\ &= 2\cos\frac{\omega T}{2} e^{-j\omega T/2} \end{aligned} \tag{5.53}$$

Thus $H(\omega)$ can be realized with a cos-shaped low-pass filter, eliminating the need for a separate digital filter. The filter characteristic $H(\omega)$ and its corresponding impulse response

$$h(t) = \frac{4}{\pi}\left(\frac{\cos \pi t/T}{1 - 4t^2/T^2}\right) \tag{5.54}$$

are sketched in Figure 5.15. If the impulse response sampling time is taken at $t = 0$, then

$$h(nT) = \begin{cases} 1 & n = 0, 1 \\ 0 & \text{otherwise} \end{cases} \tag{5.55}$$

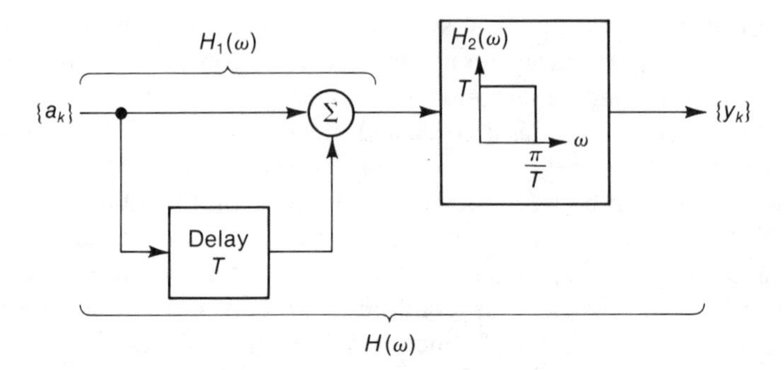

(a) Duobinary signal generation

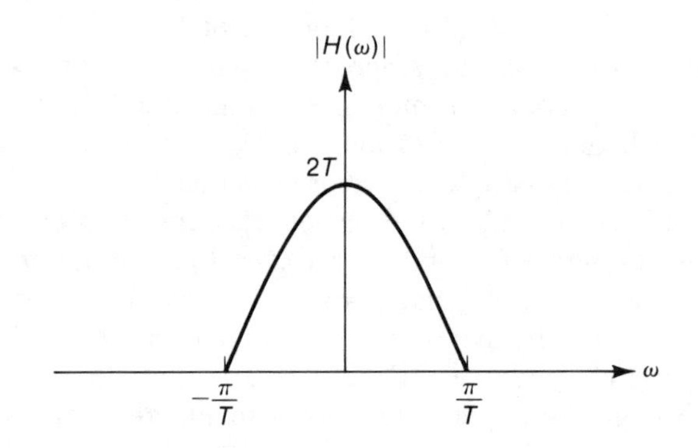

(b) Amplitude spectrum, composite filter

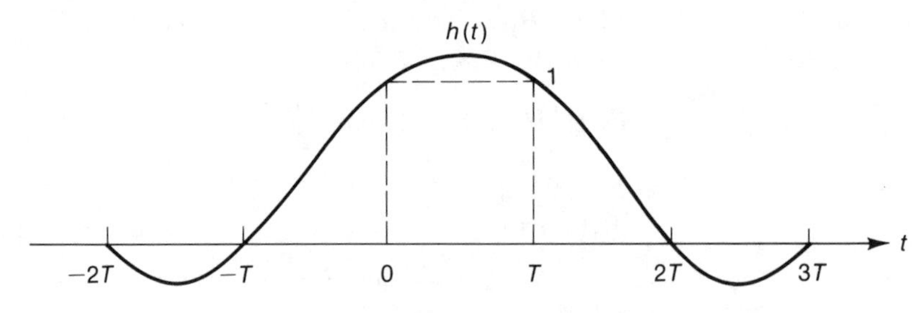

(c) Duobinary impulse response

Figure 5.15 Duobinary Signaling

The received signal can then be written as the sum of all pulse contributions

$$y_k = \sum_n a_n h_{k-n} = a_k + a_{k-1} \tag{5.56}$$

This derivation of the duobinary signal duplicates (5.51), of course, but here we see more directly the beneficial role played by intersymbol interference. The duobinary response to consecutive binary 1's is sketched in Figure 5.16*a*, which illustrates the role of ISI. Figure 5.16*b* depicts the output y_k of the duobinary filter for a random binary input $\{a_k\}$. If the binary input $\{a_k\} = \pm d$, then y_k will have three values at the sampling instants: $-2d$, 0, and $2d$. At the receiver, the three-level waveform is sliced at $\frac{1}{4}$ and $\frac{3}{4}$ of its peak-to-peak value $4d$. The slicer outputs are then decoded to recover the binary data $\hat{a}_k$, using the following three rules:

- $y_k = +d,\ \hat{a}_k = +d$
- $y_k = -d,\ \hat{a}_k = -d$ (5.57)
- $y_k = 0,\ \hat{a}_k = -\hat{a}_{k-1}$

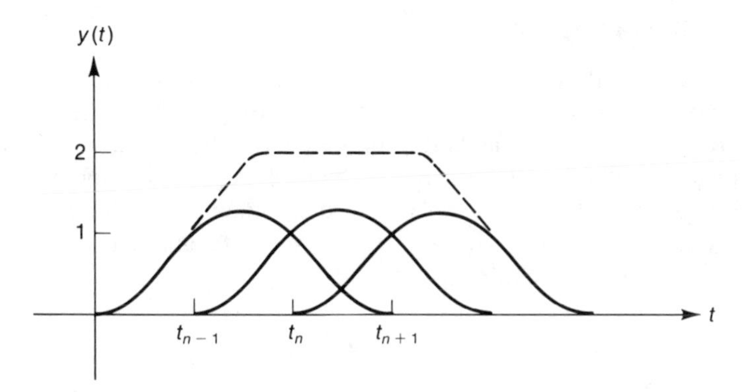

(a) Duobinary response to three consecutive 1's

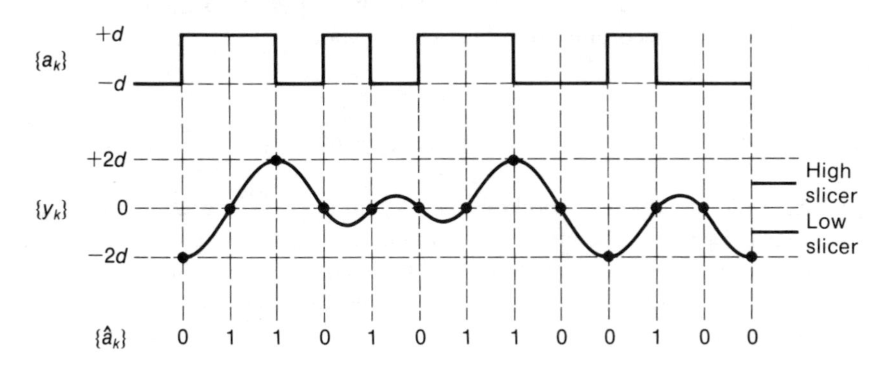

(b) Response of duobinary scheme to random binary input

Figure 5.16 Response of Duobinary Scheme

Owing to the correlation property of partial response systems, only certain types of transitions are allowed in the waveform. These level constraints can be monitored at the receiver, such that any violation of these rules results in an error indication. This feature of partial response, like error detection in bipolar coding, has proved to be valuable in performance monitoring [15]. For duobinary the following level constraints apply:

- A positive (negative) level at one sampling time may not be followed by a negative (positive) level at the next sampling time.
- If a positive (negative) peak is followed by a negative (positive) peak, they must be separated by an odd number of center samples.
- If a positive (negative) peak is followed by another positive (negative) peak, they must be separated by an even number of center samples.

Since the symbol in error cannot be determined, error correction is not possible without additional coding.

5.7.1 Duobinary Precoding

A problem with the decoder indicated in (5.57) is that errors tend to propagate. This occurs because correct decoding of $\hat{a}_k$ depends on correct decoding of $\hat{a}_{k-1}$. A method proposed by Lender eliminates this error propagation by precoding the input binary data prior to duobinary filtering. This logic operation is shown in the block diagram of Figure 5.17*a*. The binary input sequence $\{a_k\}$ is first converted to another binary sequence $\{b_k\}$ according to the rule

$$b_k = a_k \oplus b_{k-1} \tag{5.58}$$

where the symbol $\oplus$ represents modulo 2 addition or the logic operator EXCLUSIVE OR. The sequence $\{b_k\}$ is then applied to the input of the duobinary filter according to (5.51), which yields

$$\begin{aligned} y_k &= b_k + b_{k-1} \\ &= (a_k \oplus b_{k-1}) + b_{k-1} \end{aligned} \tag{5.59}$$

An example of the precoded duobinary signal is given in Figure 5.17*b* for a random binary input sequence. Examination of this example together with (5.59) indicates that $y_k = 1$ results only from $a_k = 1$ and that $y_k = 0$ or 2 corresponds only to $a_k = 0$. These values indicate that y_k can be decoded according to the simple rule

$$\hat{a}_k = y_k (\text{mod}\, 2) \tag{5.60}$$

Since each binary decision requires knowledge of only the current sample y_k, there is no error propagation.

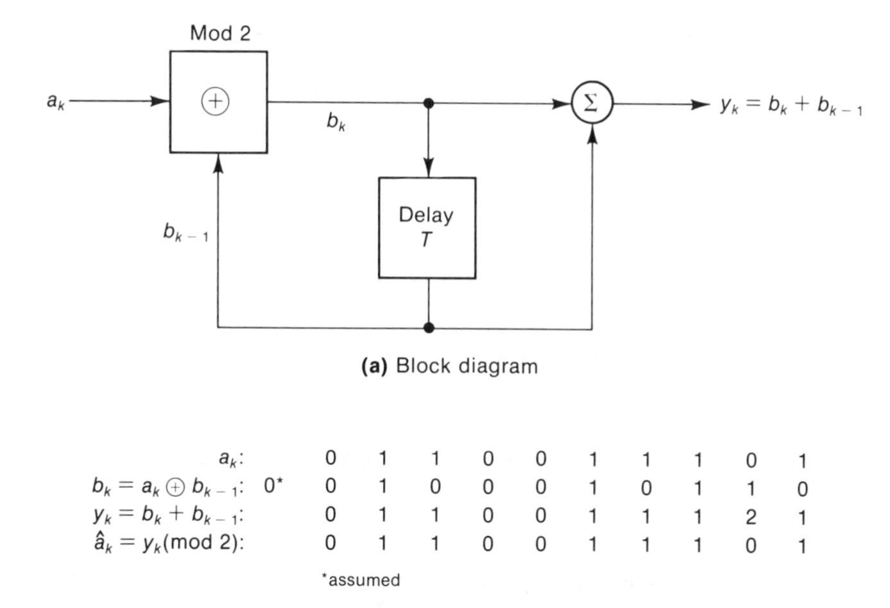

(a) Block diagram

a_k:		0	1	1	0	0	1	1	1	0	1
$b_k = a_k \oplus b_{k-1}$:	0*	0	1	0	0	0	1	0	1	1	0
$y_k = b_k + b_{k-1}$:		0	1	1	0	0	1	1	1	2	1
$\hat{a}_k = y_k(\text{mod } 2)$:		0	1	1	0	0	1	1	1	0	1

*assumed

(b) Example of signals

Figure 5.17 Precoded Duobinary Scheme

5.7.2 Generalized Partial Response Coding Systems

As indicated earlier, duobinary is just one example of the classes of partial response signaling schemes. As shown in Figure 5.18, the design of a partial response coder is based on the generalized digital filter, $H_1(\omega)$. This so-called **transversal filter** provides a weighted superposition of N digits. Appropriate choice of the weighting coefficients h_n allows the spectrum to be shaped for a particular application. From application of the Fourier series representation, the transfer function of this digital filter is known to be

$$H_1(\omega) = \sum_{n=0}^{N-1} h_n e^{-jn\omega T} \tag{5.61}$$

Example 5.2

Consider the class of partial response signaling generated by subtracting pulses spaced by two sample intervals, yielding

$$y_k = a_k - a_{k-2}$$

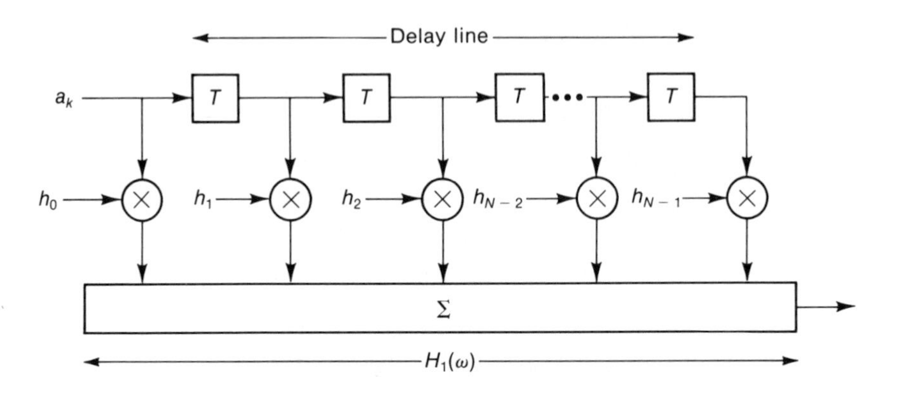

Figure 5.18 Generalized Partial Response Filter

The weighting coefficients of the generalized transversal filter are then $h_0 = 1$, $h_2 = -1$, with all other coefficients equal to zero. From (5.61), the transfer function is thus

$$H_1(\omega) = 1 - e^{-j2\omega T}$$

The amplitude characteristic is therefore

$$|H_1(\omega)| = 2\sin\omega T$$

This scheme is known both as **modified duobinary** and as class 4 partial response. It has its advantage in the spectral shape that has no dc component. This spectrum characteristic is desirable for those transmission systems, such as the telephone plant, that cannot pass dc signals because of the use of transformers. Except for the spectral shape, modified duobinary has characteristics similar to duobinary. Both schemes result in three-level waveforms and in error propagation that can be eliminated with precoding. For transmitted symbols $\{a_k\} = \pm d$, modified duobinary (without precoding) is decoded according to the following rules:

- $y_k = +d,\ \hat{a}_k = +d$
- $y_k = -d,\ \hat{a}_k = -d$
- $y_k = 0,\ \hat{a}_k = -\hat{a}_{k-2}$

Table 5.1 lists the classes of partial response signaling. The two classes that are most commonly used because of their simplicity, performance, and desirable spectral shape are *duobinary* (class 1) and *modified duobinary* (class 4). Partial response signaling is used in a variety of applications,

Table 5.1 Classes of Partial Response Signaling

Class	*Generating Function*	*Transfer Function* $H(\omega)$	*Number of Received Levels*
1	$y_k = a_k + a_{k-1}$	$2\cos(\omega T/2)$	3
2	$y_k = a_k + 2a_{k-1} + a_{k-2}$	$4\cos^2(\omega T/2)$	5
3	$y_k = 2a_k + a_{k-1} - a_{k-2}$	$2 + \cos\omega T - \cos 2\omega T$	5
4	$y_k = a_k - a_{k-2}$	$2\sin\omega T + j(\sin\omega T - \sin 2\omega T)$	3
5	$y_k = a_k - 2a_{k-2} + a_{k-4}$	$4\sin^2\omega T$	5

including baseband repeaters for T carrier [16], voice-channel modems [17], and digital radio. To enhance data rate packing, partial response schemes are also used in conjunction with other modulation schemes such as single sideband [16], FM [18], and PSK [19]. Finally, partial response has been used with existing analog microwave radios to add data above or below the analog signal spectrum [20].

5.7.3 Partial Response Techniques for *M*-ary Signals

Partial response techniques can in general be utilized with M-ary input signals to increase the data rate possible over a fixed bandwidth. Instead of a binary ($M = 2$) input to the partial response filter, the input is allowed to take on M levels. In the case of duobinary, we observed that a binary input results in a three-level received signal. The application of an M-ary input to a duobinary system results in $2M - 1$ received levels. The duobinary coder and decoder block diagrams for this M-ary application are similar to that of duobinary for binary input. Consider, for example, the case of precoded duobinary as shown in Figure 5.17. For M-ary input, the mod 2 operator must be replaced with a mod M operator. The precoder logic can then be described by the following rule:

$$b_k = a_k - b_{k-1} \qquad (\text{mod } M) \tag{5.62}$$

which is analogous to the case for binary input (Equation 5.58). The sequence $\{b_k\}$ is then applied to the input of the duobinary filter, which yields

$$y_k = b_k + b_{k-1} \qquad (\text{algebraic}) \tag{5.63}$$

The decoding rule is analogous to that of precoded duobinary for binary input, but here the operator is mod M; that is,

$$\hat{a}_k = y_k \qquad (\text{mod } M) \tag{5.64}$$

Example 5.3

Suppose the number of levels M is equal to eight. The input symbols are assumed to be $\{0, 1, 2, 3, 4, 5, 6, 7\}$. For the following random input sequence $\{a_k\}$, the corresponding precoder sequence $\{b_k\}$ and partial response sequence $\{y_k\}$ are shown in accordance with (5.62) to (5.64):

a_k:		7	5	5	0	4	6	3	6	1	4	2	2	0
b_k:	0	7	6	7	1	3	3	0	6	3	1	1	1	7
y_k:		7	13	13	8	4	6	3	6	9	4	2	2	8
$\hat{a}_k$:		7	5	5	0	4	6	3	6	1	4	2	2	0

The sequence $\{b_k\}$ is found by subtracting b_{k-1} from a_k, modulo M, so that starting from the first operation we have $7 - 0 = 7$; next, $5 - 7 = -2 \pmod 8) = 6$; next, $5 - 6 = -1 \pmod 8) = 7$; and so on. The sequence $\{y_k\}$ is simply the algebraic sum of b_{k-1} and b_k, so that the 15 levels $\{0, 1, \ldots, 14\}$ will result. Finally, $\hat{a}_k$ is obtained by taking the modulo 8 value of y_k.

The advantage of M-ary partial response signaling over zero-memory M-ary signaling is that fewer levels are needed for the same data rate. Fewer levels mean greater protection against noise and hence a better error rate performance. To show this advantage, first consider a zero-memory system based on a 100 percent raised cosine filter of bandwidth R hertz. A binary sequence ($M = 2$) input results in a data rate packing of 1 b/s per hertz of bandwidth. Therefore an M-ary sequence ($M = 2^m$) input will result in an m bps/Hz data rate packing. Now consider a duobinary scheme used on a channel with the same bandwidth of R hertz. A binary input results in 2 bps/Hz. In general, to obtain a packing of k bps/Hz, a $2^{k/2}$-level input sequence is required that results in only $2(2^{k/2}) - 1$ output levels. For $k = 4$ bps/Hz, for example, 16 levels would be required with 100 percent raised-cosine transmission but only 7 levels with duobinary. For 6 bps/Hz, 100 percent raised-cosine transmission requires 64 levels whereas duobinary requires only 15 levels.

5.7.4 Error Performance for Partial Response Coding

For duobinary, the received signal in the absence of noise has three values, $+2d$, 0, and $-2d$, and the probabilities of receiving these levels are $\frac{1}{4}$, $\frac{1}{2}$, and $\frac{1}{4}$, respectively. For the gaussian noise channel, the probability of error is then

$$P(e) = \frac{3}{2}\text{erfc}\left(\frac{d}{\sigma}\right) \tag{5.65}$$

where σ^2 is the variance of the noise. In terms of S/N ratio, the probability of error becomes [8]

$$P(e) = \frac{3}{2}\text{erfc}\left[\frac{\pi}{4}\left(\frac{S}{N}\right)^{1/2}\right] \tag{5.66}$$

where S is the average signal power, N is the noise power in the Nyquist bandwidth, and the duobinary filter is split equally between the transmitter and receiver.

Comparing duobinary error performance with that of binary ($M = 2$) transmission (Equation 5.47), we find duobinary performance to be poorer by $\pi/4$, or 2.1 dB. The binary error performance is based on the theoretical Nyquist channel, however, while duobinary performance is based on a practical channel. A fairer comparison with duobinary is to use a raised-cosine channel with $\alpha = 1$. Then for the data rate packing (R/W) to be equal, we must choose $M = 4$. For this case duobinary requires 4.9 dB less S/N ratio to attain the same error performance as the raised-cosine channel.

For M-ary partial response, the probability of error is given by [8]

$$P(e) = 2\left(1 - \frac{1}{M^2}\right)\text{erfc}\left[\frac{\pi}{4}\left(\frac{3}{M^2 - 1}\frac{S}{N}\right)^{1/2}\right] \tag{5.67}$$

Note the familiar factor $3/(M^2 - 1)$ in the S/N term, which indicates the degradation in error performance due to additional signal levels.

5.8 EYE PATTERNS

A convenient way of assessing the performance of a baseband signaling system is by means of the **eye pattern**. This pattern is obtained by displaying the received signal on an oscilloscope. The time base of the oscilloscope is set to trigger at a rate $1/T$. The resulting pattern derives its name from its resemblance to the human eye for binary data. The persistence of the cathode ray-tube blends together all allowed signal waveforms. As an example based on bipolar signaling, consider the 3-bit sequences shown in Figure 5.19*a*, where the first 1 always has positive polarity. (A corresponding set of sequences exists in which the first 1 always has negative polarity.) Figure 5.19*b* gives the eye pattern for the sequences of Figure 5.19*a* by superimposing these waveforms in the same signaling interval. Together with the other allowed sequences, Figure 5.19*c* plots the complete bipolar eye pattern. Note that this pattern traces all allowed waveforms in a given sampling interval with a resulting three-level signal at the prescribed sampling time. Moreover, the sampling time is shown at the point where the opening of the eye is greatest.

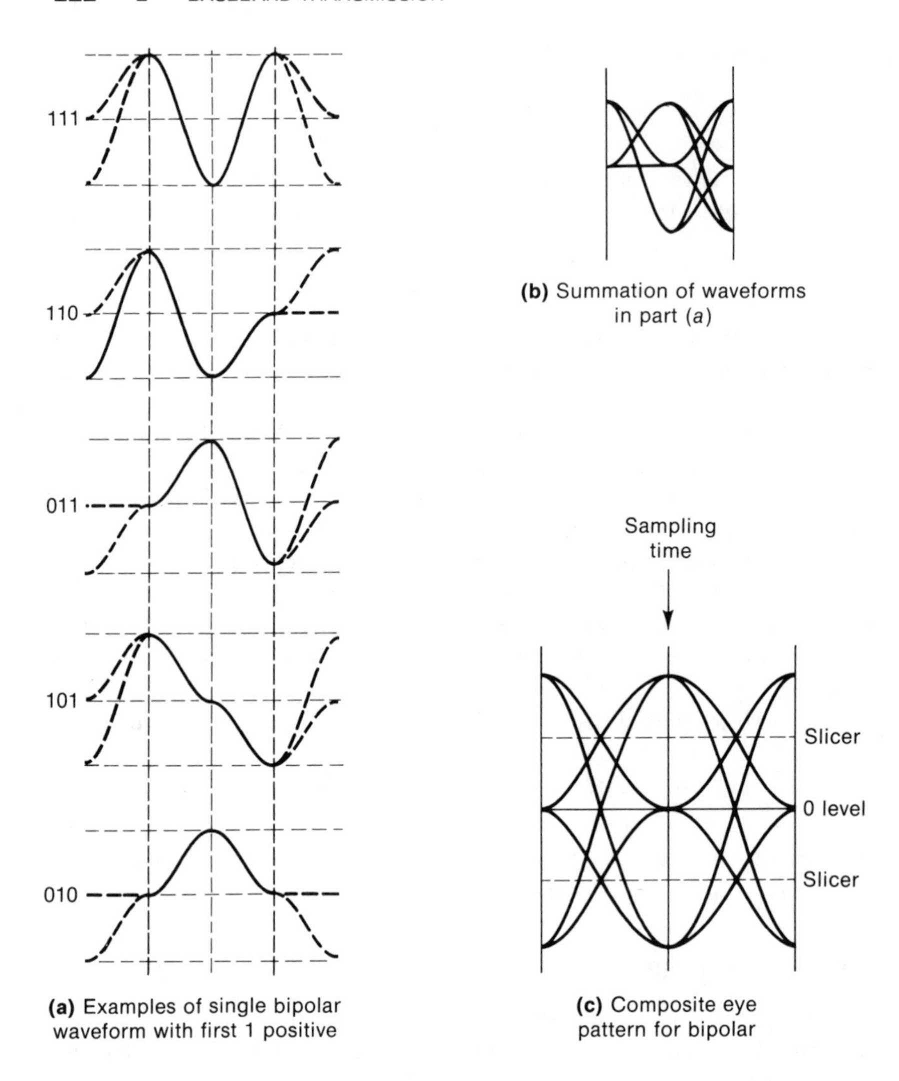

(a) Examples of single bipolar waveform with first 1 positive

(b) Summation of waveforms in part (*a*)

(c) Composite eye pattern for bipolar

Figure 5.19 Bipolar Eye Patterns

The eye opening may be used as an indication of system "health." As noise, interference, or jitter is added by the channel, the eye opening will close. The height of this opening indicates what margin exists over noise. For example, an eye closing of 50 percent indicates an equivalent S/N degradation of $-20\log(1-0.5) = 6$ dB. The three-level partial response eye patterns of Figure 5.20 illustrate the observable degradation between a zero error rate and a 10^{-3} error rate.

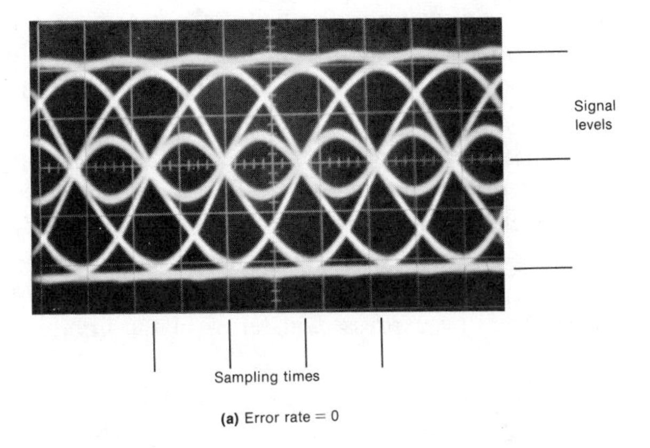

(a) Error rate = 0

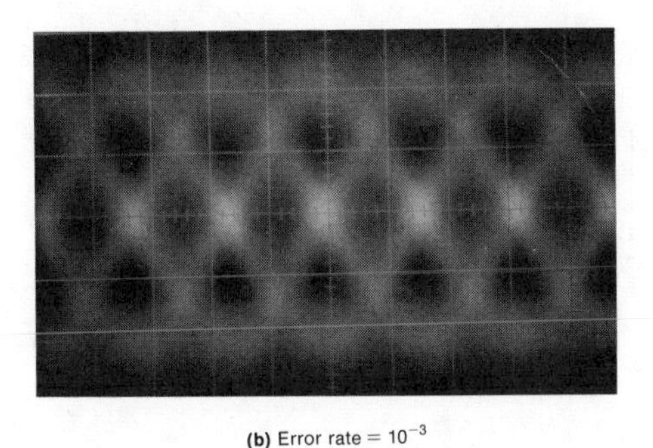

(b) Error rate = 10^{-3}

Figure 5.20 Eye Patterns of Three-Level Partial Response Signal

5.9 EQUALIZATION

Real channels introduce distortion in the digital transmission process, resulting in a received signal that is corrupted by intersymbol interference. The concept of equalization is based on the use of adjustable filters that compensate for amplitude and phase distortion. High-speed transmission over telephone circuits or radio channels often requires equalization to overcome the effects of intersymbol interference caused by dispersion. In most practical cases, the channel characteristics are unknown and may vary with time. Hence the equalizer must be updated with each new channel connection, as in the case of a voice-band modem operating in a switched network, and must also adapt the filter settings automatically to track changes in the channel with time. Today automatic adaptive equalization is used nearly universally in digital radio applications [21] and in voice-band modems for speeds higher than 2400 b/s [22].

Because of its versatility and simplicity, the transversal filter is a common choice for equalizer design. This filter consists of a delay line tapped at T-second intervals, where T is the symbol width. Each tap along the delay line is connected through an amplifier to a summing device that provides the output. The summed output of the equalizer is sampled at the symbol rate and then fed to a decision device. The tap gains, or coefficients, are set to subtract the effects of interference from symbols that are adjacent in time to the desired symbol. We assume there are $(2N + 1)$ taps with coefficients $c_{-N}, c_{-N+1}, \ldots, c_N$, as indicated in Figure 5.21. Samples of the equalizer output can then be expressed in terms of the input x_k and tap coefficients c_n as

$$y_k = \sum_{n=-N}^{N} c_n x_{k-n} \qquad k = -2N, \ldots, +2N \tag{5.68}$$

Computation of the output sequence $\{y_k\}$ is simplified by use of matrices. If we let

$$\mathbf{y} = \begin{bmatrix} y_{-2N} \\ \vdots \\ y_0 \\ \vdots \\ y_{2N} \end{bmatrix} \qquad \mathbf{c} = \begin{bmatrix} c_{-N} \\ \vdots \\ c_0 \\ \vdots \\ c_N \end{bmatrix}$$

$$\mathbf{x} = \begin{bmatrix} x_{-N} & 0 & 0 & \cdots & 0 & 0 \\ x_{-N+1} & x_{-N} & 0 & \cdots & 0 & 0 \\ \vdots & \vdots & \vdots & & \vdots & \vdots \\ x_N & x_{N-1} & x_{N-2} & \cdots & x_{-N+1} & x_{-N} \\ \vdots & \vdots & \vdots & & \vdots & \vdots \\ 0 & 0 & 0 & & x_N & x_{N-1} \\ 0 & 0 & 0 & \cdots & 0 & x_N \end{bmatrix} \tag{5.69}$$

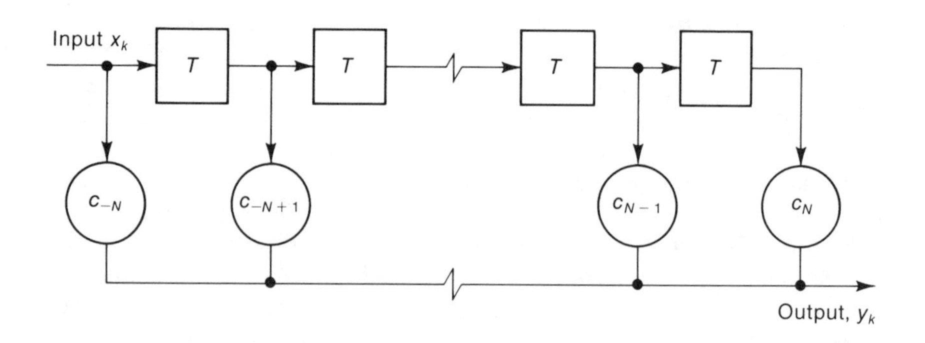

Figure 5.21 Transversal Filter

then we can write (5.68) as

$$\mathbf{y} = \mathbf{xc} \tag{5.70}$$

Example 5.4

For the channel impulse response of Example 5.1, assume that a three-tap transversal filter is used as the equalizer with tap coefficients $c_{-1} = -\frac{1}{4}$, $c_0 = 1$, and $c_1 = \frac{1}{2}$. The output sequence is found by matrix multiplication:

$$\mathbf{y} = \mathbf{xc}$$

$$= \begin{bmatrix} \frac{1}{4} & 0 & 0 \\ 1 & \frac{1}{4} & 0 \\ -\frac{1}{2} & 1 & \frac{1}{4} \\ 0 & -\frac{1}{2} & 1 \\ 0 & 0 & -\frac{1}{2} \end{bmatrix} \begin{bmatrix} -\frac{1}{4} \\ 1 \\ \frac{1}{2} \end{bmatrix}$$

$$= \left(-\tfrac{1}{16} \quad 0 \quad \tfrac{5}{4} \quad 0 \quad -\tfrac{1}{4}\right)$$

Although adjacent symbol interference has been forced to zero, some ISI has been added at sample points further removed from the center pulse. In general, a finite transversal filter cannot completely eliminate ISI.

The criterion for selection of tap coefficients is generally based on the minimization of either peak distortion or mean square distortion. The so-called **zero-forcing equalizer** has been shown to minimize peak distortion [8]. Here the tap coefficients are selected to force the equalizer output to zero at N sample points on either side of the desired pulse. We may write this in mathematical form as $2N + 1$ constraint equations:

$$y_k = \begin{cases} 1 & k = 0 \\ 0 & k = \pm 1, \pm 2, \ldots, \pm N \end{cases} \tag{5.71}$$

We can then solve for the tap coefficients c_n by combining (5.68) and (5.71), which amounts to solving $2N + 1$ simultaneous linear equations.

The mean square minimization criterion also leads to $2N + 1$ linear simultaneous equations, whose solution minimizes the sum of squares of the ISI terms. The equalizer error signal at sample point k is simply the difference between the transmitted pulse δ_k and equalized received signal y_k, or

$$e_k = y_k - \delta_k \tag{5.72}$$

The mean square error is

$$\overline{e^2} = \sum_{k=-N}^{N} (y_k - \delta_k)^2 \tag{5.73}$$

For minimum mean square error it is necessary that the gradient

$$\frac{\partial \overline{e^2}}{\partial c_j} = 0 \qquad j = -N, \ldots, N \tag{5.74}$$

From (5.73) we obtain

$$\frac{\partial \overline{e^2}}{\partial c_j} = 2 \sum_{k=-N}^{N} (y_k - \delta_k) \frac{\partial y_k}{\partial c_j} \tag{5.75}$$

Using (5.68) and (5.72) in (5.75) we obtain

$$\frac{\partial \overline{e^2}}{\partial c_j} = 2 \sum_{k=-N}^{N} e_k x_{k-j} \qquad j = -N, \ldots, N \tag{5.76}$$

Therefore the tap coefficients are optimum when the cross-correlation between the error signal and input pulse is forced to zero at each sample point within the equalizer range. In practice, the optimum tap settings are found by an iterative procedure. The tap coefficient vector $\mathbf{c}$ is updated during every symbol interval. Letting $\mathbf{c}^m$ be the mth such tap coefficient vector, we can write

$$\mathbf{c}^{m+1} = \mathbf{c}^m - \alpha \mathbf{e}^m \tag{5.77}$$

where α is the equalizer step size and $\mathbf{e}^m$ is the error vector at time m. If the sequence converges after some time m, that is,

$$\mathbf{c}^{m+1} = \mathbf{c}^m = \mathbf{c} \tag{5.78}$$

then

$$\frac{\partial \overline{e^2}}{\partial \mathbf{c}} = 0 \tag{5.79}$$

and the tap coefficients are optimized in a mean square error sense. Because of the presence of noise in the measurement of the error vector, in practice the gradient method is based on a noisy estimate and not the ideal estimate assumed above.

Two general types of automatic equalization exist. In **preset equalization**, a training sequence is transmitted and compared at the receiver with a locally generated sequence. The resulting error voltages are used to adjust the tap gains to optimum settings (minimum distortion). The training period may consist of a periodic sequence of isolated pulses or a pseudorandom sequence. After this training period the tap gains remain fixed and normal transmission ensues. Two limitations of this form of automatic equalization are immediately apparent. First, training is required initially and must be repeated with breaks in transmission; second, a time-varying channel will change intersymbol interference characteristics and hence degrade performance since tap gains are fixed.

For these reasons **adaptive equalization** techniques have been developed that are capable of deriving necessary tap gain adjustments directly from the transmitted bits. In preset equalization, since a known training sequence is transmitted, the error voltages can be directly measured with tap gains set accordingly. During normal transmission, the transmitted symbols are unknown. However, a good estimate of this sequence is produced at the receiver output. In the case of adaptive equalization the output at the receiver is used as if it were the transmitted sequence to obtain an estimate of error voltages. Such a learning procedure has been called *decision-directed* since the receiver attempts to learn by employing its own decisions. If the bit decisions are good enough initially, the equalizer can obtain a sufficient estimate of the errors to iteratively improve tap gain settings. A potential drawback to the adaptive equalizer, then, is the settling, or convergence time for poor channels. A common solution to this problem is the use of a hybrid system that uses preset equalization to establish good error performance and then switches to adaptive equalization when normal transmission commences, as illustrated in Figure 5.22.

The adaptive equalizer techniques discussed here are based on the use of linear filters. To improve performance further, the introduction of nonlinear filter techniques has led to the development of the **decision feedback equalizer** [23]. Figure 5.23 shows the basic structure of the decision feedback equalizer (DFE). A linear forward filter processes the received signal using the transversal equalizer described earlier. This filter attempts to eliminate ISI due to future symbols. Decisions made on the forward equalized signal are fed back to a second transversal filter. If the previous decisions are assumed to be correct, then the feedback equalizer cancels ISI due to previous symbols. The tap weights of the feedback equalizer are determined by the tail of the channel impulse response, including the effects of the forward equalizer. The decision feedback receiver that minimizes mean square error consists of a filter matched to the received signal followed by forward and backward transversal filters [24]. The equalizer is made adaptive through the use of the estimated gradient algorithm, in which the forward and feedback coefficients are continually adjusted to minimize mean square error. A necessary condition for convergence is that the probability of error be much less than one-half to avoid both error propa-

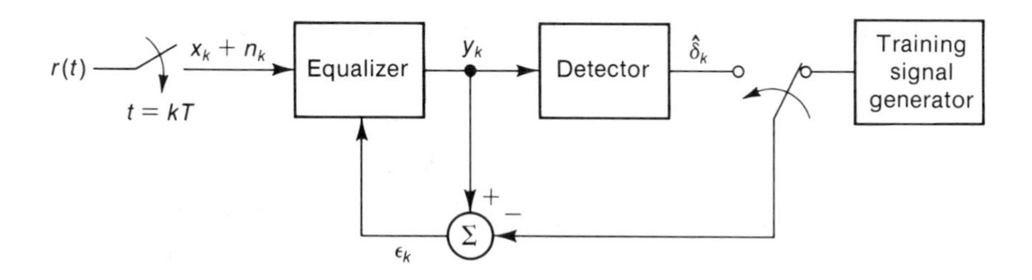

Figure 5.22 Adaptive Equalization

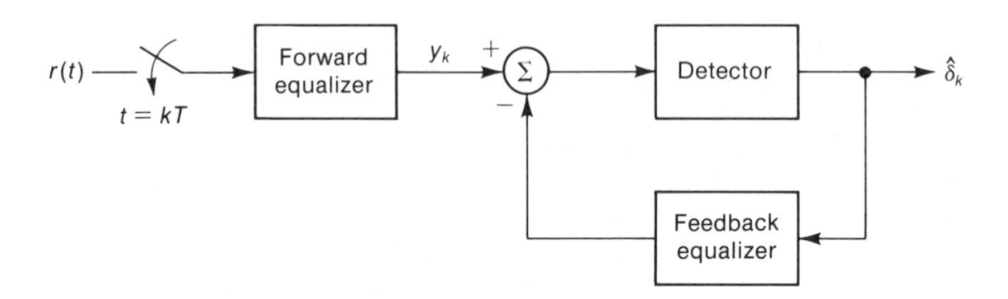

Figure 5.23 Decision Feedback Equalizer

gation in the feedback path and adaptation errors in the decision-directed process.

In practice, several conditions may limit the performance of adaptive equalizers:

1. Time-varying channels may exhibit a rate of change that exceeds the equalizer response time. This limitation can be offset by use of a larger step size in the equalizer, but at the expense of greater mean square error. For data transmission over telephone circuits, the step size is made larger for fast convergence during the training period and then reduced for operation during data transmission.
2. Practical equalizers must be based on a finite number of taps. The choice of equalizer range is based on the ratio of channel dispersion measured or estimated to data symbol width. Thus if the channel dispersion exceeds the equalizer range, performance is degraded due to the presence of ISI not cancelled by the equalizer.
3. The problem of equalizer convergence time discussed earlier is not only affected by system probability of error and equalizer step size; it is also dependent on the transmission of random data. Many data terminals transmit repetitive patterns such as all 1's during idle periods. During these periods the adaptive equalizer settings may drift due to lack of sufficient data transitions. A simple solution to this problem is afforded by scrambling of the transmitted data to eliminate periods of data idling at the receiver.

5.10 DATA SCRAMBLING TECHNIQUES

Impairments in digital transmission systems often vary with the statistics of the digital source. Timing and equalization performance usually depend on the source statistics. For example, a long string of 0's or 1's can cause a bit synchronizer to degrade or even lose synchronization. Likewise, periodic bit patterns can create discrete spectral lines that cause crosstalk in cable transmission and cochannel or adjacent channel interference in radio transmission.

Scrambling the data can minimize long strings of 0's and 1's and suppress discrete spectral components. Scrambling devices randomize or "whiten" the data by producing digits that appear to be independent and equiprobable. There are two basic classes of scramblers: techniques that scramble via logic addition (such as modulo 2) of a pseudorandom sequence with the input bit sequence and those that scramble by performing a logic addition on delayed values of the input sequence itself. In the following discussion we will restrict our attention to applications of scramblers to binary transmission, but the techniques can be generalized to M-ary transmission by use of modulo M addition.

5.10.1 Pseudorandom Sequences

Known, fixed binary sequences that exhibit properties of a random signal can be used as the basis for data scrambling. These sequences are generated by using shift registers with certain feedback connections to modulo 2 adders. A shift register is composed of a number of flip-flops cascaded in series. When the shift register receives a clock pulse, the binary state of each flip-flop is transferred to the next flip-flop. The feedback connections consist of taps at certain stages; the tapped signals are added modulo 2 and fed

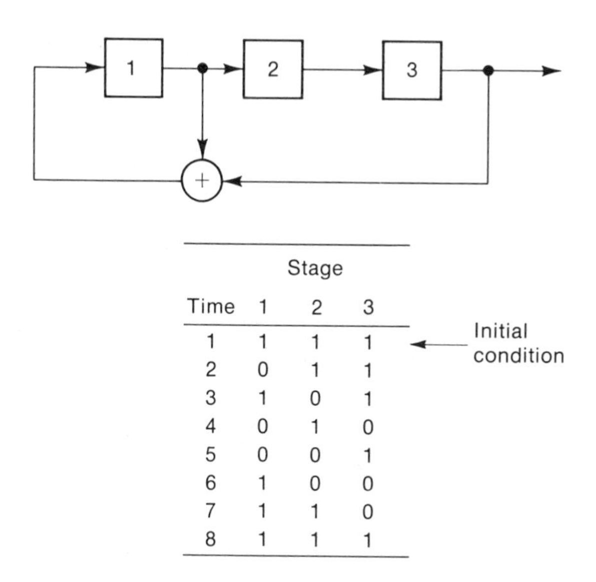

	Stage		
Time	1	2	3
1	1	1	1 ← Initial condition
2	0	1	1
3	1	0	1
4	0	1	0
5	0	0	1
6	1	0	0
7	1	1	0
8	1	1	1

Figure 5.24 Three-Stage Shift Register with Feedback for Generation of a 7-Bit Pseudorandom Sequence

back to the first flip-flop. To illustrate, consider the three-stage sequence generator shown in Figure 5.24. Suppose we initialize this generator by loading all 1's into the flip-flops. With each clock pulse, the contents of flip-flops 1 and 3 are mod 2 added, the contents of flip-flops 1 and 2 are shifted to flip-flops 2 and 3, and the mod 2 output is fed back to flip-flop 1. The contents of the shift register will cycle through seven different states and then repeat itself. The output of this three-stage sequence generator is taken from the last flip-flop of the shift register to yield the sequence 1110100. . . . This 7-bit sequence appears random to the outside observer. Because the sequence can be predicted from knowledge of the shift register length and taps, however, these sequences are known as **pseudorandom**.

The length of a pseudorandom sequence is determined by the choice of shift register length, feedback taps, and initial states of the flip-flops. As can be observed in Figure 5.24, an initial state of all zeros remains unchanged and thus is not useful for sequence generation. The *maximum-length sequence* produced by an n-stage shift register is therefore $2^n - 1$. Examples of feedback connections for maximum-length sequences are given in Table

Table 5.2 Examples of Maximum-Length Pseudorandom Sequences

Length of Shift Register	*Feedback Taps*	*Period of Sequence*
3	1, 3	7
4	1, 4	15
5	2, 5	31
6	1, 6	63
7	1, 7	127
8	1, 6, 7, 8	255
9	4, 9	511
10	3, 10	1,023
11	2, 11	2,047
12	2, 10, 11, 12	4,095
13	1, 11, 12, 13	8,191
14	2, 12, 13, 14	16,383
15	14, 15	32,767
16	11, 13, 14, 16	65,535
17	14, 17	131,071
18	11, 18	262,143
19	14, 17, 18, 19	524,287
20	17, 20	1,048,575

Sequence:	+1	+1	+1	−1	−1	+1	−1	
For $\tau = 0$								
	+1	+1	+1	−1	−1	+1	−1	
×	+1	+1	+1	−1	−1	+1	−1	
	+1	+1	+1	+1	+1	+1	+1	$= 7 = R(\tau = 0)$
For $\tau = 1$								
	+1	+1	+1	−1	−1	+1	−1	
×	+1	+1	−1	−1	+1	−1	+1	
	+1	+1	−1	+1	−1	−1	−1	$= -1 = R(\tau = 1)$
For $\tau = 2$								
	+1	+1	+1	−1	−1	+1	−1	
×	+1	−1	−1	+1	−1	+1	+1	
	+1	−1	−1	−1	+1	+1	−1	$= -1 = R(\tau = 2)$
For $\tau = -1$								
	+1	+1	+1	−1	−1	+1	−1	
×	−1	+1	+1	+1	−1	−1	+1	
	−1	+1	+1	−1	+1	−1	−1	$= -1 = R(\tau = -1)$

(a) Calculation of autocorrelation $R(\tau)$

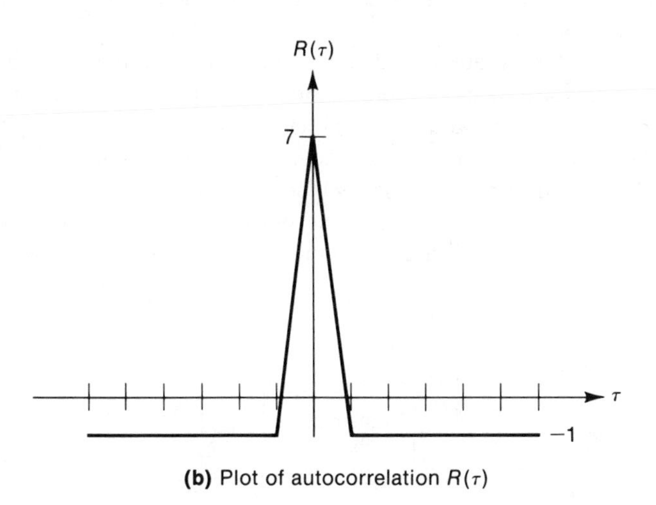

(b) Plot of autocorrelation $R(\tau)$

Figure 5.25 Example of Autocorrelation for a 7-Bit Pseudorandom Sequence

5.2 for various shift register lengths. Each connection actually specifies two feedback configurations because the inverse arrangement provides a sequence of the same length, but reversed. Thus feedback from flip-flops 2 and 3 work as well as 1 and 3 to produce a sequence of length 7. Likewise, 3 and 4 work along with 1 and 4 for length 15. Other properties of maximum-length sequences are:

- The number of 1's in one cycle of the output sequence is 1 greater than the number of 0's.
- The number of runs of consecutive 0's or 1's of length n is twice the number of runs of length $n + 1$. That is, one-half the runs are of length 1, one-fourth of length 2, one-eighth of length 3, and so on.
- The autocorrelation of the sequence has a peak equal to the sequence length $2^n - 1$ at zero shift and at multiples of the sequence length. At all other shifts, the correlation is -1. Autocorrelation for a 7-bit pseudorandom sequence is shown in Figure 5.25.

The correlation property of pseudorandom sequences results in a flat (or white) power spectral density as the sequence length increases. Because these pseudorandom sequences simulate white noise, they are sometimes called *pseudonoise* sequences.

Pseudorandom sequences can be used to scramble data by mod 2 addition of the data with the pseudorandom sequence. Figure 5.26 is a block diagram of a scrambler and descrambler that uses a prescribed pseudorandom (PR) sequence. Note that the same PR sequence generator is required at both the scrambler and descrambler. Further, a state of synchronism must exist between the encoder and decoder; that is, the same set of pseudorandom and data bits must coincide at the input to the encoder and

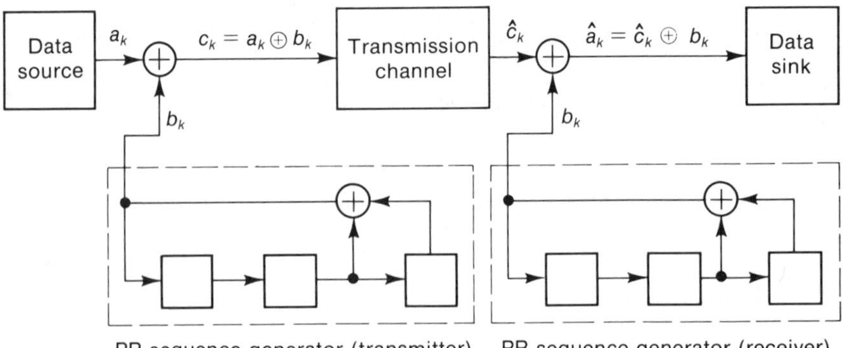

Figure 5.26 Block Diagram of Pseudorandom Scrambler and Descrambler

decoder. Initial synchronization can be accomplished by a handshaking exchange of a preamble to initiate a prestored sequence or by forward-acting operation where the initial shift register states are sent by transmitter and used to initialize the receiver shift register.

5.10.2 Self-Synchronizing Scrambler

The self-synchronizing scrambler derives its randomizing capability from a logic addition of delayed digits from the data source. Figure 5.27 shows a generalized self-synchronizing scrambler and descrambler. Each stage of the shift register represents a unit delay. For a scrambler containing M stages, the output may be written as

$$b_k = a_k \oplus \sum_{p=1}^{M} b_{k-p}\delta_p \tag{5.80}$$

where $\oplus$ and Σ denote modulo addition and δ_p is defined as

$$\delta_p = \begin{cases} 1 & \text{if stage } p \text{ is fed back and added} \\ 0 & \text{otherwise} \end{cases}$$

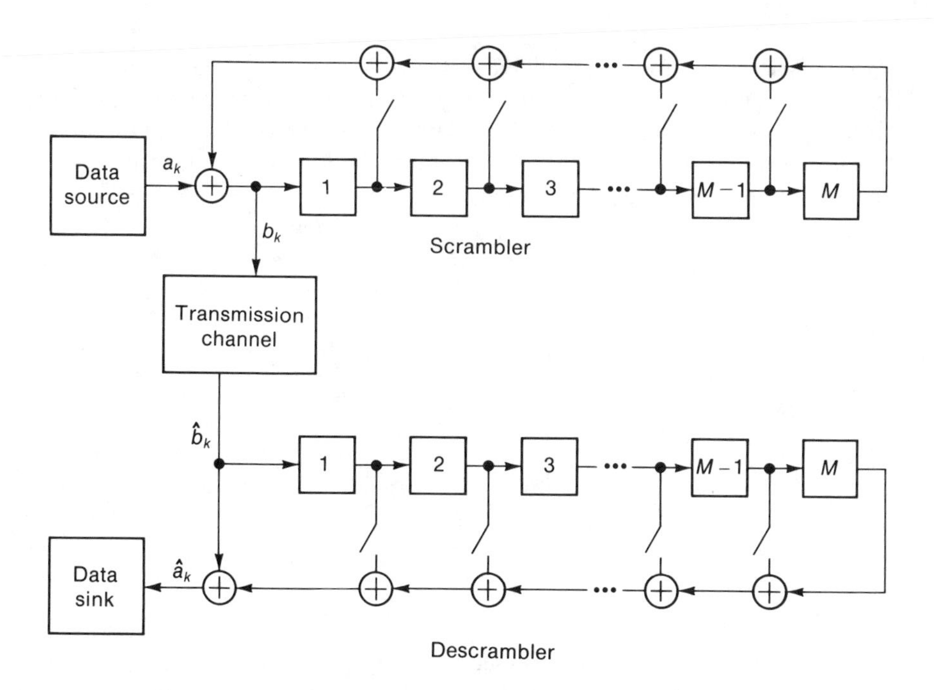

Figure 5.27 Block Diagram of Self-Synchronizing Scrambler and Descrambler

Decoding of the scrambler output b_k is possible after M error-free scrambler bits have been received, so that the shift register stages are identical at the transmitter and receiver. The equation for the decoder is then similar to that of the encoder:

$$\hat{a}_k = \sum_{p=0}^{M} \hat{b}_{k-p}\delta_p \tag{5.81}$$

The scrambled and descrambled digits are estimates, as indicated in (5.81), due to the assumption of degradation in the channel and the resulting possibility of error in digit detection.

Example 5.5

For a six-stage self-synchronizing scrambler, let δ_5 and $\delta_6 = 1$ and all other feedback connections $\delta_i = 0$ for $i = 1$ to 4. The scrambled data sequence is then

$$b_k = a_k \oplus b_{k-5} \oplus b_{k-6}$$

Assuming error-free transmission, the descrambled data sequence is

$$\begin{aligned}\hat{a}_k &= b_k \oplus b_{k-5} \oplus b_{k-6} \\ &= (a_k \oplus b_{k-5} \oplus b_{k-6}) \oplus b_{k-5} \oplus b_{k-6} \\ &= a_k\end{aligned}$$

which shows that the descrambled sequence is identical to the original data sequence. Now assume an initial loading of the binary sequence 110101 in the six stages of the scrambler. If the data source then produces an infinitely long string of zeros, the six stages will show the following sequence:

		Stage					
Time	*Input*	*1*	*2*	*3*	*4*	*5*	*6*
1	0	1	1	0	1	0	1
2	0	1	1	1	0	1	0
3	0	1	1	1	1	0	1
4	0	1	1	1	1	1	0
5	0	1	1	1	1	1	1
6	0	0	1	1	1	1	1
⋮	⋮	⋮	⋮	⋮	⋮	⋮	⋮

Development of the remaining sequences is left to the reader. For this example, the scrambled output sequence turns out to be maximum length, or $2^6 - 1 = 63$.

A potential drawback to the use of self-synchronizing scramblers is the inherent property of error extension. To observe this property, let us expand (5.81):

$$\hat{a}_k = \hat{b}_k\ \delta_0 + \hat{b}_{k-1}\ \delta_1 + \hat{b}_{k-2}\ \delta_2 + \cdots + \hat{b}_{k-M}\ \delta_M \tag{5.82}$$

By inspection we see that for each scrambled bit received in error, n other data estimates are also in error, where n is the number of feedback paths that are closed. Thus the total probability of error is

$$P_T(e) = (1 + n)P(e) \tag{5.83}$$

where $P(e)$ is the digital probability of error at the detector. In practice, the number of feedback connections required to realize a maximum-length sequence is on the order of 2 to 4. Hence the additional degradation due to error extension is usually considered negligible.

The length of the shift register in a data scrambler is selected according to the desired degree of randomness. The spectrum of the scrambler output is the appropriate measure of randomness. The power spectral density is composed of a number of discrete components within a $(\sin x/x)^2$ envelope; the discrete components are spaced by $1/T_0$, where T_0 is the period of the data sequence. Although such a demonstration is beyond the scope of this book, it can be shown that a 20-stage scrambler, arranged to provide a maximum-length sequence by choice of appropriate feedback paths, provides the desired randomness for most applications [25].

5.11 SUMMARY

Direct transmission of a signal without frequency translation is called baseband transmission. Certain codes facilitate transmission of data over baseband channels. By appropriate choice of code, certain desirable properties can be obtained, such as (1) increased data transition density to allow extraction of clock from the data signal, (2) shaping of the spectrum to minimize interference between channels, (3) reduction of bandwidth for increased spectral efficiency, and (4) error detection for on-line performance monitoring. Nonreturn-to-zero (NRZ) coding is a common choice for binary coding but lacks most of these desirable properties. Both diphase and bipolar coding result in increased data transitions and a desired reshaping of the data spectrum. Bipolar also provides inherent error detection, although the three levels used for bipolar result in a signal-to-noise (S/N) penalty compared to other two-level codes.

Because of bandwidth limitations in real channels, digital signals tend to be distorted in the transmission process. Unless compensated, this distortion causes intersymbol interference (ISI). However, certain pulse shapes called Nyquist waveforms can be used that contribute zero energy at adjacent symbol times. A $\sin x/x$ form provides the required shape but is not physically realizable; a raised-cosine shape provides zero ISI and also leads to physically realizable filters.

For a channel with bandwidth W hertz, the basic limitation in bit rate R is $2W$ bits per second for binary transmission. Hence the limitation in bandwidth efficiency (R/W) is 2 bps/Hz. With M-ary transmission, each symbol takes on one of M levels and hence represents $\log_2 M$ bits of information. M-ary transmission increases the spectral efficiency by a factor of $m = \log_2 M$ over binary transmission. However, comparison of probability of error performance indicates an S/N degradation of $3/(M^2 - 1)$ for M-ary transmission in comparison to binary $(M = 2)$ transmission.

Partial response coding uses prescribed amounts of intersymbol interference to reduce the transmission bandwidth further. There are several classes of partial response codes that permit choice of spectral shaping and number of input and output levels. The error performance is seen to be superior to that for binary and M-ary transmission.

Eye patterns are displays of random data on an oscilloscope. All allowed data transitions blend together to form a composite picture of the signal. At the receiver, the eye pattern viewed prior to data detection shows the effects of channel perturbations and is therefore an indication of system health.

Equalization of a signal distorted by a transmission channel is done by means of adjustable filters. The usual choice for equalizer design is the transversal filter, which consists of a tapped delay line. The preset equalizer uses a training sequence to initialize the tap settings. For time-varying channels, the equalizer must also be adaptive. This adaptation of the tap settings can be accomplished by using data decisions during periods of data transmission.

Data scrambling is required when performance of the communication system is dependent on the randomness of the data source. Pseudorandom sequences can be used for data scrambling by employing simple digital logic to add the data to the pseudorandom sequence. An alternative technique is the self-synchronizing scrambler, which performs a logic addition on delayed values of the data input sequence. Either technique can be used to improve performance in timing and equalization and minimize interference.

REFERENCES

1. M. R. Aaron, "PCM Transmission in the Exchange Plant," *Bell System Technical Journal* 41(January 1962):99–141.
2. V. I. Johannes, A. G. Kain, and T. Walzman, "Bipolar Pulse Transmission with Zero Extraction," *IEEE Trans. on Comm. Tech.*, vol. COM-17, no. 2, April 1969, pp. 303–310.
3. J. H. Davis, "T2: A 6.3 Mb/s Digital Repeatered Line," *1969 International Conference on Communications*, Boulder, pp. 34-9 to 34-16.
4. CCITT Yellow Book, vol. III.3, *Digital Networks—Transmission Systems and Multiplexing Equipment* (Geneva: ITU, 1981).
5. A. Croisier, "Introduction to Pseudoternary Transmission Codes," *IBM J. Res. Develop.*, July 1970, pp. 354–367.

6. W. R. Bennett, "Statistics of Regenerative Digital Transmission," *Bell System Technical Journal* 37(November 1958):1501–1542.
7. H. Nyquist, "Certain Topics in Telegraph Transmission Theory," *Trans. AIEE* 47(April 1928):617–644.
8. R. W. Lucky, J. Salz, and E. J. Weldon, *Principles of Data Communication* (*New York: McGraw-Hill*, 1968).
9. C. E. Shannon, "A Mathematical Theory of Communication," *Bell System Technical Journal* 27(July 1948):379–423.
10. C. E. Shannon, "Communications in the Presence of Noise," *Proc. IRE* 37(January 1949):10–21.
11. A. Lender, "The Duobinary Technique for High Speed Data Transmission," *IEEE Trans. Comm. and Elect.* 82(May 1963):214–218.
12. A. Lender, "Correlative Digital Communication Techniques," *IEEE Trans. Comm. Tech.*, vol. COM-12, December 1964, pp. 128–135.
13. E. R. Kretzmer, "Generalization of a Technique for Binary Data Communication," *IEEE Trans. Comm. Tech.*, vol. COM-14, February 1966, pp. 67–68.
14. A. Lender, "Correlative Level Coding for Binary Data Transmission," *IEEE Spectrum* 3(2)(February 1966):104–115.
15. D. R. Smith, "A Performance Monitoring Technique for Partial Response Transmission Systems," *1973 International Conference on Communications*, Seattle, pp. 40-14 to 40-19.
16. S. Pasupathy, "Correlative Coding: A Bandwidth-Efficient Signaling Scheme," *IEEE Comm. Mag.* 17(July 1977):4–11.
17. F. K. Becker, E. R. Kretzmer, and J. R. Sheehan, "A New Signal Format for Efficient Data Transmission," *Bell System Technical Journal* 45(5)(May–June 1966):755–758.
18. T. L. Swartz, "Performance Analysis of a Three-Level Modified Duobinary Digital FM Microwave Radio System," *1974 International Conference on Communications*, pp. 5D-1 to 5D-4.
19. C. W. Anderson and S. G. Barber, "Modulation Considerations for a 91 Mbit/s Digital Radio," *IEEE Trans. on Comm.*, vol. COM-26, no. 5, May 1978, pp. 523–528.
20. K. L. Seastrand and L. L. Sheets, "Digital Transmission over Analog Microwave Systems," *1972 International Conference on Communications*, pp. 29-1 to 29-5.
21. P. R. Hartman, "Digital Radio Technology: Present and Future," *IEEE Comm. Mag.* 19(4)(July 1981):10–15.
22. S. Qureshi, "Adaptive Equalization," *IEEE Comm. Mag.* 20(2)(March 1982): 9–16.
23. C. A. Belfiore and J. H. Parks, Jr., "Decision Feedback Equalization," *Proc. IEEE* 67(8)(August 1979):1143–1156.
24. P. Monsen, "Feedback Equalization for Fading Dispersive Channels," *IEEE Trans. Information Theory*, vol. IT-17, January 1971, pp. 56–64.
25. D. G. Leeper, "A Universal Digital Data Scrambler," *Bell System Technical Journal* 52(10)(December 1973):1851–1866.

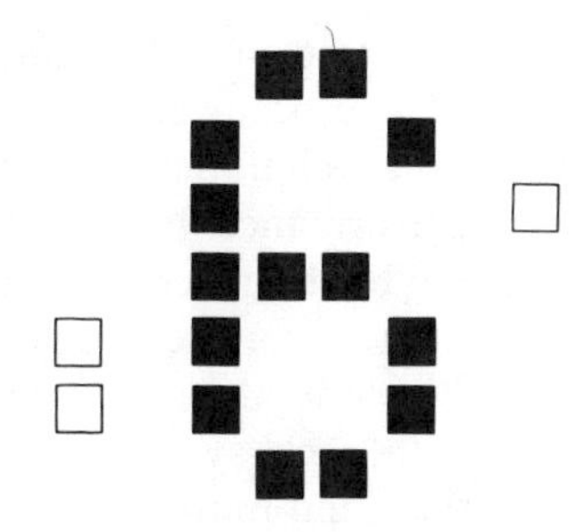

Digital Modulation Techniques

OBJECTIVES

- Considers the three basic forms of digital modulation: amplitude-shift keying, frequency-shift keying, and phase-shift keying
- Compares binary modulation systems in terms of power efficiency and spectrum efficiency
- Explains the principle of quadrature modulation used with M-ary PSK
- Describes two modulation techniques that offer certain advantages for band-limited, nonlinear channels: offset quadrature phase-shift keying and minimum-shift keying
- Explains how the technique of quadrature partial response increases the bandwidth efficiency of QAM signaling
- Compares digital modulation techniques in terms of error performance, complexity of implementation, and above all bandwidth efficiency

6.1 INTRODUCTION

In this chapter we consider modulation techniques used in digital transmission. Here we define **modulation** as the process of varying certain characteristics of a carrier in accordance with a message signal. The three basic forms of modulation are amplitude modulation (AM), frequency modulation (FM), and phase modulation (PM). Their digital representations are known as amplitude-shift keying (ASK), frequency-shift keying (FSK), and phase-shift keying (PSK). These digital modulation techniques can be characterized by their transmitted symbols, which have a discrete set of values M and occur at regularly spaced intervals T. The choice of digital modulation technique for a specific application depends in general on the error performance, bandwidth efficiency, and implementation complexity. Binary modulation schemes use two-level symbols and are therefore simple to implement, provide good error performance, but are bandwidth inefficient. M-ary modulation schemes transmit messages of length $m = \log_2 M$ bits with each symbol and are therefore appropriate for higher transmission rates and more efficient bandwidth utilization. Here we will consider binary and M-ary modulation techniques by comparing various performance and design characteristics such as waveform shape, transmitter and receiver block diagram, probability of error, and bandwidth requirements.

6.1.1 Error Performance

In comparing error performance for each modulation scheme, we will use the classic additive white gaussian noise (AWGN) channel. The AWGN channel is not always representative of the performance for real channels, particularly for radio transmission where fading and nonlinearities tend to dominate. Nevertheless, error performance for the AWGN channel is easy to derive and serves as a useful benchmark in performance comparisons.

We know that error performance is a function of the signal-to-noise (S/N) ratio. Out of the many definitions possible for S/N, the one most suitable for comparison of modulation techniques is the ratio of energy per bit (E_b) to noise density (N_0). To arrive at this definition and to obtain an appreciation for its usefulness, let us relate definitions for S/N, given earlier in Chapter 5, to E_b/N_0. Earlier we defined the average power S for a signal $s(t)$ of duration T as

$$S = \frac{1}{T}\int_{-T/2}^{T/2} s^2(t)\,dt \tag{5.5}$$

This formula can be rewritten as

$$S = \frac{1}{T}\int_{-\infty}^{\infty} s^2(t)\,dt = \frac{E_s}{T} \tag{6.1}$$

where the integral of (6.1) is defined as the energy per signal or symbol, E_s. For a transmission rate R, given by (5.42), the average energy available per bit is therefore

$$E_b = \frac{S}{R} = \frac{E_s}{TR} = \frac{E_s}{\log_2 M} \tag{6.2}$$

The noise source is assumed to be gaussian-distributed per (5.26) and to have a flat spectrum as in Figure 6.1. In many instances, the power spectral density of gaussian noise, $S_N(f)$, is indeed flat over the frequency range of the desired signal. This characteristic leads to the concept of **white noise**,

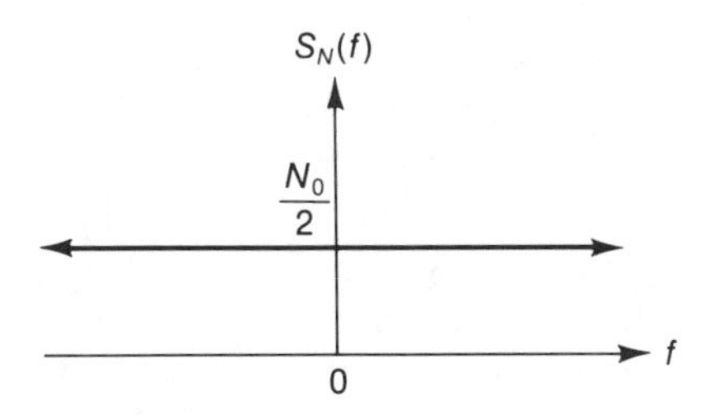

(a) Two-sided power spectral density

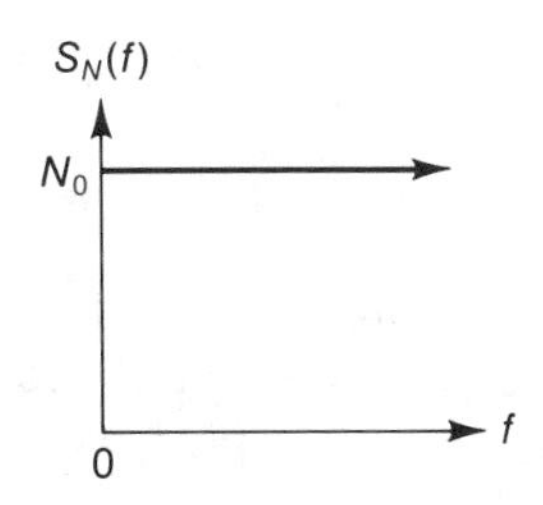

(b) One-sided power spectral density

Figure 6.1 White Noise Spectrum

which has a (two-sided) power spectral density $N_0/2$, where N_0 has dimensions of watts per hertz. For positive frequencies only, the (one-sided) power spectral density is N_0 watts/Hz. The total noise power (N) in watts is given as the product of the bandwidth (B) and power spectral density (N_0), so that $N = N_0B$.

With these relationships, we can now express the signal-to-noise ratio as a function of E_b/N_0:

$$\frac{S}{N} = \frac{E_s/T}{N_0B} \tag{6.3a}$$

Since $B = 1/2T$ is the minimum Nyquist bandwidth of the signal, we have, using (6.2),

$$\begin{aligned}\frac{S}{N} &= \frac{(E_b\log_2 M)/T}{N_0/2T} \\ &= 2\log_2 M\left(\frac{E_b}{N_0}\right)\end{aligned} \tag{6.3b}$$

Caution is advised in converting S/N measures to E_b/N_0, however, since in most practical systems the receiver noise bandwidth is larger than the Nyquist bandwidth. A more practical comparison is provided by

$$\frac{S}{N} = \frac{E_b}{N_0}\frac{R}{B_N} \tag{6.4}$$

where B_N is the noise bandwidth of the receiver.

6.1.2 Bandwidth

Generally we will make use of the power spectral density in describing bandwidth characteristics. For each digital modulation technique, there exists a power spectral density from which the bandwidth can be stated according to any given definition of bandwidth. The signal spectrum depends not only on the modulation technique but also on the baseband signal. The usual practice for specifying power spectral density is based on assumptions of random data and long averaging times. Scramblers are usually employed to guarantee a certain degree of data randomness, which tends to produce a smooth spectrum. Baseband shaping also plays a role in the modulation spectrum. For many common modulation techniques, including ASK and PSK, the spectrum is identical to that of the baseband signal translated upward in frequency.

No universal definition of bandwidth exists for modulated signals. The classic channel used by Nyquist and Shannon assumes no power outside a well-defined band. Modulated signals are not band-limited, however, thus introducing the necessity for a more practical definition of bandwidth. The definitions discussed here are those commonly used by regulatory agencies, manufacturers, and telephone and telegraph companies:

- *Null-to-null bandwidth*: The lobed nature of digital modulation spectra suggests the definition of bandwidth as null-to-null, which is equal to the width of the main lobe. This definition lacks generality, however, since not all modulation schemes have well-defined nulls. Further, side lobes may contain significant amounts of power.
- *Half-power bandwidth*: This is simply the band between frequencies at which the power spectral density has dropped to half power below the peak value. Hence the half-power bandwidth is also referred to as the 3-dB bandwidth.
- *Percentage power bandwidth*: The fraction of power located within a bandwidth B is defined as

$$P_B = \frac{\int_{-B}^{B} S(f)\,df}{\int_{-\infty}^{\infty} S(f)\,df} \tag{6.5}$$

  where $S(f)$ is the signal power spectral density. This definition requires that a certain percentage, say 99 percent, of the power be inside the allocated band B. Hence a certain percentage, say 1 percent, of the power is allowed outside the band. The 99 percent power bandwidth is somewhat larger than either the null-to-null or 3-dB bandwidth for most modulation techniques. This definition is used by many European Post, Telegraph, and Telephone (PTT) administrations.
- *Spectrum mask*: Another definition of bandwidth is to state that the power spectral density outside a spectrum mask must be attenuated to certain specified levels. The U.S. Federal Communications Commission (FCC) has defined bandwidth allocations for digital microwave radio using such a spectrum mask (see the accompanying box). For most modulation techniques, the FCC mask of Figure 6.2 results in less out-of-band power than the other bandwidth definitions considered here.

FCC Docket 19311 Spectrum Occupancy Constraints

FCC Docket 19311, adopted in 1974 by the U.S. Federal Communications Commission, established policies and procedures for the use of digital modulation techniques in microwave radio. To specify bandwidth, allowed out-of-band emissions are stated by use of a spectrum mask. For operating frequencies below 15 GHz, Docket 19311 specifies a spectrum mask as determined by the power measured in any 4-kHz band, the center frequency (f) of which is removed from the assigned carrier frequency (f_c) by 50 percent or more of the authorized transmission bandwidth. The following equation is used to define the mask:

$$A = 35 + 0.8(P - 50) + 10\log_{10}(B) \tag{6.6}$$

where A = attenuation (dB) below mean transmitted spectrum power output level

P = percentage of authorized bandwidth of which center frequency of 4-kHz power measurement bandwidth is removed from carrier frequency

B = authorized transmitted bandwidth (MHz)

A minimum attenuation (A) of 50 dB is required, and attenuation (A) of greater than 80 dB is not required. A diagram of the mask for 20, 30, and 40-MHz bandwidth allocation is shown in Figure 6.2.

6.1.3 Receiver Types

The receiver performs demodulation to recover the original digital signal. This process includes detection of the signal in the presence of channel degradation and a decision circuit to transform the detected signal back to a digital (for example, binary) signal. There are two common methods of demodulation or detection. **Synchronous** or **coherent detection** requires a reference waveform to be generated at the receiver that is matched in frequency and phase to the transmitted signal. When a phase reference cannot be maintained or phase control is uneconomical, **noncoherent detection** is used. For ASK and FSK modulation, where the binary signals are distinguished by a varying amplitude or frequency characteristic, noncoherent detection is accomplished by **envelope detection**. As the name indicates, detection is based on the presence or absence of the signal envelope. The simplest form of envelope detection is a half-wave rectifier (such as a diode) in series with an RC low-pass filter. Because PSK signals have a constant envelope, envelope detection cannot be used for PSK demodulation and some form of synchronous detection is therefore required.

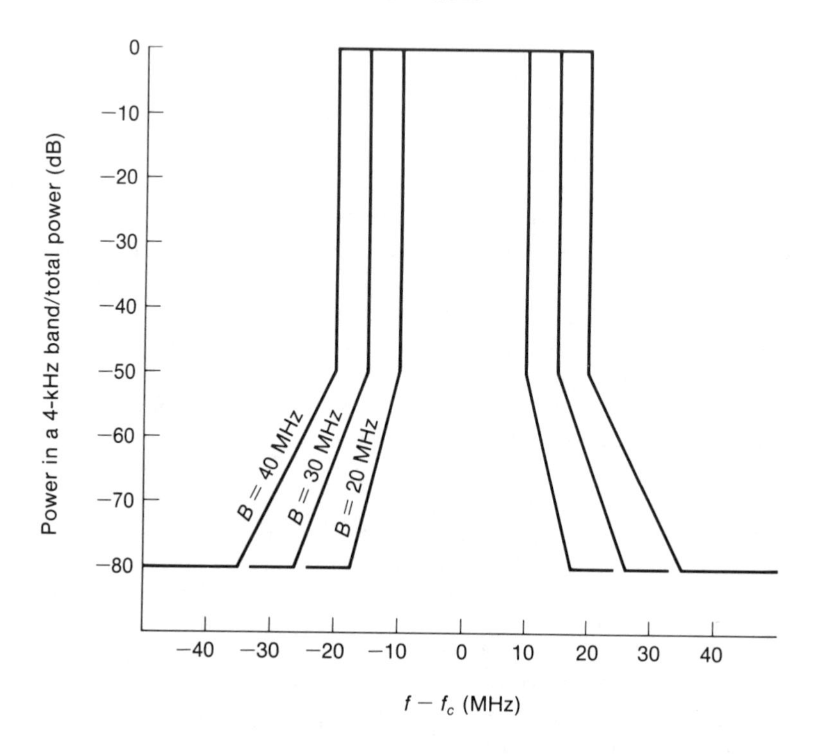

Figure 6.2 FCC Docket 19311 Spectrum Mask for 20, 30, and 40-MHz Bandwidth Allocations

For the AWGN channel the optimum detector is the matched filter, which maximizes the signal-to-noise ratio at its output [1]. For binary signals $s_1(t)$ and $s_0(t)$, the optimum detector consists of a pair of matched filters and a decision circuit. By definition, the impulse response of the matched filter is the mirror image of the transmitted signal $s(t)$ delayed by a signaling interval T. For the binary case, the matched filter impulse responses are

$$\begin{aligned} h_1(t) &= s_1(T-t) \\ h_0(t) &= s_0(T-t) \end{aligned} \tag{6.7}$$

The output signal $y(t)$ of the matched filter is simply the convolution of the received signal $r(t)$ with the filter response. In the absence of noise, the received signal is $s_i(t)$, $i = 0$ or 1, and

$$y_i(t) = \int_{-\infty}^{\infty} s_i(\alpha) h_i(t-\alpha)\, d\alpha \tag{6.8}$$

At $t = T$, the peak output of the matched filter is

$$y_i(T) = \int_{-\infty}^{\infty} s_i^2(\alpha)\, d\alpha = E_i \tag{6.9}$$

where E_i is by definition the energy of the signal $s_i(t)$. Each filter produces a maximum output only in the presence of its matched input. Therefore the filter outputs can be subtracted to produce $\pm E_i$ at the input to the appropriate decision threshold circuit.

Later in this chapter we will see that matched filter (coherent) detection yields superior error performance to noncoherent techniques. This performance advantage is theoretically quite small, however, particularly at high S/N. Hence the simplicity of the noncoherent receiver usually makes it a more popular choice over coherent methods, except for applications such as satellite transmission where the higher S/N required for noncoherent detection may be expensive to attain.

6.2 BINARY AMPLITUDE-SHIFT KEYING (ASK)

In **amplitude-shift keying** (ASK) the amplitude of the carrier is varied in accordance with the binary source. In its simplest form, the carrier is turned on and off every T seconds to represent 1's and 0's, and this form of ASK is known as on-off keying (OOK), as shown in Figure 6.3. The most general

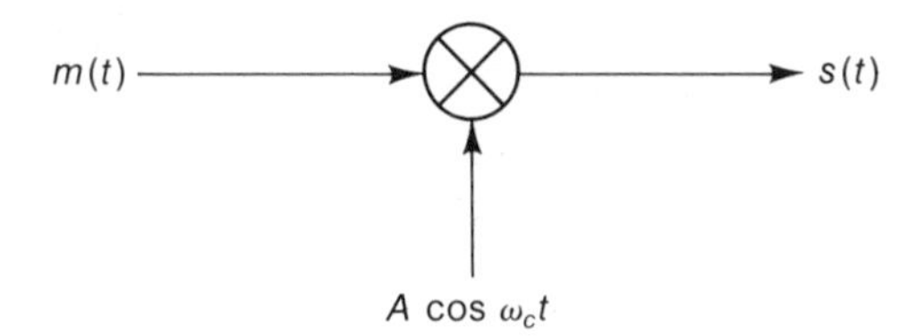

(a) Amplitude modulation

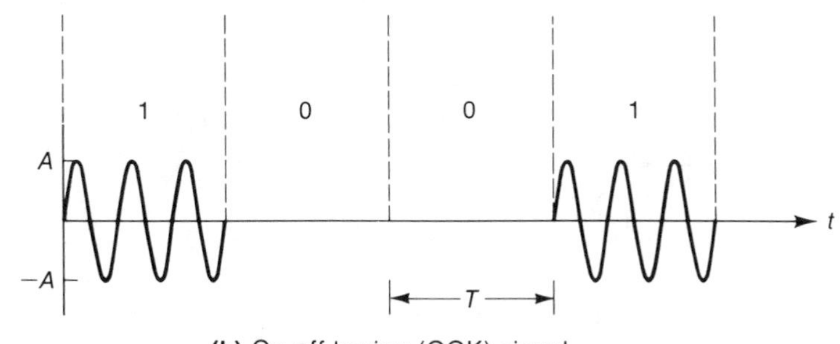

(b) On-off keying (OOK) signal

Figure 6.3 Amplitude-Shift Keying (ASK)

form of amplitude-shift keying is double sideband (DSB), represented by

$$s(t) = \frac{A}{2}[1 + m(t)]\cos \omega_c t \tag{6.10}$$

where $m(t)$ is the modulating signal (-1 and 1 for OOK) and ω_c is the carrier frequency. Since the carrier conveys no information, power efficiency is improved by suppressing the carrier and transmitting only the sidebands. The general form of the double sideband–suppressed carrier (DSB–SC) signal is

$$s(t) = Am(t)\cos \omega_c t \tag{6.11}$$

Double sideband signals contain an upper and lower sideband that are symmetrically distributed about the carrier frequency ω_c (Figure 6.4). For applications in which spectral efficiency is important, the required bandwidth may be halved by use of single sideband (SSB) modulation. The unwanted sideband is removed by a bandpass filter. The sharp cutoff required for the bandpass filter has led to vestigial sideband (VSB) in which a portion of the unwanted sideband is transmitted along with the complete other sideband. Use of vestigial sideband allows a smooth rolloff filter at the expense of only slightly more bandwidth than single sideband.

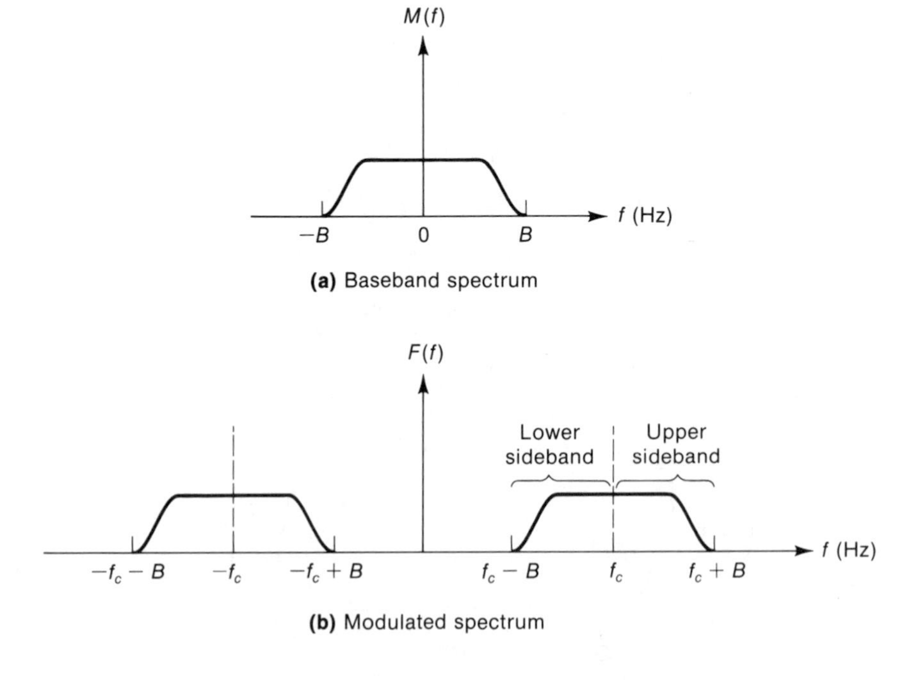

Figure 6.4 Example of ASK Spectrum for DSB-SC with Raised-Cosine Shaping

The spectrum of ASK signals is found by application of the frequency-shifting property of the Fourier transform to Equations (6.10) or (6.11). For the case of DSB–SC, the effect of multiplication by $\cos \omega_c t$ is to shift the spectrum of the original binary source up to and centered about the carrier frequency ω_c. The shaping and bandwidth of the modulated signal are determined by the baseband signal $m(t)$. If raised-cosine shaping with bandwidth B hertz is used, for example, the baseband spectrum $M(f)$ and modulated spectrum $F(f)$ for DSB–SC are as shown in Figure 6.4. Note that the bandwidth has been doubled by the modulation process to a total transmission bandwidth of $2B$ hertz.

Demodulation schemes for ASK fall into two categories, coherent and noncoherent, as illustrated in Figure 6.5. For coherent detection, matched filter detection is optimum; each filter has an impulse response that is matched to the signal being correlated. For binary transmission, assuming equally likely 1's and 0's and assuming equal energy for both signals, the matched filter detector is as shown in Figure 6.5a. For OOK, only a single matched filter is required in order to detect the presence or absence of signal energy. For OOK, with $s_0 = 0$ and $s_1 = A \cos \omega_c t$, the output of the single matched filter at the sampling time T is

$$y_1(t) = \int_0^T s_1^2(t)\, dt = \frac{A^2 T}{2} \tag{6.12}$$

or

$$y_0(t) = 0 \tag{6.13}$$

The integral in (6.12) is recognized as the energy E_s of the signal $s_1(t)$. In the presence of additive noise $N(t)$, the matched filter detector makes a

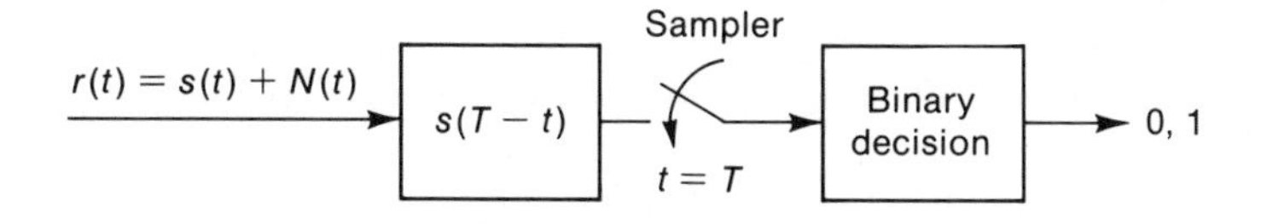

(a) Coherent detection by matched filter

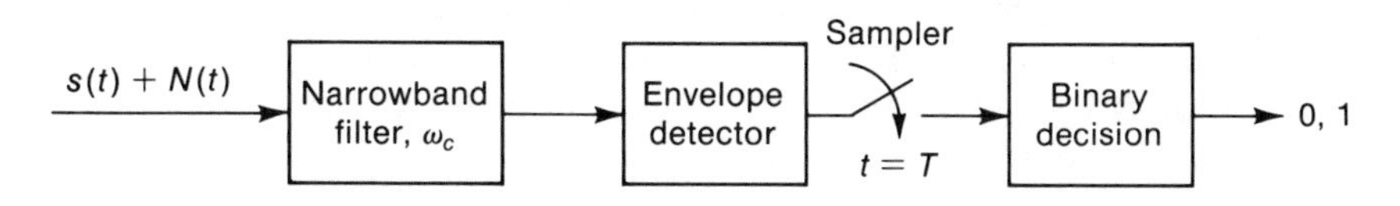

(b) Noncoherent detection by envelope detector

Figure 6.5 Receiver Structures for OOK

decision at $t = T$ based on the two signals $y_1(T) = E_s + N(T)$ and $y_0(T) = N(T)$. For the source that produces equally likely 1's and 0's and noise that has a gaussian distribution, the optimum receiver threshold is $E_s/2$. To determine the probability of error for the matched filter, we note that this case is identical to that of unipolar NRZ given by Equation (5.31), which is repeated here for convenience:

$$P(e) = \operatorname{erfc}\sqrt{\frac{S}{2N}} \tag{5.31}$$

Here the average signal power $S = (\frac{1}{2})(A^2/2) = E_s/2T$. For white noise with (one-sided) power spectral density N_0, the noise power $N = N_0 B$, where the Nyquist bandwidth $B = 1/2T$. For comparison with other modulation techniques, it is convenient to express $P(e)$ as a function of E_s and N_0. From (5.31) we have

$$P(e) = \operatorname{erfc}\sqrt{\frac{E_s}{2N_0}} \tag{6.14a}$$

The result given in (6.14*a*) pertains to the case of on-off keying, where all of the energy E_s is used to convey one binary state. This result holds for the more general case of ASK if E_s is interpreted as the total energy used for the two binary signals. Note that $E_s/2$ is then the average energy per binary signal, E_b, so that

$$P(e) = \operatorname{erfc}\sqrt{\frac{E_b}{N_0}} \tag{6.14b}$$

Comparing (6.14*a*) and (6.14*b*) we note a 3-dB difference in performance when signal energy is referenced to average value (E_b) rather than peak value (E_s).

The noncoherent demodulator for OOK consists of a narrowband filter centered at ω_c, an envelope detector, and a decision device (Figure 6.5*b*). For equiprobable 1's and 0's at the source, the probability of error is [2]

$$P(e) \approx \frac{1}{2}\exp\left(\frac{-E_s}{4N_0}\right) = \frac{1}{2}\exp\left(-\frac{E_b}{2N_0}\right) \tag{6.15}$$

A comparison of the $P(e)$ expressions in (6.15) and (6.14a) for OOK reveals that for high signal-to-noise ratios the noncoherent detector requires less than a 1-dB increase in S/N to obtain the same $P(e)$ as a coherent detector. This small performance penalty usually allows the designer to select noncoherent detection because of its simpler implementation.

6.3 BINARY FREQUENCY-SHIFT KEYING (FSK)

In frequency modulation, the frequency of the carrier varies in accordance with the source signal. For binary transmission the carrier assumes one frequency for a 1 and another frequency for a 0 as represented in Figure 6.6.

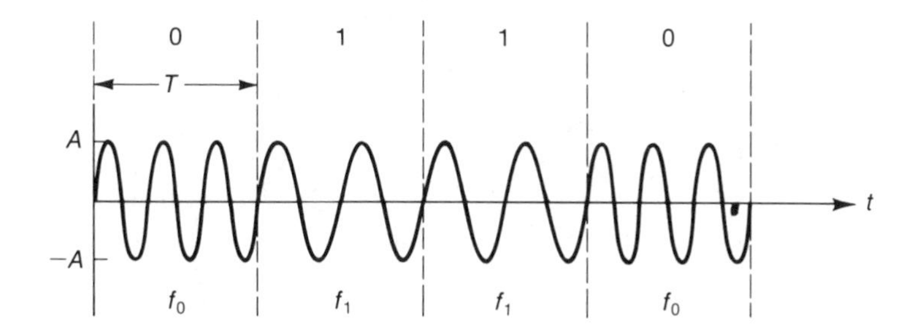

Figure 6.6 Frequency-Shift Keying (FSK) Waveform

This type of on-off modulation is called **frequency-shift keying**. The modulated signal for binary frequency-shift keying may be written

$$
\begin{aligned}
s_1(t) &= A\cos\omega_1 t \qquad \text{for binary 1} \\
s_0(t) &= A\cos\omega_0 t \qquad \text{for binary 0}
\end{aligned}
\tag{6.16}
$$

An alternative representation of the FSK waveform is obtained by letting $f_1 = f_c - \Delta f$ and $f_0 = f_c + \Delta f$, so that

$$
\begin{aligned}
s_1(t) &= A\cos(\omega_c - \Delta\omega)t \\
s_0(t) &= A\cos(\omega_c + \Delta\omega)t
\end{aligned}
\tag{6.17}
$$

where Δf is called the **frequency deviation.**

The frequency spectrum of the FSK signal is in general difficult to obtain [3]. However, transmission bandwidth (B_T) requirements can be estimated for two cases of special interest:

$$
B_T \approx \begin{cases} 2\,\Delta f & \Delta f \gg B \quad (6.18\text{a}) \\ 2B & \Delta f \ll B \quad (6.18\text{b}) \end{cases}
$$

where B is the bandwidth of the baseband modulation signal (such as NRZ). The ratio of the frequency deviation Δf to the baseband bandwidth B is called the **modulation index** m, defined as

$$
m = \frac{\Delta f}{B} \tag{6.19}
$$

A general relationship defining FM transmission bandwidth was established by Carson [4] as

$$
B_T = 2B(1 + m) \tag{6.20}
$$

Carson's rule approaches the limits given in (6.18) for both $m \gg 1$ and $m \ll 1$. These two cases are referred to as **wideband FSK** where $m \gg 1$ and **narrowband FSK** where $m \ll 1$. For $m > 1$, FSK requires more transmission bandwidth than ASK.

Coherent detection of FSK is accomplished by comparing the outputs of two matched filters, as shown in Figure 6.7*a*. The output signals of the

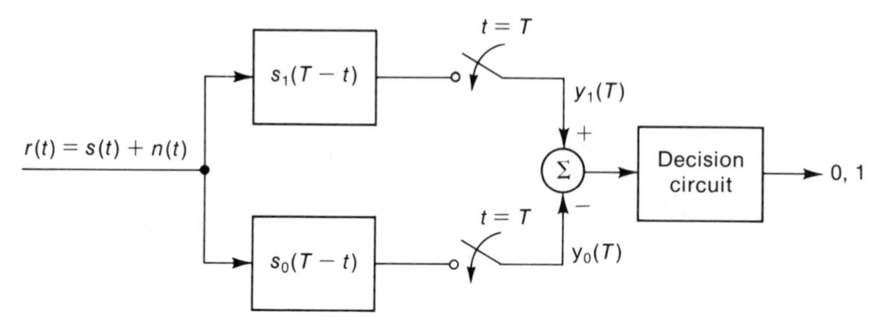

(a) Coherent detection

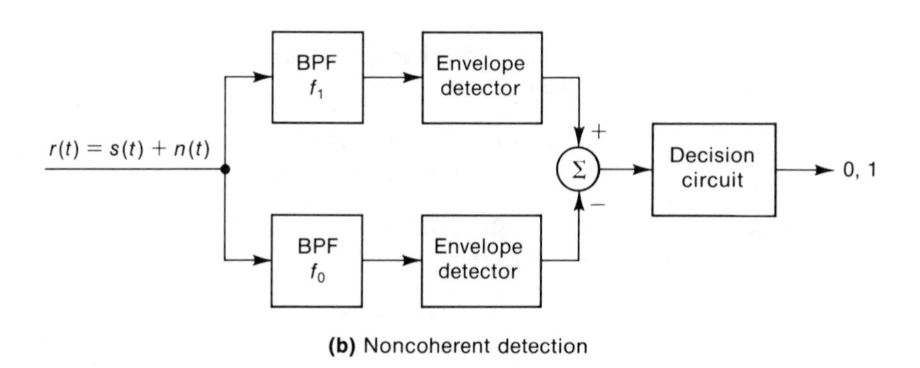

(b) Noncoherent detection

Figure 6.7 Receiver Structures for FSK

matched filters are

$$y_1(T) = y_0(T) = \int_0^T A^2 \cos^2 \omega t \, dt = \frac{A^2 T}{2} \tag{6.21}$$

The integral in (6.21) is of course the energy E_s of the signal. The summing device in Figure 6.7*a* thus produces a signal output of $+E_s$ if a 1 has been transmitted and $-E_s$ if a 0 has been transmitted (assuming no noise and orthogonal FSK signals).* The probability of error for coherent FSK would appear to be analogous to that of polar NRZ except for the fact that the noise samples are subtracted by the summing device. For orthogonal FSK signals, these (white) noise samples are independent and therefore add on a power basis [5]. The total noise power is therefore doubled to $N = 2N_0B$. The probability of error can now be written, following the case for polar

*For binary FSK, $s_1(t)$ and $s_0(t)$ are orthogonal over $(0, T)$ if the signals have zero correlation; that is,

$$\int_0^T s_1(t) s_0(t) \, dt = 0$$

signaling but with twice the noise power, so that

$$P(e) = \operatorname{erfc}\sqrt{\frac{E_s}{N_0}} = \operatorname{erfc}\sqrt{\frac{E_b}{N_0}} \tag{6.22}$$

Thus we can conclude that FSK has a 3-dB advantage over ASK on a peak power basis but is equivalent to ASK on an average power basis. The FSK waveform given in (6.16) and the receiver shown in Figure 6.7 imply other advantages of FSK over ASK. Frequency-shift keying has a constant envelope property that has merit when transmitting through a nonlinear device or channel. Further, the FSK receiver threshold is fixed at zero, independent of the carrier amplitude. For ASK, the threshold must be continually adjusted if the received signal varies in time, as happens with fading in radio transmission.

Conventional noncoherent detection for binary FSK employs a pair of bandpass filters and envelope detectors as shown in Figure 6.7*b*. The probability of error for additive white gaussian noise is [2]

$$P(e) = \frac{1}{2}\exp\left(-\frac{E_s}{2N_0}\right) = \frac{1}{2}\exp\left(-\frac{E_b}{2N_0}\right) \tag{6.23}$$

A comparison of (6.23) and (6.22) for low error rates reveals that a choice of noncoherent detection results in an S/N penalty of less than 1 dB. Therefore, because of the complexity of coherent detection, noncoherent detection is more commonly used in practice.

6.4 BINARY PHASE-SHIFT KEYING (BPSK)

In phase modulation the phase of the carrier is varied according to the source signal. For binary transmission the carrier phase is shifted by 180° to represent 1's and 0's and is called **binary phase-shift keying** (BPSK). The signal waveform for BPSK is given as

$$\begin{aligned} s(t) &= \pm A\cos\omega_c t \\ &= A\cos(\omega_c t + \phi_j) \qquad \phi_j = 0 \text{ or } \pi \end{aligned} \tag{6.24}$$

A BPSK waveform example is shown in Figure 6.8, in which the data clock and carrier frequencies are exact multiples of one another.

The form of (6.24) suggests that the BPSK signal corresponds to a polar NRZ signal translated upward in frequency. The power spectral density of PSK has a double sideband characteristic identical to that of OOK. Thus the spectrum is centered at ω_c with a bandwidth twice that of the baseband signal.

It follows that coherent detection of a PSK signal is similar to the case for ASK. Detector outputs are $\pm E_s + N(T)$, and the threshold is set at zero (assuming equiprobable 1's and 0's). The noise power $N = N_0 B$. For the case of ASK, we noted earlier that the detector outputs are $E_s + N(T)$ or $N(T)$, the noise power is $N = N_0 B$, and the optimum threshold is $E_s/2$. By

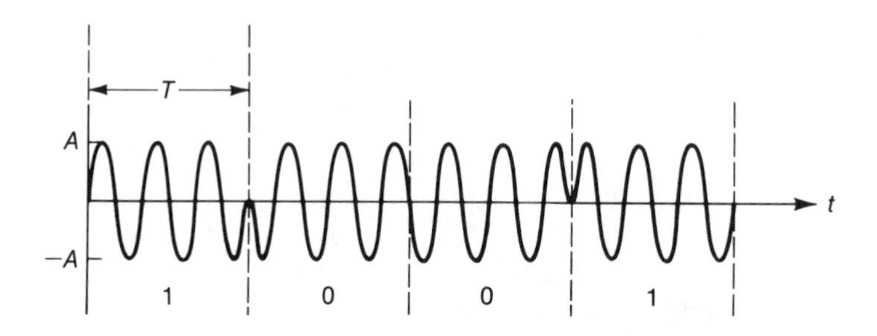

Figure 6.8 Phase-Shift Keying (PSK) Waveform

analogy with the ASK case, we can directly write the probability of error for coherent PSK as

$$P(e) = \operatorname{erfc}\sqrt{\frac{2E_s}{N_0}} = \operatorname{erfc}\sqrt{\frac{2E_b}{N_0}} \tag{6.25}$$

Comparison of (6.25) with (6.22) and (6.14) reveals that on a peak power basis PSK outperforms FSK by 3 dB and ASK by 6 dB, while on an average power basis PSK has a 3-dB advantage over both FSK and ASK.

For coherent detection, the receiver must have a phase reference available. For BPSK with the two phases separated by 180°, however, the carrier component is zero. Thus the phase reference must be generated by some other means, such as the frequency doubling phase-locked loop or the Costas loop. The frequency doubling loop shown in Figure 6.9*a* uses a square-law device to remove the modulation and create a line component in the spectrum at double the carrier frequency. A phase-locked loop (PLL) is used as a narrowband loop centered at twice the carrier frequency. Frequency division by 2 of the PLL output recovers the carrier reference. This frequency division causes a 180° phase ambiguity that can be accommodated by proper coding of the baseband signal, as in NRZ(I), which is transparent to the polarity of the carrier. The Costas loop multiplies the incoming signal by $\cos(\omega_c t + \phi)$ in one channel and $\sin(\omega_c t + \phi)$ in the other channel to generate a coherent phase reference from the suppressed carrier signal. As indicated in Figure 6.9*b*, the two phase detectors (multipliers) for sin and cos channels are used to control the phase and frequency of a voltage-controlled oscillator (VCO). The Costas loop is usually preferred over the simpler squaring loop for PSK coherent detection (Figure 6.10*a*) because of greater tolerance to shifts in the carrier frequency and the capability of wider bandwidth operation [6].

The added complexity of establishing a phase reference for coherent detection can be avoided by use of a technique known as **differential PSK** (DPSK). With DPSK, the data are encoded by means of changes in phase

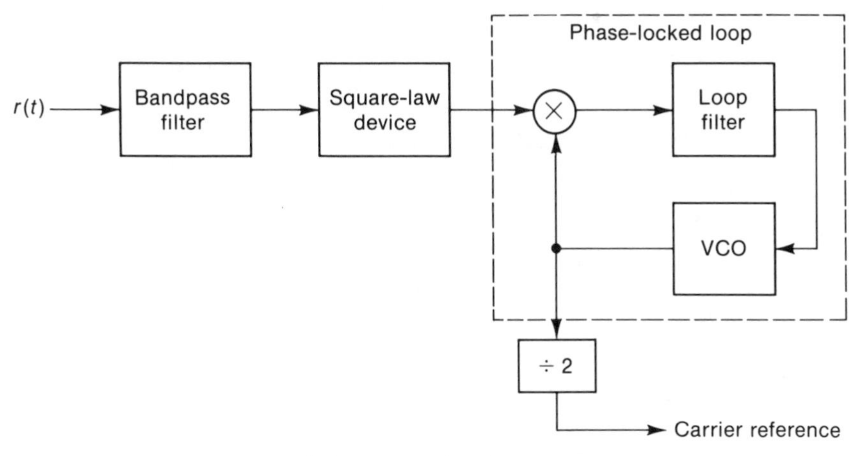

(a) Frequency doubling loop

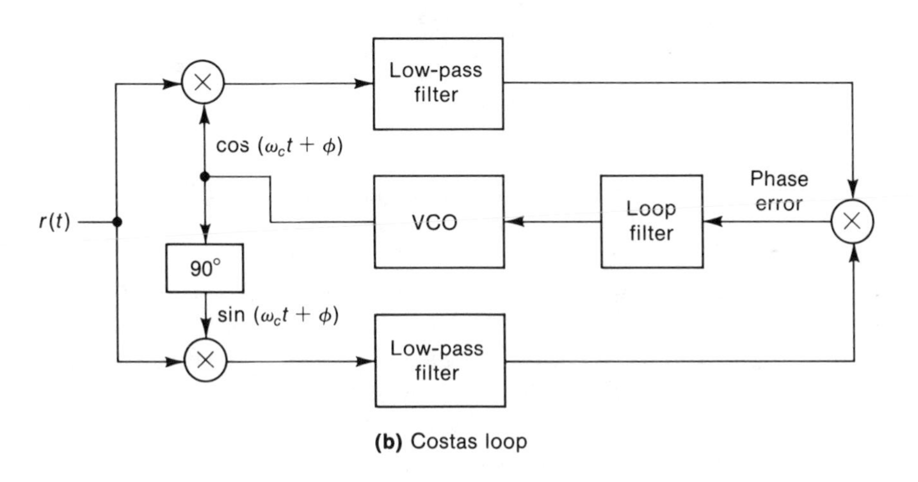

(b) Costas loop

Figure 6.9 Carrier Recovery Techniques for PSK

rather than by absolute value of phase in the carrier. For binary PSK, a 1 can be encoded as no change in the phase and a 0 encoded as a 180° change in the phase. (Note that this encoding scheme is identical to NRZ(S) shown in Figure 5.1.) The detector, shown in Figure 6.10*b*, uses the phase of the previous symbol as a reference phase to permit decoding of the current symbol. If the phases are the same, a 1 is decoded; if they differ, a 0 is decoded. The probability of error of DPSK is given by [2]

$$P(e) = \frac{1}{2}\exp\left(\frac{-E_s}{N_0}\right) = \frac{1}{2}\exp\left(\frac{-E_b}{N_0}\right) \tag{6.26}$$

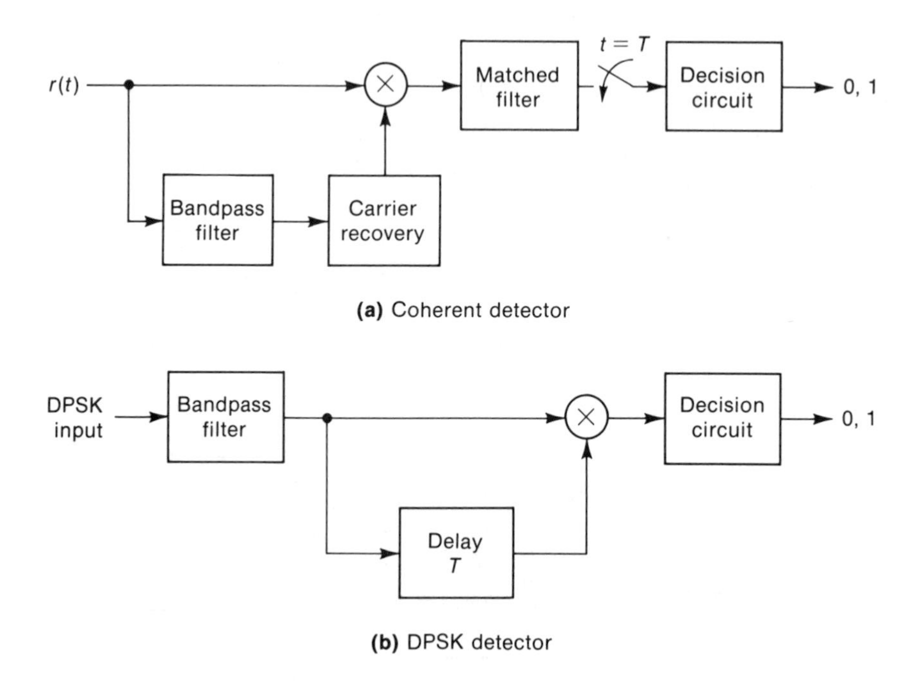

(a) Coherent detector

(b) DPSK detector

Figure 6.10 Receiver Types for PSK

Compared to coherent PSK, DPSK results in an S/N penalty of 1 dB or less for error rates of interest. However, differential encoding also causes errors to occur in pairs. Even so, DPSK is a popular alternative to coherent PSK, except for satellite applications where power limitations make coherent PSK preferable despite the added complexity of maintaining a phase reference.

6.5 COMPARISON OF BINARY MODULATION SYSTEMS

In selecting a modulation system, the two primary factors for comparison are the transmitted power and channel bandwidth that are required to achieve a specified (error rate) performance. To measure power efficiency, the parameter E_b/N_0 is most commonly used. For spectrum efficiency, the ratio of transmission rate to transmission bandwidth is used. Other factors that may influence the choice of modulation technique include effects of fading, interference, channel or equipment nonlinearities, and implementation complexity and cost.

The probability of error for ASK, FSK, and PSK binary modulation is plotted in Figure 6.11 for both coherent and noncoherent detection. As

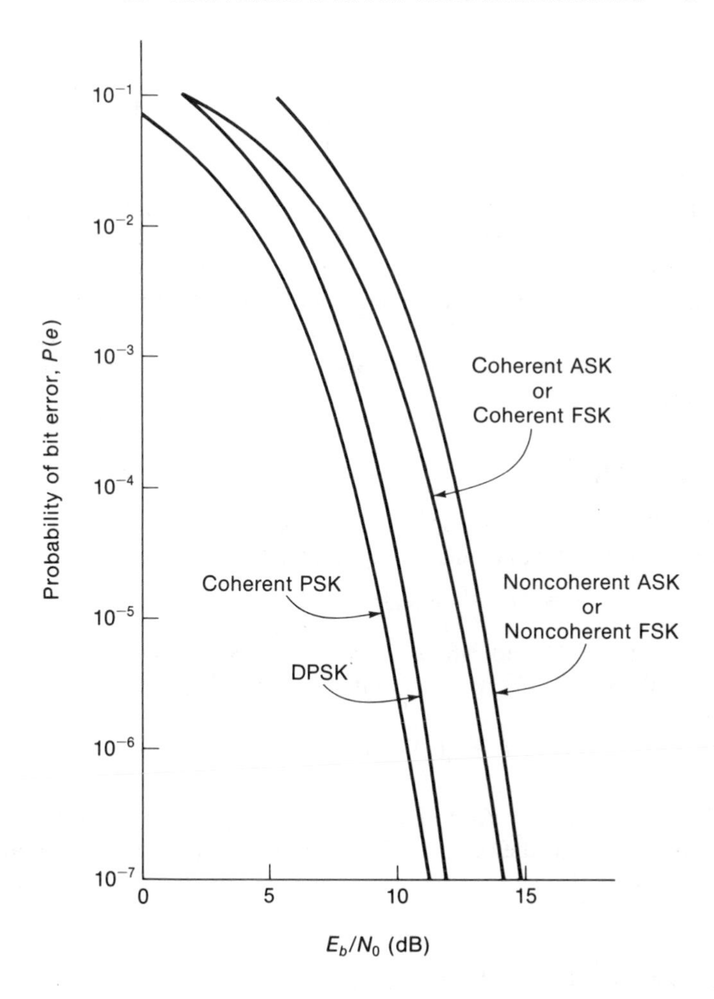

Figure 6.11 Error Performance for Binary Modulation Systems

noted earlier, PSK has a 3-dB performance advantage over both FSK and ASK, on an average power basis, for both coherent and noncoherent detection. Moreover, the difference in performance between coherent and noncoherent detection for a particular choice of modulation scheme is on the order of 1 dB for error rates of interest, that is, where $P(e) \leq 10^{-5}$.

None of the binary modulation schemes described so far is particularly efficient in bandwidth utilization. We noted earlier that ASK and PSK require the same bandwidth, whereas FSK usually (depending on choice of Δf) requires more bandwidth than ASK or PSK. The theoretical bandwidth efficiency of binary ASK and PSK is 1 bps/Hz. To obtain greater bandwidth efficiency, M-ary modulation schemes are required, but these schemes suffer an S/N penalty compared to the binary cases, as we will see later in the chapter.

In terms of implementation, the complexity and cost depend primarily on the choice of coherent versus noncoherent detection. Typically the added complexity of coherent detection is not justified by the slight improvement in error performance. Because satellite links tend to be limited in power, however, coherent detection is often used in lieu of noncoherent detection. Although ASK implementation is simple, it is not a popular choice because of its relatively poor error performance and susceptibility to fading and nonlinearities. Noncoherent FSK is commonly used for low data rates, while both coherent and differential PSK are preferred for higher data rate applications.

6.6 *M*-ARY FSK

In M-ary FSK, the M-ary symbols are represented by M-spaced frequencies selected from a set of equal energy waveforms

$$s_n(t) = A\cos\omega_n t \qquad 0 < t \le T \tag{6.27}$$

where $n = 1, 2, \ldots, M$ and where T is the symbol length. Here we will assume that the set of M-ary signals satisfies the orthogonality conditions

$$\int_0^T s_n(t)s_m(t)\,dt = \begin{cases} 0 & n \ne m \\ E_s & n = m \end{cases} \tag{6.28}$$

where E_s is the energy per symbol. Orthogonality ensures no overlap among the detector outputs at the receiver. The minimum frequency separation $\Delta\omega$ required to satisfy (6.28) is

$$\Delta\omega = \omega_m - \omega_n = \frac{\pi}{T} \tag{6.29}$$

Thus the minimum bandwidth of the signal set is

$$B \approx \frac{M}{2T} \text{ hertz} \tag{6.30}$$

Correspondingly the minimum transmission bandwidth is given by (6.18b), so that for M-ary FSK,

$$B_T \approx \frac{M}{T} \tag{6.31}$$

The optimum receiver for orthogonal M-ary FSK consists of a bank of M matched filters. At the sampling times, $t = kT$, the receiver makes decisions based on the largest filter output. Symbol s_j is selected when the jth matched filter output is the largest. For the case of the additive white gaussian noise channel, an exact probability of symbol error is given by [7]

$$P(e) = 1 - \frac{1}{\sqrt{2\pi}}\int_{-\infty}^{\infty} e^{-u^2/2}\left[1 - \operatorname{erfc}\left(u + \frac{2E_s}{N_0}\right)\right]^{M-1} du \tag{6.32}$$

A plot of this symbol error probability is given in Figure 6.12 as a function of E_b/N_0 for several values of M. (Recall that $E_s = E_b \log_2 M$.) From Figure 6.12 it is apparent that for a fixed error probability, the required E_b/N_0 may be reduced by increasing M. The required bandwidth increases as M, however, and transmitter and receiver complexity also increase with large M.

Because the integral of (6.32) cannot be evaluated exactly but rather requires numerical integration techniques, it is useful to seek an approximation of $P(e)$. An upper bound is given by [1]

$$P(e) \leq (M-1)\,\text{erfc}\left(\sqrt{\frac{E_b \log_2 M}{N_0}}\right) \tag{6.33}$$

This bound becomes increasingly tight for fixed M as the E_b/N_0 is increased. For values of $P(e) \leq 10^{-3}$, the upper bound becomes a good approximation to $P(e)$. For binary FSK ($M = 2$), the bound of (6.33) becomes an equality.

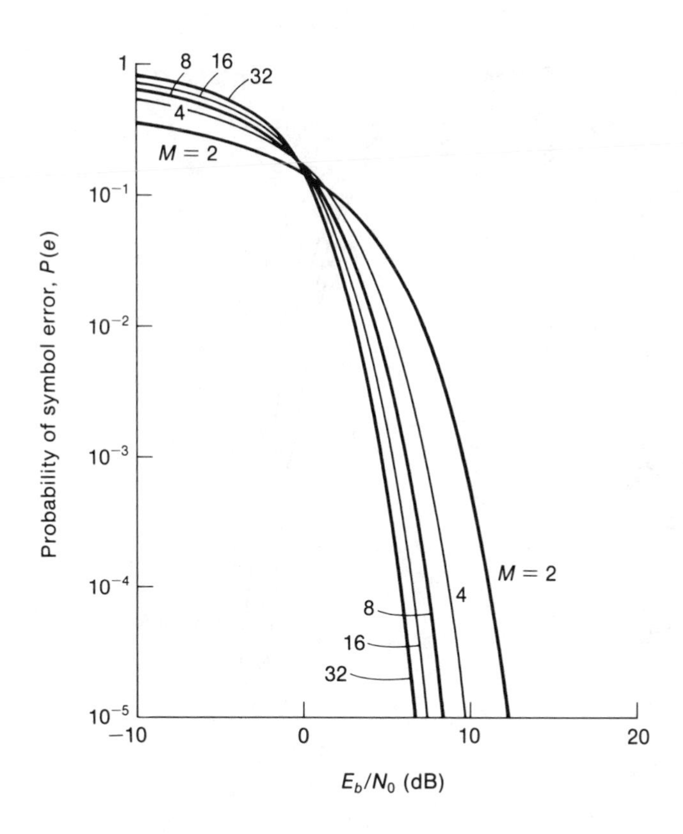

Figure 6.12 Error Performance for Coherent Detection of Orthogonal M-ary FSK

Coherent FSK detection requires exact phase references at the receiver, which can be difficult to maintain. A simpler implementation, although slightly inferior performance, results when noncoherent detection is used. An optimum noncoherent detection scheme can be realized with a bank of M filters, each centered on one of M frequencies. Each filter is followed by an envelope detector, and the decision is based on the largest envelope output. For the AWGN channel, the probability of symbol error for orthogonal signaling is given by [8]

$$P(e) = \sum_{k=1}^{M-1} \frac{(-1)^{k+1}}{k+1} \binom{M-1}{k} \exp\left[\frac{-kE_s}{(k+1)N_0}\right] \tag{6.34}$$

A plot of this symbol error probability is given in Figure 6.13 as a function of E_b/N_0 for several values of M. The set of curves in Figure 6.13 exhibit the same behavior as the case for coherent detection shown in Figure 6.12—that is, for a fixed symbol error probability, the required E_b/N_0

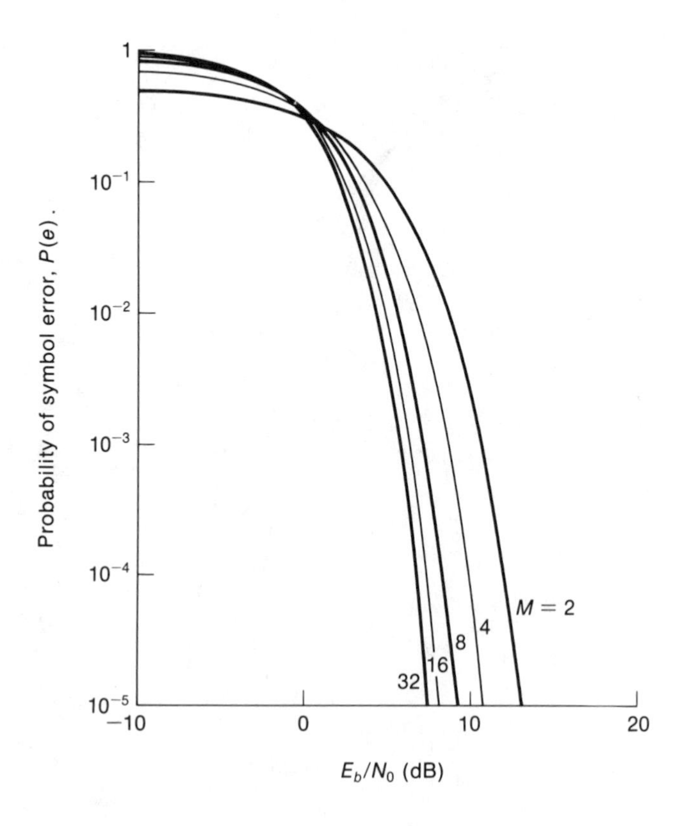

Figure 6.13 Error Performance for Noncoherent Detection of Orthogonal *M*-ary FSK

decreases with increasing M while the required bandwidth increases. A comparison of Figure 6.13 with Figure 6.12 indicates that for $P(e) \leq 10^{-4}$, the E_b/N_0 penalty for choice of noncoherent detection is less than 1 dB for any choice of M. Thus, for orthogonal codes, noncoherent detection is generally preferred because of its simpler implementation.

An upper bound on the probability of symbol error for noncoherent detection is provided by the leading term of (6.34):

$$P(e) \leq \frac{M-1}{2} \exp\left(\frac{-E_s}{2N_0}\right) \tag{6.35}$$

As the E_b/N_0 is increased, this bound becomes increasingly close to the true value of error probability. For binary FSK, the bound of (6.35) becomes an equality.

The error probability expressions presented to this point have been in terms of symbol error. This measure of error performance is appropriate where messages of length $m = \log_2 M$ bits are transmitted, such as alphanumeric characters or PCM code words. For binary transmission, however, the bit error probability is the measure of interest. For orthogonal signals, an error in detection is equally likely to be made in favor of any one of the $M - 1$ incorrect signals. For the symbol in error, the probability that exactly k of the m bits are in error is

$$P(k \text{ errors in } m \text{ bits}|\text{symbol error}) = \frac{\binom{m}{k}}{\sum_{k=1}^{m} \binom{m}{k}} \tag{6.36}$$

Hence the expected number of bits in error is

$$\frac{\sum_{k=1}^{m} k\binom{m}{k}}{\sum_{k=1}^{m} \binom{m}{k}} = \frac{m2^{m-1}}{2^m - 1} \tag{6.37}$$

The conditional probability that a given bit is in error given that the symbol is in error is then

$$P(e_b|e) = \frac{2^{m-1}}{2^m - 1} \tag{6.38}$$

In terms of the symbol error probability $P(e)$, the bit error probability, more commonly known as the bit error rate (BER), can be written

$$\text{BER} = \frac{2^{m-1}}{2^m - 1} P(e) \tag{6.39}$$

where $P(e)$ is given by (6.32) and (6.34) for coherent and noncoherent detection. Note that for large m, the BER approaches $P(e)/2$.

6.7 *M*-ARY PSK

An M-ary PSK signal may be represented by the set of signals

$$s_n(t) = A\cos(\omega_c t + \theta_n) \qquad 0 \le t \le T \tag{6.40}$$

where the M symbols are expressed as the set of uniformly spaced phase angles

$$\theta_n = \frac{2(n-1)\pi}{M} \qquad n = 1, 2, \ldots, M \tag{6.41}$$

The separation between adjacent phases of the carrier is $2\pi/M$. For BPSK, the separation is π; for 4-PSK, the separation is $\pi/2$; and for 8-PSK, the separation is $\pi/4$. A more convenient means of representing the M-ary PSK signals of (6.40) is by a phasor diagram, which displays the signal magnitude and phase angle. Examples of BPSK, 4-PSK, and 8-PSK phasors are shown in Figure 6.14. The signal points of the phasor diagram represent the **signal constellation**. As indicated in Figure 6.14, each signal has equal amplitude A, resulting in signal points that lie on a circle of radius A. Decision thresholds in the receiver are centered between the allowed phases θ_n, so that correct decisions are made if the received phase is within $\pm\pi/M$ of the transmitted phase.

Another convenient representation of M-ary PSK (and other bandwidth-efficient modulation techniques) is provided by *quadrature signal representation*. The signals of (6.40) may be expressed, by use of trigonometric expansion, as a linear combination of the carrier signals $\cos\omega_c t$ and $\sin\omega_c t$:

$$s_n(t) = A[p_n\cos\omega_c t + q_n\sin\omega_c t] \qquad 0 \le t \le T \tag{6.42}$$

where

$$\begin{aligned} p_n &= \cos\theta_n \\ q_n &= \sin\theta_n \end{aligned} \tag{6.43}$$

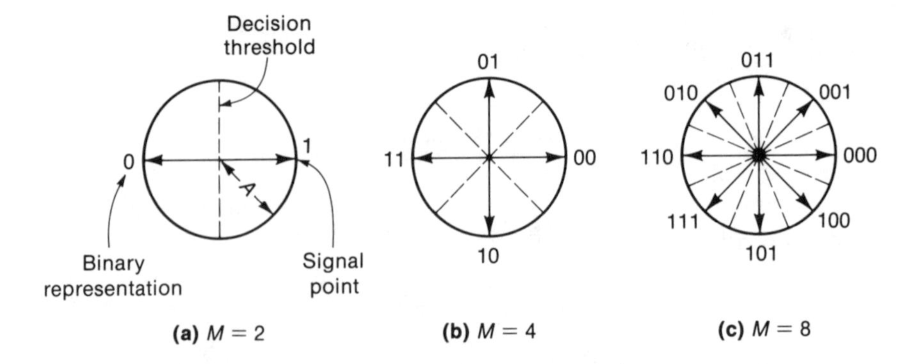

Figure 6.14 Phasor Diagrams of *M*-ary PSK Signals

Table 6.1 Quadrature Signal Coefficients for 4-PSK Modulation

Bit Values	*Quadrature Coefficients* p_n	q_n	*Phase Angle* θ_n
00	1	0	0
01	0	1	$\pi/2$
10	-1	0	π
11	0	-1	$-\pi/2$

For the case of BPSK as shown in Figure 6.14*a*, the signal phase angles of zero and π are represented in quadrature by the coefficients

$$(p_n, q_n) = (1,0), (-1,0) \qquad \text{for } 0, \pi$$

Similarly, the corresponding sets of (p_n, q_n) for the 4-PSK and 8-PSK constellations shown in Figure 6.14 are given in Tables 6.1 and 6.2. Since $\cos \omega_c t$ and $\sin \omega_c t$ are orthogonal in a phasor diagram, they are said to be in quadrature. The cosine coefficient p_n is represented on the horizontal axis and is called the *in-phase* or I signal. The sine coefficient q_n is represented on the vertical axis and is called the *quadrature* or Q signal.

Quadrature representation leads to one common method of generating M-ary PSK signals as a linear combination of quadrature signals. A block diagram of a generalized PSK modulator based on quadrature signal structure is shown in Figure 6.15. Binary data at the input to the modulator

Table 6.2 Quadrature Signal Coefficients for 8-PSK Modulation

Bit Values	*Quadrature Coefficients* p_n	q_n	*Phase Angle* θ_n
000	1	0	0
001	$1/\sqrt{2}$	$1/\sqrt{2}$	$\pi/4$
010	0	1	$\pi/2$
011	$-1/\sqrt{2}$	$1/\sqrt{2}$	$3\pi/4$
100	-1	0	π
101	$-1/\sqrt{2}$	$-1/\sqrt{2}$	$-3\pi/4$
110	0	-1	$-\pi/2$
111	$1/\sqrt{2}$	$-1/\sqrt{2}$	$-\pi/4$

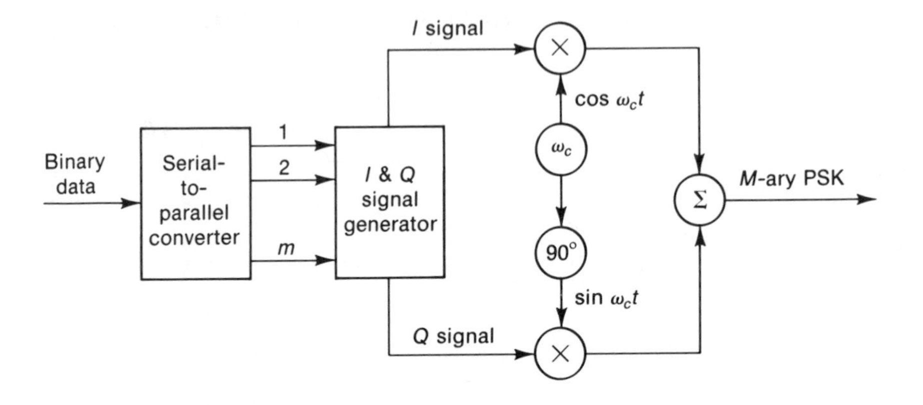

Figure 6.15 Generalized *M*-ary PSK Modulator

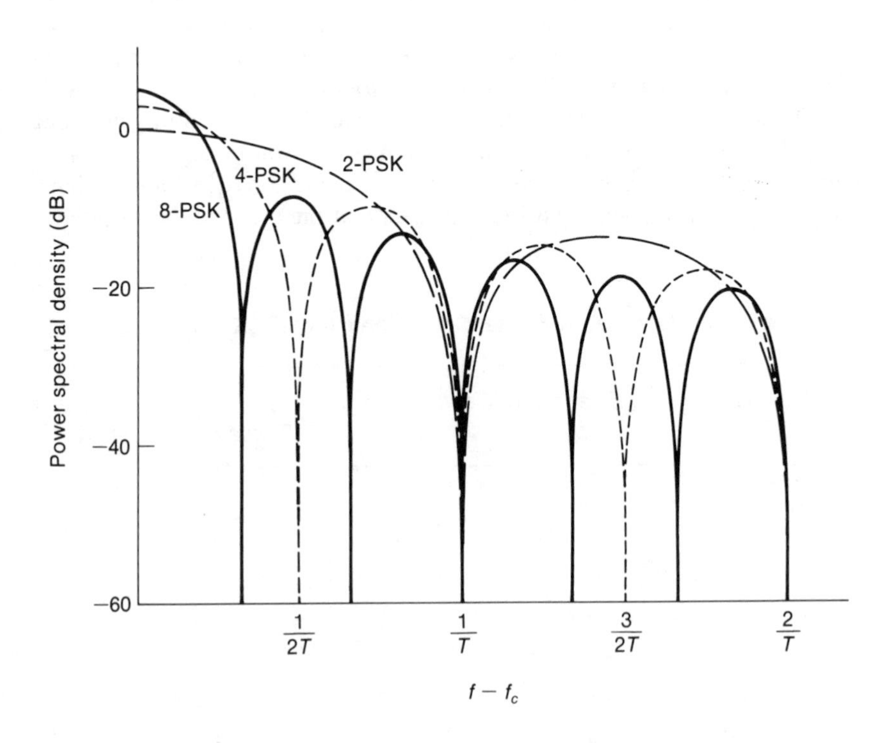

Figure 6.16 Spectra of PSK Signals for Fixed Data Rate

are serial-to-parallel converted to create $m(=\log_2 M)$ parallel bit streams, each having a bit rate of R/m. The I and Q signal generator converts each m-bit word to a pair of numbers (p_n, q_n), which are the I and Q signal coefficients. The I and Q signals are multiplied by $\cos\omega_c t$ and $\sin\omega_c t$, respectively, and the multiplier outputs are summed to create the PSK modulated signal.

As observed earlier, the spectrum of a BPSK signal corresponds to that of an NRZ signal translated upward in frequency. This relationship also holds for conventional M-ary PSK, where symbols are independent and the baseband signals applied to the quadrature channels are orthogonal, multilevel waveforms. The individual I and Q signals produce the common $\sin x/x$ spectrum, and the composite spectrum is simply the sum of the individual spectra. The resulting one-sided power spectrum for a random binary output is

$$S(f) = A^2T\left\{\frac{\sin[(f-f_c)\pi T]}{(f-f_c)\pi T}\right\}^2 \tag{6.44}$$

Figure 6.16 plots the power spectrum for 2, 4, and 8-PSK systems operating at the same data rate.

Coherent detection of M-ary PSK requires the generation of a local phase reference. For BPSK, only a single reference, $\cos\omega_c t$, is required to detect signals separated by 180°. For higher-level PSK, additional phase references are needed. One approach is to use two references that are in quadrature—that is, $\cos\omega_c t$ and $\sin\omega_c t$—along with a logic circuit to determine the linear combinations of the quadrature signals. A block diagram of a generalized PSK demodulator using two phase references is shown in Figure 6.17. A second method is to employ a set of M coherent detectors, using $\cos(\omega_c t + \theta_n)$ as the set of reference signals. In the absence of noise, the ideal phase detector measures θ_n at the sampling times. Errors due to noise will occur if the measured phase ϕ falls outside the region

$$\theta_n - \frac{\pi}{M} \le \phi < \theta_n + \frac{\pi}{M} \tag{6.45}$$

Calculation of the probability of symbol error results in a closed-form expression only when $M = 2$ and $M = 4$. For $M = 2$, the probability of error $P(e)$ was given earlier by (6.25); for $M = 4$,

$$P(e) = 2\,\mathrm{erfc}\sqrt{\frac{E_s}{N_0}}\left(1 - \frac{1}{4}\mathrm{erfc}\sqrt{\frac{E_s}{N_0}}\right) \tag{6.46}$$

For $E_s/N_0 \gg 1$, the probability of error for M-ary PSK can be approximated by [9]

$$P(e) \approx 2\,\mathrm{erfc}\sqrt{\frac{2E_s}{N_0}\sin^2\frac{\pi}{M}} \qquad M > 2 \tag{6.47}$$

This approximation becomes increasingly tight for fixed M as E_s/N_0 increases. Results for an exact calculation of symbol error probability are

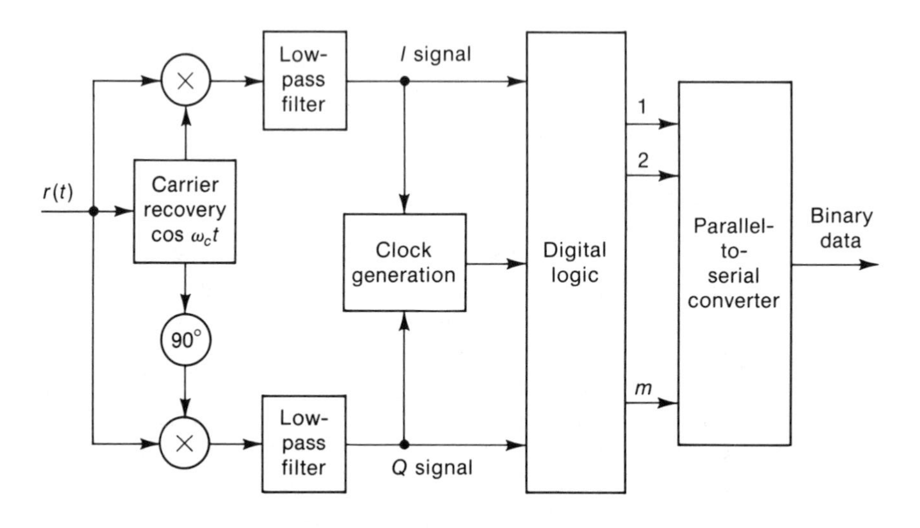

Figure 6.17 Generalized *M*-ary PSK Demodulator

plotted in Figure 6.18 as a function of $E_b/N_0 = E_s/mN_0$. The curves of Figure 6.18 show that for fixed symbol error probability, the E_b/N_0 must be increased with increasing M, while according to Figure 6.16 the required signal bandwidth is decreasing.

Coherent detection requires not only the use of multiple phase detectors but also the recovery of the carrier for phase coherence. Carrier recovery techniques were described in Section 6.4 for BPSK, and these can easily be generalized for M-ary PSK. An alternative to recovering a coherent reference is to use the phase difference between successive symbols. If the data are encoded by a phase shift rather than by absolute phase, then the receiver detects the signal by comparing the phase of one symbol with the phase of the previous symbol. This is the same technique described for BPSK in Section 6.4 and is known as differential PSK (DPSK) or differentially coherent PSK (DCPSK). Symbol errors occur when the measured phase difference differs by more than π/M in absolute phase from the phase difference of the transmitted signal. As is typical with these calculations, a closed-form expression for probability of symbol error cannot be derived for M-ary differential PSK. For large E_b/N_0, however, a good approximation to $P(e)$ is given by [9]

$$P(e) \approx 2\,\mathrm{erfc}\sqrt{\frac{2E_s}{N_0}\sin^2\frac{\pi}{\sqrt{2}\,M}} \tag{6.48}$$

A comparison of error probabilities for coherent and differentially coherent detection is given in Figure 6.18. Note that DPSK involves a tradeoff between signal power and data rate similar to that observed for coherent PSK. For BPSK, the performance of differential detection is within 1 dB of

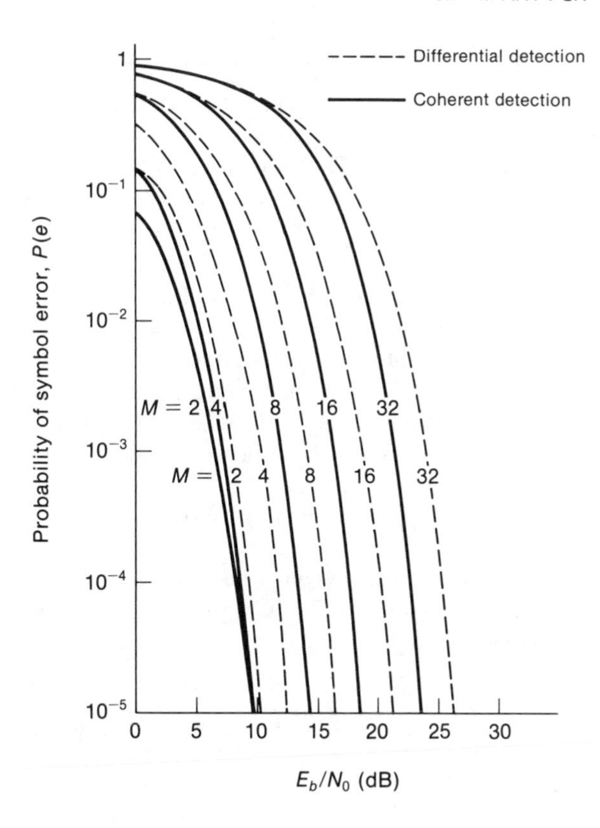

Figure 6.18 Error Performance for *M*-ary PSK

coherent detection for $P(e) < 10^{-3}$. This degradation increases with M, however, approaching 3 dB for $M \geq 8$.

The error probability expressions shown here for PSK are based on symbol error. If we assume that when a symbol error occurs the decoder randomly chooses one of the remaining $M - 1$ symbols, then the bit error probability is related to the symbol error probability by Equation (6.39). For PSK errors, however, the selected phase is more likely to be one of those adjacent to the transmitted phase. Further, Gray coding is usually employed with PSK symbols in which adjacent symbols differ in only one bit position (see Figure 6.14). Then each symbol error is most likely to cause only a single bit error. For high S/N,

$$\text{BER} \approx \frac{P(e)}{\log_2 M} \tag{6.49}$$

A comparison of BER for $M = 4$ using (6.47) and (6.49), and for $M = 2$ using (6.25), reveals that 4-PSK and BPSK have identical bit error performance.

Example 6.1

In the United States' 2-GHz common carrier band, digital radio systems are required to be able to carry 96 PCM voice channels within a 3.5-MHz bandwidth. For a choice of 4-PSK modulation with raised-cosine filtering, determine the filter rolloff factor α and the S/N required to provide $\text{BER} = 10^{-6}$.

Solution

Four PCM DS-1 (1.544 Mb/s) bit streams are required for 96 voice channels, which can be combined with a second-level multiplex into a DS-2 (6.312 Mb/s) bit stream. With a 3.5 MHz bandwidth allocation, the required bandwidth efficiency is 6.312 Mbps/3.5 MHz = 1.8 bps/Hz. For 4-PSK with raised-cosine filtering, the bandwidth efficiency is given by

$$\frac{R}{B} = \frac{(\log_2 M)/T}{(1+\alpha)/T} = 1.8 \text{ bps/Hz}$$

where α is the fraction of excess Nyquist bandwidth as defined in Chapter 5. Solving for α, we find that $\alpha = 0.1$ is required. Using (6.47) and (6.49), we obtain the following for $\text{BER} = 10^{-6}$:

$$10^{-6} = \operatorname{erfc}\sqrt{\frac{2E_s}{N_0}\sin^2\frac{\pi}{4}} = \operatorname{erfc}\sqrt{\frac{E_s}{N_0}} = \operatorname{erfc}\sqrt{\frac{2E_b}{N_0}}$$

Solving for E_b/N_0, we find

$$\left(\frac{E_b}{N_0}\right)_{\text{dB}} = 11.5 \text{ dB}$$

From Equation (6.4),

$$\left(\frac{S}{N}\right)_{\text{dB}} = \left(\frac{E_b}{N_0}\right)_{\text{dB}} + 10\log\frac{R}{B_N} = 11.5 \text{ dB} + 2.6 \text{ dB} = 14.1 \text{ dB}$$

6.8 QUADRATURE AMPLITUDE MODULATION (QAM)

The principle of quadrature modulation used with M-ary PSK can be generalized to include amplitude as well as phase modulation. With PSK, the in-phase and quadrature components are not independent. Their values are constrained in order to produce a constant envelope signal, which is a fundamental characteristic of PSK. If this constraint is removed so that the quadrature channels may be independent, the result is known as **quadrature amplitude modulation** (QAM). For the special case of two levels on each quadrature channel, QAM is identical to 4-PSK. Such systems are more popularly known as **quadrature PSK** (QPSK). The signal constellations for higher-level QAM systems are rectangular, however, and therefore distinctly different from the circular signal sets of higher-level PSK systems. Figure 6.19 compares PSK versus QAM signal constellations for $M = 16$.

A generalized block diagram of a QAM modulator is shown in Figure 6.20. The serial-to-parallel converter accepts a bit stream operating at a rate R and produces two parallel bit streams each at an $R/2$ rate. The 2-to-L ($L = \sqrt{M}$) level converters generate L-level signals from each of the two quadrature input channels. Multiplication by the appropriate phase of the carrier and summation of the I and Q channels then produces an M-ary QAM signal. The QAM demodulator resembles the PSK demodulator shown in Figure 6.17. To decode each baseband channel, digital logic is employed that compares the L-level signal against $L - 1$ decision thresholds. Finally, a combiner performs the parallel-to-serial conversion of the two detected bit streams.

The form of Figure 6.20 indicates that QAM can be interpreted as the linear addition of two quadrature DSB–SC signals. Thus it is apparent that the QAM spectrum is determined by the spectrum of the I and Q baseband signals. This same observation was made for M-ary PSK, leading us to conclude that QAM and PSK have identical spectra for equal numbers of signal points. The general expression for the QAM (or PSK) signal spectrum is given by (6.44). The transmission bandwidth B_T required for QAM is $2B$, the same as DSB–SC, where B is the (one-sided) bandwidth of the baseband spectrum (see Figure 6.4). For the Nyquist channel, the channel capacity for M-ary signaling was shown in Section 5.6 to be $2B \log_2 M$ b/s. Thus the transmission bandwidth efficiency for M-ary QAM (and PSK) is $\log_2 M$ bps/Hz. For example, 16-QAM (or 16-PSK) has a maximum transmission bandwidth efficiency of 4 bps/Hz and 64-QAM (or 64-PSK) has an efficiency of 6 bps/Hz. For practical channels, pulse shaping is employed by using specially constructed low-pass filters whose composite

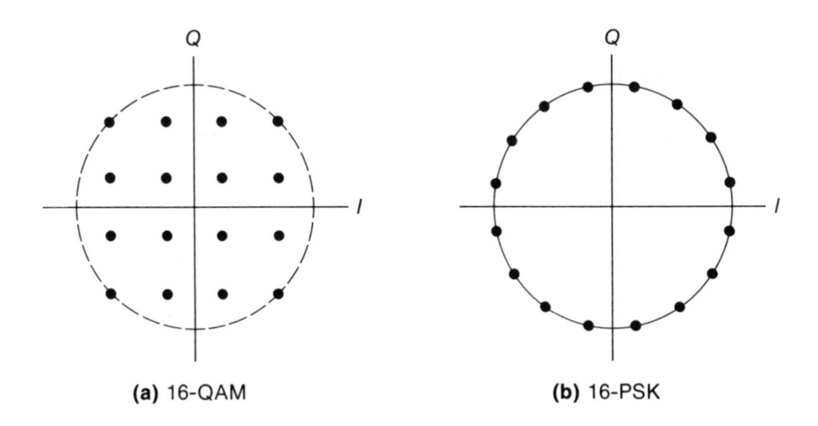

Figure 6.19 Comparison of 16-QAM and 16-PSK Signal Constellation with Same Peak Power

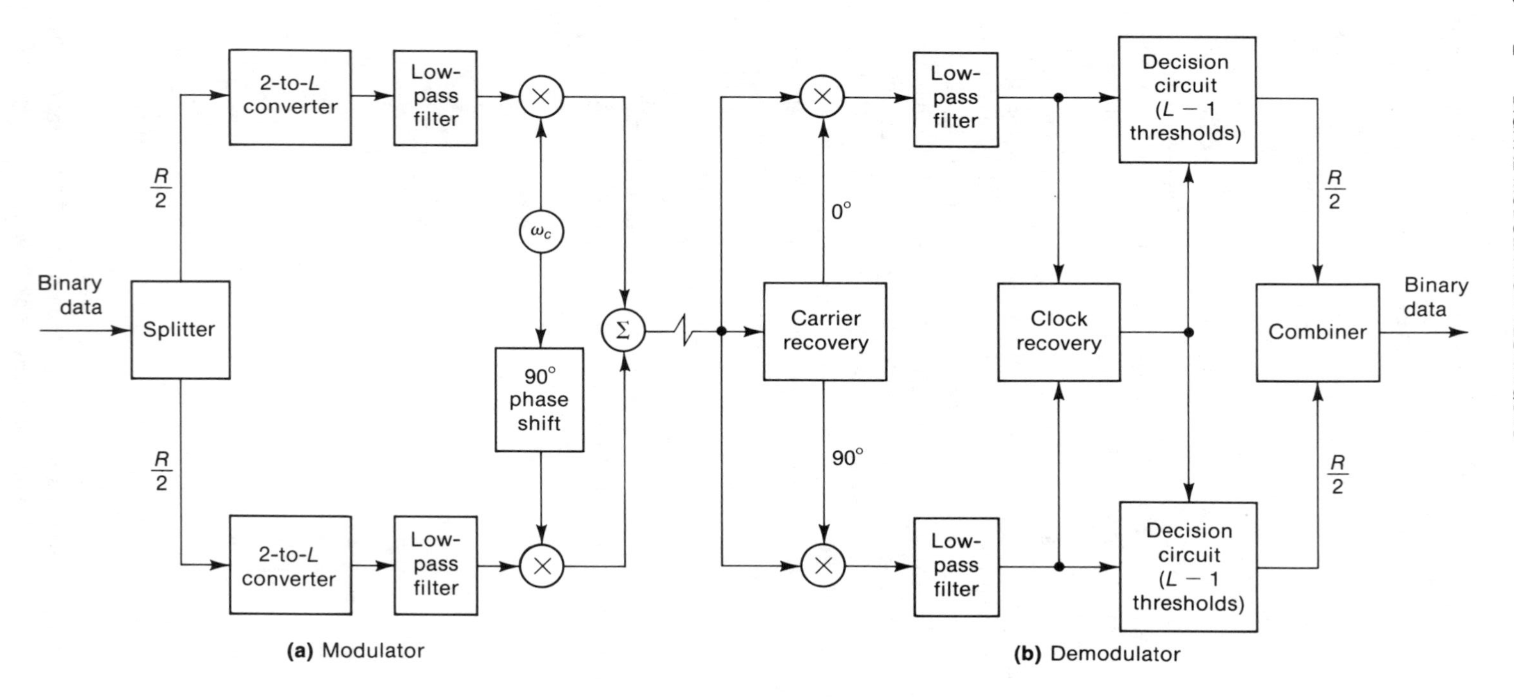

Figure 6.20 QAM Modulator and Demodulator

shape is usually split between the modulator and demodulator as indicated by Figure 6.20. For a choice of raised-cosine filtering, described in Chapter 5, the rolloff factor α determines the transmission bandwidth efficiency, given by $\log_2 M/(1 + \alpha)$ bps/Hz.

The error performance of QAM is given by the multilevel baseband error equation corresponding to each quadrature channel. Thus QAM error probability is provided by (5.47) but with M equal to the number of levels, L, on each quadrature channel. With this substitution, Equation (5.47) becomes

$$P(e) = 2\left(1 - \frac{1}{L}\right)\operatorname{erfc}\left[\left(\frac{3}{L^2 - 1}\frac{S}{N}\right)^{1/2}\right] \tag{6.50}$$

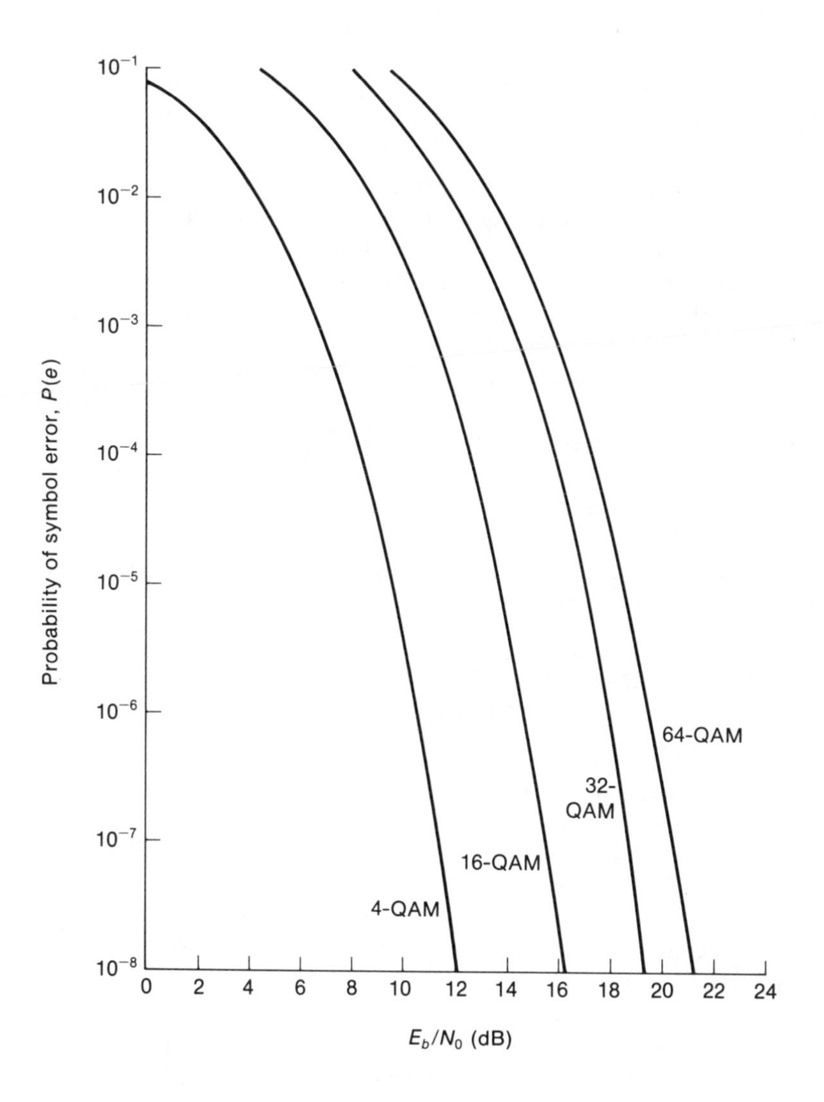

Figure 6.21 Error Performance for QAM

This error probability can be expressed as a function of E_b/N_0 by using (6.3*b*) in (6.50):

$$P(e) = 2\left(1 - \frac{1}{L}\right)\text{erfc}\left[(\log_2 L)^{1/2}\left(\frac{6}{L^2 - 1}\right)^{1/2}\left(\frac{E_b}{N_0}\right)^{1/2}\right] \quad (6.51)$$

Figure 6.21 plots the symbol error rate, given by (6.51), for M-ary QAM with $M(= L^2) = 4, 16, 32$, and 64.

To relate this symbol error rate to bit error rate, we assume that the multilevel signal is Gray coded so that symbol errors are most likely to produce only single bit errors in each quadrature channel. Hence

$$\text{BER} \approx \frac{P(e)}{\log_2 L} \quad (6.52)$$

Note that for $M = 4$ ($L = 2$), the BER equation for QAM given by (6.52) is identical to the BER equation for PSK given by (6.49), indicating the equivalence of 4-PSK and 4-QAM. For higher-level systems, however, the error performance for QAM systems is superior to that of PSK systems (compare Figure 6.21 with Figure 6.18). A comparison of the signal constellations for 16-QAM and 16-PSK, as shown in Figure 6.19, reveals the reason for this difference in error performance. The distance between signal points in the PSK constellation is smaller than the distance between points in the QAM constellation.

QAM is seen to have the same spectrum and bandwidth efficiency as M-ary PSK, but it outperforms PSK in error performance for at least the AWGN channel. These properties of QAM have made it a popular choice in current developments of high-speed digital radio [10, 11]. However, QAM is sensitive to system nonlinearities such as those found in the use of traveling wave tube (TWT) amplifiers. This effect requires that TWT amplifiers be operated below saturation for QAM systems. (See Chapter 9 for a discussion of the effects of nonlinear amplifiers in digital radios.)

6.9 OFFSET QPSK (OQPSK) AND MINIMUM-SHIFT KEYING (MSK)

For band-limited, nonlinear channels, two modulation techniques known as **offset quadrature phase-shift keying** (OQPSK) and **minimum-shift keying** (MSK) offer certain advantages over conventional QPSK [12, 13]. In a nonlinear channel, the spectral side lobes of a filtered QPSK signal tend to be restored to their initial characteristics prior to filtering. With OQPSK and MSK, the signal envelope is constant, which makes these modulation techniques impervious to channel nonlinearities. A choice of MSK further facilitates band limiting because its power spectral density decreases more rapidly than QPSK beyond the minimum bandwidth. These advantages are achieved with the same $P(e)$ performance and bandwidth efficiency (2 bps/Hz) as QPSK and with only modest increase in modulator and detector complexity.

Offset QPSK has the same phasor diagram as QPSK and thus can be represented by Equation (6.42). The difference between the two modulation techniques is in the alignment of the in-phase and quadrature bit streams, as illustrated in Figure 6.22. With OQPSK (also referred to as staggered QPSK), the I and Q bit streams are offset in time by one bit period T_b. With QPSK, the transitions of the I and Q streams coincide. This difference in quadrature signal alignment results in different characteristics in the phase changes of the carrier. The phase change in QPSK may be 0, $\pm 90°$, or 180° with each symbol interval ($2T_b$), depending on the quadrature coefficients (p_n, q_n). In OQPSK, the quadrature coefficients cannot change state simultaneously. This characteristic eliminates 180° phase changes and results in only 0° and $\pm 90°$ phase changes.

The usefulness of OQPSK is found in its application to the band-limited, nonlinear channel. In practical applications, band limiting is necessary to meet spectrum occupancy allocations. Bandpass filtering of QPSK causes the envelope of the signal to fluctuate and go to zero at each 180° phase reversal. If a nonlinear amplifier is used as the output power stage of the transmitter, envelope fluctuations are reduced and the spectrum sidebands are restored to their original level prior to filtering. These sidebands introduce out-of-band emissions that interfere with other signals in adjacent frequency bands. OQPSK when band-limited has no envelope zeros. Thus the elimination of large phase changes with OQPSK facilitates the use of nonlinear amplification, since the undesired sidebands removed by filtering are not regenerated by the amplifier.

The advantages of OQPSK result from eliminating the largest phase change (180°) associated with QPSK. This feature suggests that further suppression of out-of-band emissions is possible if phase transitions are eliminated altogether. One modulation technique that provides such a constant-envelope, continuous-phase signal is minimum-shift keying (MSK). MSK is a special case of FSK in which continuous phase is maintained at symbol transitions using a minimum difference in signaling frequencies. The general class of continuous-phase FSK (CPFSK) has a waveform defined by

$$s(t) = A\cos[\omega_c t + \phi(t)] \tag{6.53}$$

where the phase $\phi(t)$ is made continuous in time. For the case of MSK, the carrier frequency ω_c is given by

$$\omega_c = \frac{\omega_0 + \omega_1}{2} \tag{6.54}$$

and $\phi(t)$ can be written as

$$\phi(t) = \pm\frac{\Delta\omega t}{2} + \phi(0) \tag{6.55}$$

where $\Delta\omega$ is the separation between frequencies ω_0 and ω_1 used to signal binary digits 0 and 1 and where $\phi(0)$ is the initial phase. From (6.29) we know that the minimum frequency separation for orthogonal signaling is

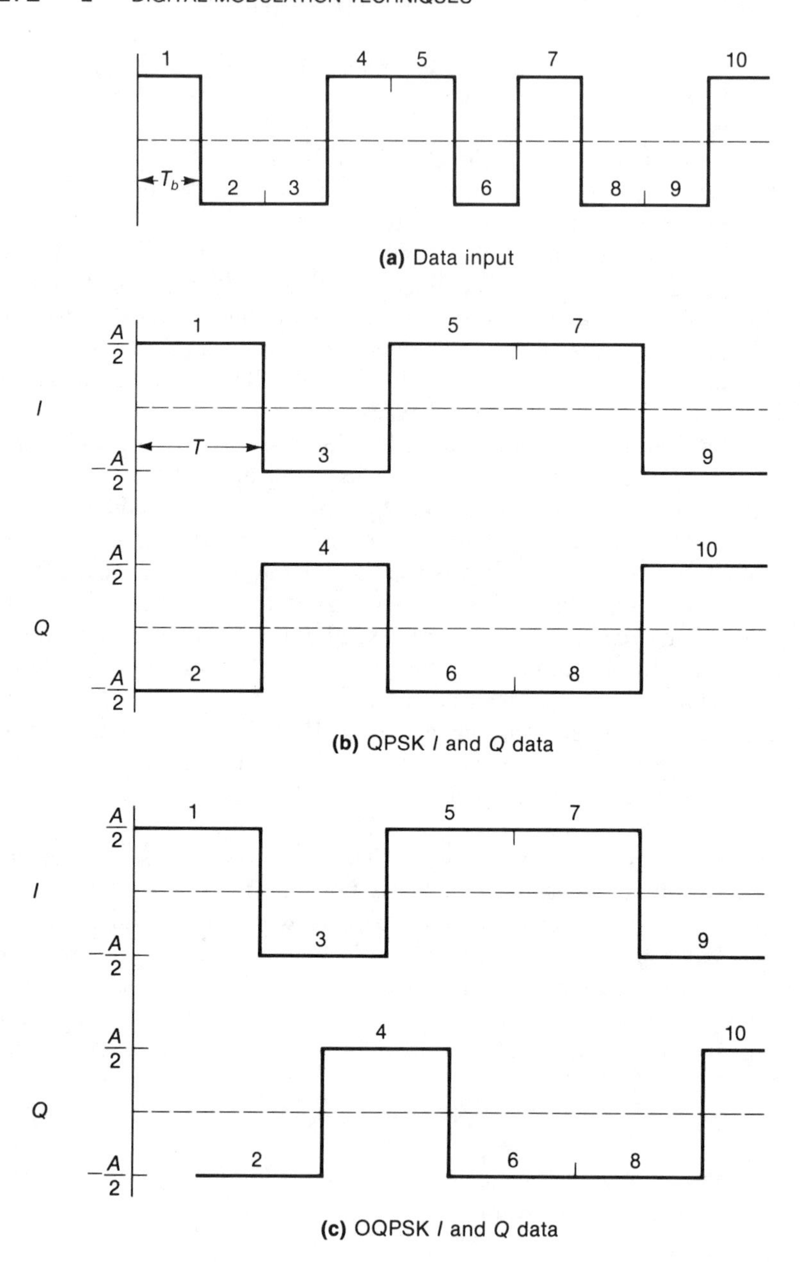

Figure 6.22 ***I* and *Q* Data Relationships in QPSK and OQPSK**

$\Delta\omega = \pi/T_b$, so that (6.53) becomes

$$s_n(t) = A\cos\left[\omega_c t + \frac{b_n \pi t}{2T_b} + \phi(0)\right] \tag{6.56}$$

where b_n is ± 1, representing binary digits 0 and 1. Thus with each bit interval T_b the phase of the carrier changes by $\pm\pi/2$, according to the value of the binary input, 0 or 1. Assuming an initial phase $\phi(0) = 0$, the phase value of $\phi(t)$ advances in time according to the phase trellis shown in Figure 6.23. Each path represents a particular combination of 1's and 0's. Since the phase change with each bit interval is exactly $\pm\pi/2$, the accumulated phase $\phi(t)$ is restricted to integral multiples of $\pi/2$ at the end of each bit interval. Further, as seen in Figure 6.23, the phase is an odd multiple of $\pi/2$ at odd multiples of T_b and an even multiple of $\pi/2$ at even multiples of T_b.

Minimum-shift keying can also be viewed as a special case of OQPSK with sinusoidal pulse shaping used in place of rectangular pulses. Using the quadrature representation of QPSK, given in (6.42), and adding sinusoidal pulse shaping, we obtain

$$s_n(t) = A\left[p_n \cos\left(\frac{\pi t}{2T_b}\right)\cos\omega_c t + q_n \sin\left(\frac{\pi t}{2T_b}\right)\sin\omega_c t\right] \tag{6.57}$$

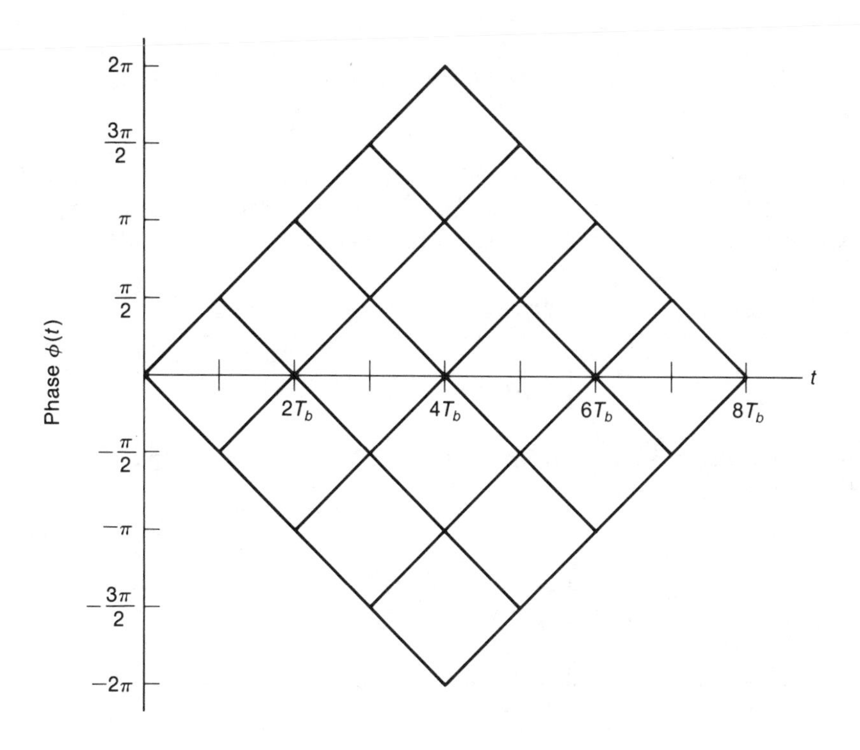

Figure 6.23 Phase Trellis for MSK

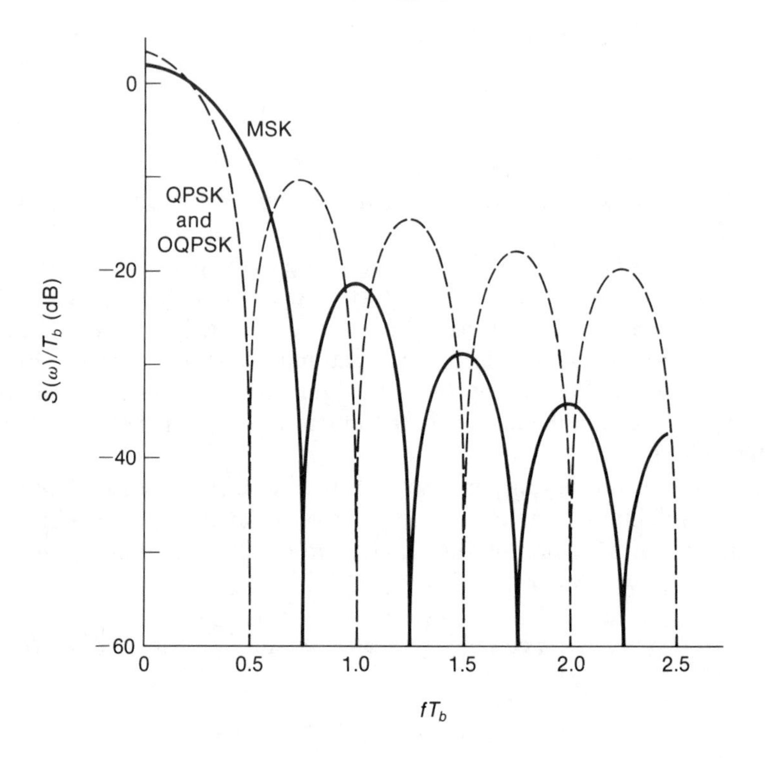

Figure 6.24 Power Spectral Density of QPSK, OQPSK, and MSK

where $\cos(\pi t/2T_b)$ is the sinusoidal shaping for the in-phase signal and $\sin(\pi t/2T_b)$ is the sinusoidal shaping for the offset quadrature signal. Applying trigonometric identities, Expression (6.57) can be rewritten as

$$s_n(t) = \begin{cases} Ap_n \cos\left(\omega_c t + \dfrac{\pi t}{2T_b}\right) & p_n \neq q_n \\ Ap_n \cos\left(\omega_c t - \dfrac{\pi t}{2T_b}\right) & p_n = q_n \end{cases} \tag{6.58}$$

By comparing (6.58) with (6.56), we see that MSK is identical to this special form of OQPSK, with

$$b_n = -p_n q_n \tag{6.59}$$

Thus MSK can be viewed as either an OQPSK signal with sinusoidal pulse shaping or as a CPFSK signal with a frequency separation (Δf) equal to one-half the bit rate.

The power spectral density for OQPSK is the same as that for QPSK, given by (6.44) with $T = 2T_b$:

$$S(\omega) = 2A^2 T_b \left\{ \frac{\sin[2(f - f_c)\pi T_b]}{2(f - f_c)\pi T_b} \right\}^2 \tag{6.60}$$

The power spectral density for MSK is

$$S(\omega) = \frac{16A^2T_b}{\pi^2}\left(\frac{\cos[2\pi(f - f_c)T_b]}{1 - 16(f - f_c)^2T_b^2}\right)^2 \tag{6.61}$$

Plots of (6.60) and (6.61) are shown in Figure 6.24. The MSK spectrum rolls off at a rate proportional to f^{-4} for large values of f whereas the OQPSK spectrum rolls off at a rate proportional to only f^{-2}. This relative difference in spectrum rolloff is expected, since MSK inherently maintains phase continuity from bit to bit. Notice that the MSK spectrum has a wider main lobe than OQPSK; the first nulls fall at $3/4T_b$ and $1/2T_b$, respectively. A better comparison of the compactness of these two modulation spectra is the 99 percent power bandwidth, which for MSK is approximately $1.2/T_b$ while for QPSK and OQPSK it is approximately $10.3/T_b$. This spectral characteristic suggests that MSK has application where a constant envelope signal and little or no filtering are desired.

The error performance of OQPSK and QPSK is identical, since the offsetting of bit streams does not change the orthogonality of the I and Q signals. Furthermore, since MSK is seen to be equivalent to OQPSK except for pulse shaping, MSK also has the same error performance as QPSK and OQPSK. Coherent detection of the QPSK, OQPSK, or MSK signals is based on an observation period of $2T_b$. If MSK is detected as an FSK signal with bit decisions made over a T_b observation period, however, MSK would be 3 dB poorer than QPSK. Hence MSK detected as two orthogonal signals has a 3 dB advantage over orthogonal FSK. MSK can also be noncoherently detected, which permits a simpler demodulator for situations where the E_b/N_0 is adequate.

6.10 QUADRATURE PARTIAL RESPONSE (QPR)

To further increase the bandwidth efficiency of QAM signaling, the in-phase and quadrature channels can be modulated with partial response coders. This technique is termed **quadrature partial response** (QPR). The quadrature addition of two partial response signals can be expressed mathematically as

$$s(t) = y_I \sin \omega_c t + y_Q \cos \omega_c t \tag{6.62}$$

As an example, assume the use of class 1 partial response coders in the I and Q channels of a 4-QAM (QPSK) modulator. Recall from Section 5.7 that the effect of the partial response coder is to produce three levels from a binary input. As shown in Figure 6.25*a*, the resulting QPR signal constellation is a 3×3 rectangle with nine signal states. Similarly, 16-QAM that has four levels on each I and Q channel produces seven levels after partial response coding, a 7×7 signal constellation, and 49 signal states, as shown in Figure 6.25*b*.

A block diagram of the QPR modulator and demodulator is shown in Figure 6.26. This QPR implementation is very similar to that of QAM

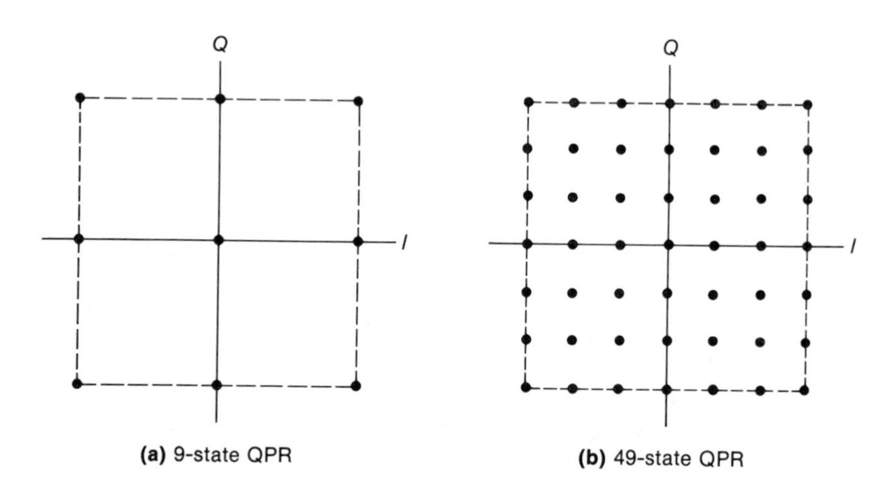

(a) 9-state QPR

(b) 49-state QPR

Figure 6.25 Phasor Diagrams of QPR

shown in Figure 6.20. The essential differences are in the filtering and detection. Partial response shaping can be accomplished at baseband with low-pass filters or after quadrature modulation with bandpass filters. In either case, the filter characteristic is usually split between the transmitter and receiver, such that the receiver filter not only finishes the required shaping but also rejects out-of-band noise. A choice of bandpass filtering allows the use of nonlinear power amplifiers prior to filtering when the signal still has a constant-envelope characteristic. After amplification, the bandpass filter converts the QAM signal to QPR. A disadvantage of bandpass filtering is the higher insertion loss compared to low-pass filtering. After final filtering in the QPR demodulator, the I and Q baseband signals are independently detected using the same partial response detection scheme described in Section 5.7.

Earlier we noted that the spectrum and error performance of M-ary QAM are determined by the corresponding characteristics of the I and Q baseband signals. Thus QPR has a spectrum defined by the partial response signal (see Table 5.1). Recall from Section 5.7 that partial response coding attains the Nyquist transmission rate. Therefore QPR permits practical realization of the transmission bandwidth efficiency theoretically possible with QAM. This efficiency is given by $\log_2 M$ bps/Hz. With class 1 partial response coding applied to the I and Q channels, for example, 4-QAM becomes 9-state QPR and achieves 2 bps/Hz whereas 16-QAM becomes 49-state QPR and achieves 4 bps/Hz efficiency. The error performance of QPR is given by the corresponding expression for partial response (Equation 5.67), but with M replaced by L, the number of levels on each quadrature channel prior to filtering. Using (6.3b) to express S/N as a function of E_b/N_0, the QPR error probability can be rewritten from (5.67)

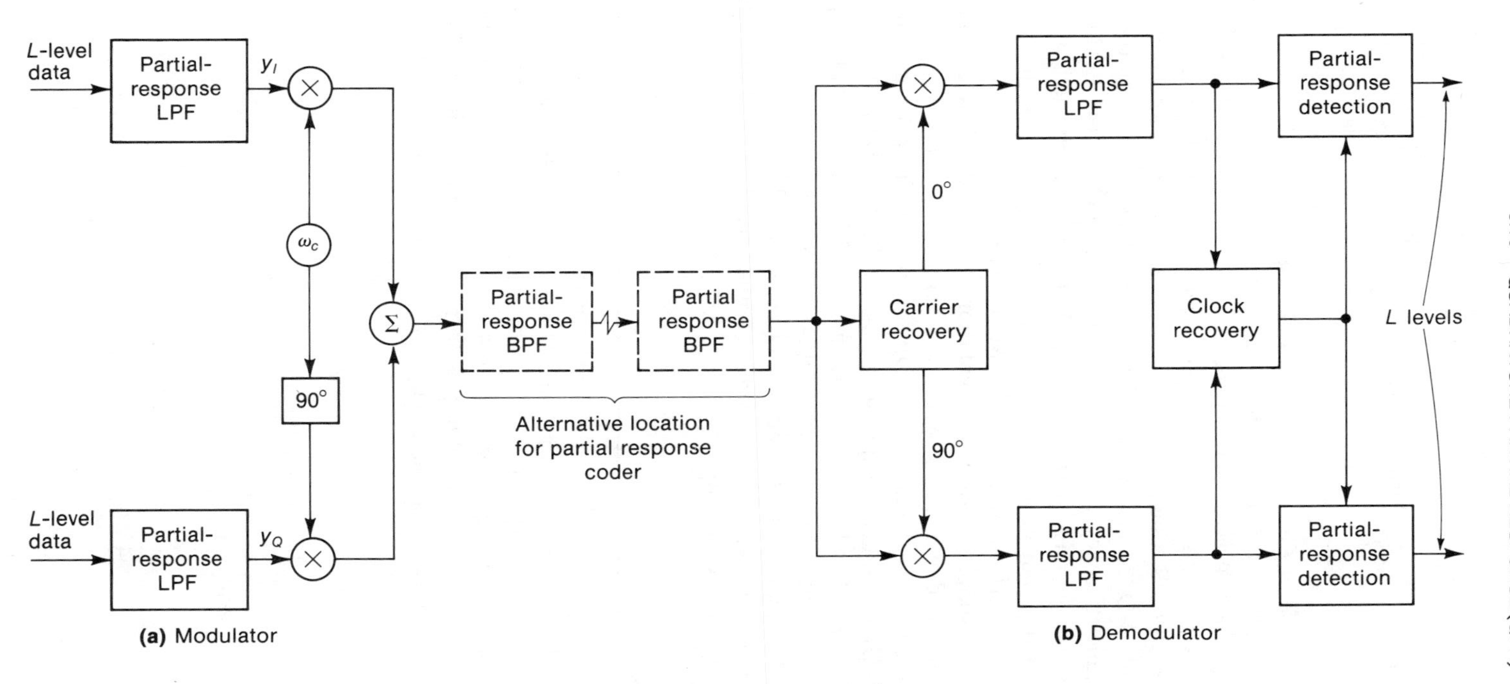

Figure 6.26 QPR Modulator and Demodulator

as

$$P(e) = 2\left(1 - \frac{1}{L^2}\right)\text{erfc}\left[\frac{\pi}{4}(\log_2 L)^{1/2}\left(\frac{6}{L^2 - 1}\right)^{1/2}\left(\frac{E_b}{N_0}\right)^{1/2}\right] \tag{6.63}$$

Assuming the use of Gray coding on each baseband signal, the bit error probability is given by (6.63) divided by $\log_2 L$.

A comparison of error performance for 9-QPR versus 4-QAM, using Expressions (6.63) and (6.51), indicates a net loss of 2 dB in E_b/N_0 for QPR. However, QPR attains a 2 bps/Hz spectral efficiency with realizable filtering, while 4-QAM requires the theoretical Nyquist channel in order to attain the same efficiency. A fairer comparison with 9-QPR is 16-QAM with 100 percent ($\alpha = 1$) raised-cosine filtering. Both systems provide 2 bps/Hz spectral efficiency, but now 9-QPR requires 1.9 dB less E_b/N_0 to provide the same error performance as 16-QAM. Its error performance, spectral efficiency, and simplicity of implementation have made QPR a popular choice in current digital radio systems. Several QPR radio systems have been described in the literature. A digital radio designed for use in the Canadian 8-GHz frequency band uses nine-state QPR to transmit 91 Mb/s in a 40-MHz RF bandwidth—an efficiency of 2.25 bps/Hz [14, 15]. A digital radio designed for the U.S. military also uses nine-state QPR to achieve an RF bandwidth efficiency of 1.9 bps/Hz [16]. Implementation of the partial response coding has been done both at baseband [16] and at RF after power amplification [14].

Example 6.2

A 9-QPR radio system is designed to transmit 90 Mb/s in a 40-MHz bandwidth. The transmit signal power is 1 watt and net system loss is 111 dB. For transmission over an additive white gaussian noise channel with one-sided power spectral density equal to 4×10^{-21} watts/Hz, compute the BER for this system. How would this performance compare with a differentially coherent 8-PSK system, assuming that the bandwidth is reduced to 30 MHz?

Solution

From (5.66) the BER for QPR modulation is

$$\begin{aligned}
\text{BER} &= \frac{3}{2}\text{erfc}\left[\frac{\pi}{4}\left(\frac{S}{N}\right)^{1/2}\right] \\
&= \frac{3}{2}\text{erfc}\left\{\frac{\pi}{4}\left[\frac{7.9 \times 10^{-12}\text{ watts}}{(4 \times 10^{-21}\text{ watts/Hz})(40 \times 10^{6}\text{ Hz})}\right]^{1/2}\right\} \\
&= \frac{3}{2}\text{erfc } 5.53 \\
&= 2.4 \times 10^{-8}
\end{aligned}$$

For 8-PSK modulation, assuming the use of Gray coding, the BER is obtained from (6.48) and (6.49) as

$$\text{BER} = \frac{P(e)}{\log_2 8} = \frac{2}{3}\text{erfc}\left[\frac{2E_b \log_2 8}{N_0}\sin^2\frac{\pi}{8\sqrt{2}}\right]^{1/2}$$

From (6.4) the S/N is given by

$$\frac{S}{N} = \frac{R}{B_N}\frac{E_b}{N_0}$$

so that

$$\begin{aligned}\text{BER} &= \frac{2}{3}\text{erfc}\left[(0.451)\left(\frac{30 \times 10^6}{90 \times 10^6}\right)\frac{7.9 \times 10^{-12}}{(4 \times 10^{-21})(30 \times 10^6)}\right]^{1/2} \\ &= \frac{2}{3}\text{erfc } 3.15 \\ &= 5.4 \times 10^{-4}\end{aligned}$$

6.11 SUMMARY

The summary for this chapter takes the form of a comparison of digital modulation techniques. The choice of digital modulation technique is influenced by error performance, spectral characteristics, implementation complexity, and other factors peculiar to the specific application (such as digital radio or telephone channel modem). Earlier we observed that binary modulation schemes provide good error performance and are simple to implement, but they lack the bandwidth efficiency required for most practical applications. Hence our emphasis here is on M-ary modulation schemes that provide the necessary bandwidth efficiency.

The symbol probability of error performance for several PSK, QAM, and QPR systems is compared in Figure 6.27. Note that with increasing M, the QAM systems hold an advantage over PSK. For example, 16-QAM requires about 4 dB less power than 16-PSK systems. Moreover, QPR systems are in general more power efficient than PSK systems for equivalent bandwidth efficiency.

Spectral efficiency among candidate modulation techniques is compared in Figure 6.28 as a function of the S/N required to provide $P(e) = 10^{-6}$. Points on this graph are labeled with the number of signal states present in the signal constellation. Two sets of curves are given for both PSK and QAM, corresponding to $\alpha = 0$ and $\alpha = 1$, where α is the fractional bandwidth required in excess of the Nyquist bandwidth. Of course, the penalty in the bandwidth efficiency of PSK or QAM systems for a choice of $\alpha = 1$ is a factor of 2 as compared to $\alpha = 0$, the minimum Nyquist bandwidth. As an example of the use of Figure 6.28, consider a requirement for 2 bps/Hz. Here QPR appears to be the best choice, when

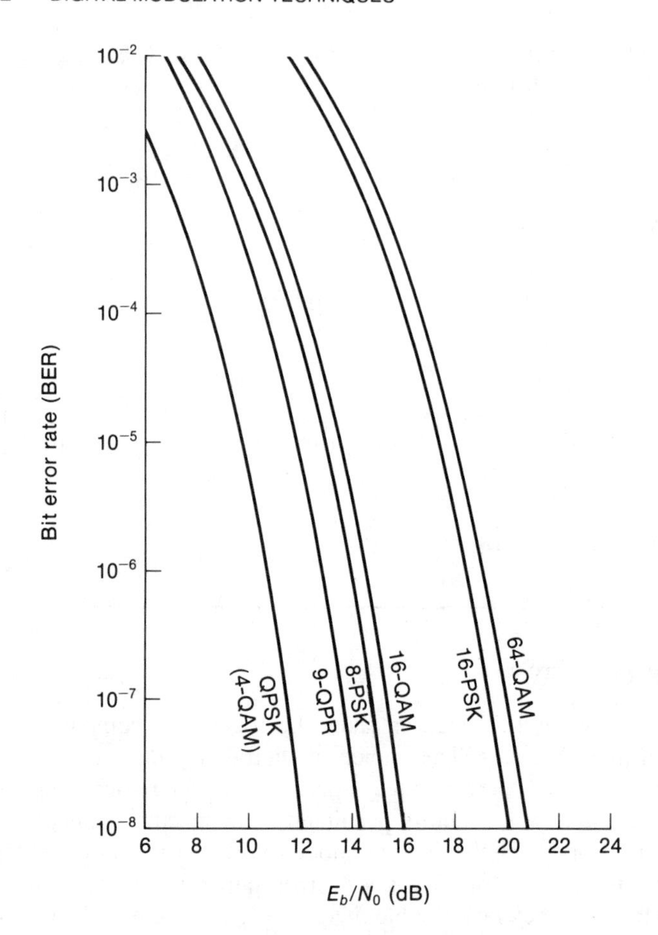

Figure 6.27 Bit Error Rate Performance for *M*-ary PSK, QAM, and QPR Coherent Systems

one trades off the required S/N against equipment complexity, considering that systems with $\alpha = 0$ are impractical.

The actual bandwidth required in digital signal transmission is a function of not only the modulation technique but also the definition of bandwidth. Table 6.3 lists the required bandwidths corresponding to three different definitions for several modulation techniques. The bandwidths are in units of R, the bit rate. The high values of 99 percent bandwidth for PSK and QAM systems are due to a slow rate of spectral rolloff. Conversely, the low value of 99 percent bandwidth for MSK is due to a greater rate of spectral rolloff, which is proportional to f^{-4} compared to the f^{-2} characteristic for PSK.

Other factors that may influence the choice of modulation technique include channel nonlinearities, radio fading, and interference from adjacent

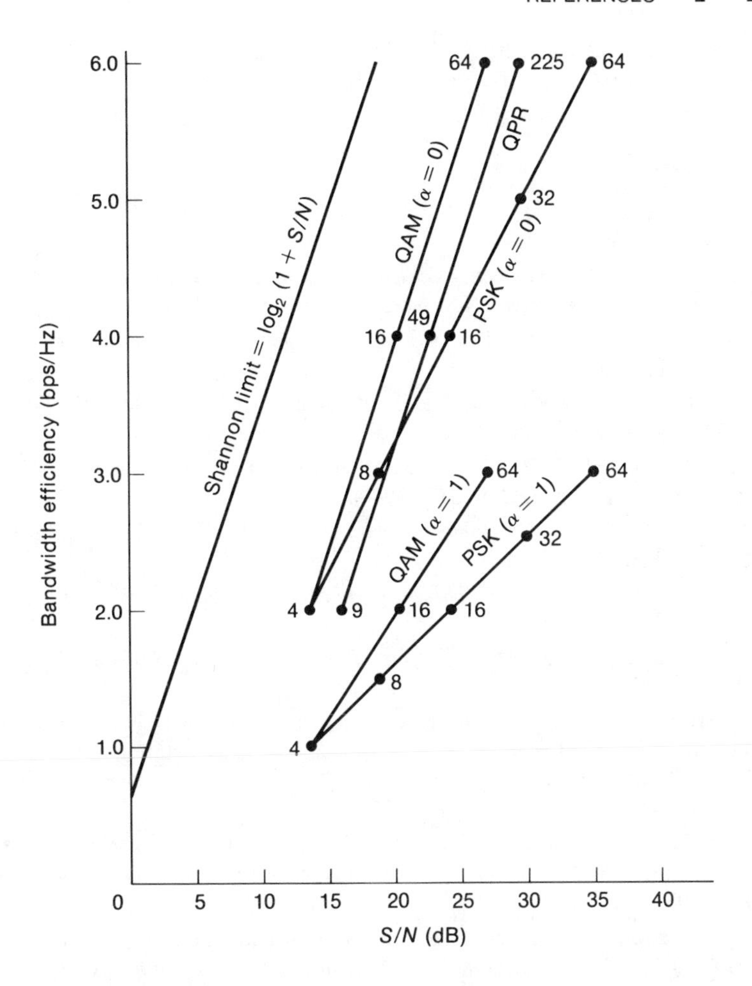

Figure 6.28 Bandwidth Efficiency for *M*-ary PSK, QAM, and QPR Coherent Systems on the Basis of S/N Required for $P(e) = 10^{-6}$

Table 6.3 Bandwidths for Digital Modulation in Units of Bit Rate *R* [17]

Modulation Technique	*Bandwidths*		
	Null-to-Null	*3 dB*	*99%*
2-PSK	2	0.88	20.56
4-PSK (QAM)	1	0.44	10.28
8-PSK	2 / 3	0.29	6.85
16-PSK (16-QAM)	1 / 2	0.22	5.14
MSK	1.5	0.59	1.18

radio channels. Discussion of these topics is deferred to the treatment of digital radio system design in Chapter 9.

REFERENCES

1. J. M. Wozencraft and I. M. Jacobs, *Principles of Communication Engineering* (New York: Wiley, 1967).
2. A. B. Carlson, *Communication Systems: An Introduction to Signals and Noise in Electrical Communication* (New York: McGraw-Hill, 1975).
3. R. W. Lucky, J. Salz, and E. J. Weldon, *Principles of Data Communication* (New York: McGraw-Hill, 1968).
4. J. R. Carson, "Notes on the Theory of Modulation," *Proc. IRE* 10 (February 1922):57–64.
5. M. Schwartz, W. R. Bennett, and S. Stein, *Communication Systems and Techniques* (New York: McGraw-Hill, 1966).
6. J. J. Spliker, *Digital Communications by Satellite* (Englewood Cliffs, N.J.: Prentice-Hall, 1977).
7. A. J. Viterbi, *Principles of Coherent Communication* (New York: McGraw-Hill, 1966).
8. W. C. Lindsey and M. K. Simon, *Telecommunication Systems Engineering* (Englewood Cliffs, N.J.: Prentice-Hall, 1973).
9. E. Arthurs and H. Dym, "On the Optimum Detection of Digital Signals in the Presence of White Gaussian Noise—A Geometric Interpretation and a Study of Three Basic Data Transmission Systems," *IRE Trans. Comm. Systems*, vol. CS-10, no. 4, December 1962, pp. 336–372.
10. P. R. Hartmann and J. A. Crossett, "135 Mb/s-6 GHz Transmission System Using 64-QAM Modulation," 1983 *International Conference on Communications*, June 1983, pp. F2.6.1 to F2.6.7.
11. J. McNicol, S. Barber, and F. Rivest, "Design and Application of the RD-4A and RD-6A 64 QAM Digital Radio Systems," 1984 *International Conference on Communications*, May 1984, pp. 646–652.
12. S. A. Gronemeyer and A. L. McBride, "MSK and Offset QPSK Modulation," *IEEE Trans. on Comm.*, vol. COM-24, no. 8, August 1976, pp. 809–820.
13. S. Pasupathy, "Minimum Shift Keying: A Spectrally Efficient Modulation," *IEEE Comm. Mag.*, July 1979, pp. 14–22.
14. C. W. Anderson and S. G. Barber, "Modulation Considerations for a 91 Mbit/s Digital Radio," *IEEE Trans. on Comm.*, vol. COM-26, no. 5, May 1978, pp. 523–528.
15. I. Godier, "DRS-8: A Digital Radio for Long-Haul Transmission," *1977 International Conference on Communications*, June 1977, pp. 5.4-102 to 5.4-105.
16. C. M. Thomas, J. E. Alexander, and E. W. Rahneberg, "A New Generation of Digital Microwave Radios for U.S. Military Telephone Networks," *IEEE Trans. on Comm.*, vol. COM-27, no. 12, December 1979, pp. 1916–1928.
17. F. Amoroso, "The Bandwidth of Digital Data Signals," *IEEE Comm. Mag.*, November 1980, pp. 13–24.

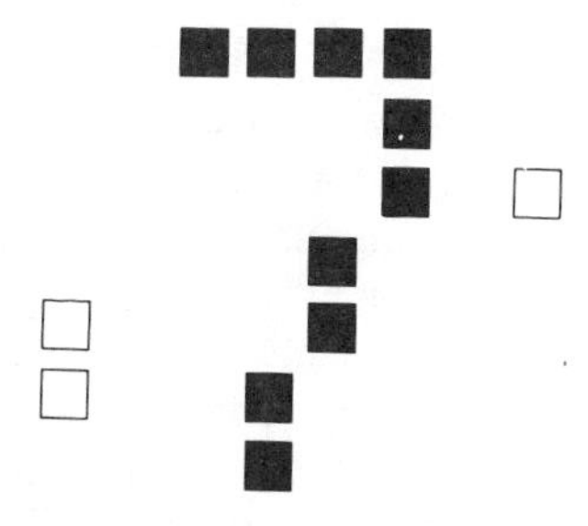

Digital Transmission Over the Telephone Network

OBJECTIVES

- Describes the techniques used to adapt telephone channels for the transmission of digital signals
- Outlines the levels of the FDM hierarchy from group to jumbogroup and supermastergroup
- Considers the transmission parameters that determine data communication performance over 4-kHz telephone circuits
- Explains how equalization can be used to correct two critical parameters: attentuation distortion and envelope delay distortion
- Discusses the use of voice-band modems and wideband modems for digital transmission
- Points out the advantages of transmultiplexers over the conventional interface between FDM and TDM systems
- Describes the new generation of hybrid transmission systems

7.1 INTRODUCTION

Current telephone networks are based predominantly on the use of analog transmission. Economics often make it impractical to build digital transmission facilities, separate from the existing telephone network, for transmission of nontelephone digital signals, such as data, facsimile, or visual telephony. This chapter deals with the techniques used to adapt telephone channels for the transmission of digital signals.

Digital signals can be converted to a form suitable for the telephone network by use of the digital modulation techniques described in the previous chapter. The equipment designed to interface data with telephone channels is termed a **modem** (an acronym for modulator/demodulator). The choice of modulation technique depends on the signal bit rate and characteristics of the channel such as bandwidth and signal-to-noise ratio. The channel characteristics found in telephone systems are well defined within the telephone industry for the basic voice channel and for the frequency-division multiplex (FDM) used to combine voice channels.

Today most data transmission takes place over individual voice channels within a bandwidth of 3 kHz or less. The characteristics of voice channels that affect modem performance can have wide-ranging values, depending on the length and routing of a call. For *dial-up*, *switched* lines, performance characteristics vary from call to call, so the modem must be designed to adapt to these varying conditions. Use of *private*, *leased* lines permits simpler modem design, since the channel conditions can be controlled by judicious routing and the use of phase and amplitude equalization.

Wideband data require a greater transmission bandwidth than the voice channel. Wideband modems provide the necessary bandwidth by using the bandwidths that occur naturally in the FDM hierarchy. If the requirement for data exceeds the total available transmission bandwidth, however, separate digital transmission facilities are required. As a compromise, hybrid transmission systems can also be used for wideband data; in hybrid systems, voice and data signals share the same transmission media. Since today's communication systems are still dominated by the analog telephone network, the techniques described in this chapter are expected to play a continuing role in future digital transmission systems.

7.2 FREQUENCY-DIVISION MULTIPLEX (FDM)

In frequency-division multiplex (FDM) each signal is allocated a discrete portion of the frequency spectrum. At the transmitter, each baseband signal is shifted in frequency by a modulation process (usually AM). Carrier frequencies are selected such that the modulated signals occupy adjacent, nonoverlapping frequency bands. Filters with sharp cutoff at the band edges are required in order to minimize interference between signals. At the receiver, each baseband signal is recovered by filtering and demodulation.

Here we focus on the use of FDM in telephone channel transmission. Several levels of multiplexing are used today that form an FDM hierarchy. The first level in the standard hierarchy is the **group**, which occupies the frequency band 60 to 108 kHz and contains 12 voice channels, each of nominal 4 kHz bandwidth. Figure 7.1 illustrates the operation of an FDM group. At the transmitter each voice signal is low-pass-filtered and then used to modulate 12 carriers spaced 4 kHz apart. The modulation process produces a double sideband signal. The outputs of the 12 modulators are

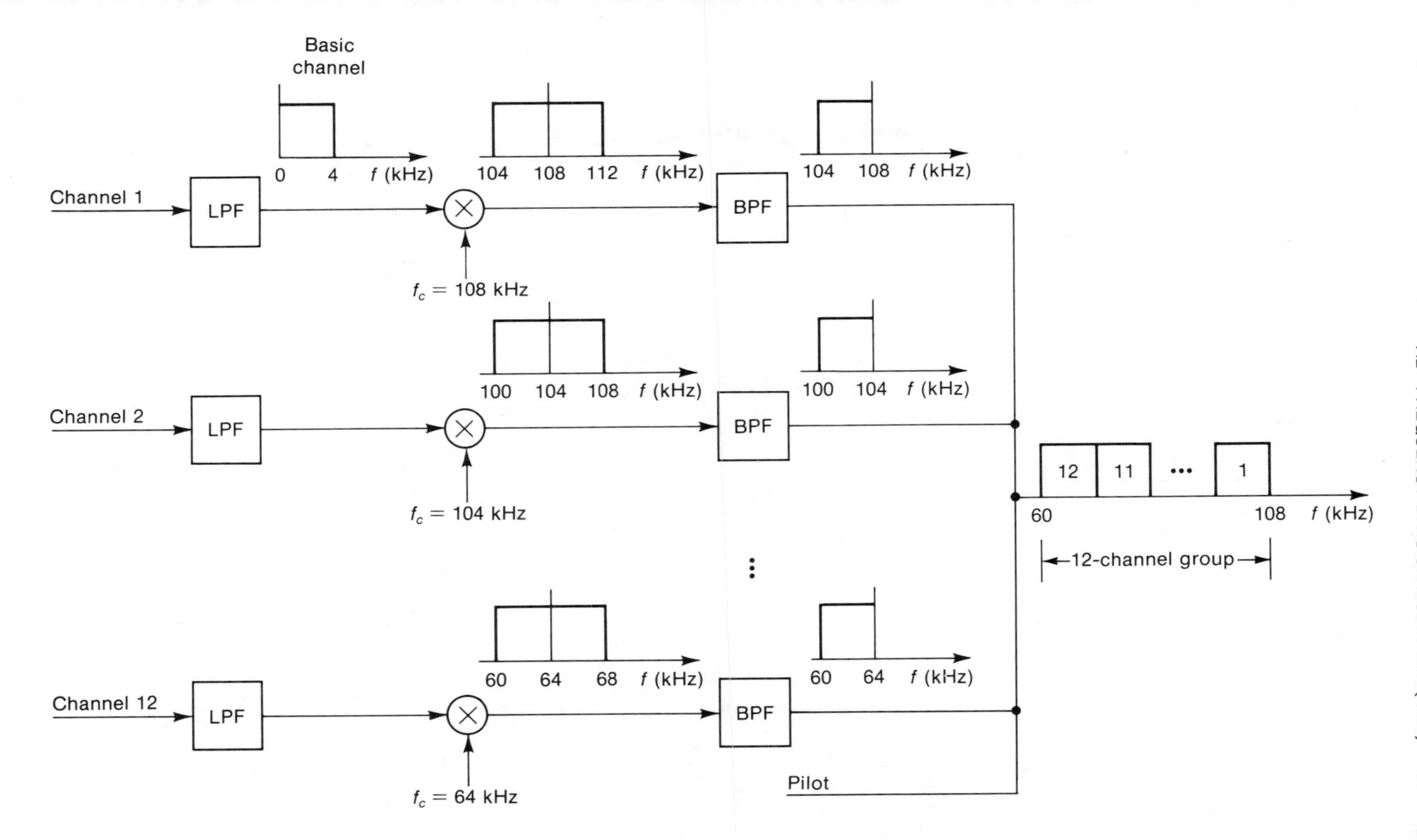

Figure 7.1 Operation of FDM Group

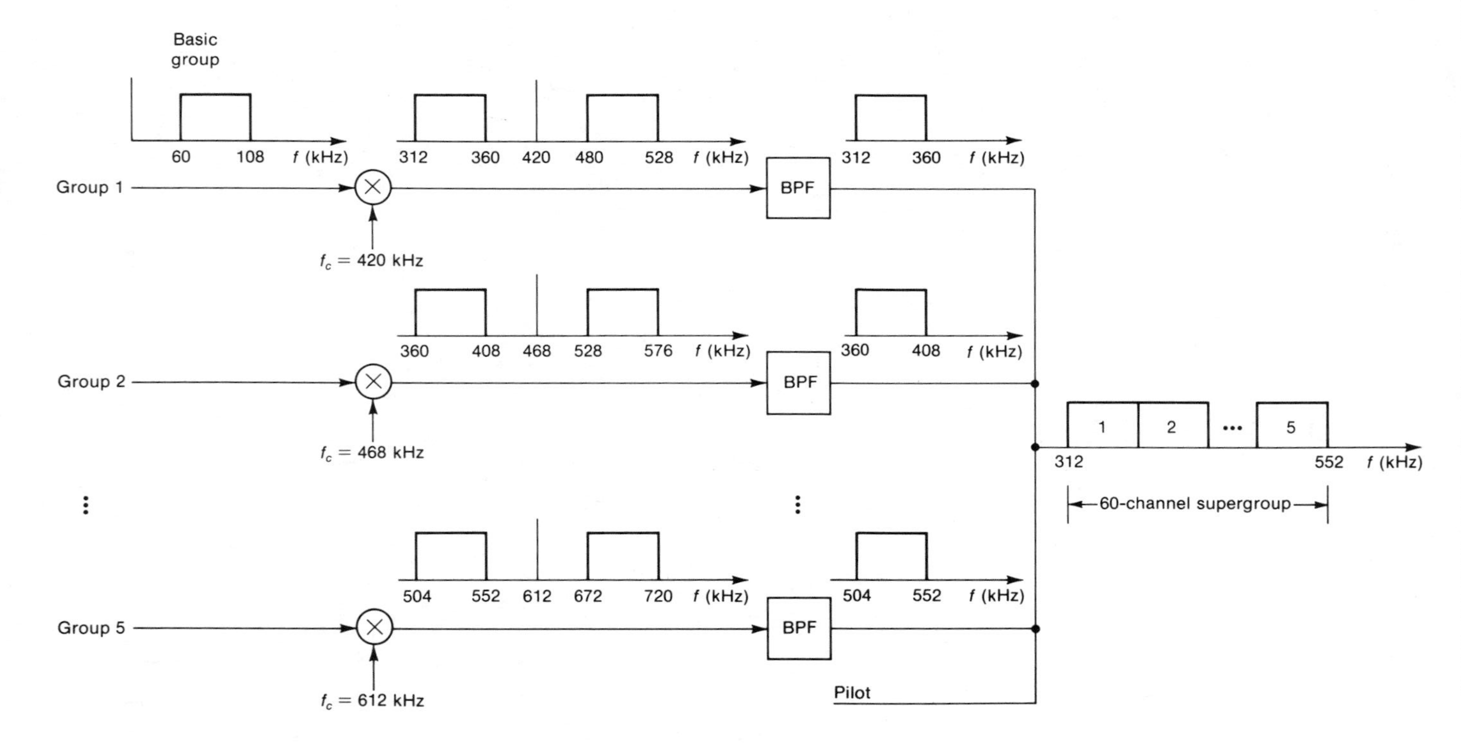

Figure 7.2 Operation of FDM Supergroup

bandpass-filtered to limit each signal to 4 kHz. By convention the lower sideband signal is selected, as illustrated in Figure 7.1. At the receiver, bandpass filters pick off each of the 12 signals for demodulation. Synchronization of the carrier frequencies between the transmitter and receiver is achieved by use of a pilot tone. The pilot is transmitted as part of the group signal, usually in one of the narrow guard bands used to separate voice channels. At the receiver, the pilot is recovered and used to synchronize the frequencies used in demodulation.

The second level in the FDM hierarchy is the **supergroup**, formed by combining five groups into a 240-kHz signal. Each group modulates a carrier with frequency 372 + 48n kHz (n = 1 to 5). A bandpass filter selects the lower sideband, and the five resulting signals are combined to form the supergroup, which occupies the frequency band 312 to 552 kHz. This modulation and filtering process is illustrated in Figure 7.2 for the supergroup signal.

The third level of the FDM hierarchy consists of the **mastergroup**, made up of five or ten supergroups, depending on the choice of the CCITT or AT&T standard. Higher levels of FDM also differ in these two standard hierarchies. Table 7.1 compares AT&T and CCITT FDM standards.

7.2.1 AT&T's FDM Hierarchy

Figure 7.3 illustrates the basic building blocks of AT&T's FDM hierarchy [1]. The basic voice channel has a 200 to 3400 Hz spectrum. Although intended primarily for voice transmission, the voice channel can also be used for data transmission provided that the data spectrum does not exceed the basic voice channel spectrum. Two examples are shown in Figure 7.3: the use of voice-band modems for rates up to 9.6 kb/s and the use of data multiplexers for combining a number of low-speed teletype channels. The 12-channel group is generated by use of single sideband (SSB) modulation by equipment known as an A-type channel bank. The 60-channel supergroup results from combining five groups using a group bank. Again using SSB modulation, the resulting supergroup signal occupies 240 kHz in the frequency range 312 to 552 kHz. The modulation process for the group and supergroup is depicted in Figures 7.1 and 7.2. Any other signal whose spectrum matches that of the group or supergroup may also be multiplexed by using this standard FDM equipment. For example, a 50-kb/s signal can be accommodated by a group bank by use of a group modem, which converts the 50-kb/s signal to a 48-kHz spectrum in the group frequency band. Another example is the transmission of a 250-kb/s signal in a 240-kHz supergroup by means of a supergroup modem.

The supergroup bank combines 10 supergroups into a 600-channel mastergroup. Single sideband modulation is again used to generate the mastergroup. As shown in Figure 7.3, two versions of the mastergroup exist. The L600 mastergroup occupies the 60 to 2788 kHz frequency band and is used for wideband transmission over the L1 coaxial cable system and

Table 7.1 CCITT vs. AT&T FDM Hierarchy

Number of Voice Channels	*Spectrum (kHz)*	*CCITT Standard [2]*	*AT&T Standard*
12	60 – 108	Group (Rec. G.232)	Group
60	312 – 552	Supergroup (Rec. G.233)	Supergroup
300	812 – 2044	Mastergroup (Rec. G.233)	
600	60 – 2788 (L600) 564 – 3084 (U600)		Mastergroup
900	8516 – 12,388	Supermastergroup (Rec. G.233)	
3600	564 – 17,548		Jumbogroup
10,800	3000 – 60,000		Jumbogroup multiplex

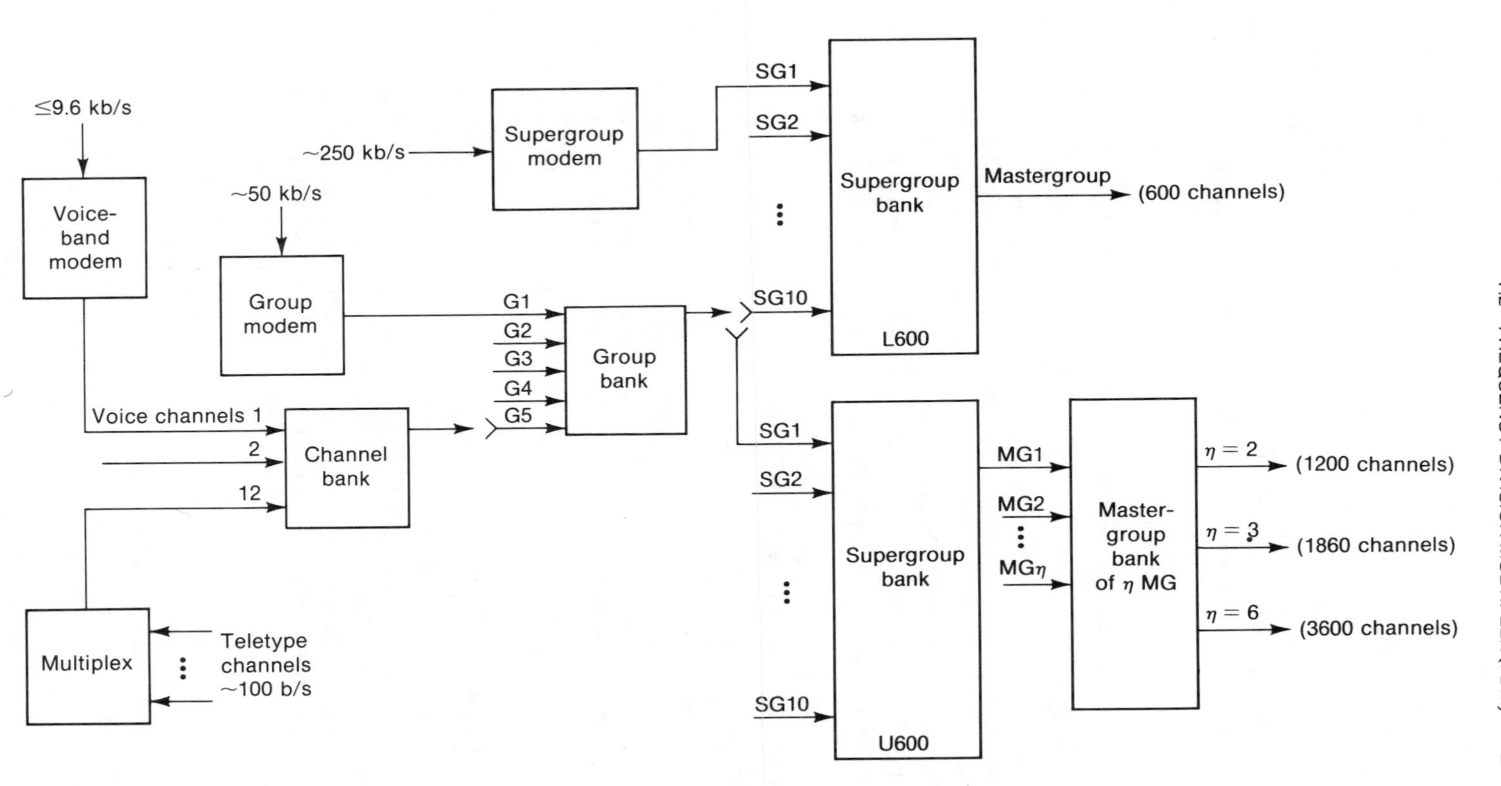

Figure 7.3 AT&T's FDM Hierarchy (Adapted from [1])

various analog microwave radio systems. The U600 mastergroup occupies the 564 to 3084 kHz frequency range and is used to build larger channel systems. One example of these larger groupings is the L3 carrier consisting of three mastergroups and one supergroup combined to provide 1860 channels. The L4 carrier is formed by combining six U600 mastergroups into a **jumbogroup** comprising 3600 voice channels and occupying the frequency band 564 to 17,548 kHz. The L5 carrier combines three jumbogroups into approximately a 60-MHz bandwidth that contains 10,800 channels.

7.2.2 CCITT's FDM Hierarchy

The standard group and supergroup as defined by the CCITT are the same as the AT&T standards [2]. However, the CCITT mastergroup contains only five supergroups comprising 300 voice channels. Using SSB modulation, the five supergroups are translated in frequency, resulting in a mastergroup that occupies the spectrum 812 to 2044 kHz. The highest level prescribed by the CCITT's FDM hierarchy is the **supermastergroup**, which contains three mastergroups and occupies the band 8516 to 12,388 kHz.

7.3 TRANSMISSION PARAMETERS

In this section we consider the transmission parameters that determine data communications performance over 4-kHz telephone circuits [3, 4]. Depending on the modem design, some of these parameters may have little effect on performance. For most modems, however, the amplitude and phase characteristics are of primary importance due to the intersymbol interference that results from amplitude and phase distortion.

7.3.1 Attenuation Distortion

Ideally, all frequencies of a signal experience the same loss (or gain) in traversing the transmission channel. Typical channels exhibit variation in loss with frequency, however, a form of distortion known as **attenuation** or **amplitude distortion**. Wire-line systems, for instance, are characterized by greater attenuation at high frequencies than at lower ones. Moreover, filters used to band-limit a signal cause greater attenuation at the band edges than at band center.

Attenuation distortion is specified by a limit placed on the loss at any frequency in the passband relative to the loss at a reference frequency. CCITT Rec. G.132 shown in Figure 7.4 allows up to 9 dB variation from the value expected at an 800 Hz reference frequency. This recommendation is for the case of a four-wire international connection comprising up to 12 circuits in tandem. AT&T uses a 1004 Hz reference frequency. Attenuation characteristics of the basic AT&T telephone circuit are given in Table 7.2.

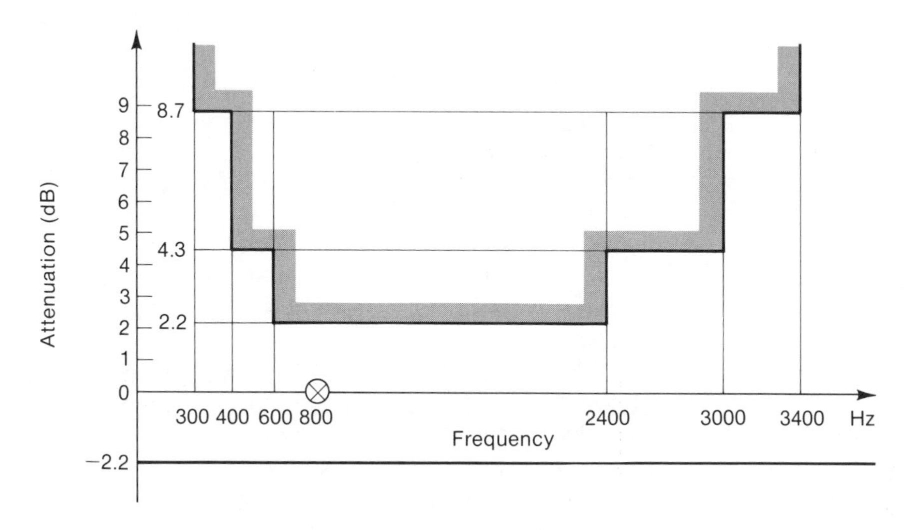

Figure 7.4 Attenuation Distortion Allowed by CCITT for a Four-Wire International Telephone Circuit (Courtesy CCITT [2])

7.3.2 Envelope Delay Distortion

The time required for signal transmission through a channel is finite and therefore produces a delay in the arrival time of the signal. For example, a waveform $\sin \omega t$ applied to the input of a transmission channel has an output waveform $\sin(\omega t - \beta)$. The phase shift β can be a function of frequency, so that different phase shifts may occur at various frequencies. The time delay between input and output waveforms is called **phase delay**, defined as

$$T_p = \frac{\beta}{\omega} \frac{\text{rad}}{\text{rad/s}} \tag{7.1}$$

If the phase delay varies with frequency, the output signal will be disturbed because of the difference in arrival time of each frequency. The difference between phase delay at two frequencies is termed **delay distortion**, defined by

$$T_d = \frac{\beta_2}{\omega_2} - \frac{\beta_1}{\omega_1} \tag{7.2}$$

For modulated waveforms, the envelope of the signal may also suffer distortion due to differences in propagation time between any two specified frequencies. This **envelope delay** or **group delay distortion** is defined as the variation in the slope of the phase shift characteristic:

$$T_e = \frac{d\beta}{d\omega} \tag{7.3}$$

Table 7.2 AT&T Conditioning Specifications

Channel Conditioning[a]	Attenuation Distortion (Frequency Response) Relative to 1004 Hz		Envelope Delay Distortion	
	Frequency Range (Hz)[b]	Variation (dB)[c]	Frequency Range (Hz)[b]	Variation (μs)
Basic	500–2500	−2 to +8	800–2600	1750
	300–3000	−3 to +12		
C1	1000–2400[d]	−1 to +3	1000–2400[d]	1000
	300–2700[d]	−2 to +6	800–2600	1750
	300–3000	−3 to +12		
C2	500–2800[d]	−1 to +3	1000–2600[d]	500
	300–3000[d]	−2 to +6	600–2600[d]	1500
			500–2800[d]	3000
C3 (access line)	500–2800[d]	−0.5 to +1.5	1000–2600[d]	110
	300–3000[d]	−0.8 to +3	600–2600[d]	300
			500–2800[d]	650
C3 (trunk)	500–2800[d]	−0.5 to +1	1000–2600[d]	80
	300–3000[d]	−0.8 to +2	600–2600[d]	260
			500–2800[d]	500
C4	500–3000[d]	−2 to +3	1000–2600[d]	300
	300–3200[d]	−2 to +6	800–2800[d]	500
			600–3000[d]	1500
			500–3000[d]	3000
C5	500–2800[d]	−0.5 to +1.5	1000–2600[d]	100
	300–3000[d]	−1 to +3	600–2600[d]	300
			500–2800[d]	600

[a]C conditioning applies only to the attenuation and envelope delay characteristics.
[b]Measurement frequencies will be 4 Hz above those shown. For example, the basic channel will have −2 to +8 dB loss, with respect to the 1004 Hz loss, between 504 and 2504 Hz.
[c](+) means loss with respect to 1004 Hz; (−) means gain with respect to 1004 Hz.
[d]These specifications are FCC-tariffed items. Other specifications are AT&T objectives.
Source: Reprinted from Reference [3] by permission of AT&T.

Envelope delay distortion is expressed relative to a reference frequency. For example, CCITT Rec. G.133 uses a 800 Hz reference frequency and AT&T uses 1004 Hz. Envelope delay found in a voice channel of an AT&T A5 channel bank is shown in Figure 7.5. The parabolic shape of envelope delay for the A5 channel bank is typical for FDM equipment, where maximum delays occur at the band edges and minimum delays at the center of the

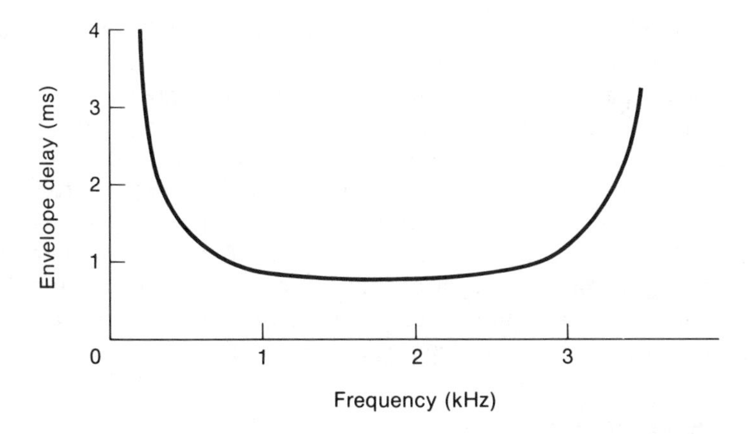

Figure 7.5 Envelope Delay Distortion of A5 Channel Bank [3]

band. Envelope delay distortion is specified by the maximum variation in envelope delay permitted over a band of frequencies. Table 7.2 indicates the limits on envelope delay distortion for the basic AT&T voice channel.

7.3.3 Noise

When amplitude and delay distortion are controlled with equalization, noise often becomes the fundamental limitation to data transmission performance. Noise found in telephone circuits can be classified as gaussian, impulse, or single frequency. Gaussian noise is due to thermal agitation of electrons in a conductor. Its spectrum is uniform over the whole frequency range; hence the noise is also white. The effects of gaussian white noise on probability of error in data transmission over telephone circuits has already been presented in Chapter 6, which described digital modulation techniques used in data modems.

Impulse noise is characterized by large peaks that are much higher than the normal peaks of channel (thermal) noise. Thus with the occurrence of impulse noise, the probability of data error is even higher than that caused by gaussian noise. Measurement is made by counting the number of impulses above a selected threshold in a certain test period. Impulse noise can be caused by natural sources, such as lightning, but most impulse noise can be traced to such sources as switching transients and normal maintenance, repair, and installation activities in the telephone plant. Because of the nature of impulse noise, mathematical characterization is not practical. Instead, direct measurement of a particular channel is used. The AT&T specification for voice channels limits impulse noise to 15 counts in 15 min, using a threshold of -22 dBm [3].

Examples of single-frequency interference are crosstalk and intermodulation. Crosstalk occurs when an unwanted signal path is coupled with the desired signal path. Intermodulation noise results from the presence of nonlinearities that cause intermodulation products, some of which appear in the passband. For data transmission, these single-frequency interferers have greatest effect on multiplexers that transmit several low-speed sources, such as teletype signals, over a single voice channel. Whereas gaussian noise is uniformly distributed over the voice band, a single-frequency interferer may present a higher noise level at a particular slot in the voice channel. The AT&T standard voice channel specifies single-frequency interference to be at least 3 dB below the thermal noise level [3].

7.3.4 Phase Jitter

Phase jitter arises from incidental frequency or phase modulation about the desired transition times of the data signal. Phase jitter contributes to data errors by reducing the receiver margin to other impairments. Sources of phase jitter include instabilities in power supplies and in oscillators used to generate carrier frequencies in FDM equipment. The most common frequency components of jitter are found at power frequencies (say 50 or 60 Hz), ringing currents (say 20 Hz), and at harmonics of these frequencies. The AT&T limit on jitter in the voice channel is 10° peak-to-peak for the frequency band of 20 to 300 Hz and 15° peak-to-peak for the band 4 to 300 Hz [5].

7.3.5 Level Variation

Standard practice in FDM systems is to set transmission levels to control crosstalk and intermodulation. Allowed level variations are also prescribed. For example, the channel loss for a 1004 Hz reference frequency is specified as 16 dB $\pm$ 1 dB by AT&T [3]. However, variations in the prescribed level occur due to short-term and long-term effects. Hence a data modem receiver must incorporate automatic gain control.

Short-term level variations occur typically over a period of a few seconds and may be due to maintenance actions such as patching, automatic switching to standby equipment, or fading in microwave radio systems. Longer-term level variations are caused primarily by temperature changes, component aging, and amplifier drift. AT&T limitations on level variations are ± 3 dB for short-term variations and ± 4 dB for long term, relative to the nominal 16-dB channel loss specification [3].

7.4 CONDITIONING

Of the transmission parameters described in Section 7.3, two critical parameters, attenuation distortion and envelope delay distortion, can be corrected by equalization. The remaining characteristics described in Section 7.3 can be corrected only by careful selection of the circuit configuration, which involves routing the circuit to avoid certain types of media or equipment. Equalization may be done by separate amplitude and envelope delay compensation networks or simultaneously in a single network. In either case, the objective is to flatten the amplitude and envelope delay response to prescribed levels.

An equalizer thus performs two functions. Using the amplitude of the reference frequency, the equalizer adds attenuation or gain to other frequencies until the entire amplitude response is flat. The second function of the equalizer is to add delay to center frequencies in order to match the delay found at the band edge. The overall result is to flatten the amplitude response and envelope delay characteristic to within a specified tolerance relative to values at the reference frequency. In practice, it is not possible to equalize amplitude and envelope delay completely over the entire channel bandwidth.

Equalization can be performed by the user's equipment, by the telephone company's equipment, or by a combination of the two. A user-provided modem is likely to include a form of equalization, which may be fixed or adjustable. Fixed or "compromise" equalizers are based on the average characteristics of a large number of telephone circuits. Adjustable equalizers are of two types: manual and automatic. For high-speed data transmission at rates above 4800 b/s, automatic adaptive equalization is now a standard feature in modems. Thus newer modems are designed to operate over basic (unconditioned) lines.

Equalization provided by the telephone company is known as **conditioning** and is available in different grades for voice-band channels. These grades correspond to different sets of allowed attenuation and envelope delay distortion. The required grade of conditioning depends on the bandwidth of the data signal; larger bandwidths require a greater degree of equalization. Table 7.2 shows the grades of circuit conditioning available on AT&T facilities. The various levels of conditioning are specified as C1 through C5. The attenuation and envelope delay characteristics of the basic voice channel are shown for comparison with conditioned circuits. Conditioned circuits are also prescribed by the CCITT for international leased circuits according to the following recommendations [6]:

- Rec. M.1020 (equivalent to AT&T C2): conditioning for use with modems that do not contain equalizers
- Rec. M.1025: conditioning for use with modems that contain equalizers
- Rec. M.1040: ordinary quality for applications that do not require conditioning

Circuit conditioning may be performed at various points in the transmission system. Most commonly, circuits are equalized at the receiver, which is termed **postequalization**. If the line characteristics are known, **preequalization** may be employed in which the transmitted signal is predistorted. If the predistortion amounts to the inverse characteristic of the line, then the overall amplitude and envelope delay response will be flat at the receiver. For long-haul international circuits, additional equalization may be required at the gateway nodes. This form of equalization, which combines the effects of postequalization and preequalization, is known as **midpoint equalization.**

A new type of conditioning, high-performance data conditioning or type D, specifies performance of two parameters: signal-to-noise ratio and harmonic distortion. Type D conditioning is independent of C conditioning and is achieved by careful selection of transmission facilities. The need for D conditioning arises with high-speed modems ($\geq$ 9.6 kb/s) whose performance is limited by noise and distortion.

7.5 VOICE-BAND MODEMS

Digital transmission over existing FDM telephone facilities was initially limited to modems operating over single voice-band channels. The earliest forms of voice-band modems used binary FSK. Rates up to 1800 b/s can be conveniently transmitted in a nominal 4-kHz channel using FSK. For binary FSK, two frequencies are required within the voice band, one used for a mark (1) and the other for a space (0). For example, CCITT Rec. V.23 specifies binary FSK for a 600/1200-b/s modem operating in the general switched telephone network, as indicated here:

	Mark Frequency	*Space Frequency*
Mode 1: up to 600 b/s	1300 Hz	1700 Hz
Mode 2: up to 1200 b/s	1300 Hz	2100 Hz

For slower rates, the voice-channel bandwidth can be divided into separate frequency subbands to accommodate several data signals. This form of FDM is known as voice frequency telegraph (VFTG) or voice frequency carrier telegraph (VFCT). Each data channel is allocated a certain frequency subband. FSK modulation is most commonly used, where the space and mark frequencies are contained within a specific subband. Typical applications range from multiplexing two 600-b/s channels into a composite 1200 b/s to multiplexing twenty-four 75-b/s channels into a composite 1800 b/s. Standards for VFCT modulation format have been developed for several applications. In these modulation plans, frequencies are usually spaced uniformly across the voice-channel bandwidth. For 75-b/s telegraph channels, 12 channels can be handled with 120 Hz spacing, 16 channels with 85 Hz spacing, and 24 channels with 60 Hz spacing.

Differential phase-shift keying (DPSK) has been commonly used to extend modem transmission rates beyond the limitations of FSK modems. Today, four-phase DPSK is a universal standard for 2400 b/s, as described by CCITT Rec. V.26; similarly, eight-phase DPSK has become a universal standard for 4800 b/s, as described by CCITT Rec. V.27. Table 7.3 and Figure 7.6 present characteristics of these two and other standard high-speed voice-band modems. DPSK is generally limited to rates up to 4800 b/s because of the susceptibility of PSK systems to phase jitter found on voice channels.

Further increases in bandwidth efficiency were introduced with the use of single sideband (SSB) and vestigial sideband (VSB) modulation. Modems designed with SSB or VSB proved to be complex and vulnerable to channel perturbations, however, especially amplitude and envelope delay distortion. The use of these linear modulation techniques introduced the need for automatic equalization. Both preset and adaptive modes of equalization have been used. (See Chapter 5 for a detailed description of equalization techniques.) Preset equalization uses a training sequence to fix the equalizer settings; adaptive equalization provides continuous updating. The adaptive equalizer is preferable and has become a universal feature in high-speed modems. To ensure convergence of the equalizer, data scrambling is required and this feature too has become standard in high-speed modems.

Advanced modulation and spectrum shaping techniques are today used to achieve up to 16-kb/s transmission rates. Filtering schemes such as partial response and raised cosine, described in Chapter 5, are used to control or eliminate intersymbol interference. Such spectrum shaping is often used in combination with the modulation technique in high-speed modems. For rates at 9600 b/s and higher, quadrature amplitude modulation (QAM) and various forms of amplitude/phase modulation (AM/PM) have been used. At 9600 b/s, AM/PM with a 16-state constellation is part of a universal standard, CCITT Rec. V.29, as indicated in Table 7.3 and Figure 7.6*c*. A 64-state constellation that has been implemented with a 16-kb/s modem also uses AM/PM as illustrated in Figure 7.6*d* [8]. Attendant with these advances in modulation techniques, sophisticated equalization and carrier recovery have been incorporated into modems to the point where conditioned lines may not be required.

Other important transmission characteristics of voice-band modems are briefly described here:

- *Synchronization mode*: Modems operate with either asynchronous or synchronous interface with the data terminal equipment (DTE). Asynchronous data terminals do not use a clock signal to define data transitions; synchronous data terminals use a clock provided by either the DTE or modem. In general, low-speed modems (below 1200 b/s) operate asynchronously and high-speed modems synchronously.
- *Transmission mode*: Most modems operate with simultaneous two-way data exchange (**duplex**). When the modulator and demodulator share the

Table 7.3 Transmission Characteristics of Standard High-Speed Voice-Band Modems [7]

Characteristic	*CCITT Rec. V.26 bis*	*CCITT Rec. V.27 ter*	*CCITT Rec. V.29*
Data rate	2400 b/s $\pm$ 0.01%	4800 b/s $\pm$ 0.01%	9600 b/s $\pm$ 0.01%
Modulation rate	1200 symbols/s $\pm$ 0.01%	1600 symbols/s $\pm$ 0.01%	2400 symbols/s $\pm$ 0.01%
Carrier frequency	1800 $\pm$ 1 Hz	1800 $\pm$ 1 Hz	1700 $\pm$ 1 Hz
Spectrum shaping	N / A	50% raised cosine	N / A
Modulation type[a]	4-DPSK	8-DPSK	16-state AM / PM
Equalization	Fixed compromise	Automatic adaptive	Automatic adaptive
Scrambler	N / A	Self-synchronizing with length $2^7 - 1$	Self-synchronizing with length $2^{23} - 1$

[a]For signal constellations see Figure 7.6.

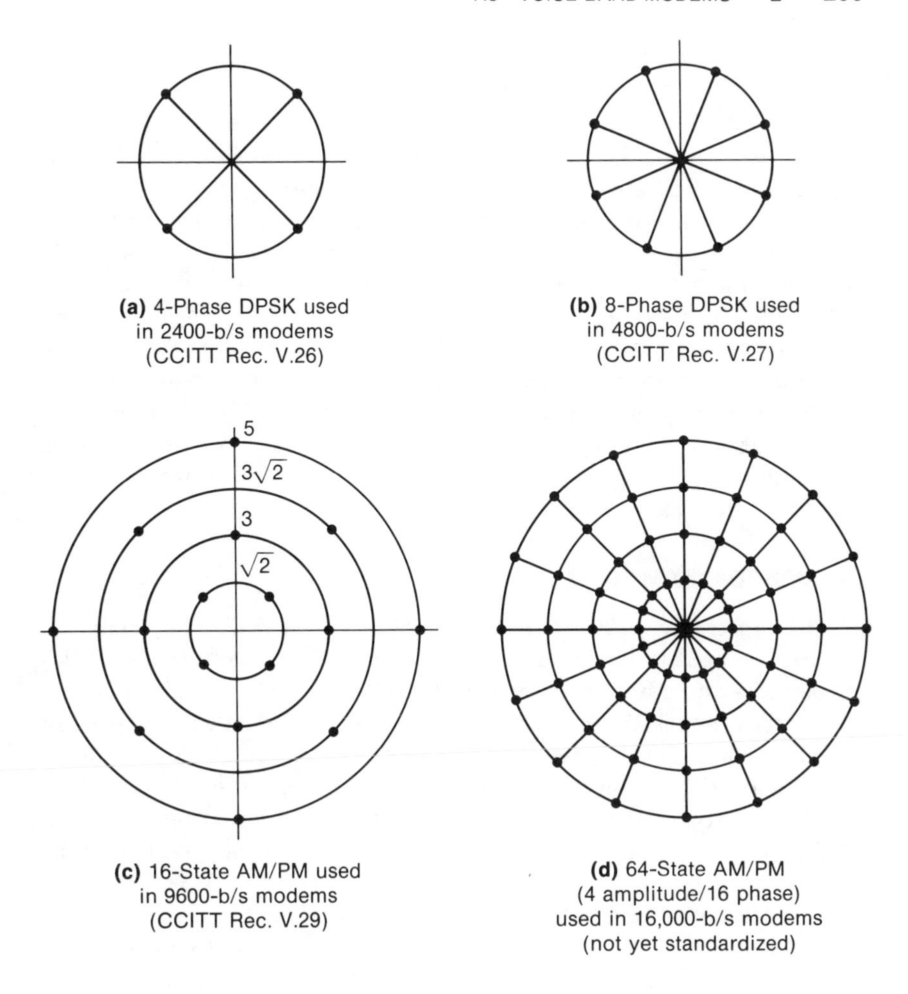

(a) 4-Phase DPSK used in 2400-b/s modems (CCITT Rec. V.26)

(b) 8-Phase DPSK used in 4800-b/s modems (CCITT Rec. V.27)

(c) 16-State AM/PM used in 9600-b/s modems (CCITT Rec. V.29)

(d) 64-State AM/PM (4 amplitude/16 phase) used in 16,000-b/s modems (not yet standardized)

Figure 7.6 Standard Signal Constellations for Voice-Band Modems

same line, however, **half-duplex** transmission is required in which the modem can transmit and receive data but not simultaneously. **Simplex** modems can only transmit or receive data.

- *Calling mode*: *Originate* modems can access the telephone line and initiate transmission. *Answer* modems can respond to a call but cannot initiate one. *Originate/answer* modems can perform both functions.
- *Terminal interface*: Compatibility between the DTE and modem requires that data, timing, and control lines be specified. Standard interfaces have been developed by the EIA and CCITT (see Table 2.1).
- *Line interface*: In general, half-duplex modems interface with two-wire lines and duplex modems interface with four-wire lines. For low-speed

transmission, however, modems can be designed to operate duplex over two-wire lines by use of separate frequency bands between transmit and receive carriers.

- *Echo suppressors*: Echoes are a problem on long-distance telephone connections, especially by satellite. Echo suppressors solve this problem by detecting the direction of transmission of the talking party and introducing loss into the return transmission path. For duplex data transmission, however, echo suppressors must be disabled or removed. In the AT&T network, echo suppressors are disabled by a single-frequency tone within the band 2010 to 2240 Hz applied for at least 400 ms. The echo suppressors remain disabled as long as there is continuous energy in the channel. Any interruption in the data signal over 100 ms in duration reactivates the echo suppressors. Echo suppressors, however, are being replaced with echo cancellers that do not impair modem performance and eliminate the requirement for disabling.

7.6 WIDEBAND MODEMS

For data rates above 20 kb/s, higher levels in the FDM hierarchy are required for modem operation. The FDM group with 48 kHz bandwidth can be accessed with a group modem to provide wideband data transmission. Earliest applications of group modems allowed data rates of 19.2 or 38.4 kb/s using simple ASK, FSK, or PSK modulation. These choices of data rate were extensions of the teletype hierarchy given by 75×2^n b/s, with n equal to an integer. Recent standards for group modems have been based on an $8n$-kb/s standard, which has evolved from the 8-kHz sampling rate for digitized voice. CCITT Recs. V.35, V.36, and V.37 for group modems are based on this $8n$-kb/s standard, as shown in Table 7.4. For data rates above 72 kb/s, automatic adaptive equalization techniques have been applied to group modem design, resulting in rates of up to 168 kb/s, as specified by CCITT Rec. V.37.

The next level in the FDM hierarchy is the 240-kHz bandwidth supergroup. Data rates on the order of 250 kb/s can be transmitted via a supergroup modem, such as the Western Electric 303 data set. Using efficient modulation types and automatic adaptive equalization, supergroup modems have been demonstrated for rates up to 800 kb/s, although there appears to be insufficient demand for the introduction of such rates.

An increasing demand for wideband digital transmission over existing microwave systems has led to the development of wideband modems that interface directly with an analog radio. Frequency modulation has long been a worldwide standard in microwave radio systems in conjunction with standard FDM equipment for multichannel voice transmission. With increased use of PCM repeatered lines, however, modems have been developed to allow direct interface of PCM and TDM equipment with analog FM radios, thereby eliminating the need for converters between FDM and TDM systems.

Table 7.4 Transmission Characteristics of CCITT-Recommended Group Modems [7]

Characteristic	*CCITT Rec. V.35*	*CCITT Rec. V.36*	*CCITT Rec. V.37*
Data rates	48 kb/s	48, 56, 64, and 72 kb/s	96, 112, 128, 144, and 168 kb/s
Modulation type	AM, suppressed carrier	3-level partial response	7-level partial response
Equalization	N/A	N/A	Automatic adaptive
Scrambler	Self-synchronizing with length $2^{20} - 1$	Self-synchronizing with length $2^{20} - 1$	Self-synchronizing with length $2^{20} - 1$

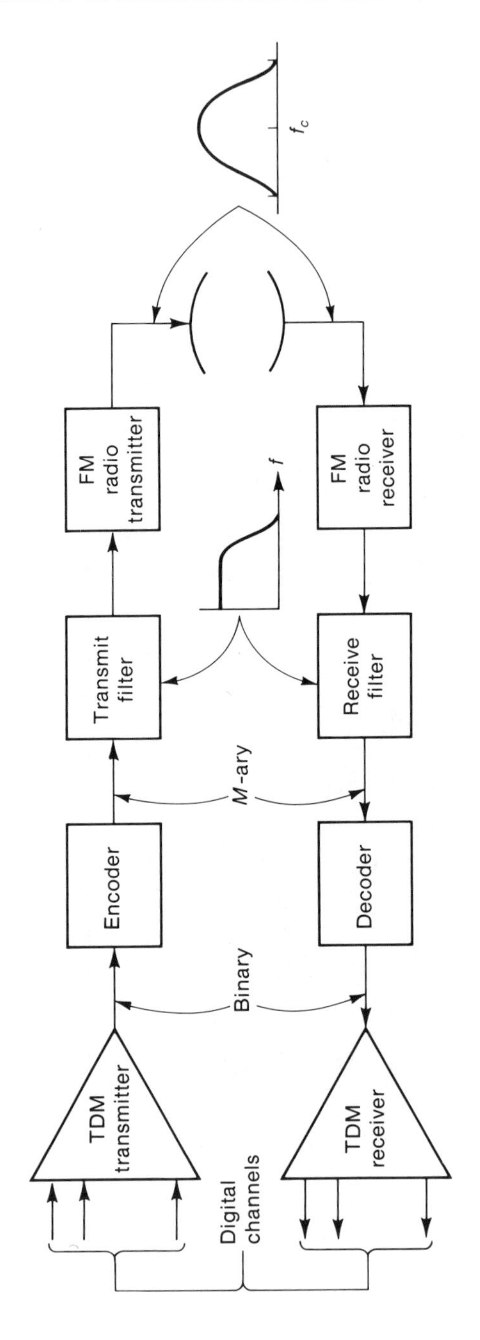

Figure 7.7 Digital Transmission Over FM Microwave

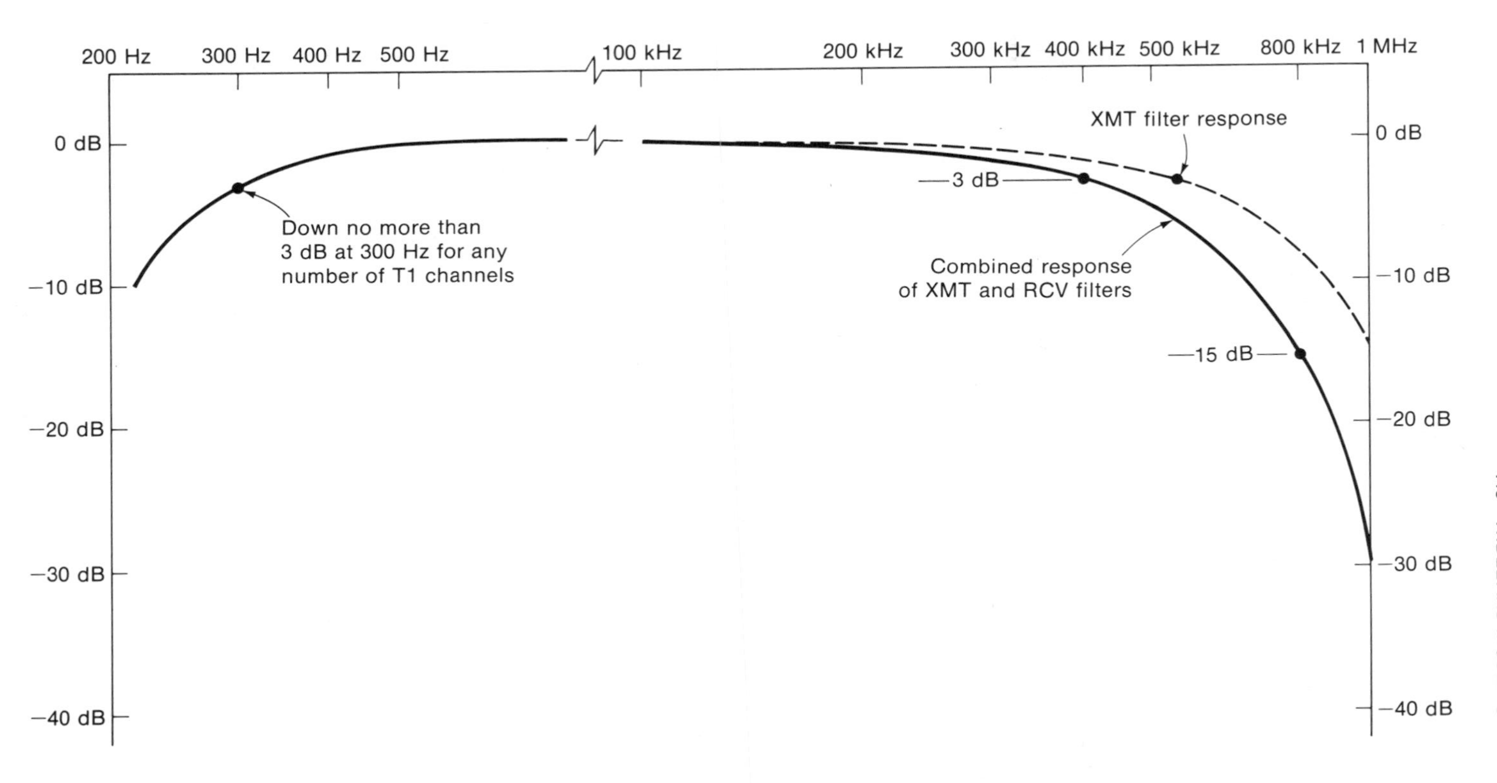

Figure 7.8 Partial Response Filter Characteristic for Digital FM Application (Reprinted by permission of California Microwave, Inc.)

The basic functions of a digital FM microwave radio system are shown in Figure 7.7. The transmitting TDM combines digital channels into a single serial bit stream. The multiplexer output is encoded into an M-ary baseband signal, which is then shaped by the transmit filter. After transmission over the FM radio link, the demodulated signal is passed through the receive filter and decoder to recover the TDM signal. Typical choices of filter characteristic are raised cosine or partial response, split equally between the transmitter and receiver as shown in Figure 7.7. Implementation alternatives are to build the encoder and filter into the radio or multiplexer [9] or as separate equipment [10].

Digital FM has been used to provide radio transmission of multiple T1 and T2 digital channels. For example, the TD-2 FM radio used by AT&T has been configured for digital transmission of three T2 signals, at a total bit rate of 20.2 Mb/s, using four-level encoding with 50 percent raised-cosine filtering [11]. Partial response filtering is also commonly used in digital FM

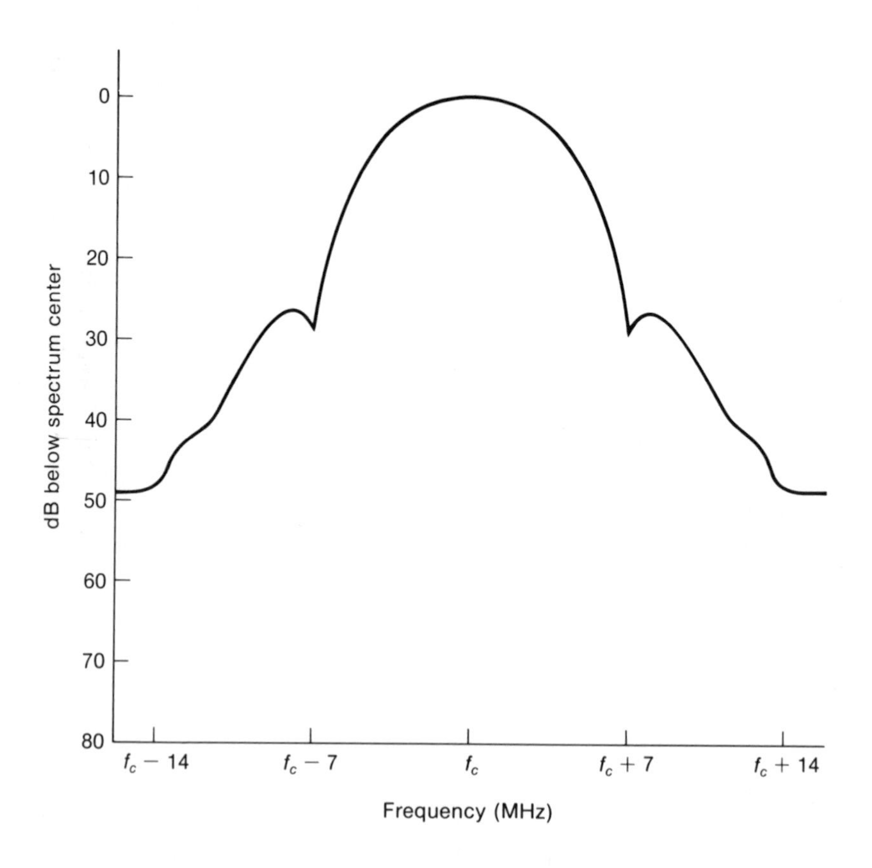

Figure 7.9 RF Spectrum of 12.6-Mb/s Digital FM Radio Using Partial Response

implementations in both three-level [12] and seven-level [13] versions. Figure 7.8 plots the transmit and combined filter characteristic, normalized to a single T1 (1.544 Mb/s) channel, for one implementation of three-level partial response.* The RF spectrum depends on the baseband spectrum and the frequency deviation of the radio. For example, using the partial response filter of Figure 7.8, Figure 7.9 indicates the RF spectrum occupied by an eight-T1-channel multiplexer (12.6 Mb/s) with 4-MHz FM deviation of the radio. For this application, 99 percent of the power is contained within a 14 MHz bandwidth. This configuration of an eight-T1-channel TDM with built-in partial response filtering operating over FM radio links with a 14 MHz bandwidth has been commonly used in U.S. military communication systems [14].

7.7 TRANSMULTIPLEXERS

The interface between FDM and TDM systems can be accomplished by use of back-to-back analog and digital channel banks with individual voice channels connected in between. An alternative to this conventional approach is the **transmultiplexer**, which directly translates FDM into PCM signals and vice versa. The advantage of transmultiplexers stems from the use of a single piece of equipment versus two channel banks. These advantages are realized while meeting or exceeding the performance for a tandem connection of PCM and FDM channel banks. The primary applications of transmultiplexers are the interface of digital switches with analog transmission facilities and the interface of analog with digital transmission facilities.

In the design and application of transmultiplexers, three standard configurations have been adopted:

1. Translation between the 60-channel supergroup and two 30-channel, European standard, PCM multiplexers
2. Translation between two 12-channel groups and the 24-channel, North American standard, PCM multiplexer
3. Translation between two 60-channel supergroups and five 24-channel, North American standard, PCM multiplexers

The first two configurations have been recognized by the CCITT in Recs. G.793 and G.794, respectively [15]. Salient characteristics described in these CCITT recommendations are digital and analog interfaces, correspondence between analog (3 kHz) and digital (64 kb/s) channels, synchronization of

*In Figure 7.8 the high end of the frequency scale is normalized. To obtain the actual filter response, multiply the frequency scale by the number of T1 channels to be serviced by the multiplexer. For example, a four-channel multiplexer requires a combined filter response that is down 3 dB at four times 400 kHz or 1.6 MHz.

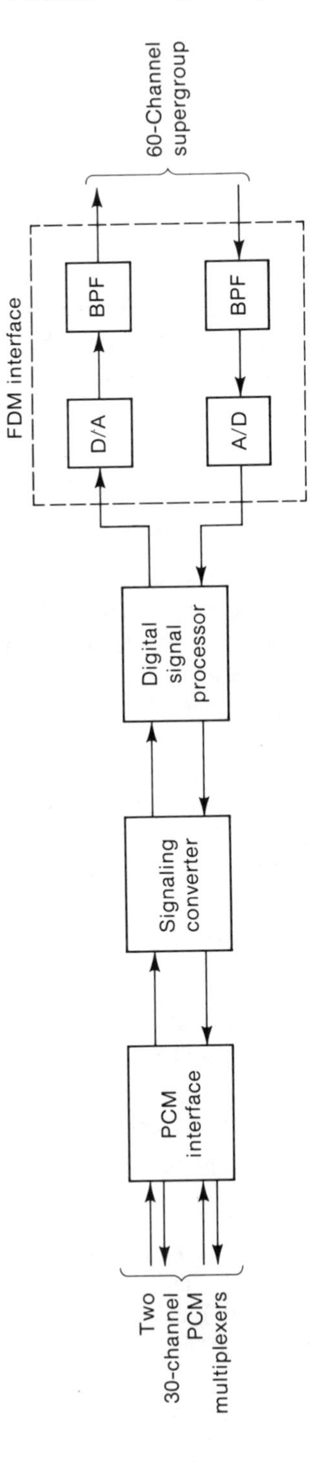

Figure 7.10 Block Diagram of 60-Channel Transmultiplexer

the transmultiplexer with PCM and FDM equipment, and operation with different types of signaling.

The algorithms used in FDM–TDM translation in general are based on the use of digital signal processing as illustrated in Figure 7.10. The FDM signal is digitized, and processing such as filtering, modulation, and amplification is performed on the digital representation of the FDM signal to produce a PCM signal. In the reverse direction, the PCM signal is digitally processed to produce a digital version of a FDM signal. A digital-to-analog converter is then used to produce the conventional FDM signal. Many design approaches have been used [16], and no single technique is considered standard practice in the telephone industry today.

7.8 HYBRID TRANSMISSION SYSTEMS

With the growing need for digital transmission of nontelephone signals, such as data, facsimile, and visual telephony, hybrid transmission systems have been developed that carry FDM voice and digital services on the same transmission media. Existing FDM transmission systems can be adapted to simultaneously carry a digital signal within the baseband of a cable or microwave system. Three basic methods have been used. Their names indicate where the digital spectrum is located with respect to the FDM spectrum: **data under voice** (DUV), **data in voice** (DIV), and **data above voice** (DAV), which is sometimes called **data over voice** (DOV).

As part of its Digital Data Service, AT&T uses its existing microwave radio systems to carry 1.544-Mb/s data via the DUV technique [17]. The AT&T U600 mastergroup occupies the band from 564 to 3084 kHz, so that the lower 564 kHz of baseband spectrum is available for data transmission. Use of the AT&T L600 mastergroup requires removal of the lower two supergroups (60 to 564 kHz) to provide the same 564-kHz spectrum for data. Figure 7.11 indicates the scheme used to transmit 1.544-Mb/s data within the lower 564-kHz frequency band of the baseband spectrum. A clock recovery circuit extracts a 1.544-MHz clock from the 1.544-Mb/s bipolar input and distributes it with a 772-kHz half-rate clock to the appropriate transmitter circuits. A scrambler converts the bipolar signal to a unipolar format and then scrambles the signal to prevent discrete spectral components that could interfere with the FDM channels. The encoder converts the binary input to a four-level rate of 772 kilosymbols per second. The class 4 partial response filter shapes the digital signal, resulting in a seven-level signal with a spectral null at 386 kHz. The DUV receiver reconstructs the 1.544-Mb/s bipolar signal from the seven-level partial response signal by the inverse functions of the transmitter.

DUV systems have been employed with other FDM microwave systems, such as those operated by Western Union in the United States and by the Canadian Dataroute System. In the Western Union application of DUV, the baseband of a 6-GHz, 1200-channel microwave system is divided into two frequency bands by filtering techniques. Using three-level partial

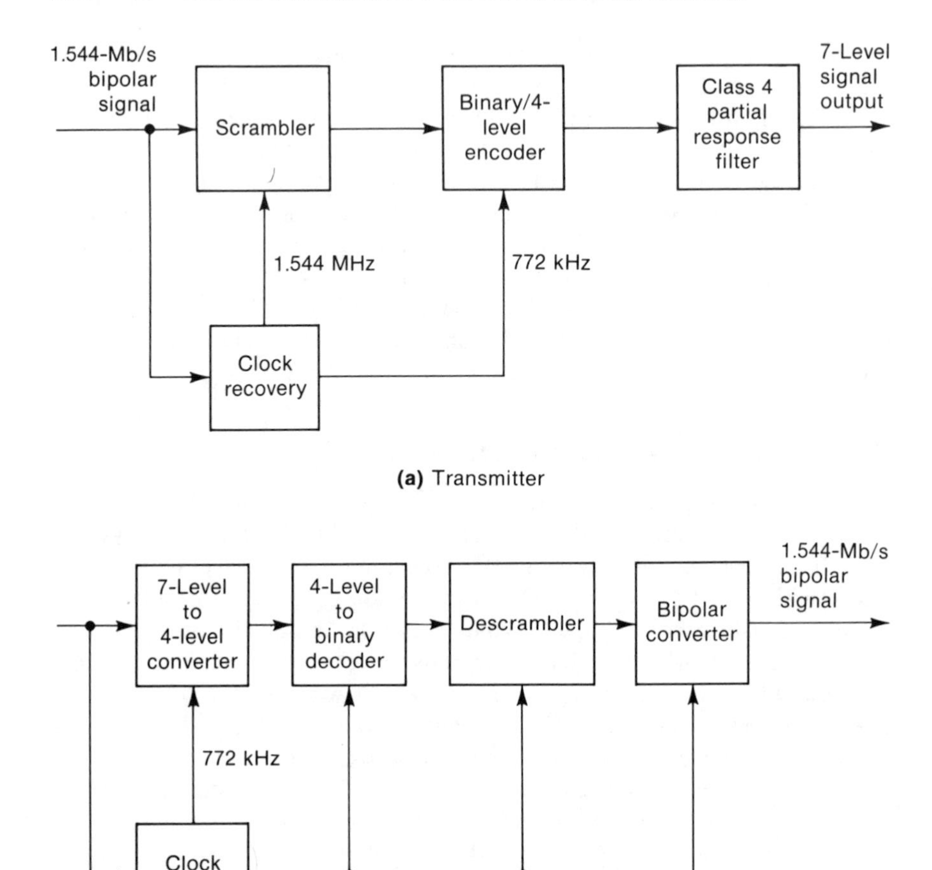

Figure 7.11 Data Under Voice (DUV) System Used for Transmission of 1.544 Mb/s

response, a scrambled 6.312-Mb/s digital signal is compressed into the lower 3.2 MHz of baseband spectrum. The FDM spectrum occupies the remaining baseband spectrum between 4.0 and 6.8 MHz and provides 600 voice channels. Complementary transmit and receive filters are used to achieve the necessary partial response characteristic. This hybrid baseband signal is shown in Figure 7.12. Note that the service channel subcarrier and the pilot, two necessary components of a radio transmission system, must be placed above the data and voice spectra.

The use of data above voice (DAV) requires that the data spectrum be translated above the FDM spectrum. FSK, PSK, and digital AM can be

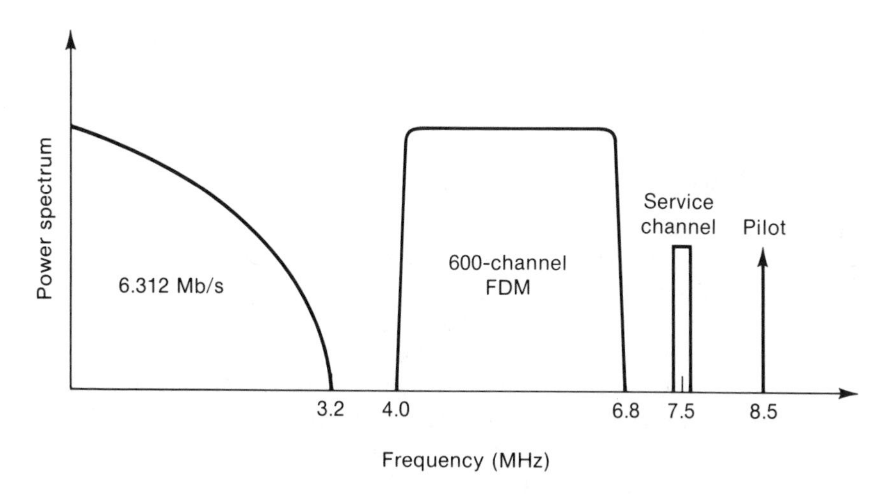

Figure 7.12 Baseband Spectrum of Western Union Hybrid Transmission System

used to place the data in the desired frequency spectrum. An advantage of the DAV technique over DUV is that FDM signals and any service channels, which normally may be located at the lower end of the spectrum, do not require translation. DAV has been adopted by Canadian National and Canadian Pacific for transmission of 1.544 Mb/s over a 6-GHz FDM microwave system [18]. A DAV system for 2.048-Mb/s transmission over FDM systems has been recommended by CCITT Rec. G.941 [15] and has been developed in hardware [19].

The data in voice (DIV) technique has not been as commonly used as DUV or DAV. Nevertheless, a provision for 6.312-Mb/s DIV transmission in place of a 1232-kHz mastergroup is given in CCITT Rec. G.941 [15].

The advantage of hybrid digital/analog transmission is the ability to carry digital and analog signals over a single microwave or cable system when there is not sufficient analog or digital traffic to justify separate systems. Hybrid transmission has greater equivalent digital capacity and performance than an analog system using modems to carry data within FDM channelization. The AT&T version of DUV described here provides approximately 4 bps/Hz spectral efficiency, for example, while the Western Union approach provides 2 bps/Hz. Hybrid techniques also allow regeneration of the digital signal with every radio or cable repeater. Finally, hybrid transmission is more efficient than using PCM coding for voice channels. Using standard 64-kb/s PCM, 600 voice channels would require 40 Mb/s compared to the 2.5-MHz bandwidth required for a 600-channel FDM mastergroup. Until long-haul digital links become pervasive, hybrid transmission techniques will be a popular compromise between existing analog transmission systems and growing requirements for digital services.

7.9 SUMMARY

With the predominance of the telephone network in today's communication systems, the application of digital transmission must often coexist with analog transmission facilities. Frequency division multiplex (FDM), in which telephone channels are allocated different portions of the spectrum, is the backbone of most telephone networks. Each of the different multiplex levels in the FDM hierarchy is used to combine lower-level FDM signals, but digital signals can also be accommodated via modems, devices that convert a digital signal into a format compatible with FDM transmission. The AT&T FDM hierarchy is composed of the 12-channel group that occupies a 48 kHz bandwidth, the 60-channel supergroup that occupies a 240 kHz bandwidth, the 600-channel mastergroup that occupies approximately a 2.5 MHz bandwidth, the 3600-channel jumbogroup that occupies a 17 MHz bandwidth, and the 10,800-channel jumbogroup multiplex that occupies a 60 MHz bandwidth. The CCITT and AT&T hierarchies differ above the supergroup level—the CCITT hierarchy has a 300-channel mastergroup and a 900-channel supermastergroup (see Table 7.1).

Data transmission over 4-kHz telephone channels is affected by attenuation distortion, envelope delay distortion, noise, phase jitter, and level variation. For most voice-band modems, attenuation and group delay distortion are the dominant sources of degradation. Attenuation distortion is specified by a limit placed on the loss at any frequency relative to the loss at a reference frequency. Envelope delay distortion is specified by the maximum variation in envelope delay. Both attenuation and envelope delay distortion can be corrected by use of equalization, which adds attenuation or gain to flatten the amplitude response and adds delay to flatten the group delay. Equalization may be included in the modem or supplied by the telephone company. Equalizers found in modems may be fixed or adaptive, although automatic adaptive equalization is standard at rates above 4800 b/s. Equalization provided by the telephone company is termed conditioning and is available in different grades (see Table 7.2).

Voice-band modems provide transmission of data rates up to 20 kb/s over 4-kHz telephone channels. Early modem applications, however, were limited to about 1800 b/s and used frequency-shift keying (FSK). Rates up to 4.8 kb/s are provided by use of differential phase-shift keying (DPSK). Higher data rates have been achieved through use of combined amplitude/phase modulation (AM/PM). Modem characteristics for data rates to 9.6 kb/s have now been standardized by the CCITT (see Table 7.3). Rates above 20 kb/s require the use of higher levels in the FDM hierarchy. Group modems, for example, provide transmission of rates up to 168 kb/s in a 48-kHz group bandwidth. Demands for even higher data rates have led to the replacement of the entire FDM equipment with a modem that interfaces directly with analog radio or cable systems.

Alternative approaches to conventional modems have been developed to provide greater efficiency in combining analog and digital transmission. The transmultiplexer, for example, directly translates FDM into PCM signals and vice versa, thereby avoiding the use of back-to-back PCM and FDM channel banks. Existing FDM systems have been adapted to carry data by insertion of the data signal under, in the middle, or above the FDM spectrum. These hybrid transmission systems are known respectively as data under voice (DUV), data in voice (DIV), and data above voice (DAV). When both digital and analog signals are to be carried over a single microwave or cable system, hybrid transmission systems provide a more bandwidth-efficient approach than the use of individual modems for data signals or PCM for voice channels.

REFERENCES

1. Members of the Technical Staff, Bell Telephone Laboratories, *Transmission Systems for Communications* (Winston-Salem: Western Electric Company, 1971).
2. CCITT Yellow Book, vol. III.2, *International Analogue Carrier Systems. Transmission Media—Characteristics* (Geneva: ITU, 1981).
3. *Data Communications Using Voiceband Private Line Channels*, Bell System Tech. Ref. Pub. 41004 (New York: AT&T, 1973).
4. *Transmission Parameters Affecting Voiceband Data Transmission—Description of Parameters*, Bell System Tech. Ref. Pub. 41008 (New York: AT&T, 1974).
5. *Notes on the Network* (New York: AT&T, 1980).
6. CCITT Yellow Book, vol. IV.2, *Maintenance; International Voice-Frequency Telegraphy and Facsimile, Internationally Leased Circuits* (Geneva: ITU, 1981).
7. CCITT Yellow Book, vol. VIII.1, *Data Communication Over the Telephone Network* (Geneva: ITU, 1981).
8. RADC TR-76-311, "16 kb/s Data Modem Techniques," Rome Air Development Center, Rome, N.Y., October 1976.
9. W. E. Fleig, "A Stuffing TDM for Independent T1 Bit Streams," *Telecommunications* 6(7)(July 1972):23–32.
10. J. L. Osterholz and M. K. Klukis, "Spectrally Efficient Digital Transmission Using Analog FM Radios," *IEEE Trans. on Comm.*, vol. COM-27, no. 12, December 1979, pp. 1837–1841.
11. C. W. Broderick and R. W. Gutshall, "A 20 Mbps Digital Terminal for TD-2 Radio," *Conference Record ICC 1969*, June 1969, pp. 27-21 to 27-26.
12. T. L. Swartz, "Performance Analysis of a Three-Level Modified Duobinary Digital FM Microwave Radio System," *Conference Record ICC 1974*, June 1974, pp. 5D-1 to 5D-4.
13. A. Lender, "Seven Level Correlative Digital Transmission Over Radio," *Conference Record ICC 1976*, June 1976, pp. 18-22 to 18-26.
14. *PCM/TDM System Design Verification Test Program*, U.S. Defense Communications Agency, Reston, Va., February 1972.

15. CCITT Yellow Book, vol. III.3, *Digital Networks—Transmission Systems and Multiplexing Equipment* (Geneva: ITU, 1981).
16. S. L. Freeny, "TDM/FDM Translation as an Application of Digital Signal Processing," *IEEE Comm. Mag.* 18(1)(January 1980):5–15.
17. K. L. Seastrand and L. L. Sheets, "Digital Transmission Over Analog Microwave Radio Systems," *Conference Record ICC 1972*, June 1972, pp. 29-1 to 29-5.
18. K. Feher, R. Goulet, and S. Moris, "1.544 Mbit/s Data Above FDM Voice and Data Under FDM Voice Microwave Transmission," *IEEE Trans. on Comm.*, vol. COM-23, no. 11, November 1975, pp. 1321–1327.
19. H. Panschar and O. Ringelhaan, "Data Above Baseband Modem for Analog Radio Relay Systems," Siemens Telecom Report 2, Special Issue, "Digital Transmission," 1979, pp. 142–143.

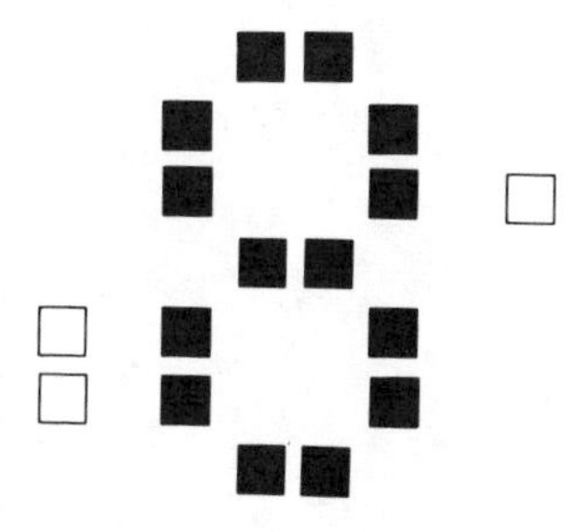

Digital Cable Systems

OBJECTIVES

- Provides the theory and practices necessary for designing digital cable systems
- Describes twisted-pair, coaxial, and optical cable, including physical and propagation characteristics
- Explains the design and operation of regenerative repeaters in detail
- Assesses the performance of tandemed repeaters in terms of bit error rate, jitter, and crosstalk
- Presents methods for calculating repeater spacing for each of the three cable types
- Discusses additional implementation factors including power feeding, fault location, automatic protection switching, and orderwires

8.1 INTRODUCTION

Digital transmission over cable employs the basic elements shown in Figure 8.1. At each end office the digital signal to be transmitted is converted to a form suitable for cable transmission using the baseband coding techniques described in Chapter 5. In the case of metallic cable systems, the digital signal can then be directly interfaced with the cable, whereas for optical cable systems the electrical signal must first be converted into light pulses by an optical transmitter. After propagation along the cable, the received signal is an attenuated and distorted version of the transmitted signal. Digital

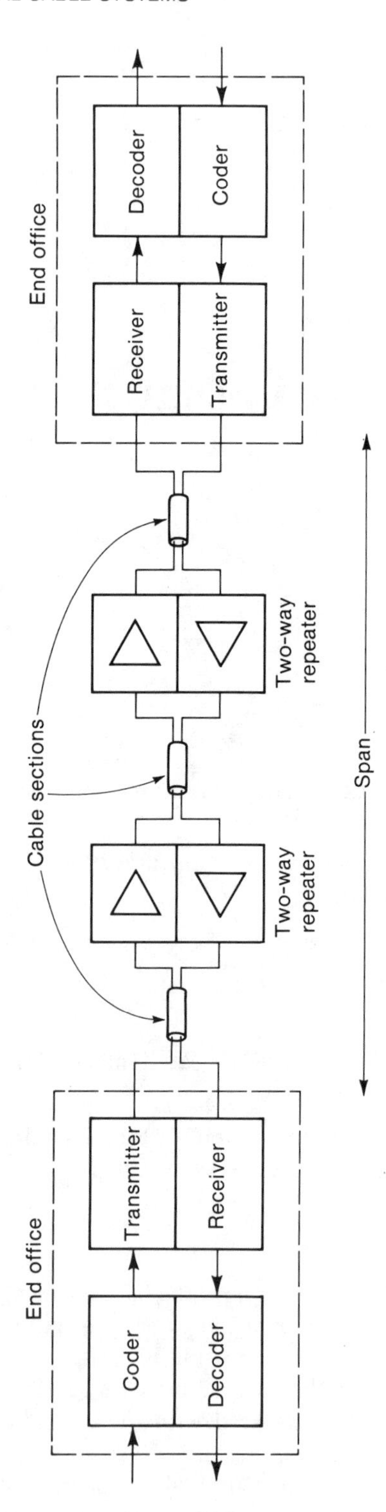

Figure 8.1 Digital Transmission Over Cable

repeaters are stationed at regular intervals along the transmission path to provide reshaping, retiming, and regeneration of the received signal. In the case of optical cable systems a repeater includes an optical receiver and transmitter. The receiver in each end office is identical to a one-way repeater, although for optical cable transmission the receiver must first provide optical-to-electrical signal conversion.

8.2 CABLE CHARACTERISTICS

Three cable types are considered here: twisted-pair, coaxial, and optical. Historically, twisted-pair was the first of these cable types to be used for digital transmission, dating back to the early 1960s with the first T-carrier systems. Limitations in bandwidths, repeater spacing, and crosstalk performance, however, have led to use of coaxial and more recently to optical cable systems. When compared to metallic cable systems, optical cable systems provide greater bandwidths, lower attenuation, freedom from crosstalk and electrical interference, and potentially lower costs.

8.2.1 Twisted-Pair Cable

Twisted-pair cable consists of two insulated conductors twisted together. The standard configuration uses copper conductors and plastic (such as polyethylene) or wood pulp insulation. For multipair cable assemblies, the individual cable pairs are grouped in units or concentric layers to form the core of the cable. An outer sheathing, of lead or plastic, is applied over the core for protection. Adjacent pairs are twisted at a different pitch to minimize interference (crosstalk) between pairs in multipair cable.

The four primary electrical characteristics of twisted-pair cable are its series resistance R, series inductance L, shunt capacitance C, and shunt conductance G per unit length of cable. Transmission characteristics, such as characteristic impedance Z and propagation constant γ, can be calculated from these primary electrical characteristics. The impedance is given by

$$Z = \sqrt{\frac{R + j\omega L}{G + j\omega C}} \tag{8.1}$$

For polyethylene cable, the conductance G is small, so that at low frequencies, where $\omega \ll R/L$,

$$Z \approx \sqrt{\frac{R}{j\omega C}} \tag{8.2}$$

while at high frequencies, where $\omega \gg R/L$,

$$Z \approx \sqrt{\frac{L}{C}} \tag{8.3}$$

The propagation constant γ describes the attenuation α and phase shift β as

$$\gamma = \alpha + j\beta = \sqrt{(R + j\omega L)(G + j\omega C)} \tag{8.4}$$

At low frequencies, the phase and attenuation characteristics are proportional to ω, as seen by

$$\alpha = \beta \approx \sqrt{\frac{\omega RC}{2}} \tag{8.5}$$

For high frequencies, the attenuation becomes a constant and is given by

$$\alpha = \frac{R}{2}\sqrt{\frac{C}{L}} \tag{8.6}$$

while the phase has a linear dependence on frequency. Other factors that influence the attenuation characteristic are the skin effect, where attenuation exhibits a $\sqrt{\omega}$ dependence for very high frequencies, and temperature change, where an additional factor proportional to $\sqrt{R}$ appears in the expression for attenuation. Typical attenuation values for polyethylene-insulated cable (PIC) of various sizes are plotted in Figure 8.2 for typical carrier frequencies.

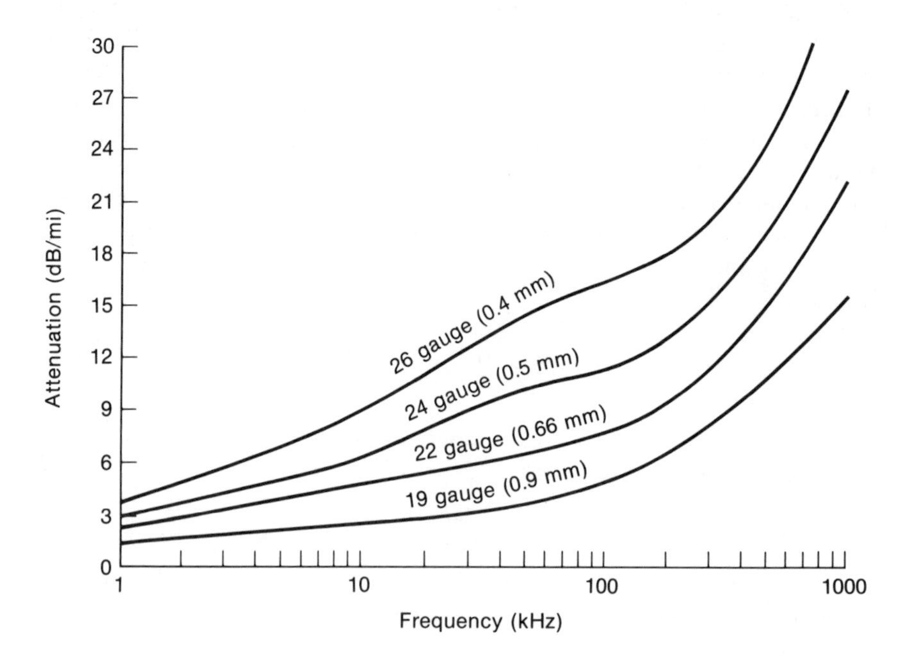

Figure 8.2 Typical Attenuation Characteristic for PIC Cable at 68°F

Twisted-pair cable has a long history as a major form of transmission for voice frequency (VF) and FDM signals. In the last 20 years, twisted-pair cable has also been used for digital transmission, beginning in the early 1960s with T1 carrier in the Bell System. Engineering design for T carrier was based on the capabilities of standard VF cable [1]. Using standard 22-gauge cable, T1 repeaters were designed for 6000-ft spacings, which coincided with the distance between loading coils (inductors) used in VF transmission to improve the frequency response. Because loading coils introduce a nonlinear phase response, they can cause intersymbol interference and must therefore be removed for T1 service. With the growing use of T carrier in the United States, cables have been designed specifically for T1 and T2 carrier. The use of low-capacitance PIC cable reduces loss at high frequencies, allowing greater distance between repeaters. To reduce crosstalk when both directions of transmission are in one sheath, *screened* cable has been introduced in which an insulated metallic screen separates the two directions of transmission within the cable core. In North America, T-carrier systems operate over twisted-pair cable at rates of 1.544 Mb/s (T1), 3.152 Mb/s (T1C), and 6.312 Mb/s (T2). Similar applications exist in Europe and elsewhere at standard rates of 2.048 and 8.448 Mb/s.

At transmission rates above 10 Mb/s, the use of twisted-pair cable becomes undesirable because of higher loss and greater crosstalk. As the next section explains, coaxial cable is preferred at such transmission rates.

8.2.2 Coaxial Cable

A **coaxial cable** consists of an inner conductor surrounded by a concentric outer conductor. The two conductors are held in place and insulated by a dielectric material, usually polyethylene disks that fit over the inner conductor and are spaced at approximately 1-in. intervals. The inner conductor is solid copper; the outer conductor consists of a thin copper tape covered by steel tape. A final covering of paper tape insulates coaxial pairs from each other. The standard cable size for long-haul transmission has an inner conductor diameter of 0.104 in. (2.6 mm) and an outer conductor diameter of 0.375 in. (9.5 mm), usually expressed as 2.6/9.5-mm cable. Other sizes commonly used are 1.2/4.4-mm and 0.7/2.9-mm coaxial cable.

Coaxial cables have low attenuation at carrier frequencies, where loss is directly proportional to resistance R and is relatively independent of the other primary characteristics. Attenuation due to resistive loss is found to be proportional to the square root of frequency and inversely proportional to the cable diameter, as seen in Figure 8.3, which is based on data extracted from CCITT recommendations [2]. The attenuation can be expressed as a function of frequency (in megahertz) by

$$\alpha = a + b\sqrt{f} + cf \qquad \text{dB/km} \tag{8.7}$$

where the set of constants (a, b, c) are peculiar to a particular size of cable.

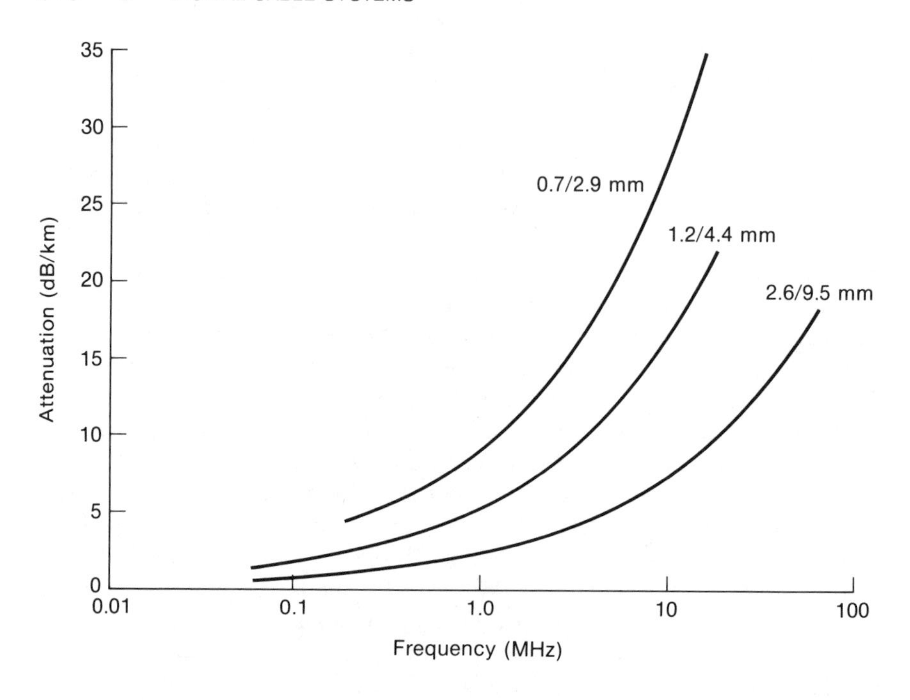

Figure 8.3 Attenuation Characteristic for Standard Coaxial Cable at 10°C

With 1.2/4.4-mm coaxial cable, for example,

$$a = 0.066$$
$$b = 5.15$$
$$c = 0.047$$

The characteristic impedance of these cables is also a function of frequency and conductor diameter ratio and can be expressed as

$$Z = Z_\infty + \frac{A}{\sqrt{f}}(1 - j) \tag{8.8}$$

where Z_∞ is the impedance at infinite frequency and f is given in megahertz. The constant A is peculiar to each size of cable. For the cable sizes mentioned here, the characteristic impedance varies by only a few ohms over the frequency range of interest. A typical value for the real part of the impedance is 75 ohms at 1 MHz for 0.7/2.9-mm or 1.2/4.4-mm cable.

Coaxial cable is a standard transmission medium for many FDM and analog video systems. A good example of coaxial cable application is the AT&T L carrier, which provides capacities up to 10,800 voice channels (Chapter 7). Because digital coaxial cable systems are still relatively new, their application is not as widespread or standardized as analog coaxial

Table 8.1 Transmission Rates and Cable Types for Digital Transmission Using Twisted-Pair and Coaxial Cable

Transmission Rate (Mb/s)	*Application*	*Cable Type*
1.544	North America (T1) and Japan	VF cable with paper or polyethylene insulation
2.048	Europe	
3.152	North America (T1C)	
6.312	North America (T2) and Japan	Low-capacitance PIC
8.448	Europe	0.7/2.9-mm coaxial cable
34.368	Europe	1.2/4.4-mm or 2.6/9.5-mm coaxial cable
139.264	Europe	
97.772	Japan (PCM-100M)	2.6/9.5-mm coaxial cable
274.176	North America (T4)	
400.352	Japan (PCM-400M)	
564.992	Europe	
800	Japan	

cable. Nevertheless, there are digital coaxial cable systems in use today, particularly at standard rates of 8.448 Mb/s and above. The CCITT has under study the characteristics of digital line systems on coaxial cable for rates up to 564.992 Mb/s [3]. Most of the digital cable systems currently in use take advantage of the characteristics of existing analog cable plant and repeater spacing. For example, North American 274-Mb/s systems [4], European 565-Mb/s systems [5], and Japanese 400 and 800-Mb/s systems [6, 7] all are compatible with 60-MHz FDM systems, which operate on 2.6/9.5-mm coaxial cable with a nominal repeater spacing of 1.5 km. Table 8.1 lists the standard data rates and cable types used with twisted-pair and coaxial cable.

8.2.3 Fiber Optic Cable

An **optical fiber** is a cylindrical waveguide made of two transparent materials each with a different index of refraction. The two materials, usually high-quality glass, are arranged concentrically to form an inner core and an outer cladding, as shown in Figure 8.4. Fiber size is designated by two numbers: core diameter and cladding diameter. For example, 50/125 μm is a standard fiber size specified by CCITT Rec. G.651 [2]. Light transmitted through the core is partly reflected and partly refracted at the boundary with the cladding. This property of optics, illustrated in Figure 8.5, is known

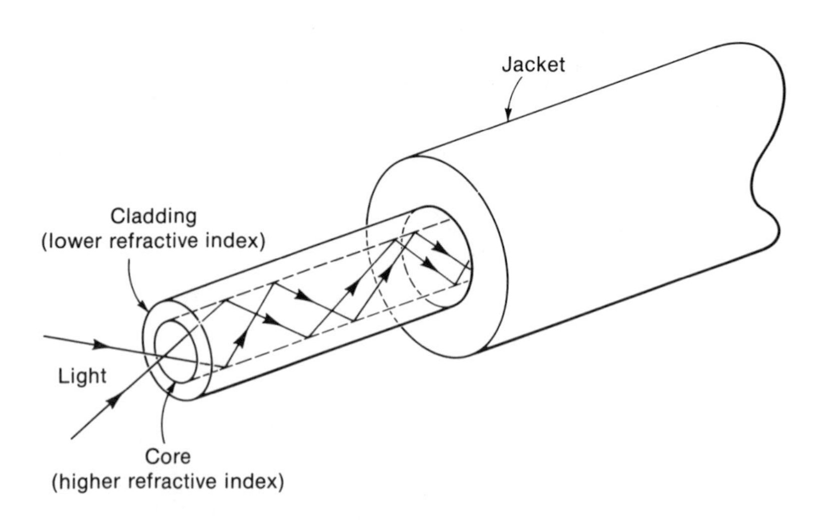

Figure 8.4 Geometry of an Optical Fiber

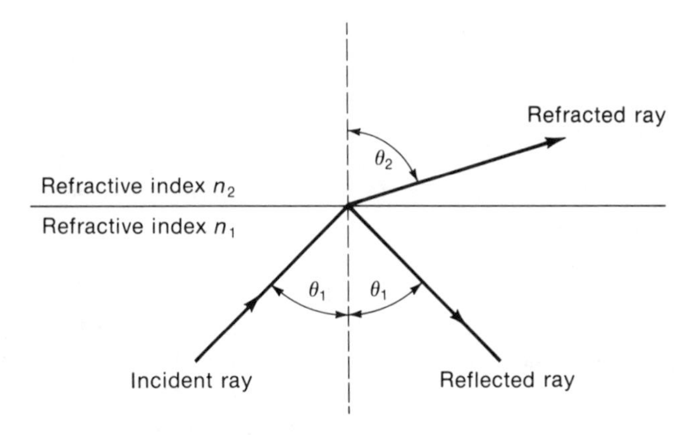

Figure 8.5 Optical Fiber Reflection and Refraction

as Snell's law:

$$n_1 \sin\theta_1 = n_2 \sin\theta_2 \tag{8.9}$$

where θ_1 = angle of incidence = angle of reflection

θ_2 = angle of refraction

n_1 = refractive index of medium 1

n_2 = refractive index of medium 2

If $n_1 > n_2$ there is a critical angle θ_c in which $\sin\theta_2 \geq 1$ when

$$\theta_1 \geq \theta_c = \sin^{-1}\left(\frac{n_2}{n_1}\right) \tag{8.10}$$

For all $\theta_1 \geq \theta_c$ the incident ray is totally reflected with no refractive loss at the boundary of n_1 and n_2. Thus fiber optic cables are designed with the core's index of refraction higher than that of the cladding.

The maximum angle θ_m at which a ray may enter a fiber and be totally reflected is known as the **numerical aperture** (NA) and is given by

$$\text{NA} = \sin\theta_m = \sqrt{n_1^2 - n_2^2} \tag{8.11}$$

Numerical aperture indicates the fiber's ability to accept light, where the amount of light gathered by the fiber increases with θ_m. Different entry angles of the light source result in multiple modes of wave propagation. The total number of modes that can occur in an optical fiber is determined by the core size, wavelength, and numerical aperture. Propagation can be restricted to a single mode by using a small-diameter core, by reducing the numerical aperture, or both. The choice between **single-mode** and **multimode fiber** depends on the desired repeater spacing or transmission rate; single mode is the preferred choice for long-haul or high-data-rate systems. The range of NA for most multimode fiber is 0.20 to 0.31.

The bandwidth of a fiber optic system is limited by pulse spreading due to dispersion. **Multimode dispersion** (also called **intermodal dispersion**) arises because of the different propagation velocities and path lengths of the different rays (modes) in a multimode cable. These differences result in different arrival times for rays launched into the fiber coincidentally. Although this form of dispersion may be eliminated by use of single-mode fibers, **material dispersion** (also called **intramodal dispersion**) may be produced by a variation in propagation velocity within the optical fiber. This form of dispersion is dependent on the wavelength and spectral width of the light source. Of the two types of dispersion, multimode dispersion when present is predominant. The effects of dispersion can thus be minimized by using a single-mode fiber with a light source of narrow spectrum. Using state-of-the-art lasers and single-mode fibers, material dispersion of less than 0.1 ns/km has been achieved at a wavelength of 850 nm. The use of a longer wavelength source—for example at 1300 nm—further reduces material dispersion to allow even greater bandwidths.

Dispersion effects may also be remedied by constructing a fiber whose index of refraction increases toward the axis. This type of fiber, called **graded-index**, is illustrated in Figure 8.6. With such an arrangement, rays that travel longer paths have greater velocity than rays traveling the shorter paths. The various modes then tend to have the same arrival time, such that dispersion is minimized and greater bandwidths become possible for multimode fibers. This is a significant improvement over the fiber described previously, called **step-index** fiber, where the core has a uniform index of refraction. In this case the propagation velocity is constant, so that rays traveling a longer path arrive behind rays traveling a shorter path, thus producing mode dispersion.

A second propagation characteristic of fiber optics is attenuation, measured in decibels per kilometer. Whereas the dispersion characteristic determines the achievable bandwidth, attenuation determines the available signal-to-noise ratio. Attenuation in an optical fiber has two sources, absorption and scattering, as illustrated in Figure 8.7 for a typical fiber. Absorption is caused by the natural presence of impurities in glass, such as OH ions at 945, 1240, and 1390 nm. The fiber manufacturing process must control the concentration of such impurities to ensure that attenuation is kept low. Rayleigh scattering arises because of small variations in density that are introduced in the glass during the manufacturing process. Rayleigh scattering is known to be inversely proportional to the fourth power of wavelength and represents the fundamental limit of low-loss fibers.

The repeater spacing is a function not only of attenuation but of bandwidth as well. Because of dispersion, the bandwidth of a fiber optic system is inversely proportional to the cable length. This relationship has

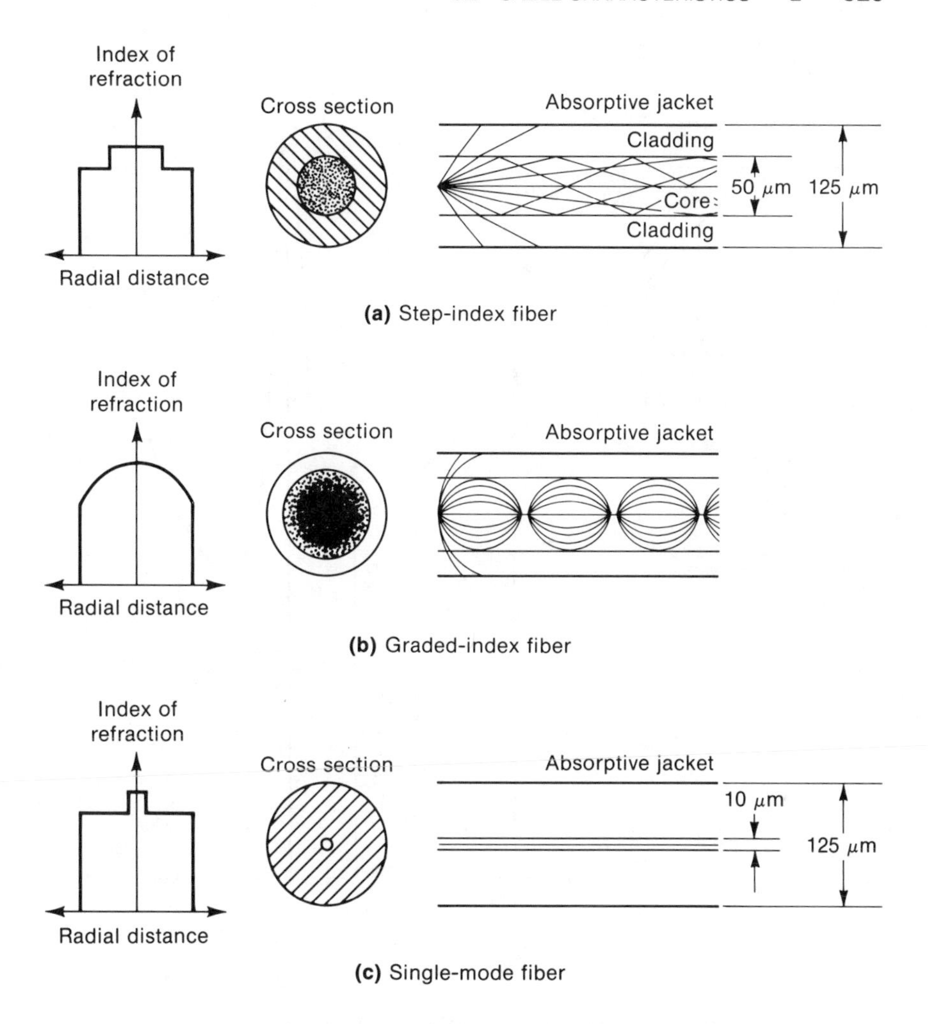

Figure 8.6 Types of Optical Fiber

led to the use of a characteristic known as the **bandwidth-distance product**, given in terms of MHz • km. Typical figures are 10 MHz • km for step-index fibers and 400 to 1200 MHz • km for graded-index fibers.

There are currently two types of optical sources used with fiber optic systems: light-emitting diodes (LEDs) and injection laser diodes. Where applicable, the LED is preferred because of lower cost, higher reliability, less temperature sensitivity, and greater stability. Compared to lasers, however, LEDs produce less optical power output, have a wider emission angle, and have smaller bandwidth, which limits their application to short-haul, low-data-rate systems. Lasers with their greater power output, narrower emission angle and higher operating speed are favored in long-haul, high-data-rate systems.

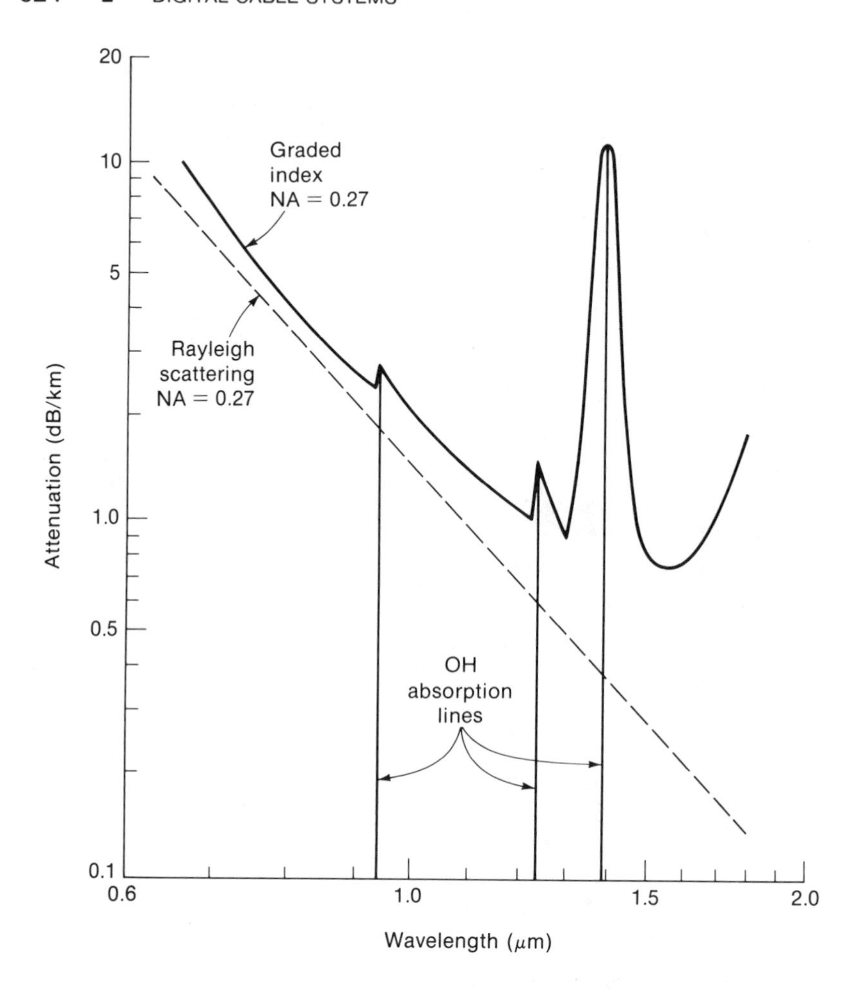

Figure 8.7 Spectral Attenuation for Low-Loss Fibers

The two detectors used in fiber optic systems are the PIN diode (containing positive, intrinsic, and negative layers) and avalanche photo diode (APD). The PIN diode is more commonly used because of its lower cost, higher reliability, and greater temperature range. Although it is more costly and requires a high-voltage power supply, the APD is favored in long-haul, high-data-rate systems because of its greater receiver sensitivity.

Two wavelength bands are of current interest for fiber optic systems: the short-wavelength band of roughly 800 to 860 nm and the long-wavelength band of 1200 to 1600 nm. Most current equipment operates in the short-wavelength band. Operation in the longer-wavelength band, particularly at 1300 and 1550 nm, is attractive because of improved attenuation and dispersion characteristics at these wavelengths. Typically the longer-wavelength band is used in combination with lasers and APDs to achieve a long-haul, high-data-rate capability.

The principal application of fiber optics is with digital transmission, where rates in excess of 1 Gb/s have been realized. In the United States, fiber optic systems are available from several manufacturers for rates at DS-3 (44.736 Mb/s) and higher. The CCITT is currently studying the application of digital transmission to fiber optic systems [3]. Typical characteristics of fiber optic communication systems are listed in Table 8.2 for both short-haul, low-data-rate and long-haul, high-data-rate applications.

The advantages of fiber optic cable over metallic cable are illustrated here with a series of comparisons:

- *Size*: Since individual optic fibers are typically only 125 μm in diameter, a multiple-fiber cable can be made that is much smaller than corresponding metallic cables. For example, cable manufactured by AT&T Technologies provides 144 individual fibers along with strengthening materials (twine and steel) encased in a protective sheath with outer diameter of 12 mm ($\frac{1}{2}$ in.). To equal that bandwidth capacity with coaxial cable would require a cable with several hundred times the fiber cable's cross-sectional area.
- *Weight*: The weight advantage of fiber cable over metallic cable is small for single-fiber, low-rate systems (such as T1) but increases dramatically for multiple-fiber, high-rate systems. As a result of this weight advantage, the transporting and installation of fiber optic cable is much easier than for other types of communication cable.
- *Bandwidth*: Fiber optic cables have bandwidths that can be orders of magnitude greater than metallic cable. Low-data-rate systems can be easily upgraded to higher-rate systems without the need to replace the

Table 8.2 Typical Characteristics of Fiber Optic Communication Systems

Component	*Short Haul, Low Data Rate*	*Long Haul, High Data Rate*
Transmitter	Light-emitting diode	Injection laser diode
Receiver	PIN diode	Avalanche photodiode
Optical fiber	Step index	Graded index or single mode
Wavelength	800–850 nm	1300 nm
Attenuation	$\leq$ 6 dB/km	$\leq$ 2 dB/km
Bandwidth	$\leq$ 20 MHz · km	$\geq$ 400 MHz · km
Multimode dispersion	$\leq$ 1 ns/km	$\leq$ 0.5 ns/km
Material dispersion	$\leq$ 4 ns/km	$\leq$ 0.2 ns/km

fibers. Upgrading can be achieved by changing light sources (LED to laser), improving the modulation technique, improving the receiver, or using wavelength-division multiplexing.

- *Repeater spacing*: With low-loss fiber optic cable, the distance between repeaters can be significantly greater than in metallic cable systems. Moreover, losses in optical fibers are independent of bandwidth, whereas with coaxial or twisted-pair cable the losses increase with bandwidth. Thus this advantage in repeater spacing increases with the system's bandwidth.
- *Electrical isolation*: Fiber optic cable is electrically nonconducting, which eliminates all electrical problems that now beset metallic cable. Fiber optic systems are immune to power surges, lightning-induced currents, ground loops, and short circuits. Fibers are not susceptible to electromagnetic interference from power lines, radio signals, adjacent cable systems, or other electromagnetic sources.
- *Crosstalk*: Because there is no optical coupling from one fiber to another within a cable, fiber optic systems are free from crosstalk. In metallic cable systems, by contrast, crosstalk is a common problem and is often the limiting factor in performance.
- *Radiation*: Unlike metallic cable systems, fiber optic cable does not radiate electromagnetic energy, which is important in applications involving military security. This means that the signal can be detected only if the cable is physically accessed.
- *Environment*: Properly designed fiber optic systems are relatively unaffected by adverse temperature and moisture conditions and therefore have application to underwater cable. For metallic cable, however, moisture is a constant problem particularly in underground (buried) applications, resulting in short circuits, increased attenuation, corrosion, and increased crosstalk.
- *Reliability*: The reliability of optical fibers, optical drivers, and optical receivers has reached the point where the limiting factor is usually the associated electronics circuitry.
- *Cost*: The numerous advantages listed here for fiber optic systems have resulted in dramatic growth in their application with attendant reductions in cost due to technological improvements and sales volume. Today fiber optic systems are more cost-effective than metallic cable for long-haul, high-bit-rate applications. Fiber optic cable is also expected eventually to overtake metallic cable in short-haul applications, including metro facilities and local networks. One final cost factor in favor of fiber optics is the choice of material—copper, which may someday be in short supply, versus silicon, one of the earth's most abundant elements.

It should be pointed out here that fiber optic cables, in common with other types of cable, have certain advantages over radio systems. For instance, cable systems do not require frequency allocations from the already crowded frequency spectrum. Moreover, cable systems do not have the terrain clearance problems encountered in line-of-sight microwave systems. Fiber optic systems also avoid the multipath fading and interference problems common to radio systems. However, fiber optic and other cable types have the distinct disadvantage of requiring right-of-way along the physical path.

8.3 REGENERATIVE REPEATERS

The purpose of a regenerative repeater is to construct an accurate reproduction of the original digital waveform. Its application is described here for cable systems, but regenerative repeaters are common to all digital transmission systems. The functions performed in a digital repeater can be grouped into three essential parts—reshaping, retiming, and regeneration, as illustrated in Figure 8.8 and described in the following paragraphs.

After transmission over a section of cable, each received digital pulse is flattened and spread over several symbol times due to the effects of attenuation and dispersion. Before these pulses can be sampled for detection and regeneration, reshaping is necessary; it is accomplished by equalization and amplification. The equalizer compensates for the frequency-dependent nature of attenuation in cable. Ideally the equalizer provides a characteristic that is inverse to the cable characteristic, so that the overall response is independent of frequency. In practice, however, perfect equalization is neither possible nor necessary. Equalized pulses still tend to appear "rounded off," but this shape is adequate for proper regeneration. The equalizer and amplifier combination is designed for a standard length of cable. Variations in cable length are compensated by a network that builds out the cable to the equivalent of a standard section of cable. This line build-out feature is automatic (ALBO) and typically covers a wide range of line loss. The ALBO network is controlled by a peak detector, whose output controls a variable resistance in the ALBO.

Retiming in a regenerative repeater is accomplished by deriving timing from the data. This self-timing method proves to be acceptable if the data maintain a sufficient density of transitions, or zero crossings. A discussion of suitable transmission codes and scramblers for proper clock recovery is given in Chapter 5. The alternatives to this retiming scheme include use of highly accurate oscillators at each repeater or separate transmission of a clock signal, but such schemes are costly and therefore seldom used. Typically the clock recovery circuit consists of a resonant tank circuit tuned to the timing frequency (Figure 8.8). This circuit is preceded by a rectifier that converts the reshaped pulses into unipolar marks. Each mark stimulates the tank circuit, which produces a sine wave at the desired timing frequency. This sine-wave signal decays exponentially until another mark restimulates the tank circuit. So long as these marks arrive often enough, the tank circuit can be forced to reproduce the correct frequency of the incoming data

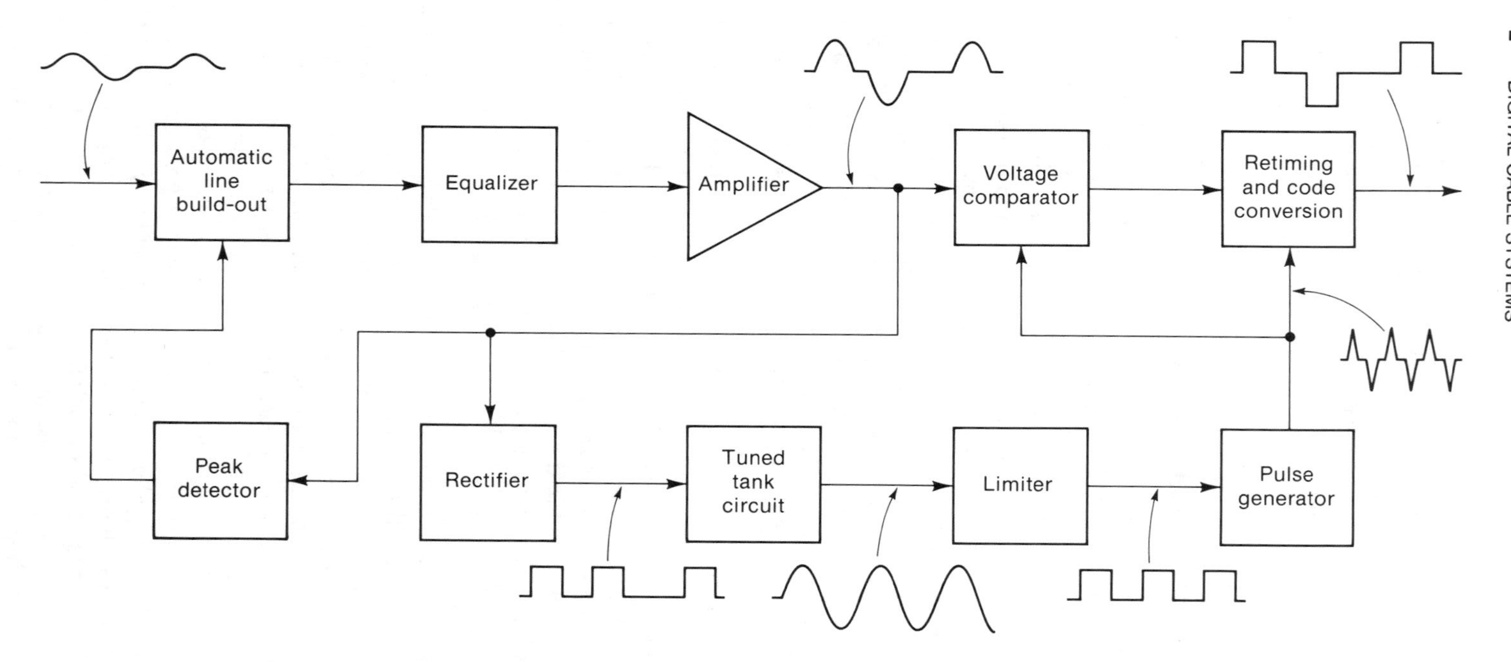

Figure 8.8 Block Diagram of a Regenerative Repeater

signal. The sinusoidal timing signal is then limited to produce a constant-amplitude square wave. A pulse generator yields positive and negative timing spikes at the zero crossings of the square wave.

The regenerator performs voltage comparison, clock sampling, and any necessary code conversion. Using the timing spikes generated by the clock recovery circuit, the reshaped signal is sampled near the center of each symbol interval. The result is compared with a threshold by a decision circuit. The width of each regenerated pulse is controlled by the recovered clock. Finally, code conversion is applied as would be required, for example, with a bipolar format.

8.4 CLOCK RECOVERY AND JITTER

In a regenerative repeater, a form of short-term phase modulation known as **jitter** is introduced during the clock recovery process. The sampled and regenerated data signal then has its pulses displaced from their intended position, as illustrated by the photograph in Figure 8.9 for a binary signal. This display in Figure 8.9 is generated by use of an oscilloscope in which many transitions of the data signal are superimposed to indicate the

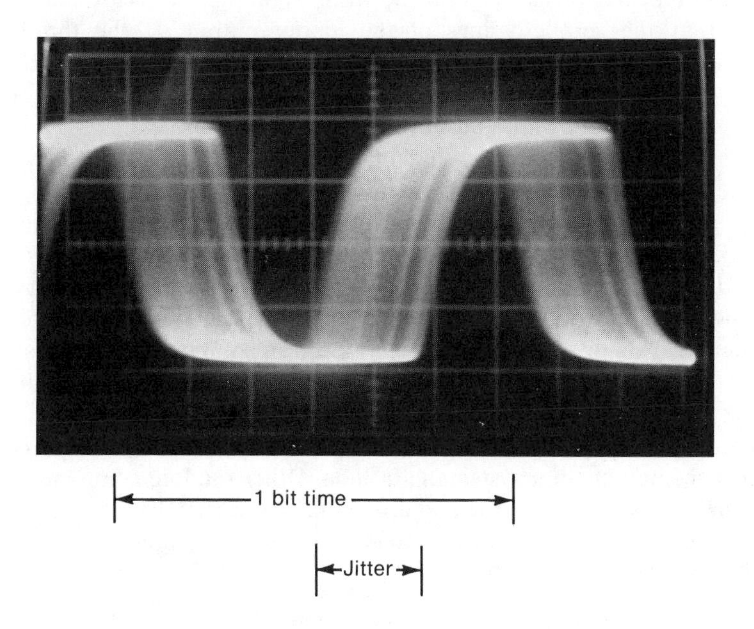

Figure 8.9 Photograph of Jitter for Binary Signal

magnitude of jitter. Jitter leads to sampling offsets and potential data errors in a repeater and also causes distortion in the reconstructed analog waveform for PCM systems.

Sources of jitter may be classified as systematic or random depending on whether or not they are pattern-dependent. Since each repeater in a long repeater chain operates on the same data pattern, any jitter source related to the data pattern produces the same effect at each repeater and hence is systematic. Random sources of jitter are uncorrelated at each repeater and should not cause a significant buildup of jitter in a repeater chain. Thus accumulation of jitter in a long repeater chain is caused primarily by systematic sources.

8.4.1 Sources of Jitter

Major sources of jitter include intersymbol interference, threshold misalignment, and tank circuit mistunings. These sources are difficult to analyze individually, which complicates complete jitter analysis for a typical repeater.

Imperfect equalization during pulse reshaping causes a skewing of the pulses due to residual intersymbol interference. The effective center of each pulse varies from pulse to pulse depending on the surrounding data pattern. The resulting shifts in peak pulse position cause a corresponding phase shift in the timing circuit output.

In the repeater block diagram shown in Figure 8.8, the clock signal is generated at the zero crossings of the limiter output. If the threshold detector is offset from the ideal zero crossings, the timing spikes used for retiming will incur a phase shift. Variation in the timing signal amplitude leads to corresponding variation in the triggering time.

The frequency of the tank circuit, f_T, is tuned to the expected line frequency, f_0. Mistuning results from initial frequency offset plus changes in f_T and f_0 with time. Both static and dynamic phase shifts in the timing signal occur from mistuning. Static phase shifts, however, can be compensated by phase adjustment in the retiming portion of the regenerator. Dynamic phase shifts are caused by variation in the transition density of the data signal. The presence of a strong spectral component at f_0 causes the tank circuit to converge on f_0, while a weak spectral component allows the tank circuit to drift toward its natural frequency, f_T.

These major jitter components are all to some extent pattern-dependent and thus accumulate in a systematic fashion. Other random components of jitter, due for example to noise and crosstalk, also exist. These components are uncorrelated at successive repeaters and do not accumulate in a systematic manner.

8.4.2 Timing Jitter Accumulation

A model for the analysis of timing jitter accumulation in a chain of repeaters has been developed by Byrne and colleagues [8] and provides the basis of our discussion (see Figure 8.10). To make the model tractable, several assumptions must be made:

1. The same jitter is injected at each repeater.
2. All significant jitter sources can be represented by an equivalent jitter at the input to each clock recovery circuit.
3. Jitter adds linearly from repeater to repeater.
4. Clock recovery is performed by a tank circuit tuned to the pulse repetition frequency.
5. Since the rate of change of jitter is small compared to the recovered clock rate, the tank circuit is equivalent to a low-pass filter with a single pole corresponding to the half-bandwidth of the tank circuit.

The input and output jitter amplitude spectra, $\theta_i(f)$ and $\theta_o(f)$, are related by the jitter transfer function at each repeater $Y(f)$:

$$\begin{aligned} \theta_o(f) &= \frac{1}{(1+j)(f/B)}\theta_i(f) \\ &= Y(f)\theta_i(f) \end{aligned} \tag{8.12}$$

where $B = \dfrac{f_0}{2Q}$ = half-bandwidth of tank circuit

f_0 = timing frequency of data

Q = quality factor of tank circuit

The result of (8.12) is that low-frequency jitter is passed by the tank circuit while higher-frequency jitter is attenuated and shifted in phase.

The jitter at the end of a chain of N repeaters is the sum of jitter introduced at each repeater and operated upon by all succeeding repeaters. Since the jitter introduced in each repeater is assumed to be identical, the

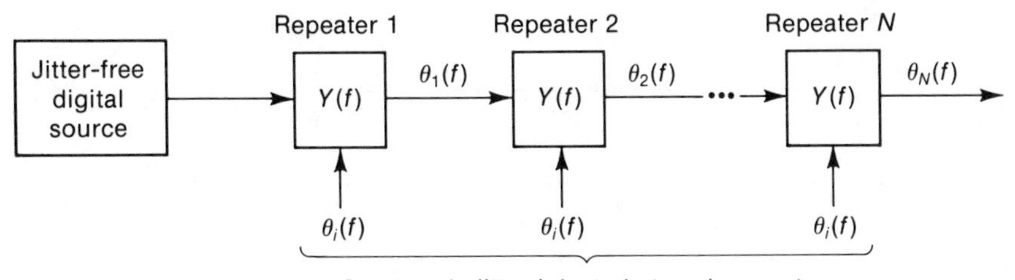

Figure 8.10 Model of Jitter Accumulation Along a Chain of Repeaters

accumulated jitter spectrum can be expressed as

$$\theta_N(f) = \theta_i(f)\left[Y(f) + Y^2(f) + \cdots + Y^N(f)\right] \tag{8.13}$$

The right-hand side of (8.13) is the sum of a geometric series that is given by

$$\theta_N(f) = \theta_i(f)\frac{B}{jf}\left[1 - \left(\frac{1}{1 + jf/B}\right)^N\right] \tag{8.14}$$

The power density of jitter at the Nth repeater output is obtained by squaring the magnitude of the transfer function:

$$S_N(f) = \left(\frac{B}{f}\right)^2 \left|1 - \left(\frac{1}{1 + jf/B}\right)^N\right|^2 S_i(f) \tag{8.15}$$

where $S_i(f)$ is the power density of the jitter injected in each repeater. The normalized jitter power spectrum, obtained from (8.15), has been plotted in Figure 8.11 for $N = 1$, 10, 100, and 1000 repeaters. At very low frequencies the normalized jitter power spectrum is proportional to the square of N. For higher frequencies the power gain falls off as the inverse square of frequency.

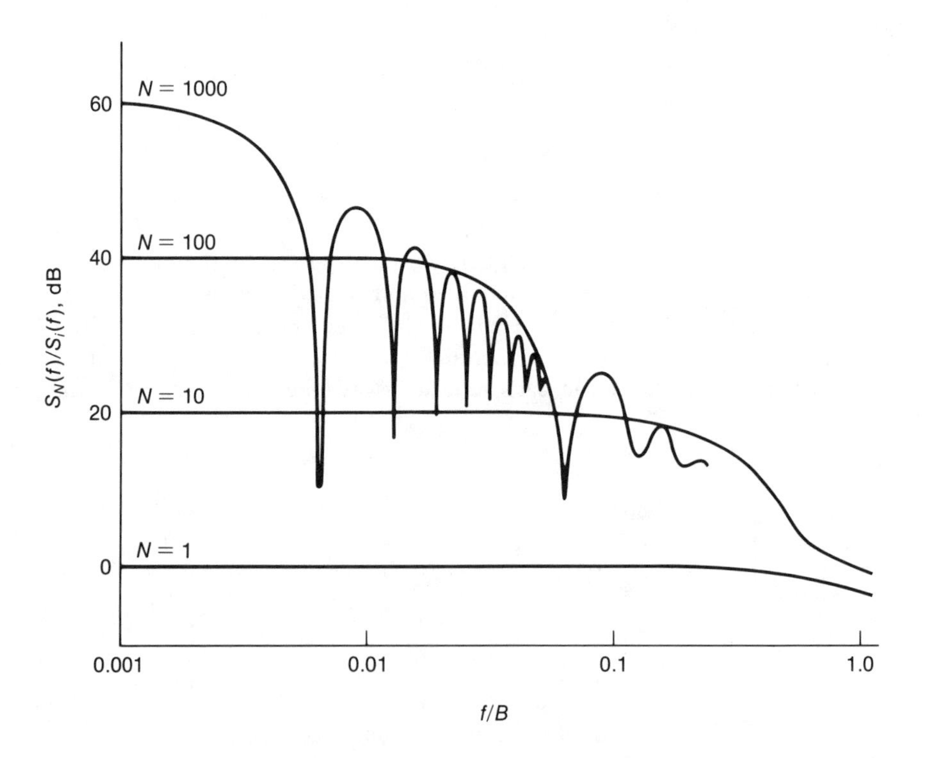

Figure 8.11 Normalized Jitter Power Spectrum for *N* Repeaters in Tandem

As a result of this spectrum shape, the bulk of jitter gain for large N is at low frequency where jitter increases with an amplitude proportional to N.

Random Patterns

In normal operation, the data transmitted along a chain of repeaters have a nearly random bit pattern. For this case, Byrne and colleagues have determined that the low-frequency power density of jitter injected in each repeater is independent of frequency, such that $S_i(f)$ in Equation (8.15) is a constant.

The mean square value of jitter can be found by integrating the jitter spectrum (Equation 8.14) over all frequencies. The result is

$$\overline{J_N^2} = \frac{S_i B}{2}\left(N - \frac{1}{2}\left\{\frac{(2N-1)!}{4^{N-1}[(N-1)!]^2}\right\}\right) \tag{8.16}$$

For $N > 100$, this expression reduces to the approximation

$$\overline{J_N^2} \approx \frac{1}{2} S_i B N \tag{8.17}$$

Thus the mean square jitter in a long repeater chain increases as N, and the rms amplitude increases as the square root of N. Moreover, Expression (8.17) indicates that the jitter power is directly proportional to the timing circuit bandwidth B and inversely proportional to the Q of the circuit.

Note that analysis of rms jitter amplitude due to uncorrelated sources indicates an accumulation according to only the fourth root of N [9]. Therefore, as suggested earlier, systematic jitter dominates in long repeater chains.

Repetitive Patterns

Any data bit pattern **m** can be shown to have a characteristic jitter, $\theta_{\mathbf{m}}(t)$. A change from pattern **m** to pattern **n** causes a jump in jitter amplitude from $\theta_{\mathbf{m}}$ to $\theta_{\mathbf{n}}$. This change in jitter stems from the response of the timing circuit to the new pattern. The rate of change is controlled by the bandwidth B of the timing circuit and is given by $B(\theta_{\mathbf{m}} - \theta_{\mathbf{n}})$. The total change in jitter at the end of N repeaters accumulates in a linear fashion and is given by $N(\theta_{\mathbf{m}} - \theta_{\mathbf{n}})$. These characteristics of accumulated jitter for N repeaters due to transition in the data pattern are illustrated in Figure 8.12.

The worst-case jitter for all data pattern changes is characterized by the maximum value of $(\theta_{\mathbf{m}} - \theta_{\mathbf{n}})$ for all possible pairs of repetitive patterns. Experimental results indicate that the jitter magnitude due to transitions between two fixed patterns is affected by the 1's density (for bipolar) and intersymbol interference. In practice such repetitive data patterns are rare, except when the line is idle, test patterns are being transmitted, or a PCM system is being used to sample slowly changing analog signals.

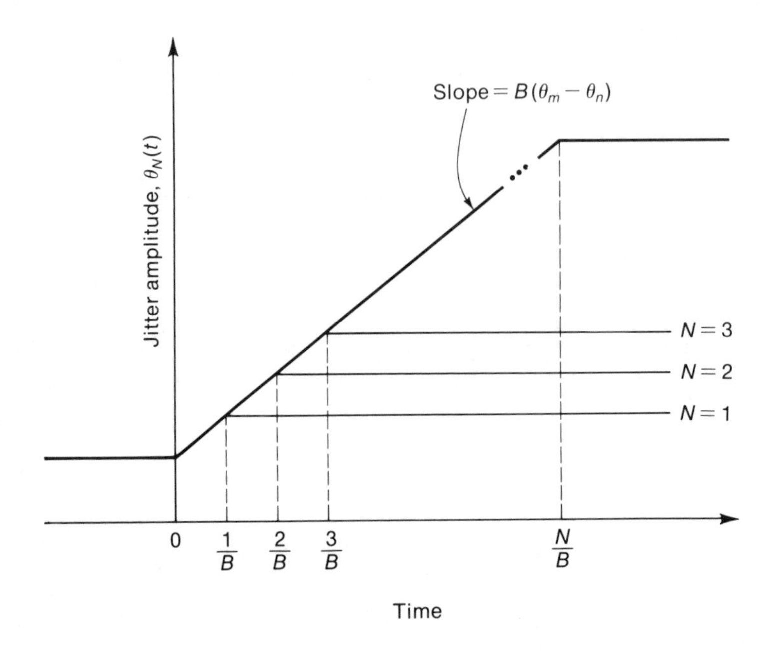

Figure 8.12 Jitter at End of *N* Repeaters Due to Change in Data Pattern

8.4.3 Alignment Jitter

So far we have not related jitter to the error performance of a regenerative repeater. Here we will see that the misalignment caused by jitter between the data signal and sampling times does not contribute significantly to bit errors. The alignment jitter at the Nth repeater, Δ_N, is defined as the difference between the input and output jitter, so that for the Nth repeater we have

$$\Delta_N(t) = \theta_N(t) - \theta_{N-1}(t) \tag{8.18}$$

From (8.14) we can rewrite (8.18) as

$$\Delta_N(f) = \theta_i(f)\left(\frac{1}{1 + jf/B}\right)^N \tag{8.19}$$

Using the Fourier transform, Equation (8.19) can be rewritten in the time domain as

$$\begin{aligned} \Delta_N(t) &\leq M_i(1 - e^{-Bt})^N \\ &\leq M_i \end{aligned} \tag{8.20}$$

where M_i is the maximum value of $\theta_i(t)$, the jitter injected at each repeater. Thus the maximum alignment jitter does not increase with the length of the chain and is bounded by the maximum jitter injected at each repeater. This property can also be seen from Figure 8.12, where the constant slope of jitter change indicates that the alignment jitter is the same between any two successive repeaters.

8.5 CROSSTALK

Crosstalk is defined as the coupling of energy from one circuit to another, resulting in a form of interference. In cable transmission, crosstalk takes the form of electrical coupling between twisted pairs in a multipair cable or between coaxials in a multicoaxial cable. As noted before, crosstalk is negligible in fiber optic cable. Depending on the cable arrangement, crosstalk may be the limiting factor in determining transmission distance between repeaters. Crosstalk is eliminated by a regenerative repeater, however, so that it does not accumulate in a chain of repeaters.

Figure 8.13 illustrates the two principal types of crosstalk. **Near-end crosstalk** (NEXT) refers to the case where the disturbing circuit travels in a direction opposite to that of the disturbed circuit. **Far-end crosstalk** (FEXT) is the case where the disturbed circuit travels in the same direction as the disturbing circuit. Of the two types, NEXT is the dominant source of interference because a high-level disturbing signal is coupled into the disturbed signal at the receiver input where signal level is normally low. For this reason, opposite directions of transmission are often isolated by using separate cables or a screened cable to avoid NEXT so that only FEXT is of concern.

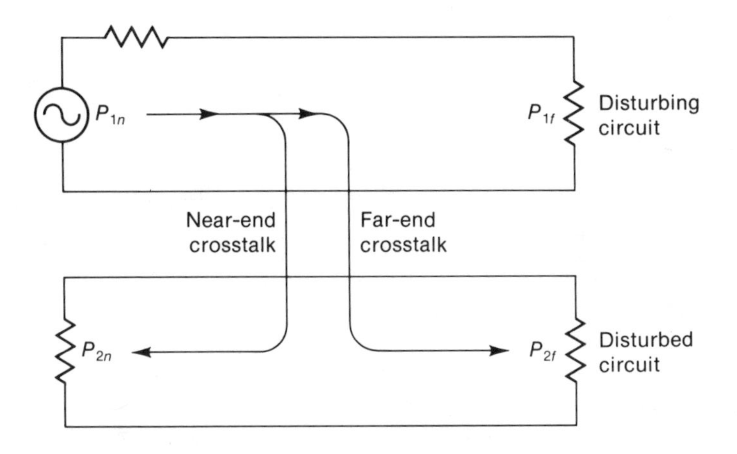

Figure 8.13 Types of Crosstalk

Crosstalk is usually measured and specified by a ratio of the power in the disturbing circuit to the induced power in the disturbed circuit. Specifically, using the notation shown in Figure 8.13, the NEXT ratio is given by

$$\text{NEXT} = \frac{P_{2n}}{P_{1n}} \tag{8.21}$$

and the FEXT by

$$\text{FEXT} = \frac{P_{2f}}{P_{1f}} \tag{8.22}$$

where P_{1n} = transmit power of disturbing circuit

P_{1f} = receiver power of disturbing circuit

P_{2n} = near-end power in disturbed circuit due to coupling from disturbing circuit

P_{2f} = far-end power in disturbed circuit due to coupling from disturbing circuit

For long transmission lines and high (carrier) frequencies, the NEXT ratio is given by

$$\frac{P_{2n}}{P_{1n}} \approx k_n f^{3/2} \tag{8.23}$$

or, since NEXT is usually given in decibels,

$$\text{NEXT} = 10 \log k_n + 15 \log f \tag{8.24}$$

where k_n is the NEXT loss per unit frequency and f is frequency in the same units as k_n. Note that NEXT increases with frequency at a rate of 15 dB per frequency decade and is independent of cable length. For like pairs or coaxials, the FEXT ratio is given by

$$\frac{P_{2f}}{P_{1f}} \approx k_f f^2 L \tag{8.25}$$

or, in decibels,

$$\text{FEXT} = 10 \log k_f + 20 \log f + 10 \log L \tag{8.26}$$

where k_f is the FEXT loss per unit frequency and L is the cable length. Note that FEXT increases with frequency at a rate of 20 dB per frequency decade and with length at a rate of 10 dB per length decade.

For multipair cable, the total crosstalk in a given pair is the power sum of individual contributions from the other pairs. Measurements of total crosstalk indicate that the distribution of crosstalk values, in decibels, is normal (gaussian). Additionally, the distribution of pair-to-pair crosstalk, in decibels, has been found to be normal [10].

The effect of crosstalk on error rate depends on its mean square value, analogous to the case of additive gaussian noise used in modeling thermal noise. The total interference at a repeater is then the sum of crosstalk interference σ_c^2 and thermal noise σ_n^2:

$$\sigma_T^2 = \sigma_c^2 + \sigma_n^2 \tag{8.27}$$

If the noise is gaussian and the crosstalk is known to be gaussian, the total interference σ_T^2 must also be gaussian, thus allowing a straightforward calculation of error rate using the appropriate expressions of the next section.

8.6 ERROR PERFORMANCE FOR TANDEM REPEATERS

Here we are interested in the behavior of the error rate after n tandem repeaters, where each repeater has probability of error p. The probability of error for various digital transmission codes has already been shown in Chapter 5. Each transmitted digit may undergo cumulative errors as it passes from repeater to repeater. If the total number of errors is even, they cancel out. A final error is made only if an odd number of errors is made in the repeater chain. For a given bit the probability of making k errors in n repeaters is given by the binomial distribution:

$$P_k = \binom{n}{k} p^k (1-p)^{n-k} \tag{8.28}$$

The net probability of error for the n repeater chain, $P_n(e)$, is obtained by summing over all odd values of k, yielding

$$\begin{aligned} P_n(e) &= \sum_{k=1}^{n} \binom{n}{k} p^k (1-p)^{n-k} \qquad (k \text{ odd}) \\ &= np(1-p)^{n-1} + \frac{n(n-1)(n-2)}{3!} p^3 (1-p)^{n-3} + \cdots \\ &\approx np \end{aligned} \tag{8.29}$$

where the approximation holds if $p \ll 1$ and $np \ll 1$. This approximation indicates that if the probability of a single error is small enough, the probability of multiple errors is negligible. In this case, the net error probability increases linearly with the number of repeaters.

It is interesting to note here the advantage of digital repeaters over analog repeaters. An analog repeater is nothing more than an amplifier that amplifies both signal and noise. Assuming the presence of additive gaussian noise in each repeater, the noise accumulates in a linear fashion for a chain of repeaters. Consequently, the signal-to-noise ratio progressively decreases with each repeater so that after n repeaters the final signal-to-noise ratio is

$$(S/N)_n = \frac{S/N}{n} \tag{8.30}$$

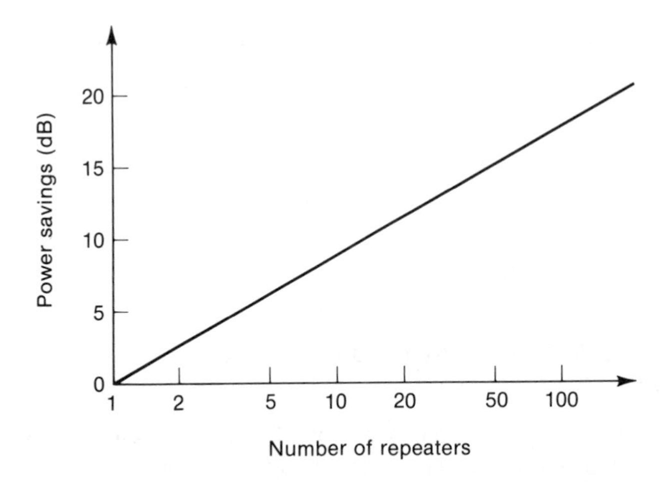

Figure 8.14 Power Savings Gained by Digital Repeaters Over Analog Repeaters for a Fixed $P(e) = 10^{-5}$

where S/N is the value after the first repeater. Choosing NRZ as the transmission code for binary data, the error probability for n digital repeaters in tandem is obtained from (8.29) and (5.30) as

$$P_n(e) \approx n \operatorname{erfc}\sqrt{S/N} \tag{8.31}$$

For n analog repeaters in tandem,

$$P_n(e) = \operatorname{erfc}\sqrt{\frac{S/N}{n}} \tag{8.32}$$

Figure 8.14 indicates the power savings in using digital (regenerative) repeaters for a fixed 10^{-5} probability of error.

8.7 REPEATER SPACING IN MULTIPAIR CABLE SYSTEMS

Repeaters are designed to provide a certain error rate objective on the basis of each repeater section and for the end-to-end system. The error rate in multipair cable systems is affected by the transmitted signal level, cable loss, crosstalk, and the required receiver signal-to-noise ratio. The required spacing between repeaters can in general be written as

$$R_s = \frac{L_D}{L_c} \tag{8.33}$$

where R_s = repeater spacing (units of distance)

L_D = maximum section loss (dB)

L_c = cable attenuation (dB/unit distance)

The maximum section loss L_D is the loss a signal may undergo and still provide an adequate signal-to-noise ratio at the repeater input to meet the error rate objective.

Repeatered lines for multipair cable may be designed with both directions of transmission in the same cable or with each direction in a separate cable. In one-cable operation, near-end crosstalk is the limiting factor in determining L_D and hence repeater spacing. In two-cable (or one-cable T-screen) operation, the physical separation of transmit and receive directions eliminates NEXT. The limiting factor in determining repeater spacing is then a combination of FEXT and thermal noise, which permits wider repeater spacing than in one-cable operation.

While crosstalk can dictate the repeater spacing, particularly in one-cable systems, there is a maximum spacing determined by cable attenuation alone. Cable attenuation characteristics depend on the gauge (size), construction, insulation, and operating temperature. Temperature changes in multipair cable affect the resistance and hence the attenuation characteristic. Therefore the maximum operating temperature must be known in determining the maximum section loss. Cable attenuation and section loss values are often given for a standard temperature, such as 55°F. Then a temperature correction factor, f_T, can be used to account for other operating temperatures, where we define

$$f_T = \frac{\text{loss at } T^\circ\text{F}}{\text{loss at } 55^\circ\text{F}} \tag{8.34}$$

Cable attenuation for a given maximum operating temperature is given by

$$L_c(T^\circ\text{F}) = L_c(55^\circ\text{F}) f_T \tag{8.35}$$

The maximum section loss corresponding to a given maximum operating temperature is then given by

$$L_D(T^\circ\text{F}) = \frac{L_D(55^\circ\text{F})}{f_T} \tag{8.36}$$

For the design of T1 carrier systems (1.544 Mb/s), values of f_T and L_c are listed in Table 8.3 for various types of cable. Based on U.S. climate, the maximum ambient temperature is assumed to be 100°F for buried cable and 140°F for aerial cable.

Table 8.3 Cable Losses at 772 kHz for T1 Carrier Design

Cable Gauge	Cable Type: Construction	Cable Type: Insulation	L_c at 55°F (dB/1000 ft)	Buried Cable (100°F Max. Temp.) f_T	Buried Cable (100°F Max. Temp.) L_c (dB/1000 ft)	Aerial Cable (140°F Max. Temp.) f_T	Aerial Cable (140°F Max. Temp.) L_c (dB/1000 ft)
19	Unit or layer	Paper	3.00	1.038	3.11	1.071	3.21
	Unit	PIC	3.18	1.043	3.31	1.080	3.43
	T-screen, unfilled	PIC	3.28	1.056	3.46	1.108	3.63
	T-screen, filled to be watertight	PIC	2.94	1.056	3.10	1.108	3.26
22	Unit or layer	Paper	5.10	1.042	5.31	1.078	5.49
	Unit	PIC	4.39	1.044	4.58	1.083	4.76
	T-screen, unfilled	PIC	4.47	1.056	4.72	1.108	4.95
	T-screen, filled to be watertight	PIC	3.99	1.056	4.20	1.108	4.39
24	Unit or layer	Paper	6.80	1.044	7.09	1.083	7.36
	Unit	PIC	5.58	1.026	5.72	1.052	5.87
	T-screen, unfilled	PIC	5.52	1.056	5.83	1.108	6.12
	T-screen, filled to be watertight	PIC	4.92	1.056	5.19	1.108	5.40
26	Unit or layer	Pulp	6.79	1.054	7.16	1.101	7.48
	Unit	PIC	7.48	1.024	7.66	1.046	7.82
	T-screen, filled to be watertight	PIC	6.65	1.056	7.02	1.098	7.30

f_T = temperature correction factor
L_c = cable loss at 772 kHz

For T1 carrier systems, repeaters are spaced at a nominal distance of 6000 ft, where the loss for 22-gauge cable is in the range of 27 to 33 dB. Hence T1 repeaters are designed to operate with a maximum transmission loss of typically 32 to 35 dB. The exact choice of maximum allowed loss is based on the expected maximum loss plus a margin to account for variation in loss. The range of repeater operation is extended to lower-loss sections by using automatic line build-out techniques.

8.7.1 One-Cable Systems

In one-cable systems, near-end crosstalk depends on the type of cable, the frequency, the physical separation of the pairs in the two directions of transmission, and the number of carrier systems contained within the cable. Engineering of a one-cable system first requires detailed knowledge of the cable construction. Multipair cable is constructed with pairs divided into units, layers, or groups. To minimize NEXT, it is standard practice to use pairs in nonadjacent units (layers or groups). Greater separation decreases the interference from NEXT. For small cables, however, it may be necessary to use pairs in adjacent units, or even pairs in the same unit, for opposite directions of transmission. In smaller cables, therefore, crosstalk interference increases significantly with the number of carrier systems.

Table 8.4 presents typical figures for NEXT on 22-gauge plastic insulated cable at 772 kHz [11]. This choice of frequency applies to T1 systems using bipolar coding. The frequency correction factor of 15 dB per decade can be used to adjust these figures to other frequencies. Some differences in crosstalk values exist between plastic, pulp, and paper-insulated cable. For a highly filled system, however, it has been found that the values do not differ noticeably. Therefore, as a first approximation, the values presented in Table 8.4 can be used for any type of insulation [12]. NEXT values do vary with cable gauge, though, according to the following expression:

$$\text{NEXT}_1 = \text{NEXT}_2 + 10 \log \frac{\alpha_1}{\alpha_2} \tag{8.37}$$

where subscripts 1 and 2 denote the wire gauge being compared and α is attenuation per unit length. The values given in Table 8.4 are **NEXT coupling loss** defined as the ratio of power in the disturbing circuit to the induced power in the disturbed circuit. Because of variations in the basic physical and electrical parameters of individual cable pairs, crosstalk coupling loss varies from pair to pair. The distribution of crosstalk coupling loss values is known to be normal (gaussian) so that only the mean (m) and standard deviation (σ) are needed to characterize the distribution. As crosstalk becomes worse, the crosstalk coupling loss values get smaller. Note that the values in Table 8.4 are affected both by the pair count and by the cable layout. The smaller cable sizes, such as 12 and 25-pair cable, have inherently greater crosstalk because of the close proximity of pairs. With

Table 8.4 Typical Near-End Crosstalk Coupling Loss at 772 kHz for 22-Gauge Cable [11]

Cable Size	*Location of Opposite-Direction Pairs*	*Mean (m), dB*	*Standard Deviation (σ), dB*
12-Pair cable	Adjacent pairs (same layer)	67	4.0
	Alternate pairs (same layer)	78	6.5
25-Pair cable	Adjacent pairs (same layer)	68	7.2
	Alternate pairs (same layer)	74	6.5
50-Pair cable	Adjacent pairs (same layer)	67	9.8
	Alternate pairs (same layer)	70	6.4
	Adjacent units	83	8.2
	Alternate units	94	8.7
100-Pair cable	Adjacent pairs (same layer)	66	5.5
	Alternate pairs (same layer)	73	8.2
	Adjacent units	85	8.6
	Alternate units	101	6.7
200-Pair cable	Adjacent pairs (same layer)	65	5.8
	Alternate pairs (same layer)	73	7.1
	Adjacent units	84	9.0
	Alternate units	103	6.7
Screened cable			
25-Pair	Opposite halves	103	7.5
50-Pair	Opposite halves	107	7.5
100-Pair	Opposite halves	112	7.0
200-Pair	Opposite halves	117	6.5

larger cable sizes, it is possible to select widely separated pairs, in separate units, and thus provide tolerable levels of crosstalk. Screened cable may be used to improve crosstalk performance further, especially if a small pair cable is involved.

The relationship between maximum section loss (L_D), near-end crosstalk coupling loss (m and σ), and number of carrier systems (n) can be written in the general form

$$L_D \leq am - b\sigma - c \log n - d \qquad \text{(decibels)} \tag{8.38}$$

where a, b, c, and d are constants whose values depend on the system. The U.S. industry standard for T1 cable engineering, for example, is given as [1]

$$L_D \leq (m - \sigma) - 10 \log n - 32 \tag{8.39}$$

This expression for T1 repeater section engineering is based on a requirement that at least 99 percent of the lines in a repeater section should have an error rate below 1×10^{-7}. The constant term, $d = 32$, in Equation (8.39) includes a 6-dB margin of safety factor to allow for variations in cable manufacturing, additional interference from far-end crosstalk, and other sources of degradation such as jitter. For convenience, Expression (8.39) can be put in the form of a chart as in Figure 8.15. For selected values of $(m - \sigma)$ and n, the value of L_D can be read directly off this chart. This value of L_D may then be used in (8.33) to determine repeater spacing.

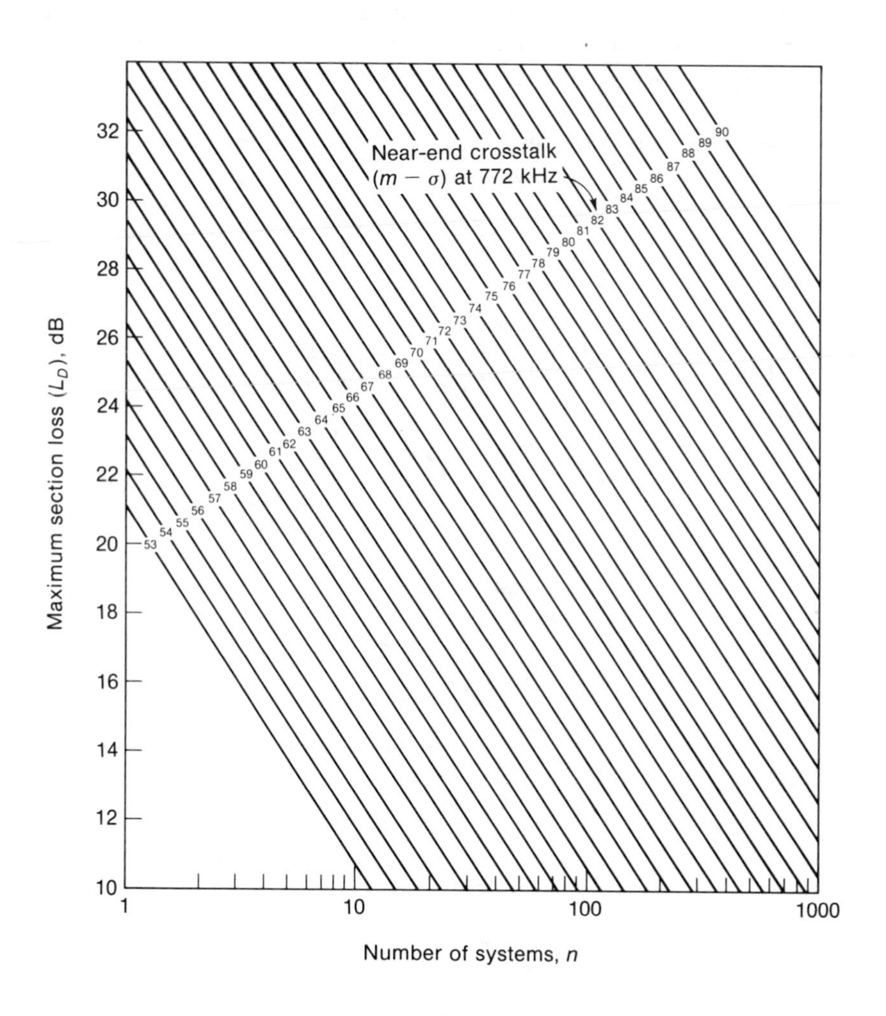

Figure 8.15 Maximum Section Loss vs. Crosstalk Coupling Loss $(m - \sigma)$ and Number of Systems

A value for L_D from (8.38) or (8.39) that would lead to a section loss greater than the basic loss due to transmission attenuation simply means that the allowed section loss is set equal to the attenuation loss. Thus for T1 carrier the allowed section loss cannot exceed 32 to 35 dB, the fundamental limit due to cable attenuation.

As noted earlier, the number of carrier systems in a cable affects crosstalk performance. Using Expression (8.39), this effect is illustrated for T1 carrier in Figure 8.16, which shows the crosstalk performance required to support a given number of carrier systems, assuming a certain allowed repeater section loss. Figure 8.16 also includes two plots of actual crosstalk performance values extracted from Table 8.4 for two cases: (1) transmit and receive cable pairs that are in adjacent units and (2) pairs that are separated by screening. A comparison of the required and actual NEXT performance indicates the improvement needed (case 1) or margin provided (case 2) by actual crosstalk performance to match the performance required for 100 percent cable fill. Conversely, use of Figure 8.16 and Table 8.4 can show the

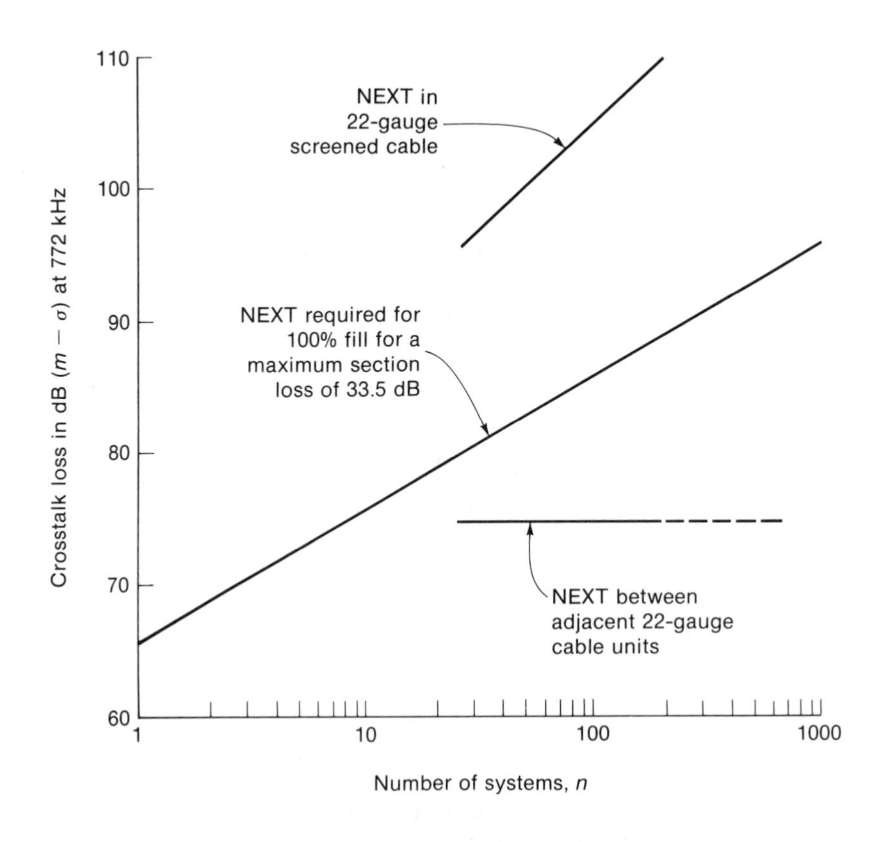

Figure 8.16 Near-End Crosstalk Loss vs. Number of Systems

backoff from 100 percent cable fill required to allow use of existing cable. For 50-pair cable, for example, the near-end crosstalk coupling loss $(m - \sigma)$ for adjacent units is 75 dB from Table 8.4. From Figure 8.16 the near-end crosstalk loss required to support 25 (two-way) systems at maximum allowed section loss is 80 dB. Therefore the improvement in crosstalk loss required for maximum fill is 5 dB. Such an improvement is not easily realized even by redesign of the cable. Faced with this predicament, the system designer must therefore reduce the number of carrier systems in the cable or use a screened or two-cable system in place of a one-cable system.

Example 8.1

A capacity for 50 T1 carrier systems is to be provided in a buried 200-pair cable of unit construction and 22-gauge plastic insulated conductors. The 50-pair unit used for one direction of transmission will be adjacent to the 50-pair unit used for the opposite direction. Calculate the required repeater spacing. Repeat the calculation but for only 10 carrier systems.

Solution

The maximum section loss due to NEXT is found from (8.39), where from Table 8.4 we have $m - \sigma = 75$. Therefore

$$L_D = 75 - 32 - 10 \log 50 = 26 \text{ dB}$$

(Note that the L_D can also be read directly from Figure 8.15.) Repeater spacing is found by using (8.33), where from Table 8.3 we have $L_c(100°\text{F}) = 4.58$ dB/1000 ft. Therefore

$$R_s = \frac{L_D}{L_c} = \frac{26 \text{ dB}}{4.58 \text{ dB}/1000 \text{ ft}} = 5677 \text{ ft}$$

For only 10 carrier systems,

$$L_D = 75 - 32 - 10 \log 10 = 33 \text{ dB}$$

so that the repeater spacing can be increased to

$$R_s = \frac{33 \text{ dB}}{4.58 \text{ db}/1000 \text{ ft}} = 7205 \text{ ft}$$

Example 8.2

Find the maximum number n of T1 carrier systems that can be accommodated by the cable type given in Example 8.1 if the repeater spacing is fixed at 6000 ft.

Solution

The maximum section loss due to NEXT is found from (8.33) as

$$L_D = R_s L_c = (6000 \text{ ft})(4.58 \text{ dB}/1000 \text{ ft})$$
$$= 27.5 \text{ dB}$$

Then from (8.39), or using Figure 8.15,

$$10 \log n = 75 - 32 - 27.5 = 15.5$$

Therefore the maximum number of T1 carrier systems is

$$n \leq 35$$

8.7.2 Two-Cable Systems

For the engineering of two-cable systems, an expression for the maximum section loss (L_D) or maximum cable fill (n) can be obtained by using the general form given by (8.38). Here, however, the statistical parameters m and σ apply to FEXT performance. For many two-cable system applications, losses due to far-end coupling do not exceed the level established by cable losses alone. It follows, therefore, that the maximum section loss is usually determined by only cable attenuation and not crosstalk, even for a large fill (n) in the cable.

Example 8.3

Calculate the required repeater spacing for a T1 carrier system, given the use of 19-gauge PIC cable constructed in units and a maximum section loss designed at 33.5 dB for:

(a) Aerial cable (140°F maximum temperature)

(b) Buried cable (100°F maximum temperature)

Solution

Using (8.33) and Table 8.3, we obtain

(a) $R_s = \dfrac{33.5 \text{ dB}}{3.43 \text{ dB}/1000 \text{ ft}} = 9767 \text{ ft}$

(b) $R_s = \dfrac{33.5 \text{ dB}}{3.31 \text{ dB}/1000 \text{ ft}} = 10{,}121 \text{ ft}$

8.8 REPEATER SPACING IN COAXIAL CABLE SYSTEMS

For coaxial cable systems, repeater spacing is determined by transmitter signal level, cable attenuation, and the required receiver signal-to-noise ratio. Whereas crosstalk is usually the major contributor to section loss in multipair cable, the use of coaxial cable virtually eliminates crosstalk as a factor in repeater spacing design. Thus repeater spacing is calculated by using Expression (8.33), which in the case of coaxial cable amounts to dividing the allowed attenuation between repeaters (L_D) by the attenuation of the cable per unit length (L_c).

The attenuation characteristic for several standard coaxial cables was given in Figure 8.3 for a fixed temperature, 10°C. However, attenuation increases with temperature because of the increase in conductor resistance with temperature. The attenuation characteristic at temperature T°C is referenced to 10°C by

$$\alpha(T°\text{C}) = \alpha(10°\text{C})[1 + K(T°\text{C} - 10°\text{C})] \tag{8.40}$$

The temperature coefficient, K, depends on both cable size and frequency. A typical value for frequencies above 1 MHz is $K = 2 \times 10^{-3}$ per degrees Celsius for any of the cable sizes shown in Figure 8.3.

Example 8.4

A 34.368-Mb/s system is to be applied to a 2.6/9.5-mm coaxial cable system. Maximum operating temperature is anticipated as 40°C for buried cable. Assuming an allowed attenuation between repeaters of 75 dB, calculate the required repeater spacing. Repeat the calculation for 1.2/4.4-mm coaxial cable.

Solution

The attenuation of 2.6/9.5-mm cable is found from Figure 8.3 and Equation (8.40). Assuming the use of bipolar signaling and a temperature coefficient of $K = 2 \times 10^{-3}$/°C, the attenuation at 17.184 MHz and 40°C is

$$\alpha(40°\text{C}) = 10[1 + (0.002)(30)] = 10.6 \text{ dB/km}$$

The repeater spacing is found from (8.33) as

$$R_s = \frac{75 \text{ dB}}{10.6 \text{ dB/km}} = 7.1 \text{ km}$$

For 1.2/4.4-mm cable, the attenuation is given by

$$\alpha(40°\text{C}) = 22[1 + (0.002)(30)] = 23.3 \text{ dB/km}$$

and the repeater spacing is then $R_s = 3.2$ km.

8.9 REPEATER SPACING IN FIBER OPTIC SYSTEMS

The maximum spacing of repeaters in a fiber optic system is determined by two transmission parameters, dispersion and attenuation, both of which increase with cable distance. Dispersion reduces the bandwidth available due to time spreading of digital signals, while attenuation reduces the signal-to-noise ratio available at the receiver. The distance between repeaters is then determined as the smaller of the distance limitation due to dispersion and that due to attenuation. In general, path length is limited by attenuation at lower data rates and limited by bandwidth at higher data rates.

To calculate repeater distance as determined by the attenuation characteristic, a power budget is required. Figure 8.17 illustrates the composition of a repeater section together with the concept of a power budget. Each component of the repeater section—transmitter, connectors, fiber, splices, and receiver—has a loss or gain associated with it. For repeater section design, it is convenient to divide the power budget into a system gain and loss. Each component has a characteristic gain or loss, usually expressed in decibels, which may be expressed in statistical terms (such as mean and standard deviation) or as a worst case, depending on the component supplier's specifications. The system gain G accounts for signal power characteristics in the transmitter and receiver of the repeater section and is given by

$$G = P_T - C_T - L_T - P_R - C_R - L_R \tag{8.41}$$

where P_T = average transmitter power

P_R = minimum input power to receiver required to achieve the error rate objective

C_T, C_R = connector loss of transmitter and receiver

L_T, L_R = allowance for degradation of transmitter and receiver with temperature variation and aging

The average transmitter power, P_T, depends on the peak output power of the light source, the duty cycle or modulation technique, and the loss in coupling the light source into the fiber. Coupling loss depends on the fiber numerical aperture and source emission characteristic. Of the two standard sources, lasers couple typically 10 dB more power into the fiber than LEDs because of greater output power and narrower emission angle. At both the transmitter and receiver, connector losses are included to account for joining of the fibers. The receiver sensitivity and choice of modulation technique then determine P_R, the minimum required input power. Because source output power and receiver sensitivity may degrade with temperature excursions or with time, an allowance is made for these additional losses.

The loss L takes into account cable losses, including fiber attenuation and splice losses, and is given by

$$L = DL_D + L_F + N_S L_S \tag{8.42}$$

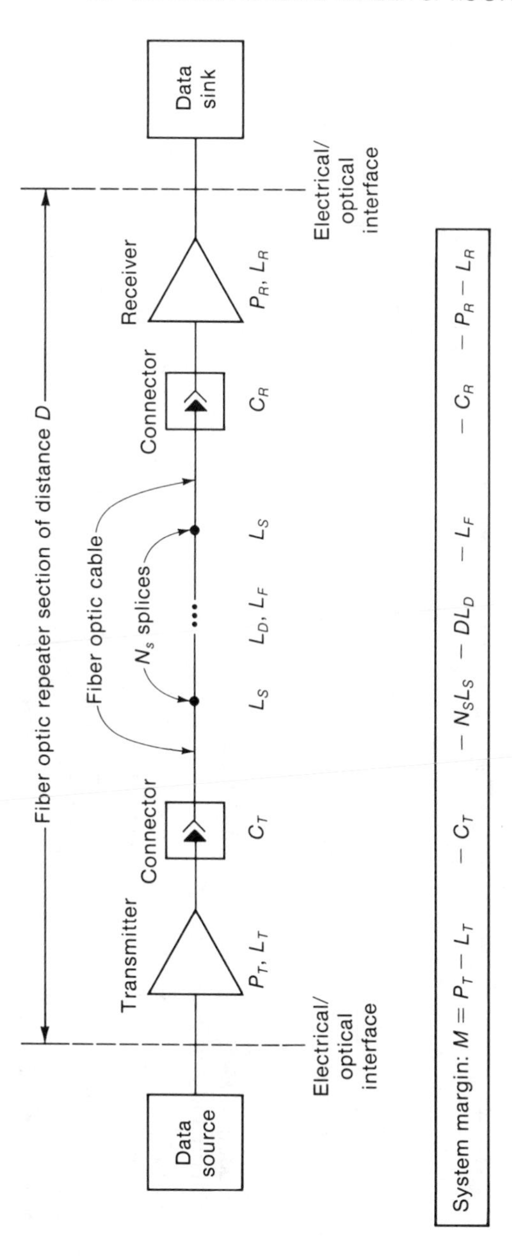

Figure 8.17 Fiber Optic Repeater Section Composition and Power Budget

Table 8.5 Power Budget Calculation for Fiber Optic System

Parameter	*Source*	*Sample Value for Low-Data-Rate, Short-Haul System*	*Sample Value for High-Data-Rate, Long-Haul System*
Average transmitter power, P_T	Manufacturer's specifications	−10 dBm (LED)	0 dBm (laser diode)
Allowance for transmitter degradation with temperature variation and aging, L_T	Manufacturer's specifications	2 dB	2 dB
Transmitter connector loss, C_T	Manufacturer's specifications	1.5 dB	1.5 dB
Receiver connector loss, C_R	Manufacturer's specifications	1.5 dB	1.5 dB
Minimum receiver power input, P_R	Manufacturer's specifications	−40 dBm[a] (PIN diode)	−58 dBm[a] (APD)
Allowance for receiver degradation with temperature variation and aging, L_R	Manufacturer's specifications	2 dB	2 dB

System gain, G	$P_T - L_T - C_T - C_R - P_R - L_R$	23 dB	51 dB
Average loss per unit length of fiber, L_D	Manufacturer's specifications	5 dB/km	1.5 dB/km
Repeater section distance, D	System design	2 km	25 km
Allowance for fiber degradation with temperature variation and aging, L_F	Manufacturer's specifications	3 dB	3 dB
Total fiber loss	$DL_D + L_F$	13 dB	40.5 dB
Average loss per splice, L_S	Manufacturer's specifications	0.2 dB	0.2 dB
Number of splices, N_S	System design	5	25
Total splice loss	$N_S L_S$	1 dB	5 dB
Total loss, L	$DL_D + L_F + N_S L_S$	14 dB	45.5 dB
System margin, M	$G - L$	9 dB	5.5 dB

[a]For BER $= 10^{-6}$.

where D = distance

L_D = average loss per unit length of fiber

L_F = allowance for degradation of fiber with temperature variation and aging

N_S = number of splices

L_S = average loss per splice

The loss in fiber optic cable consists of two components: intrinsic loss, which is due to absorption and scattering introduced in the fiber manufacturing process, and cabling loss caused by small bends in the fiber (called **microbending**) introduced during cable manufacturing. Total cable loss is found by multiplying cable distance (D) by the attenuation characteristic (L_D) and adding a loss factor (L_F) to account for the worst-case effects of temperature and age on cable loss. Losses due to splices are added to cable loss in order to yield the total loss L. Splices are necessary to join fibers when repeater spacing exceeds the standard manufacturing lengths (such as 1 km). Because splices are normally permanent junctions, the loss per splice is typically 0.5 dB or less, which is lower than typical connector losses of 1.5 dB or less.

After calculation of system gain G and loss L, the system margin M is obtained as $G - L$. To assure that error rate objectives are met, the system margin must be a nonnegative value. Often the system designer sets a minimum acceptable margin such as 3 or 6 dB that then accounts for additional sources of degradation not included in the gain and loss calculations. If the required system margin is not met, a system redesign is required, involving the selection of an improved component (such as the transmitter, fiber, connectors, splices, or receiver) or a reduction in repeater spacing.

The calculation of fiber optic link parameters for the power budget is facilitated by the form shown in Table 8.5. Here the system designer may determine the required value for a component, such as fiber loss or receiver sensitivity, by fixing the values of all other parameters including the system margin and working back to arrive at the value of the component in question. Conversely, the system designer may fix all component values and work forward to calculate the system margin or allowed link distance. To illustrate the use of the form in Table 8.5, sample values are given for two hypothetical fiber optic systems:

1. Case 1 is a low-data-rate (DS-2), short-haul (2 km) system based on an LED transmitter, a PIN diode receiver, and 5-dB/km fiber attenuation.
2. Case 2 is a high-data-rate (DS-3), long-haul (25 km) system based on a laser diode transmitter, avalanche photodiode (APD) receiver, and 1.5-dB/km fiber attenuation.

Repeater spacing must also be determined as a function of dispersion, since dispersion increases with distance and limits the available bandwidth.

The bandwidth required to support a given data rate depends on the baseband coding scheme. For NRZ, which is commonly used in fiber optic systems, the required bandwidth is equal to the data rate. (See Figure 5.7*a* for NRZ bandwidth requirements.) The dependence of bandwidth on length varies with fiber type and wavelength and in general is not linear. The relationship for calculating the bandwidth of an optical fiber section is

$$B_e = \frac{B_d}{D^\gamma} \tag{8.43}$$

where B_e = end-to-end bandwidth in megahertz

B_d = bandwidth-distance product in megahertz-kilometers

D = distance in kilometers

γ = length dependence factor (typically $0.7 \le \gamma \le 1$, where γ varies with fiber size, numerical aperture, and wavelength)

Table 8.6 provides a convenient form for calculating the required bandwidth-distance product, B_d, according to (8.43). Sample values for the two cases described above are given in Table 8.6. This table may also be used to

Table 8.6 Bandwidth Calculation for Fiber Optic System

Parameter	*Source*	*Sample Value for Low-Data-Rate, Short-Haul System*	*Sample Value for High-Data-Rate, Long-Haul System*
Data rate, R	System design	6.312 Mb/s	44.736 Mb/s
End-to-end bandwidth, B_e	Dependent on choice of baseband coding scheme	6.3 MHz for NRZ	45 MHz for NRZ
Repeater section distance, D	System design	2 km	25 km
Fiber size	System design	50/125 μm	50/125 μm
Numerical aperture, NA	Manufacturer's specifications	0.23	0.29
Wavelength	System design	825 nm	1300 nm
Length dependence factor, γ	Manufacturer's specifications	0.88	0.72
Bandwidth-distance product, B_d	$B_e D^\gamma$	11.6 MHz-km	765 MHz-km

Table 8.7 Rise Time Calculation for Fiber Optic System

Parameter	*Source*	*Sample Value for Low-Data-Rate, Short-Haul System*	*Sample Value for High-Data-Rate, Long-Haul System*
Data rate, R	System design	6.312 Mb / s	44.736 Mb / s
Maximum allowed rise time, t_s	$0.7 / R$	111 ns	15.6 ns
Repeater section distance, D	System design	2 km	25 km
Transmitter rise time, t_t	Manufacturer's specifications	6 ns (LED)	1.5 ns (laser diode)
Rise time per unit length fiber due to multimode dispersion, $t_{mo/km}$	Manufacturer's specifications	10 ns / km[a]	0.5 ns / km[b]
Total fiber rise time due to multimode dispersion, t_{mo}	$Dt_{mo/km}$	20 ns	12.5 ns
Rise time per unit length fiber due to material dispersion, $t_{ma/km}$	Manufacturer's specifications	4 ns / km (LED)	0.2 ns / km (laser diode)
Total fiber rise time due to material dispersion, t_{ma}	$Dt_{ma/km}$	8 ns	5.0 ns
Receiver rise time, t_r	Manufacturer's specifications	10 ns (PIN diode)	4 ns (APD)
System rise time, t_s	$1.1\,(t_t^2 + t_r^2 + t_{ma}^2 + t_{mo}^2)^{1/2}$	27 ns	15.5 ns

[a] For step-index fiber with bandwidth 15 MHz · km.
[b] For graded-index fiber with bandwidth 800 MHz · km.

determine the maximum repeater spacing, D, for the required bandwidth B_e.

Another measure of an optical fiber system's dispersive properties and bandwidth is the **system rise time**, usually defined as the time for the voltage to rise from 0.1 to 0.9 of its final value. For detection of NRZ data, the system rise time is typically specified to be no more than 70 percent of the bit interval. Rise time is determined by the dispersive properties of each system component—transmitter, fiber, and receiver. The overall rise time is approximately 1.1 times the square root of the sum of the squares of the rise times of the individual system components [13]:

$$t_s = \text{system rise time} = 1.1\left(t_t^2 + t_r^2 + t_{\text{ma}}^2 + t_{\text{mo}}^2\right)^{1/2} \tag{8.44}$$

where t_t = transmitter rise time

t_r = receiver rise time

t_{ma} = material dispersion

t_{mo} = multimode dispersion

A budget for rise time is derived in much the same manner as the power budget. Once again a form such as Table 8.7 is handy in calculating the system rise time. Sample values are shown for the two cases: a low-data-rate, short-haul system and a high-data-rate, long-haul system. The system rise time for the first case is negligible compared to the maximum allowed rise time, a result that is expected for low-data-rate, short-haul systems. However, the small margin in rise time for the second case indicates that the rise time budget may place limits on the distance and data rate or require improved performance in the transmitter, fiber, or receiver.

8.10 IMPLEMENTATION CONSIDERATIONS

Figure 8.18 shows representative implementations of wire and fiber optic PCM repeatered lines, both of which provide the same 672 voice channel capacity. Figure 8.18 also illustrates the savings in equipment and increase in repeater spacings afforded by fiber optic systems over conventional T-carrier cable. The fiber optic system operates at the T3 rate, 44.736 Mb/s, with one operating line and one spare line (1:1 redundancy). The T1 wire-line system uses 28 operating and 4 spare lines (1:7 redundancy). At the end offices, transmitted and received signals are fed through a protection switch that contains monitoring and control logic and allows a failed line to be switched off-line and replaced with a spare line. Repeaters for regeneration are placed at regular intervals according to the repeater spacing calculations. These repeaters are housed in protective cases and may be above or below ground, depending on the application. Access to the repeater is then typically by telephone pole or manhole. Because repeater sites are frequently in remote locations, power is often unavailable locally

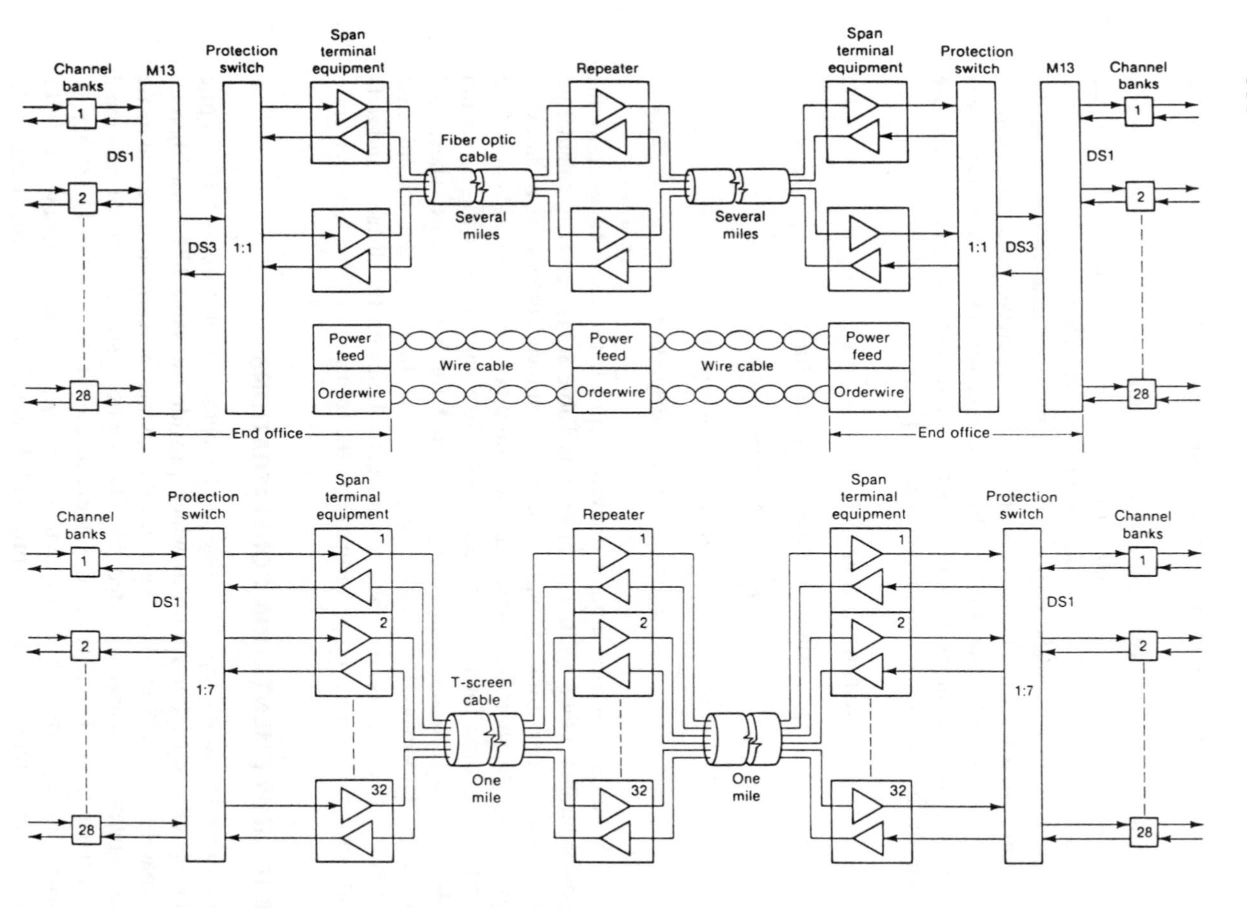

Figure 8.18 Fiber Optics vs. T-Carrier for PCM Repeatered Lines (Reprinted by permission of ITT)

and must be fed along the communication lines. For troubleshooting, fault isolation devices and an orderwire channel are also standard features in repeatered line equipment.

8.10.1 Power Feeding

In metallic cable systems, line repeaters are powered by using the cable itself for dc power transmission. In multipair cable, power from the end offices is passed through the pairs and decoupled from the signal through transformers. The required voltage V_M for a cable section is determined by

$$V_M = RV_R + LV_L \tag{8.45}$$

where V_R = voltage drop of one repeater

R = number of repeaters

V_L = voltage drop per unit length of cable

L = cable length

For a specified maximum voltage V_M and for given voltage drops V_R and V_L, Equation (8.45) determines the maximum number of repeaters allowed. Typical values supplied by the end office power supply are 50 to 100 mA constant current and 48 to 260 volts maximum voltage.

In coaxial and fiber optic cable systems, standard practice is to include a metallic pair for dc power transmission. In the case of coaxial systems, this pair is included within the cable sheath containing the coaxials; for fiber optic cable, the power line is separate from the fiber cable.

8.10.2 Automatic Protection Switching

The use of spare lines with automatic switching provides increased availability and facilitates maintenance, especially at remote unattended locations. The ratio of spare lines to operating lines depends on the required availability, the failure rate of individual line components (repeaters, cable), and the failure rate of the protection switch itself. The usual practice is to drive spare lines with a dummy signal (from a pseudorandom sequence generator, for example) and monitor the received signal for errors. At the same time, operational lines are monitored for failure—for example, by counting bipolar violations or consecutive zeros and comparing the count to a threshold. Logic circuitry controls the transfer of operating lines to spare lines.

8.10.3 Fault Location

When a fault occurs, a means of locating the fault from a remote location or end office is desirable and usually standard practice in cable systems. Fault location techniques can be categorized as **in-service**, where the traffic is not interrupted during testing, or as **out-of-service**, in which case operational traffic is replaced by special test signals. A faulty repeater section can be identified by monitoring the error rate at each repeater. This may be done on an in-service basis by use of parity error or bipolar violation checking. Or it may be done out-of-service by transmitting a pseudorandom sequence from the end office, looping back this signal successively from one repeater to the next, and detecting errors in the sequence at the end office. In theory, the preferred method of fault location is in-service, since the techniques employed are usually simple to implement, provide continuous and automatic monitoring, and do not result in user outage. Out-of-service techniques are more accurate, however, and are more commonly used for fault location.

For 24 and 30-channel PCM transmission over multipair cable, fault location is performed out-of-service by use of a pseudorandom test sequence applied to the line under test. A particular repeater is accessed for test by use of an address transmitted as one of several audio frequencies or bit sequences. The addressed repeater responds by closing a loop from the transmit path to the receive path, so that the test signal returns to the fault location unit. A comparison of detected versus transmitted patterns indicates the exact bit error rate. Each repeater may be progressively tested by use of the appropriate address until the faulty section is discovered. The number of addresses is typically 12 to 18 for a fault location unit in an end office. Since this fault location can be carried out at both end offices, a line span may contain a maximum of 24 to 36 repeaters for proper fault location.

8.10.4 Orderwire

Orderwires are voice circuits used by maintenance personnel for coordination and control. In cable transmission, the orderwire is usually provided by a separate pair in parallel with the communication lines. Alternatively, the orderwire can be modulated onto the same cable by using frequencies below or above the digital signal spectrum. Using the orderwire, maintenance personnel can dial any number from a repeater location.

8.11 SUMMARY

Digital transmission via cable has been developed for twisted-pair, coaxial, and optical cables. The performance of twisted-pair cable generally limits its application to below 10 Mb/s, but nevertheless twisted-pair cable has been used extensively for T-carrier systems. Coaxial cable is a standard transmission medium for FDM and has been adapted for digital transmission at rates approaching 1 Gb/s (see Table 8.1). Fiber optic cable offers greater bandwidths, lower attenuation, and no crosstalk or electrical interference compared to metallic cable, and these advantages have led to dramatic growth in fiber optic systems. Low-data-rate, short-haul fiber optic systems tend toward multimode cable, LED transmitters, and PIN diode receivers; high-data-rate, long-haul systems tend toward single-mode cable, laser diode transmitters, and avalanche photodiode receivers.

Regenerative repeaters are required in a digital cable system at regularly spaced intervals to reshape, retime, and regenerate the digital signal. Reshaping involves amplification and equalization; retiming requires recovery of clock from the data signal; and regeneration performs sampling, voltage comparison, and code conversion. Each regenerative repeater contributes jitter and data errors in the process of regenerating the digital signal. Jitter amounts to a displacement of the intended data transition times, which leads to sampling offsets and data errors. Sources of jitter may by systematic or random. Jitter related to the data pattern is systematic and tends to accumulate along a chain of repeaters. Jitter from uncorrelated sources is random and does not accumulate in a repeater chain. The error rate in a chain of n repeaters is approximately equal to np, where p is the probability of error for a single repeater.

Repeater spacing in multiple twisted-pair (multipair) cable is determined largely by attenuation and crosstalk. Crosstalk occurs when energy is coupled from one circuit to another. Near-end crosstalk (NEXT) refers to the case when the disturbing circuit travels in a direction opposite to that of the disturbed circuit. Far-end crosstalk (FEXT) is the case where the disturbed circuit travels in the same direction as the disturbing circuit. The degree of crosstalk depends on cable size and construction. When both directions of transmission are present in the same cable, NEXT is the limiting factor in determining repeater spacing. When two separate cables are used for the two directions, NEXT is eliminated and repeater spacing is determined by FEXT and noise.

For coaxial cable systems, crosstalk is virtually eliminated and repeater spacing becomes a function of only the attenuation characteristic. Repeater spacing is calculated by dividing the allowed attenuation between repeaters by the attenuation of the cable per unit length.

Repeater spacing in fiber optic systems is determined by dispersion and attenuation characteristics. A power budget accounts for attenuation effects by summing the loss or gain of each system component, including transmitter, connectors, fiber, splices, and receiver. The net difference between gains and losses is equivalent to a system margin, which provides protection against degradation not included in the power budget. Like attenuation, dispersion increases with distance; thus dispersion reduces the available bandwidth and allowed repeater spacing. The calculation of repeater spacing as a function of dispersion is based on the required bandwidth (see Table 8.6) or system rise time (see Table 8.7).

Implementation of metallic or fiber optic cable systems must also include consideration of power feeding, redundancy with automatic protection switching, fault location using in-service or out-of-service techniques, and voice orderwire for maintenance personnel.

REFERENCES

1. H. Cravis and T. V. Crater, "Engineering of T1 Carrier System Repeatered Lines," *Bell System Technical Journal* 42(2)(March 1963):431–486.
2. CCITT Yellow Book, vol. III.2, *International Analog Carrier Systems. Transmission Media—Characteristics* (Geneva: ITU, 1981).
3. CCITT Yellow Book, vol. III.3, *Digital Networks—Transmission Systems and Multiplexing Equipment* (Geneva: ITU, 1981).
4. Members of the Bell Telephone Laboratories, *Engineering and Operations in the Bell System* (Bell Telephone Laboratories, 1977).
5. J. Legras and C. Paccaud, "Development of a 565 Mbit/s System on 2.6/9.5 mm Coaxial Cables," *1981 International Conference on Communications*, pp. 75.3.1 to 75.3.5.
6. N. Inoue, H. Kasai, T. Miki, and N. Sakurai, "PCM-400M Digital Repeatered Line," *1975 International Conference on Communications*, pp. 24.11 to 24.15.
7. H. Kasai, K. Ohue, and T. Hoshino, "An Experimental 800 Mbit/s Digital Transmission System Over Coaxial Cable," *1981 International Conference on Communications*, pp. 75.5.1 to 75.5.5.
8. C. J. Byrne, B. J. Karafin, and D. B. Robinson, "Systematic Jitter in a Chain of Digital Regenerators," *Bell System Technical Journal* 42(6)(November 1963):2679–2714.
9. H. E. Rowe, "Timing in a Long Chain of Binary Regenerative Repeaters," *Bell System Technical Journal* 37(6)(November 1958):1543–1598.
10. Bell Telephone Laboratories, *Transmission Systems for Communications* (Winston-Salem: Western Electric Company, 1970).
11. L. Jachimowicz, J. A. Olszewski, and I. Kolodny, "Transmission Properties of Filled Thermoplastic Insulated and Jacketed Telephone Cables at Voice and Carrier Frequencies," *IEEE Trans. on Comm.*, vol. COM-21, no. 3, March 1973, pp. 203–209.
12. T. C. Henneberger and M. D. Fagen, "Comparative Transmission Characteristics of Polyethylene Insulated and Paper Insulated Communication Cables," *Trans. AIEE Communications Electronics* 59(March 1962):27–33.
13. D. L. Baldwin and colleagues, "Optical Fiber Transmission System Demonstration Over 32 km with Repeaters Data Rate Transparent up to 2.3 Mbit/s," *IEEE Trans. on Comm.*, vol. COM-26, no. 7, July 1978, pp. 1045–1055.

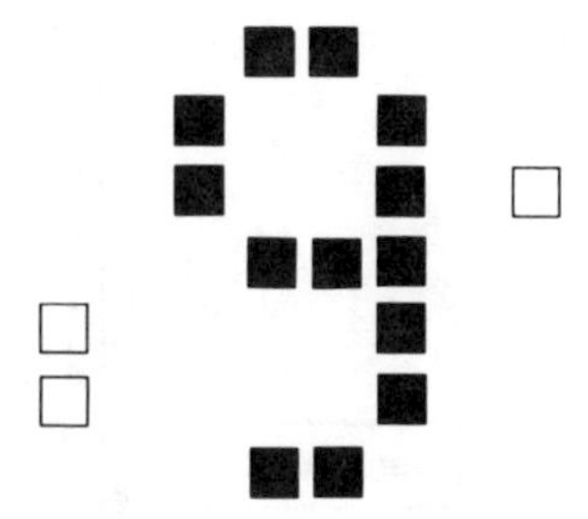

Digital Radio Systems

OBJECTIVES

- Considers free-space propagation and discusses deviations from it due to effects of terrain, atmosphere, and precipitation
- Describes multipath fading due to ground reflection or atmospheric refraction and characterizes it with a statistical model
- Explains how the effects of multipath fading on digital radio performance can be mitigated by such techniques as diversity and adaptive equalization
- Discusses the effects of frequency allocations on digital radio capacity and system planning
- Points out techniques for minimizing intrasystem and intersystem RF interference
- Describes the components of a digital radio—including transmitter, receiver, diversity combiner, adaptive equalizer, power amplifiers, and filters
- Lists the procedures for calculating link performance or determining design parameters

9.1 INTRODUCTION

The basic components required for operating a radio over a line-of-sight (LOS) link are the transmitter, towers, antennas, and receiver (Figure 9.1). Transmitter functions typically include multiplexing, encoding, modulation, up-conversion from baseband or intermediate frequency (IF) to radio frequency (RF), power amplification, and filtering for spectrum control.

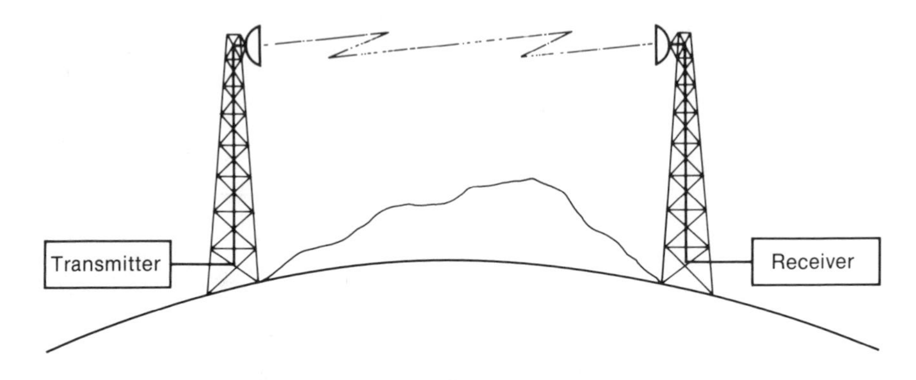

Figure 9.1 Line-of-Sight Radio Configuration

Antennas are placed on a tower or other tall structure at sufficient height to provide a direct, unobstructed path between the transmitter and receiver sites. Receiver functions include RF filtering, down-conversion from RF to IF, amplification at IF, equalization, demodulation, decoding, and demultiplexing. Some of these transmitter and receiver functions have been described in earlier chapters, including multiplexing (Chapter 4), coding (Chapter 5), and modulation (Chapter 6). The other functions that are unique to digital radio are described in detail in this chapter. Further, various phenomena associated with line-of-sight propagation are described here as they affect digital radio performance.

In describing radio and link design, the focus is on digital line-of-sight microwave systems (from about 1 to 30 GHz). The design of millimeter-wave (30 to 300 GHz) radio links is also considered, however. Further, much of the material on line-of-sight propagation, including multipath and interference effects, and link design methodology also applies to the design of analog radio systems.

9.2 LINE-OF-SIGHT PROPAGATION

The modes of propagation between two radio antennas may include a direct, line-of-sight (LOS) path but also a ground or surface wave that parallels the earth's surface, a sky wave from signal components reflected off the troposphere or ionosphere, a ground reflected path, and a path diffracted from an obstacle in the terrain. The presence and utility of these modes depend on the link geometry, both distance and terrain between the two antennas, and the operating frequency. For frequencies in the microwave band, the LOS propagation mode is the predominant mode available for use; the other modes may cause interference with the stronger LOS path. Line-of-sight links are limited in distance by the curvature of the earth, obstacles along the path, and free-space loss. Average distances for conservatively designed LOS links are 25 to 30 mi, although distances up to 100 mi have been used. The performance of the LOS path is affected by several phenomena addressed in this section, including free-space loss, terrain, atmosphere, and precipitation. The problem of fading due to multiple paths is addressed in the following section.

9.2.1 Free-Space Loss

Consider a radio path consisting of isotropic antennas at the transmitter and receiver. An isotropic transmitting antenna radiates its power P_{ta} equally in all directions. In the absence of terrain or atmospheric effects (that is, free space), the radiated power density is equal at points equidistant from the transmitter. One can imagine a sphere of radius D where the power density on the surface of the sphere is given by

$$P_D = \frac{P_{\text{ta}}}{4\pi D^2} \tag{9.1}$$

If a receiving antenna has an effective area A_r and is located a distance D from the transmitting antenna, the received power P_{ra} is equal to

$$P_{\text{ra}} = \frac{P_{\text{ta}} A_r}{4\pi D^2} \tag{9.2}$$

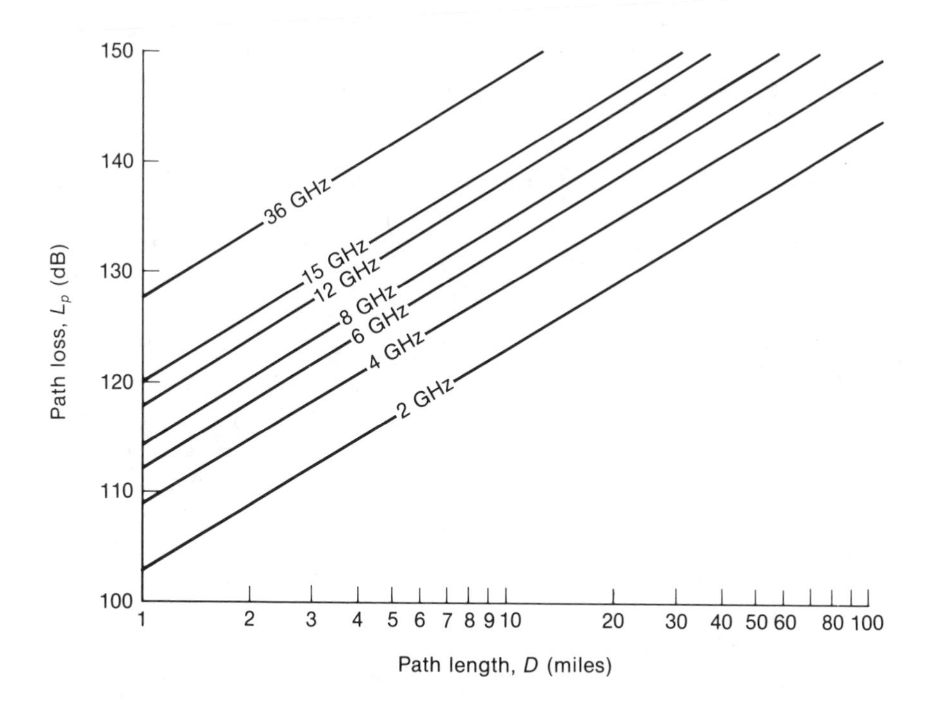

Figure 9.2 Free-Space Loss Between Isotropic Antennas

The effective area of an isotropic antenna is known to be [1]

$$A_r = A_{\text{isotropic}} = \frac{\lambda^2}{4\pi} \tag{9.3}$$

where λ is the wavelength of the radio signal. Combining (9.2) and (9.3), we can express the free-space path loss between isotropic antennas as

$$\frac{P_{\text{ta}}}{P_{\text{ra}}} = \left(\frac{4\pi D}{\lambda}\right)^2 \tag{9.4}$$

When expressed in decibels, path loss L_p is

$$L_p = 96.6 + 20\log f + 20\log D \tag{9.5}$$

where f is radio frequency in gigahertz and D is path length in miles. Plots of L_p for representative frequencies and distances are shown in Figure 9.2. Note that the doubling of either frequency or distance causes a 6-dB increase in path loss.

9.2.2 Terrain Effects

Obstacles along a line-of-sight radio path can cause the propagated signal to be reflected or diffracted, resulting in path losses that deviate from the free-space value. This effect stems from electromagnetic wave theory, which postulates that a wave front diverges as it advances through space. A radio beam that just grazes the obstacle is diffracted, with a resulting obstruction loss whose magnitude depends on the type of surface over which the diffraction occurs. A smooth surface, such as water or flat terrain, produces the maximum obstruction loss at grazing. A sharp projection, such as a mountain peak or even trees, produces a knife-edge effect with minimum obstruction loss at grazing. Most obstacles in the radio path produce an obstruction loss somewhere between the limits of smooth earth and knife edge.

Reflections

When the obstacle is below the optical line-of-sight path, the radio beam can be reflected to create a second signal at the receiving antenna. Reflected signals can be particularly strong when the reflection surface is smooth terrain or water. Since the reflected signal travels a longer path than the direct signal, the reflected signal may arrive out of phase with the direct signal. The degree of interference at the receiving antenna from the reflected signal depends on the relative signal levels and phases of the direct and reflected signals.

At the point of reflection, the indirect signal undergoes attenuation and phase shift, which is described by the reflection coefficient R, where

$$R = \rho e^{j\phi} \tag{9.6}$$

The magnitude ρ represents the change in amplitude and ϕ is the phase shift on reflection. The values of ρ and ϕ depend on the wave polarization (horizontal or vertical), angle of incidence, dielectric constant of the reflection surface, and the wavelength λ of the radio signal. The mathematical relationship has been developed elsewhere [2] and will not be covered here. For microwave frequencies, however, two general cases should be mentioned:

1. For horizontally polarized waves with small angle of incidence, $R = -1$ for all terrain, such that the reflected signal suffers no change in amplitude but has a phase change of 180°.
2. If the polarization is vertical with grazing incidence, $R = -1$ for all terrain. With increasing angle of incidence, the reflection coefficient magnitude decreases, reaching zero in the vicinity of $\phi = 10°$.

To examine the problem of interference from reflection, we first simplify the analysis by neglecting the effects of the curvature of the earth's surface. Then, when the reflection surface is flat earth, the geometry is as illustrated in Figure 9.3*a*, with transmitter (Tx) at height h_1 and receiver

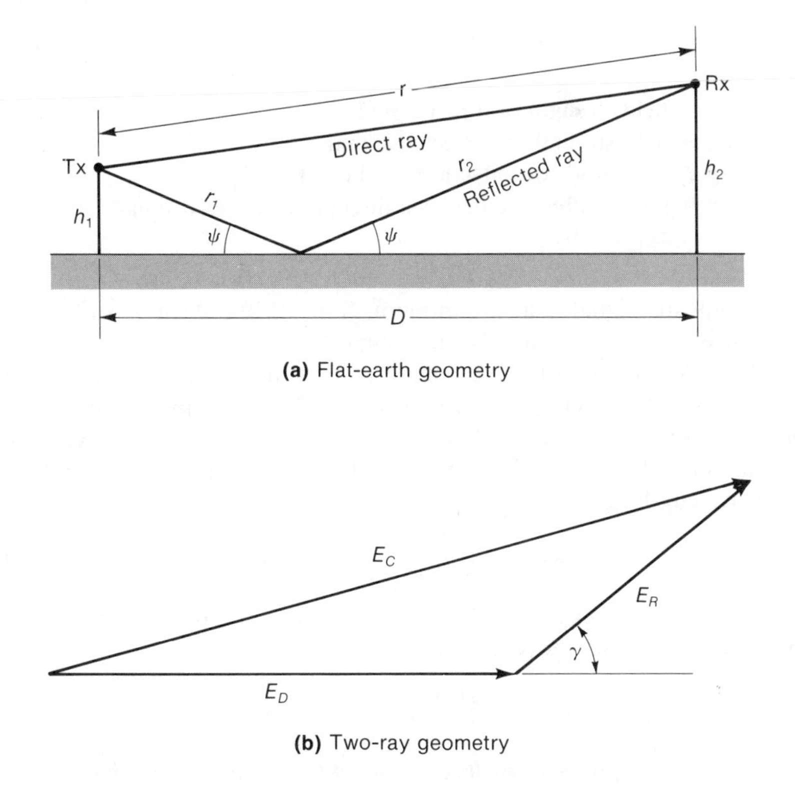

Figure 9.3 Geometry of Two-Path Propagation

(Rx) at height h_2 separated by a distance D and the angle of reflection equal to the angle of incidence ψ. Using plane geometry and algebra, the path difference δ between the reflected and direct signals can be given by

$$\begin{aligned}\delta &= (r_1 + r_2) - r \\ &\approx \frac{4h_1h_2}{D}\end{aligned} \tag{9.7}$$

The overall phase change experienced by the reflected signal relative to the direct signal is the sum of the phase difference due to the path length difference δ and the phase ϕ due to the reflection. The total phase shift is therefore

$$\gamma = \frac{2\pi\delta}{\lambda} + \phi \tag{9.8}$$

At the receiver, the direct and reflected signals combine to form a composite signal with field strength E_C. By the simple geometry of Figure 9.3*b* and using the law of cosines,

$$E_C^2 = E_D^2 + E_R^2 + 2E_DE_R\cos\gamma \tag{9.9}$$

or

$$E_C = E_D\sqrt{1 + \rho^2 + 2\rho\cos\gamma} \tag{9.10}$$

where E_D = field strength of direct signal
E_R = field strength of reflected signal
ρ = magnitude of reflection coefficient = E_R/E_D
γ = phase difference between direct and reflected signal as given by (9.8)

The composite signal is at a minimum, from (9.10), when $\gamma = (2n + 1)\pi$, where n is an integer. Similarly, the composite signal is at a maximum when $\gamma = 2n\pi$. As noted earlier, the phase shift ϕ due to reflection is usually around 180° for microwave paths since the angle of incidence upon the reflection surface is typically quite small. For this case, the received signal minima, or nulls, occur when the path difference is an even multiple of a half-wavelength, or

$$\delta = 2n\left(\frac{\lambda}{2}\right) \qquad \text{for minima} \tag{9.11}$$

The maxima, or peaks, for this case occur when the path difference is an odd multiple of a half-wavelength, or

$$\delta = \frac{(2n + 1)}{2}\lambda \qquad \text{for maxima} \tag{9.12}$$

When the spherical surface is substituted for the plane reflecting surface of Figure 9.3*a*, the reflected signal diverges at a greater rate such that the field strength is further reduced. To allow for the effects of round earth, it is necessary to introduce a *divergence factor*, defined as the ratio of the field strength obtained after reflection from a spherical surface to that obtained after reflection from a plane surface. The analysis required in calculating the divergence factor is not simple but has been done for smooth earth [2]. This calculated value is then multiplied by the reflection coefficient in (9.6) to describe the reflected signal characteristics.

Fresnel Zones

The effects of reflection and diffraction on radio waves can be more easily seen by using the model developed by A. Fresnel for optics. Fresnel accounted for the diffraction of light by postulating that the cross section of an optical wave front is divided into zones of concentric circles separated by half-wavelengths. These zones alternate between constructive and destructive interference, resulting in a sequence of dark and light bands when diffracted light is viewed on a screen. When viewed in three dimensions, as necessary for determining path clearances in line-of-sight radio systems, the Fresnel zones become concentric ellipsoids. The first Fresnel zone is that locus of points for which the sum of the distances between the transmitter and receiver and a point on the ellipsoid is exactly one half-wavelength longer than the direct path between the transmitter and receiver. The nth Fresnel zone consists of that set of points for which the difference is n half-wavelengths. The radius of the nth Fresnel zone at a given distance along the path is given by

$$F_n = 17.3\sqrt{\frac{nd_1d_2}{fD}} \quad \text{meters} \tag{9.13}$$

where d_1 = distance from transmitter to a given point along the path (km)
d_2 = distance from receiver to the same point along the path (km)
f = frequency (GHz)
D = path length (km)($D = d_1 + d_2$)

As an example, Figure 9.4a shows the first three Fresnel zones for an LOS path of length (D) 40 km and frequency (f) 8 GHz. The distance h represents the clearance between the LOS path and the highest obstacle along the terrain.

Using Fresnel diffraction theory, the effects of path clearance on transmission loss can be calculated as shown in Figure 9.4b. The three cases shown correspond to different reflection coefficient values as determined by differences in terrain roughness. The curve marked $R = 0$ represents the case of knife-edge diffraction, where the loss at a grazing angle (zero clearance) is equal to 6 dB. The curve marked $R = -1.0$ illustrates diffraction from a smooth surface, which produces a maximum loss equal to 20 dB at grazing. In practice, most microwave paths have been found to have a reflection coefficient magnitude of 0.2 to 0.4; thus the curve marked $R = -0.3$ represents the ordinary path [3]. For most paths, the signal attenua-

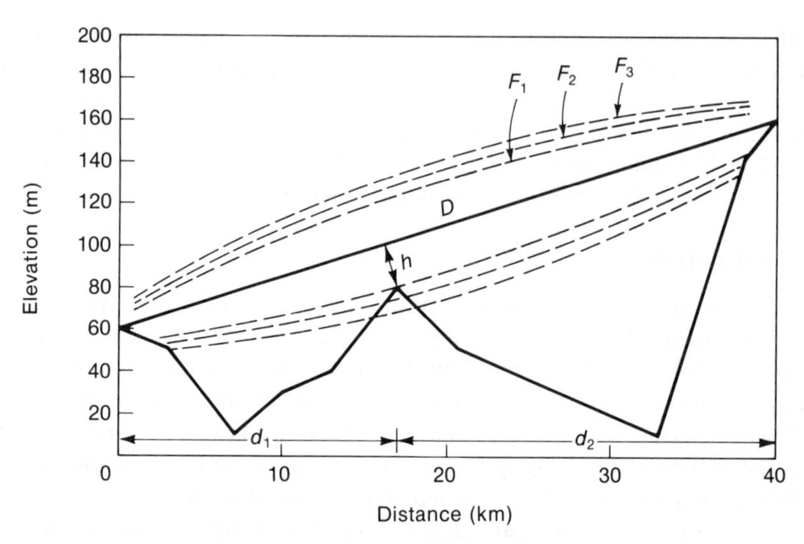

(a) Fresnel zones for an 8-GHz, 40-km LOS path

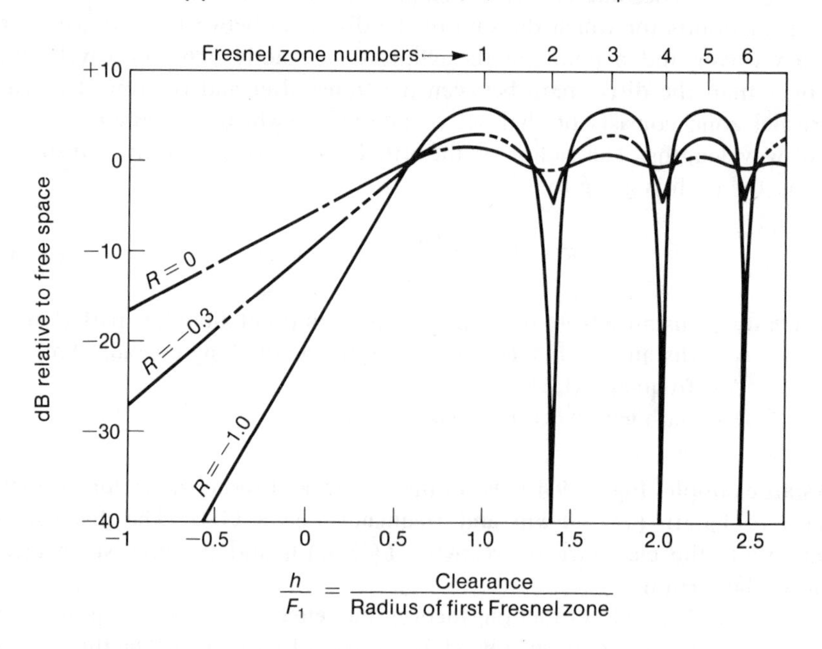

(b) Attenuation vs. path clearance (adapted from [8])

Figure 9.4 Fresnel Zones

tion becomes small with a clearance of 0.6 times the first Fresnel zone radius. Thus microwave paths are typically sited with a clearance of at least $0.6F_1$.

The fluctuation in signal attenuation observed in Figure 9.4*b* is due to alternating constructive and destructive interference with increasing clearance. Clearance at odd-numbered Fresnel zones produces constructive interference since the delayed signal is in phase with the direct signal; with a reflection coefficient of -1.0, the direct and delayed signals sum to a value 6 dB higher than free-space loss. Clearance at even-numbered Fresnel zones produces destructive interference since the delayed signal is out of phase with the direct signal by a multiple of $\lambda/2$; for a reflection coefficient of -1.0, the two signals cancel each other. As indicated in Figure 9.4*b*, the separation between adjacent peaks or nulls decreases with increasing clearance, but the differences in signal strength decrease with increasing Fresnel zone numbers.

9.2.3 Atmospheric Effects

Radio waves travel in straight lines in free space, but they are bent, or *refracted*, when traveling through the atmosphere. Bending of radio waves is caused by changes with altitude in the index of refraction, defined as the ratio of propagation velocity in free space to that in the medium of interest. Normally the refractive index decreases with altitude, meaning that the velocity of propagation increases with altitude, causing radio waves to bend downward. In this case, the radio horizon is extended beyond the optical horizon.

The index of refraction n varies from a value of 1.0 for free space to approximately 1.0003 at the surface of the earth. Since this refractive index varies over such a small range, it is more convenient to use a scaled unit, N, which is called **radio refractivity** and defined as

$$N = (n - 1)10^6 \tag{9.14}$$

Thus N indicates the excess over unity of the refractive index, expressed in millionths. When $n = 1.0003$, for example, N has a value of 300. At microwave frequencies, the radio refractivity is given by

$$N = \frac{77.6p}{T} + \frac{3.73 \times 10^5 w}{T^2} \tag{9.15}$$

where p = atmospheric pressure (millibars)
w = partial pressure of water vapor (millibars)
T = absolute temperature (°K)

Owing to the rapid decrease of pressure and humidity with altitude and the slow decrease of temperature with altitude, N normally decreases with altitude and tends to zero.

To account for atmospheric refraction in path clearance calculations, it is convenient to replace the true earth radius a by an **effective earth radius**

a_e and to replace the actual atmosphere with a uniform atmosphere in which radio waves travel in straight lines. The ratio of effective to true earth radius is known as the ***k* factor**:

$$k = \frac{a_e}{a} \tag{9.16}$$

By application of Snell's law in spherical geometry, it may be shown that as long as the change in refractive index is linear with altitude, the k factor is given by

$$k = \frac{1}{1 + a(dn/dh)} \tag{9.17}$$

where dn/dh is the rate of change of refractive index with height. It is usually more convenient to consider the gradient of N instead of the gradient of n. Making the substitution of dN/dh for dn/dh and also entering the value of 6370 km for a into (9.17) yields the following:

$$k = \frac{157}{157 + (dN/dh)} \tag{9.18}$$

where dN/dh is the N gradient per kilometer. Under most atmospheric conditions, the gradient of N is negative and constant and has a value of approximately

$$\frac{dN}{dh} = -40 \text{ units/km} \tag{9.19}$$

Substituting (9.19) into (9.18) yields a value of $k = 4/3$, which is commonly used in propagation analysis. An index of refraction that decreases uniformly with altitude resulting in $k = 4/3$ is referred to as **standard refraction**.

Anomalous Propagation

Weather conditions may lead to a refractive index variation with height that differs significantly from the average value. In fact, atmospheric refraction and corresponding k factors may be negative, zero, or positive. The various forms of refraction are illustrated in Figure 9.5 by presenting radio paths over both true earth and effective earth. Note that radio waves become straight lines when drawn over the effective earth radius. Standard refraction is the average condition observed and results from a well-mixed atmosphere. The other refractive conditions illustrated in Figure 9.5—including subrefraction, superrefraction, and ducting—are observed a small percentage of the time and are referred to as **anomalous propagation**.

Subrefraction ($k < 1$) leads to the phenomenon known as inverse bending or earth bulge illustrated in Figure 9.5*b*. This condition arises because of an increase in refractive index with altitude and results in an upward bending of radio waves. The effect produced is likened to the bulging of the earth into the microwave path that reduces the path clearance

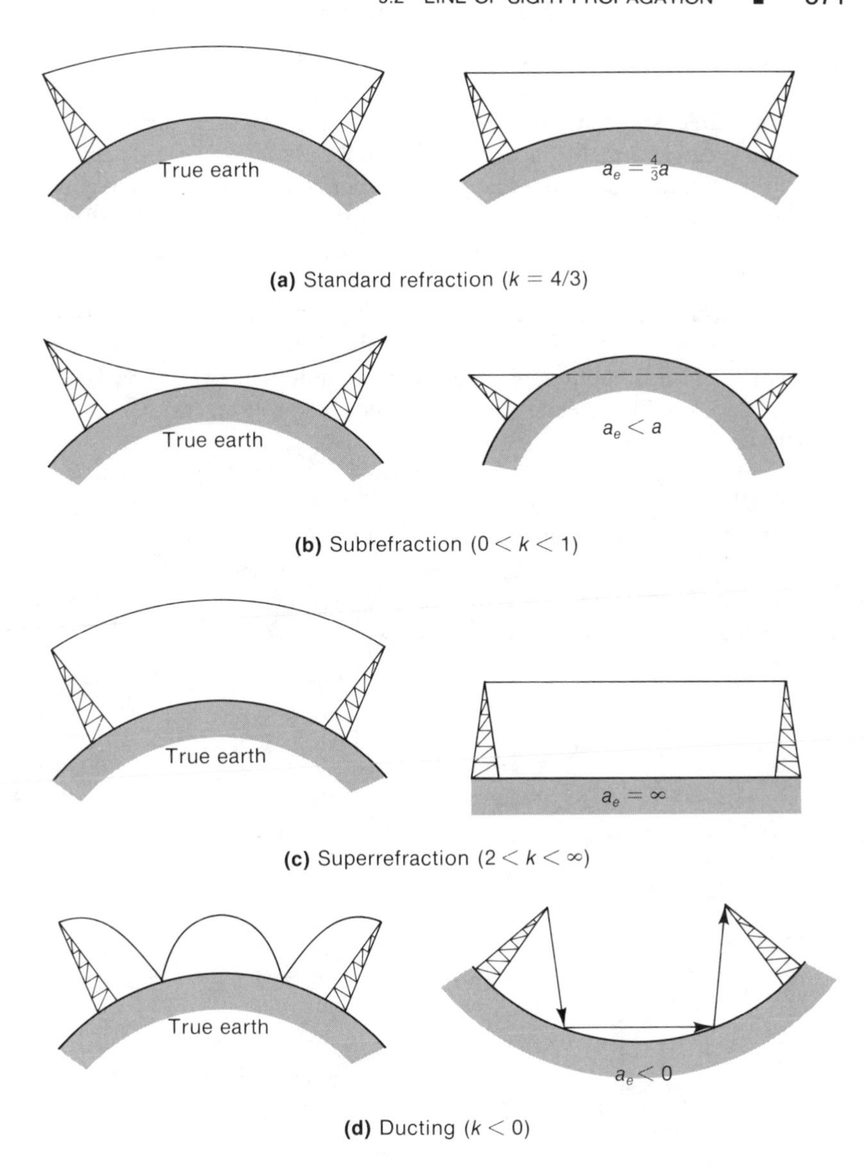

Figure 9.5 Various Forms of Atmospheric Refraction

or obstructs the line-of-sight path. Substandard atmospheric refraction may occur with the formation of fog or as cold air passes over a warm earth.

Superrefraction ($2 < k < \infty$) causes radio waves to refract downward with a curvature greater than normal. The result is an increased flattening of the effective earth. For the case illustrated in Figure 9.5*c* the effective earth radius is infinity ($k = \infty$)—that is, the earth reduces to a plane. From

Equation (9.18) it can be seen that an N gradient of -157 units per kilometer yields a k equal to infinity. Under these conditions radio waves are propagated at a fixed height above the earth's surface, creating unusually long propagation distances and the potential for overreach interference with other signals occupying the same frequency allocation. Superrefractive conditions arise when the index of refraction decreases more rapidly than normal with increasing altitude, which is produced by a rise in temperature with altitude, a decrease in humidity, or both. An increase in temperature with altitude, called a temperature inversion, occurs when the temperature of the earth's surface is significantly less than that of the air, which is most commonly caused by cooling of the earth's surface through radiation on clear nights or by movement of warm dry air over a cooler body of water.

A more rapid decrease in refractive index gives rise to more pronounced bending of radio waves, in which the radius of curvature of the radio wave is smaller than the earth's radius. As indicated in Figure 9.5*d* the rays are bent to the earth's surface and then reflected upward from it. With multiple reflections, the radio waves can cover large ranges far beyond the normal horizon. In order for the radio wave's bending radius to be

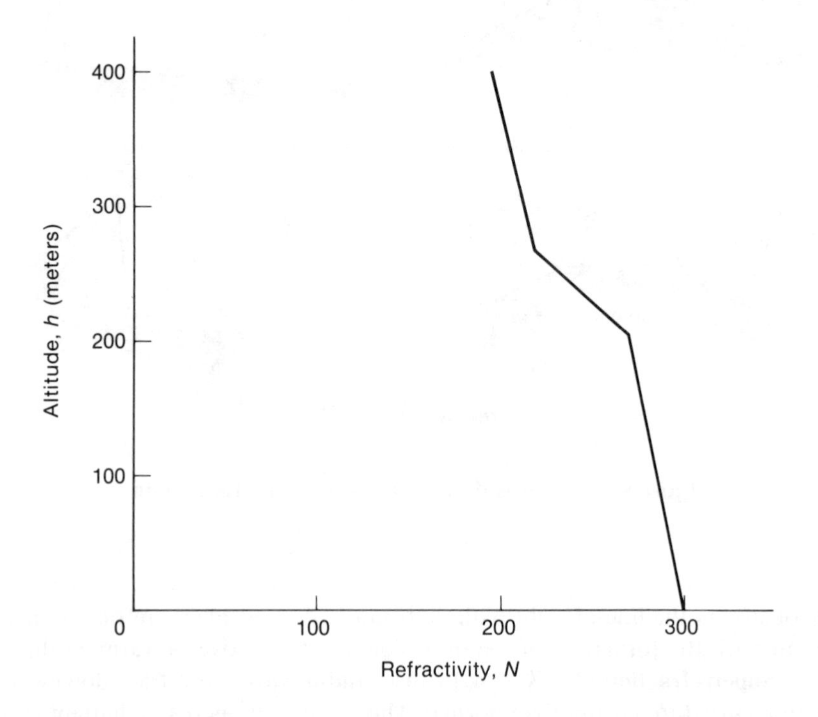

Figure 9.6 Model of Refractive Index Variation When Elevated Ducting Is Present

smaller than the earth's radius, the N gradient must be less than -157 units per kilometer. Then, according to (9.18), the k factor and effective earth radius both become negative quantities. As illustrated in Figure 9.5*d* the effective earth is approximated by a concave surface. This form of anomalous propagation is called **ducting** because the radio signal appears to be propagated through a waveguide, or duct. A duct may be located along the earth's surface or it may be elevated above the earth's surface. The meteorological conditions responsible for either surface or elevated ducts are similar to conditions causing superrefractivity. With ducting, however, a transition region between two differing air masses creates a trapping layer. In ducting conditions, refractivity N decreases with increasing height in an approximately linear fashion above and below the transition region, where the gradient departs from the average. In this transition region, the gradient of N becomes steep (Figure 9.6).

Atmospheric Absorption

For frequencies above 10 GHz, attenuation due to atmospheric absorption becomes an important factor in radio link design. The two major atmospheric gases contributing to attenuation are water vapor and oxygen. Studies have shown that absorption peaks occur in the vicinity of 22.3 and 187 GHz due to water vapor and in the vicinity of 60 and 120 GHz for oxygen [4]. These peaks with their attenuation are illustrated in Figure 9.7. While absorption by oxygen is nearly constant, absorption due to water vapor varies according to the humidity of the air. The plot of Figure 9.7 has been drawn for an absolute humidity of 7.5 g/m^3.

At millimeter wavelengths (30 to 300 GHz), atmospheric absorption becomes a significant problem. To obtain maximum propagation range, frequencies around the absorption peaks are to be avoided. As can be seen in Figure 9.7, certain frequency bands have relatively low attenuation. In the millimeter wave range, the first two such bands, or *windows*, are centered at approximately 36 and 85 GHz.

Rain Attenuation

Attenuation due to rain and suspended water droplets (fog) can be a major cause of signal loss, particularly for frequencies above 10 GHz. Rain and fog cause a scattering of radio waves that results in attenuation. Moreover, for the case of millimeter wavelengths where the raindrop size is comparable to the wavelength, absorption occurs and increases attenuation. The degree of attenuation on an LOS link is thus a function of radio frequency, rainfall rate, and the distribution of rainfall along the path. Figure 9.8 indicates calculated rainfall attenuation as a function of frequency for several rainfall rates [5]. Note that heavy rain, as found in thunderstorms, produces significant attenuation particularly for frequencies above 10 GHz. To relate rainfall rates to path attenuation, measurements must be made by use of

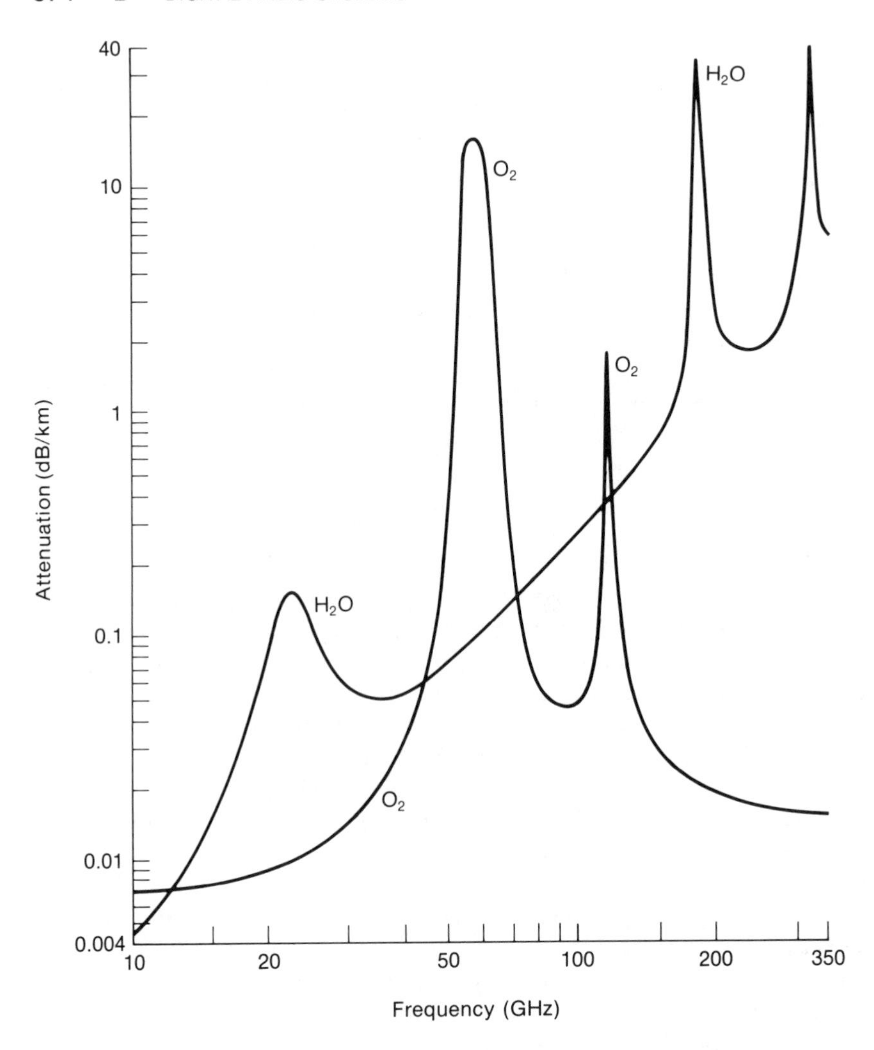

Figure 9.7 Absorption by Water Vapor and Oxygen

rain gauges placed along the propagation path. As an example, Figure 9.9 shows the path average rainfall rate in Washington, D.C., as a function of path length for various probability levels [6]. In the absence of such rainfall data along a specific path, it becomes necessary to use maps of rain climate regions, such as those provided by CCIR Rep. 563.2 [7].

A comparison of Figures 9.8 and 9.9 with Figure 9.7 indicates that rainfall attenuation often exceeds the combined oxygen and water vapor attenuation. To counter the effects of rain attenuation, it should first be noted that neither space nor frequency diversity is effective as protection

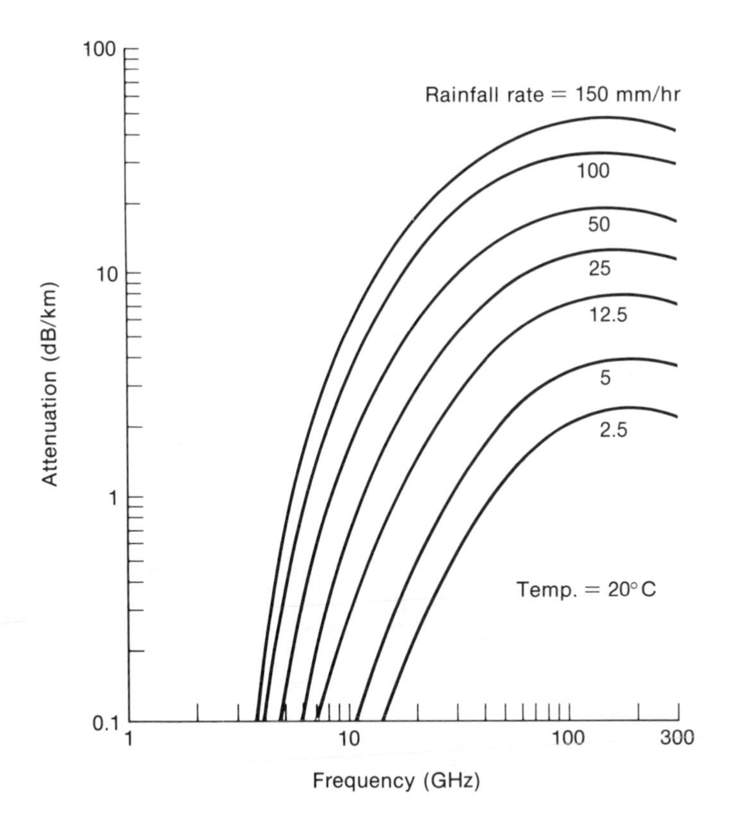

Figure 9.8 Calculated Values of Rain Attenuation in dB / km [5]

against rainfall effects. Measures that are effective, however, include increasing the fade margin, shortening the path length, and using a lower frequency band.

Example 9.1

Consider a millimeter-wave radio link to operate at 36 GHz over a path of length 5 km (3.1 mi) in the Washington, D.C., area. Calculate the total propagation losses allowed to meet a path availability of 99.99 percent.

Solution

For millimeter-wave radio links, the factors that affect path availability are free-space loss and atmospheric absorption (which are here assumed to be constant in time) and rainfall (which is time-variable). For short paths at 36 GHz, varying atmospheric conditions (multipath fading) have negligible effect on path loss variation compared to rainfall attenuation.

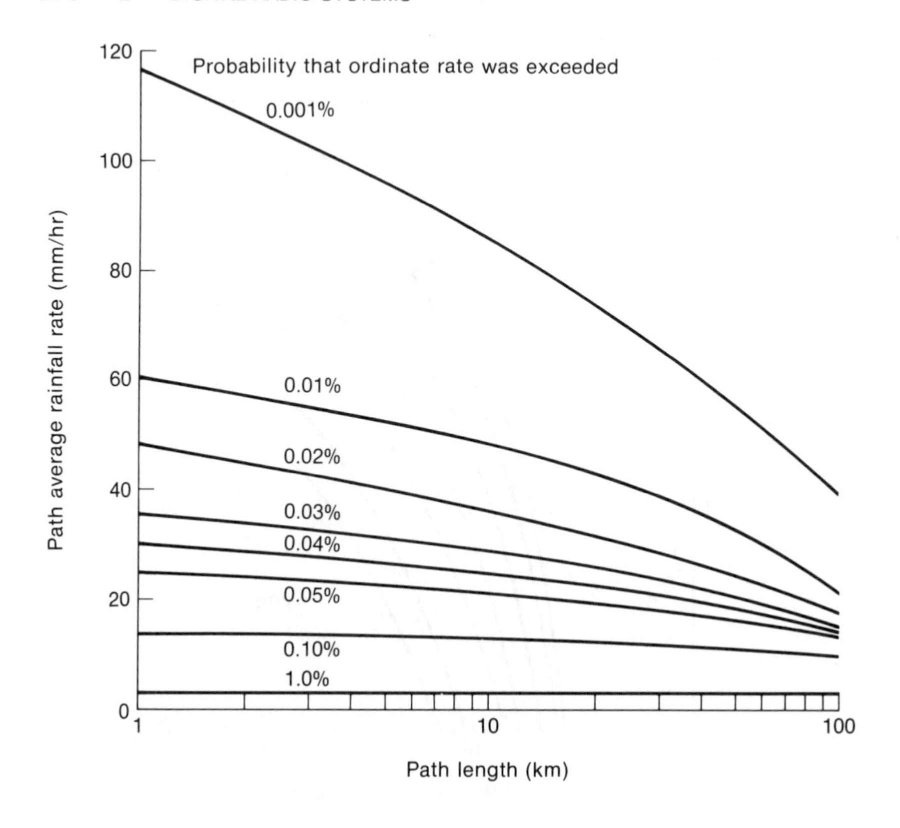

Figure 9.9 Path Average Rainfall Rate in Washington, D.C., vs. Path Length [6]

From Equation (9.5) or Figure 9.2, the free-space loss is

$$\begin{aligned} L_p &= 96.6 + 20\log(36) + 20\log(3.1) \\ &= 137.5 \text{ dB} \end{aligned}$$

From Figure 9.7, atmospheric absorption due to the presence of oxygen and water vapor is

$$\begin{aligned} O_2: &\quad (0.02 \text{ dB/km})(5 \text{ km}) = 0.1 \text{ dB} \\ H_2O: &\quad (0.05 \text{ dB/km})(5 \text{ km}) = 0.25 \text{ dB} \end{aligned}$$

To provide a path availability of 99.99 percent we find from Figure 9.9 that the rainfall rate will not exceed 50 mm/hr for 0.01 percent of the time. The corresponding attenuation due to 50 mm/hr rainfall is found from Figure 9.8 to be 12.6 dB/km, or 63 dB for a 5-km path. Summing up propagation losses, we find the total propagation loss to be 201 dB.

9.2.4 Path Profiles

In order to determine tower heights for suitable path clearance, a profile of the path must be plotted. The path profile is obtained from topographical maps that should have a scale of 1:50,000 or less. For line-of-sight links under 70 km in length, a straight line may be drawn connecting the two endpoints. For longer links, the great circle path must be calculated and plotted on the map. The elevation contours are then read from the map and plotted on suitable graph paper, taking special note of any obstacles along the path.

The path profile may be plotted on special graph paper that depicts the earth as curved and the transmitted ray as a straight line or on rectilinear graph paper that depicts the earth as flat and the transmitted ray as a curved line. The use of linear paper is preferred because it eliminates the need for special graph paper, permits the plotting of rays for different effective earth radius, and simplifies the plotting of the profile. Figure 9.10 is an example of a profile plotted on linear paper for a 40-km, 8-GHz radio link; Figure 9.11 is the same example plotted on $k = \frac{2}{3}$ earth radius paper.

The use of rectilinear paper, as suggested, requires the calculation of the earth bulge at a number of points along the path, especially at obstacles. This calculation then accounts for the added elevation to obstacles due to curvature of the earth. Earth bulge in meters may be calculated as

$$h = \frac{d_1 d_2}{12.76} \tag{9.20}$$

where d_1 = distance from one end of path to point being calculated (km)
d_2 = distance from same point to other end of the path (km)

As indicated earlier, atmospheric refraction causes ray bending, which can be expressed as an effective change in earth radius by using the k factor. The effect of refraction on earth bulge can be handled by adding the k factor to (9.20):

$$h = \frac{d_1 d_2}{12.76k} \tag{9.21}$$

or, for d_1 and d_2 in miles and h in feet,

$$h = \frac{d_1 d_2}{1.5k} \tag{9.22}$$

To facilitate path profiling, Equation (9.21) or (9.22) may be used to plot a curved ray template for a particular value of k and for use with a flat earth profile. Alternatively, the earth bulge can be calculated and plotted at selected points that represent the clearance required below a straight line drawn between antennas; when connected together, these points form a smooth parabola whose curvature is determined by the choice of k.

In path profiling, the choice of k factor is influenced by its minimum value expected over the path and the path availability requirement. With

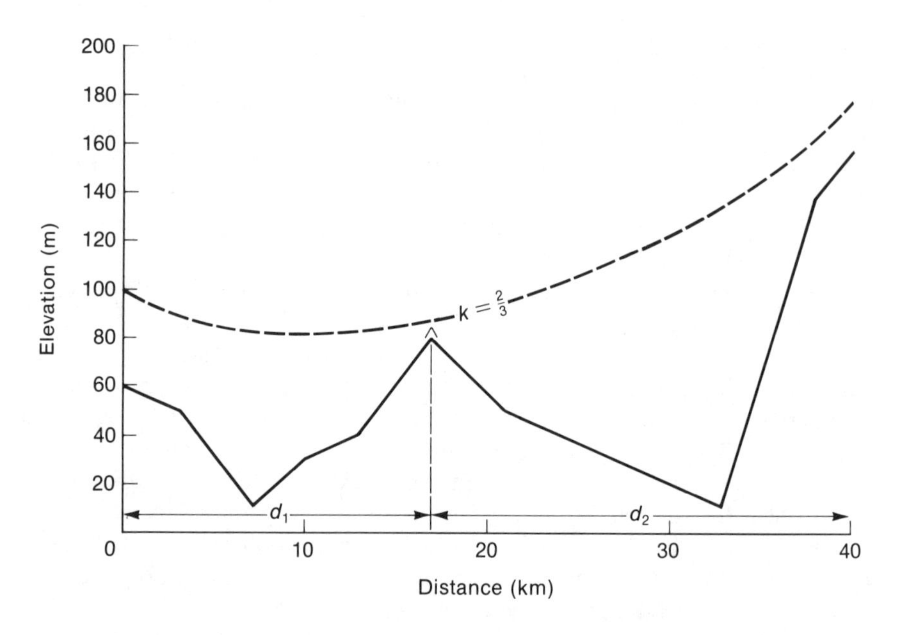

Figure 9.10 Example of an LOS Path Profile Plotted on Linear Paper

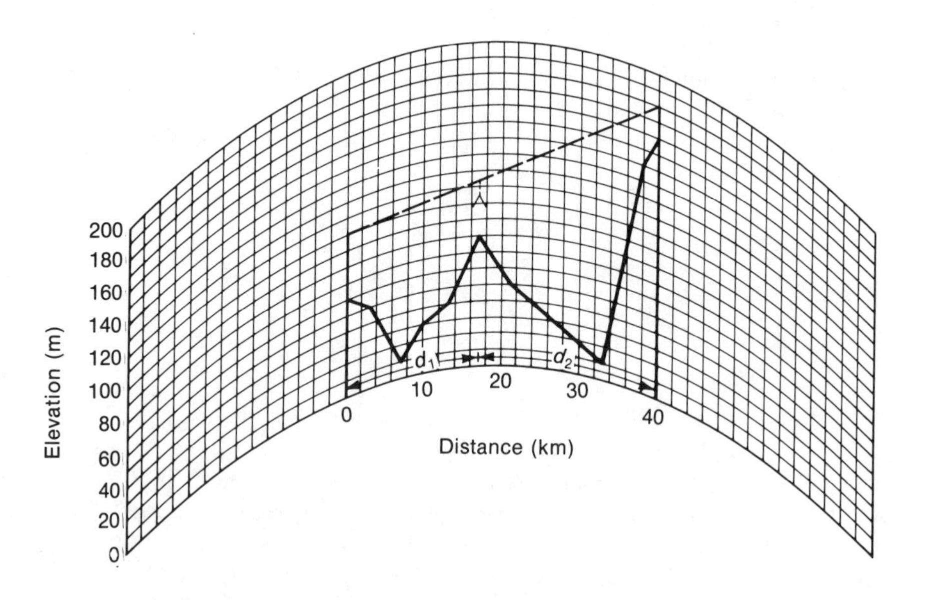

Figure 9.11 Example of an LOS Path Profile Plotted on $k = \frac{2}{3}$ Paper

lower values of k, earth bulging becomes pronounced and antenna height must be increased to provide clearance. To determine the clearance requirements, the distribution of k values is required; it can be found by meteorological measurements [7]. This distribution of k values can be related to path availability by selecting a k whose value is exceeded for a percentage of the time equal to the availability requirement.

Apart from the k factor, the Fresnel zone clearance must be added. Desired clearance of any obstacle is expressed as a fraction, typically 0.3 or 0.6, of the first Fresnel zone radius. This additional clearance is then plotted on the path profile, shown as a small tick mark on Figures 9.10 and 9.11, for each point being profiled. Finally, clearance should be provided for trees (nominally 15 m) and additional tree growth (nominally 3 m) or, in the absence of trees, for smaller vegetation (nominally 3 m).

The clearance criteria can thus be expressed by specific choices of k and fraction of first Fresnel zone. Here is one set of clearance criteria that is commonly used for highly reliable paths [8]:

1. Full first Fresnel zone clearance for $k = \frac{4}{3}$
2. 0.3 first Fresnel zone clearance for $k = \frac{2}{3}$

whichever is greater. Over the majority of paths, the clearance requirements of criterion 2 will be controlling. Even so, the clearance should be evaluated by using both criteria along the entire path.

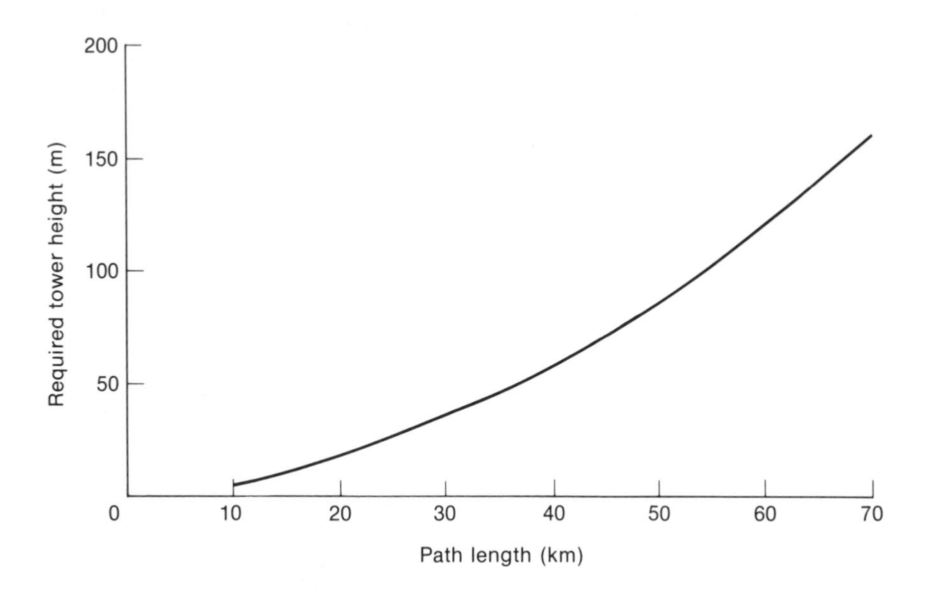

Figure 9.12 Tower Height Required for Smooth Earth Clearance for $k = \frac{2}{3}$, 0.6 First Fresnel Zone Clearance, and Equal Antenna Heights

Adequate tower heights may now be determined by plotting the radio ray for the proper value of k and superimposing this ray on the terrain profile such that proper first Fresnel zone and any vegetation clearance is achieved. Figure 9.12 shows the relationship between tower height and path length for $k = \frac{2}{3}$ and 0.6 first Fresnel zone clearance. Note that the tower height required for clearance over a smooth earth increases as the square of the path length.

9.3 MULTIPATH FADING

Fading is defined as variation of received signal level with time due to changes in atmospheric conditions. The propagation mechanisms that cause fading include refraction, reflection, and diffraction associated with both the atmosphere and terrain along the path. The two general types of fading are referred to as **multipath** and **power** fading and are illustrated by the recordings of RF received signal levels shown in Figure 9.13.

Power fading, sometimes called attenuation fading, results mainly from anomalous propagation conditions, such as (see Figure 9.5) *subrefraction* ($k < 1$), which causes blockage of the path due to the effective increase in earth bulge, *superrefraction* ($k > 2$), which causes pronounced ray bending and decoupling of the signal from the receiving antenna, and *ducting* ($k < 0$) in which the radio beam is trapped by atmospheric layering and directed away from the receiving antenna. Rainfall also contributes to power fading, particularly for frequencies above 10 GHz. Power fading is characterized as slowly varying in time, usually independent of frequency, and causing long periods of outages. Remedies include greater antenna

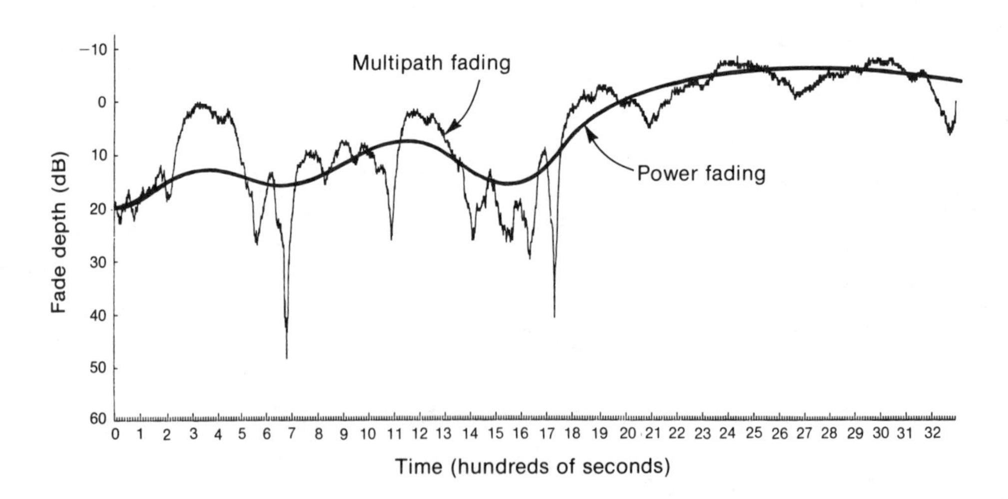

Figure 9.13 Example of Multipath and Power Fading for LOS Link

heights for subrefractive conditions, antenna realignment for superrefractive conditions, and added link margin for rainfall attenuation.

Multipath fading arises from destructive interference between the direct ray and one or more reflected or refracted rays. These multiple paths are of different lengths and have varied phase angles on arrival at the receiving antenna. These various components sum to produce a rapidly varying, frequency-selective form of fading. Deep fades occur when the primary and secondary rays are equal in amplitude but opposite in phase, resulting in signal cancellation and a deep amplitude null. Between deep fades, small amplitude fluctuations are observed that are known as **scintillation**; these fluctuations are due to weak secondary rays interfering with a strong direct ray.

Multipath fading is observed during periods of atmospheric stratification, where layers exist with different refractive gradients. The most common meteorological cause of layering is a temperature inversion, which commonly occurs in hot, humid, still, windless conditions, especially in late evening, at night, and in early morning. Since these conditions arise during the summer, multipath fading is worst during the summer season. Multipath fading can also be caused by reflections from flat terrain or a body of water. Hence multipath fading conditions are most likely to occur during periods of stable atmosphere and for highly reflective paths. Multipath fading is thus a function of path length, frequency, climate, and terrain. Techniques used to deal with multipath fading include the use of diversity, increased fade margin, and adaptive equalization.

9.3.1 Statistical Properties of Fading

The random nature of multipath fading suggests a statistical approach to its characterization. The statistical parameters commonly used in describing fading are:

- Probability (or percentage of time) that the line-of-sight link is experiencing a fade below threshold
- Average fade duration and probability of fade duration greater than a given time
- Expected number of fades per unit time

The terms to be used are defined graphically in Figure 9.14. The threshold L is the signal level corresponding to the minimum acceptable signal-to-noise ratio or, for digital transmission, the maximum acceptable probability of error. The difference between the normal received signal level and threshold is the **fade margin**. A fade is defined as the downward crossing of the received signal through the threshold. The time spent below threshold for a given fade is then the fade duration.

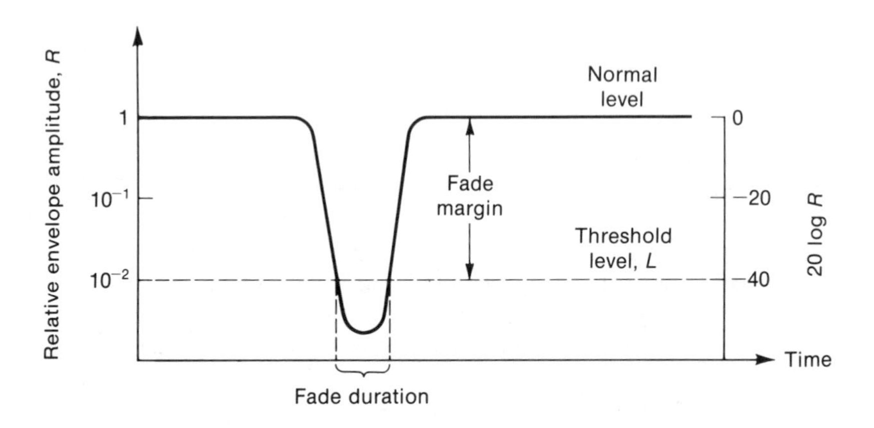

Figure 9.14 Definition of Fading Terms

For line-of-sight links, the probability distribution of fading signals is known to be related to and limited by the Rayleigh distribution, which is well known and is found by integrating the curve shown in Figure 9.15. The Rayleigh probability density function is given by

$$p(r) = \frac{r}{\sigma^2} e^{-(r^2/2\sigma^2)} \tag{9.23}$$

for envelope amplitude r ($r \geq 0$) and mean square amplitude σ^2. The

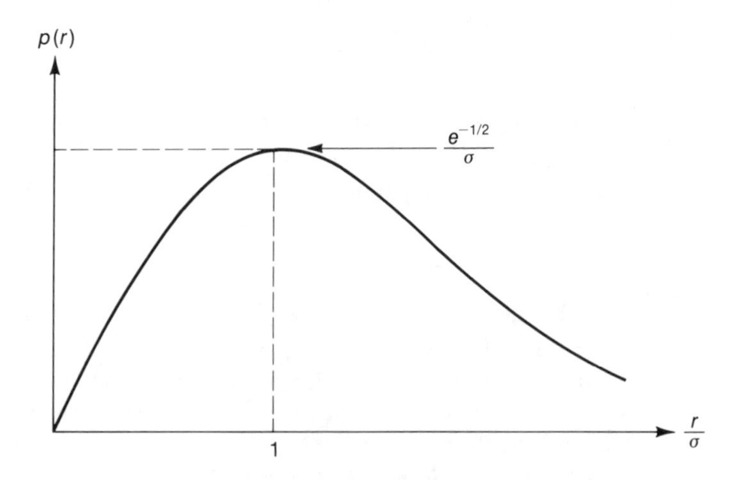

Figure 9.15 Rayleigh Probability Density Function

Rayleigh distribution function has the form

$$P(r \le r_0) = \int_0^{r_0} p(r)\,dr = \int_0^{r_0} \frac{r}{\sigma^2} e^{-(r^2/2\sigma^2)}\,dr$$
$$= 1 - e^{-r_0^2/2\sigma^2} \qquad r_0 \ge 0 \tag{9.24}$$

For relative envelope amplitude $R = r/\sigma\sqrt{2}$ and relative threshold amplitude $L = r_0/\sigma\sqrt{2}$, the distribution function becomes

$$P(R < L) = 1 - e^{-L^2} \tag{9.25}$$

An approximation to (9.25) valid for small values of L (representing deep fades) is

$$P(R < L) \approx L^2 \qquad \text{for } L < 0.1 \tag{9.26}$$

Fading probabilities are more conveniently expressed in terms of the fade margin F in decibels by letting $F = -20 \log L$. Then

$$P(R < L) = 10^{-F/10} \tag{9.27}$$

Actual observations of multipath fading indicate that in the region of deep fades, amplitude distributions have the same slope as the Rayleigh distribution but displaced. This characteristic corresponds to the special case of the Nakagami–Rice distribution where a direct (or *specular*) component exists that is equal to or less than the Rayleigh fading component [9]. Thus, for deep fading, the distribution function becomes

$$P(R < L) = d(1 - e^{-L^2}) \tag{9.28}$$

The parameter d that modifies the Rayleigh distribution has been termed a multipath occurrence factor. Experimental results of Barnett [10] show that

$$d = \frac{abD^3 f}{4} \times 10^{-5} \tag{9.29}$$

where D = path length in miles

f = frequency in GHz

a = terrain factor

$$= \begin{cases} 4 \text{ for overwater or flat terrain} \\ 1 \text{ for average terrain} \\ \frac{1}{4} \text{ for mountainous terrain} \end{cases}$$

b = climate factor

$$= \begin{cases} \frac{1}{2} \text{ for hot, humid climate} \\ \frac{1}{4} \text{ for average, temperate climate} \\ \frac{1}{8} \text{ for cool, dry climate} \end{cases}$$

Combining this factor with the basic Rayleigh probability of Equation (9.27) results in the following overall expression for probability of outage due to fading deeper than the fade margin:

$$P(o) = d10^{-F/10} = \left(\frac{abD^3f}{4} \times 10^{-5}\right)(10^{-F/10}) \tag{9.30}$$

As an example, values of $P(o)$ as a function of fade margin are plotted in Figure 9.16 for a 30-mi path with average terrain and climate. The Rayleigh

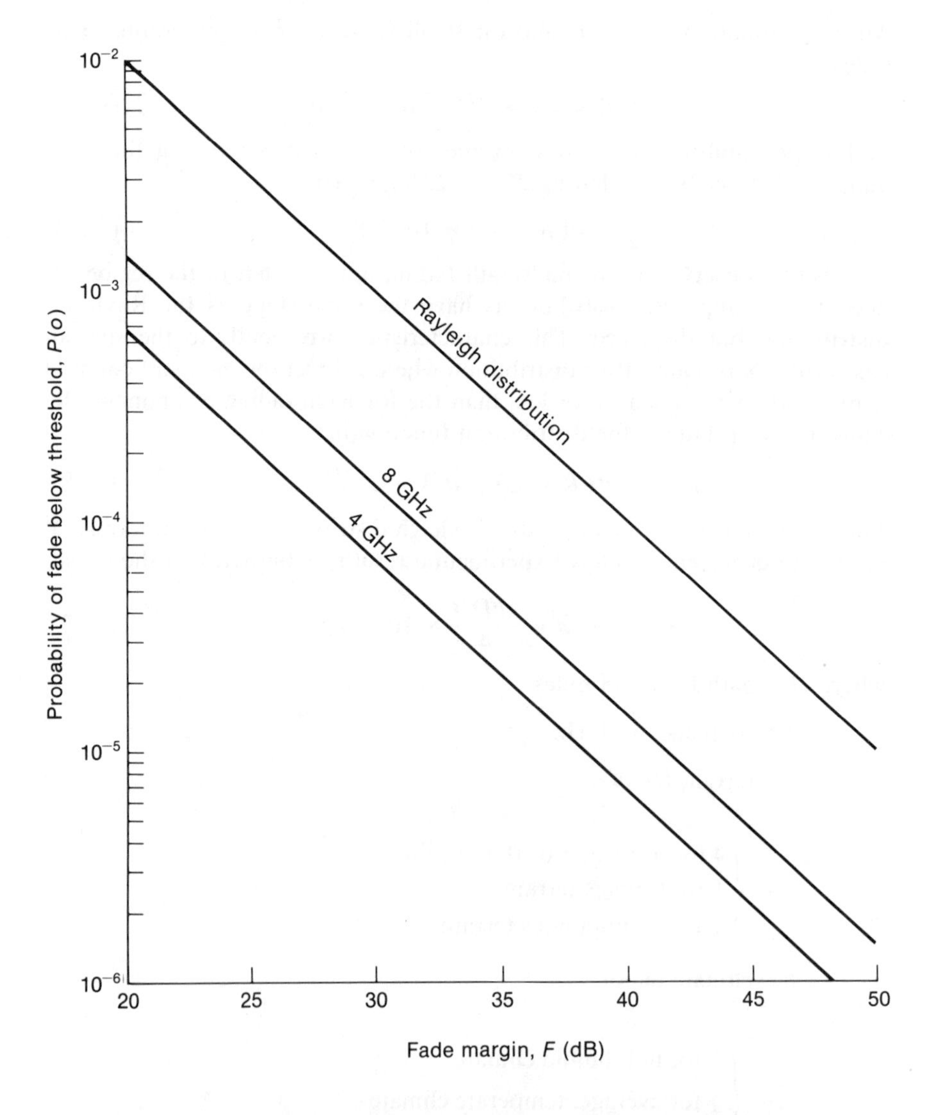

Figure 9.16 Probability of Fading Below the Fade Margin on Typical LOS Link (30-mi Path with Average Terrain and Climate)

distribution given by (9.27) is also shown to indicate the limiting value for multipath fading. Note that the distributions all have a slope of 10 dB per decade of probability.

Vigants [11, 12] observed that fade durations have an average value $\bar{t}$, proportional to L and independent of frequency, given by

$$\bar{t} = \frac{L}{c} \quad \text{for } L < 0.1$$
$$= \frac{10^{-F/20}}{c} \tag{9.31}$$

where c is an experimental constant equal to approximately 2.22×10^{-3} s^{-1}. The probability distribution function for fade durations is also based on empirical results from Vigants and is given by

$$P(t) = e^{-1.15(t/\bar{t})^{2/3}} \tag{9.32}$$

An expression for expected number of fades per unit time, N, can now be written directly in terms of $P(o)$ and $\bar{t}$:

$$N = \frac{P(o)}{\bar{t}} \tag{9.33}$$

Substituting (9.30) and (9.31) into (9.33), N becomes

$$N = (2.5 \times 10^{-6})(abcD^3f)10^{-F/20} \tag{9.34}$$

For digital transmission applications, it is often more useful to express the fading statistics in terms of error rate—for example as:

- Probability of exceeding a specified error rate
- Average duration of error rate exceeding a specified level
- Frequency that a specified error rate is reached

These error statistics are derivable by a change of variables, from signal level to error rate, in the Rayleigh or Nakagami–Rice distribution functions given in (9.25) and (9.28). The modulation technique must be specified, however, to state the error rate as a function of signal level or signal-to-noise ratio first. D. J. Kennedy has provided these error rate statistics for various modulation techniques [13]. For the Nakagami–Rice channel, the probability distribution function was found to be

$$P(o) = d\left[1 - (2p)^{1/\alpha\gamma_0}\right] \tag{9.35}$$

where d = multipath occurrence factor
p = bit error rate
γ_0 = signal-to-noise (E_b/N_0) averaged over fading channel

$$\alpha = \begin{cases} 0.25 & \text{for binary noncoherent ASK} \\ 0.5 & \text{for binary noncoherent FSK} \\ 1.0 & \text{for binary differential PSK} \end{cases}$$

The expression for $P(o)$ is plotted in Figure 9.17 as a function of γ_0 for

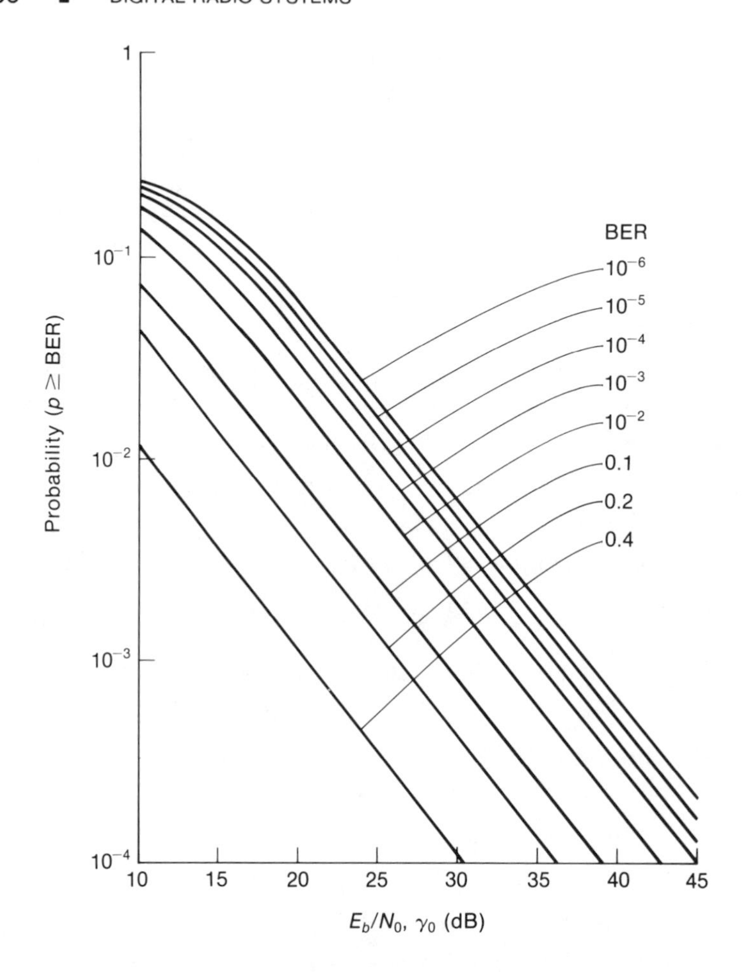

Figure 9.17 Probability of Specified BER or Worse for Typical LOS Link [13] © 1973 IEEE.

$\alpha = 0.5$ and $d = 0.25$. For other values of α, the factor of $-10\log(2\alpha)$ must be added to the abscissa, while for other values of d the ordinate must be multiplied by the factor $d/0.25$. A similar set of curves can be derived to show the rate of fading or average duration of fading to a specified error rate. Caution is advised in using these error statistics in predicting performance of wideband digital LOS links, however, since intersymbol interference due to multipath dispersion may be the dominant source of degradation unless adaptive equalization is employed (see Section 9.3.3).

9.3.2 Diversity Improvement

Diversity is used in line-of-sight radio links to protect against either equipment failure or multipath fading. Here we consider the improvement in multipath fading afforded by the two most commonly used diversity techniques:

- **Space diversity**, which provides two signal paths by use of vertically separated receiving antennas
- **Frequency diversity**, which provides two signal frequencies by use of separate transmitter-receiver pairs

The degree of improvement provided by diversity depends on the degree of correlation between the two fading signals. In practice, because of limitations in allowable antenna separation or frequency spacing, the fading correlation tends to be high. Fortunately, improvement in link availability remains quite significant even for high correlation. To derive the diversity improvement, we begin with the joint Rayleigh probability distribution function to describe the fading correlation between diversity signals, given by

$$P(R_1 < L, R_2 < L) \approx \frac{L^4}{1 - k^2} \qquad \text{(for small } L\text{)} \tag{9.36}$$

where R_1 and R_2 are signal levels for diversity channels 1 and 2 and k^2 is the correlation coefficient. By experimental results, empirical expressions for k^2 have been established that are a function of antenna separation or frequency spacing, wavelength, and path length.

Space Diversity Improvement Factor

Vigants [14] has developed the following expression for k^2 in space diversity links:

$$k^2 = 1 - \frac{S^2}{2.75D\lambda} \tag{9.37}$$

where S = antenna separation, D = path length, λ = wavelength, and S, D, and λ are in the same units. A more convenient expression is the space diversity improvement factor, given by

$$\begin{aligned} I_{sd} &= \frac{P(R_1 < L)}{P(R_1 < L, R_2 < L)} \approx \frac{L^2}{L^4/(1 - k^2)} \\ &= \frac{1 - k^2}{L^2} \end{aligned} \tag{9.38}$$

Using Vigants' expression for k^2 (Equation 9.37), we obtain

$$I_{sd} = \frac{S^2}{2.75D\lambda L^2} \tag{9.39}$$

When D is given in miles, S in feet, and λ in terms of carrier frequency f in gigahertz, I_{sd} can be expressed as

$$I_{sd} = \frac{(7.0 \times 10^{-5}) f S^2}{D} 10^{F/10} \tag{9.40}$$

where F is the fade margin associated with the second antenna.

Frequency Diversity Improvement Factor

Using experimental data and a mathematical model, Vigants and Pursley [15] have developed an improvement factor for frequency diversity, given by

$$I_{\mathrm{fd}} = \left(\frac{50}{fD}\right)\left(\frac{\Delta f}{f}\right)10^{F/10} \tag{9.41}$$

where f = frequency (GHz)

Δf = frequency separation (GHz)

F = fade margin (dB)

D = path length (mi)

Effect of Diversity on Fading Statistics

The effect of diversity improvement, I_d, on probability of outage due to fading can be expressed as

$$P_d(o) = \frac{P(o)}{I_d} \tag{9.42}$$

where $P_d(o)$ is the probability of simultaneous fading in the two diversity signals. For space diversity, substituting Equations (9.30) and (9.40) into (9.42) we obtain

$$P_{\mathrm{sd}}(o) = \frac{abD^4}{28S^2}10^{-F/5} \tag{9.43}$$

where the fade margins on the two antennas are assumed equal. Likewise, for frequency diversity we obtain

$$P_{\mathrm{fd}}(o) = (5 \times 10^{-8})\left(\frac{abD^4 f^3}{\Delta f}\right)10^{-F/5} \tag{9.44}$$

Vigants [11] and Lin [16] observed in their experimental results that diversity reduces the average fade duration by a factor of 2, so that Equation (9.31) is modified to read

$$\bar{t}_d = \frac{L}{2c} = \frac{10^{-F/20}}{2c} \tag{9.45}$$

The probability distribution function for fade durations has the same form as the expression for nondiversity (Equation 9.32), but with t_d replacing t so that

$$P(t_d) = e^{-1.15(t_d/\bar{t}_d)^{2/3}} \tag{9.46}$$

The expected number of fades per unit time is also reduced with diversity, according to

$$N_d = \frac{P_d(o)}{\bar{t}_d} \tag{9.47}$$

For the case of space diversity, N_{sd} is found by substituting (9.43) and (9.45) into (9.47) to yield

$$N_{sd} = \frac{abD^4}{(6.3 \times 10^3)S^2} 10^{-3F/20} \tag{9.48}$$

where the fade margin F is assumed to be equal for both antennas. Similarly, for frequency diversity we find

$$N_{fd} = 2.22 \times 10^{-10} \left(\frac{abD^4 f^3}{\Delta f} \right) 10^{-3F/20} \tag{9.49}$$

Example 9.2

Consider the design of an 8-GHz line-of-sight microwave link for a path with length 30 mi, average terrain, and moderate climate. You are to characterize the effects of fade margin and diversity on the probability of fade outage, fade duration, and expected number of fades per unit time. Assume the use of space diversity with equal fade margins on the two receiving antennas.

Solution

To obtain general insight into the effect of fade margin and antenna separation, the fading statistics can be plotted by using the equations given earlier. For the link conditions given, and using Equation (9.43), Figure 9.18 shows the probability of fading below threshold as a function of the fade margin. Plots are provided for antenna separations of 20, 30, and 40 ft. Using Equation (9.48), Figure 9.19 shows the expected number of fades per second, again as a function of fade margin and antenna separation. Using Equations (9.45) and (9.46), the average fade duration is given in Figure 9.20; the probability distribution of fade durations is given in Figure 9.21 for fade margins of 20, 30, and 40 dB. Note that the fading duration statistics are applicable to both space and frequency diversity and are independent of the path parameters.

Example 9.3

Using the same link characteristics given in Example 9.2, compare the fade outage probabilities and link availabilities for (a) nondiversity, (b) space diversity with antenna separation of 30 ft, and (c) frequency diversity with 5 percent spacing. Assume a fade margin of 40 dB.

Solution

(**a**) Using Equation (9.30) or Figure 9.16, we obtain

$$P_{nd}(o) = 1.35 \times 10^{-5}$$

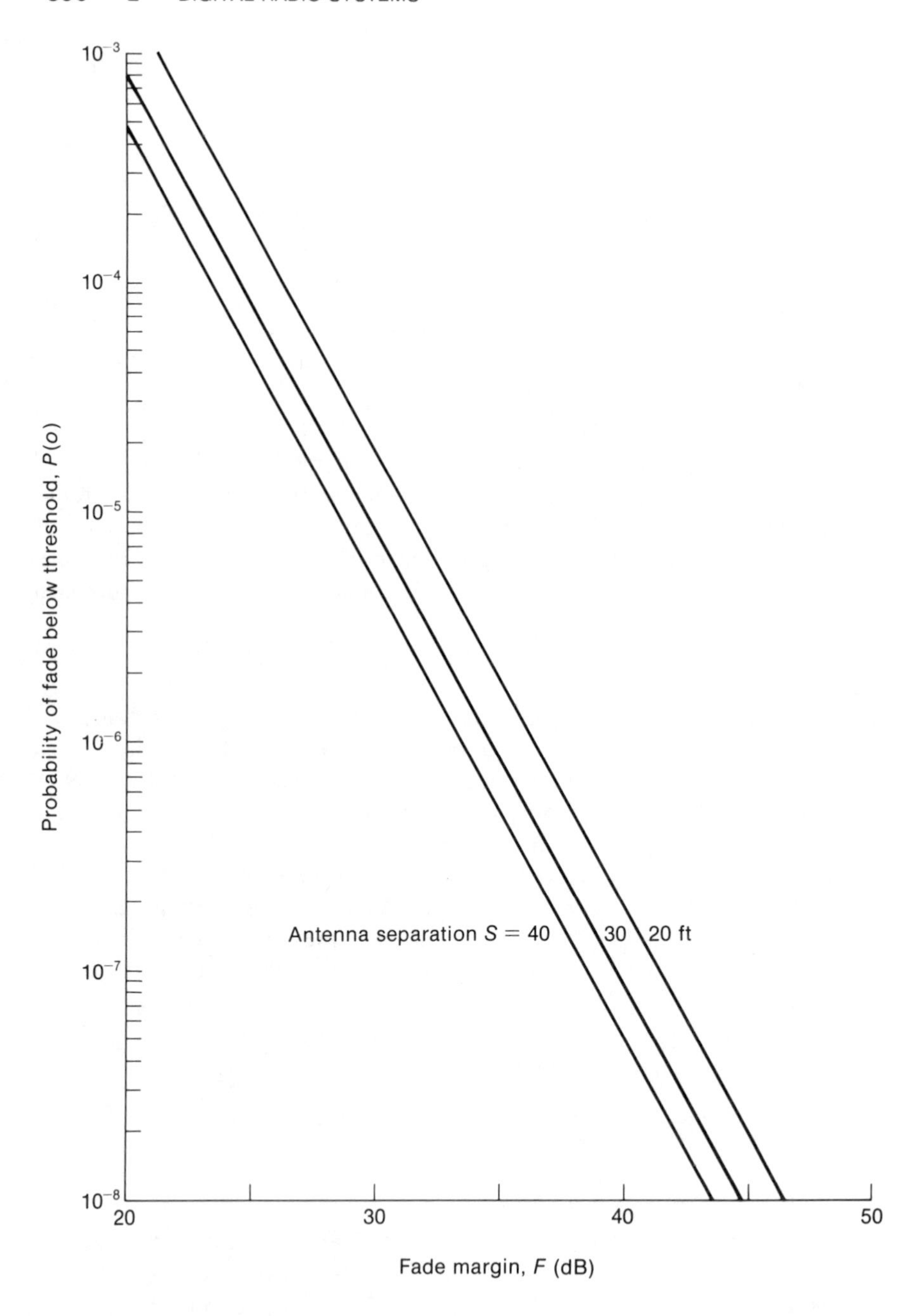

Figure 9.18 Probability of Fading for LOS Space Diversity Link vs. Fade Margin (30-mi Path with Average Terrain and Climate)

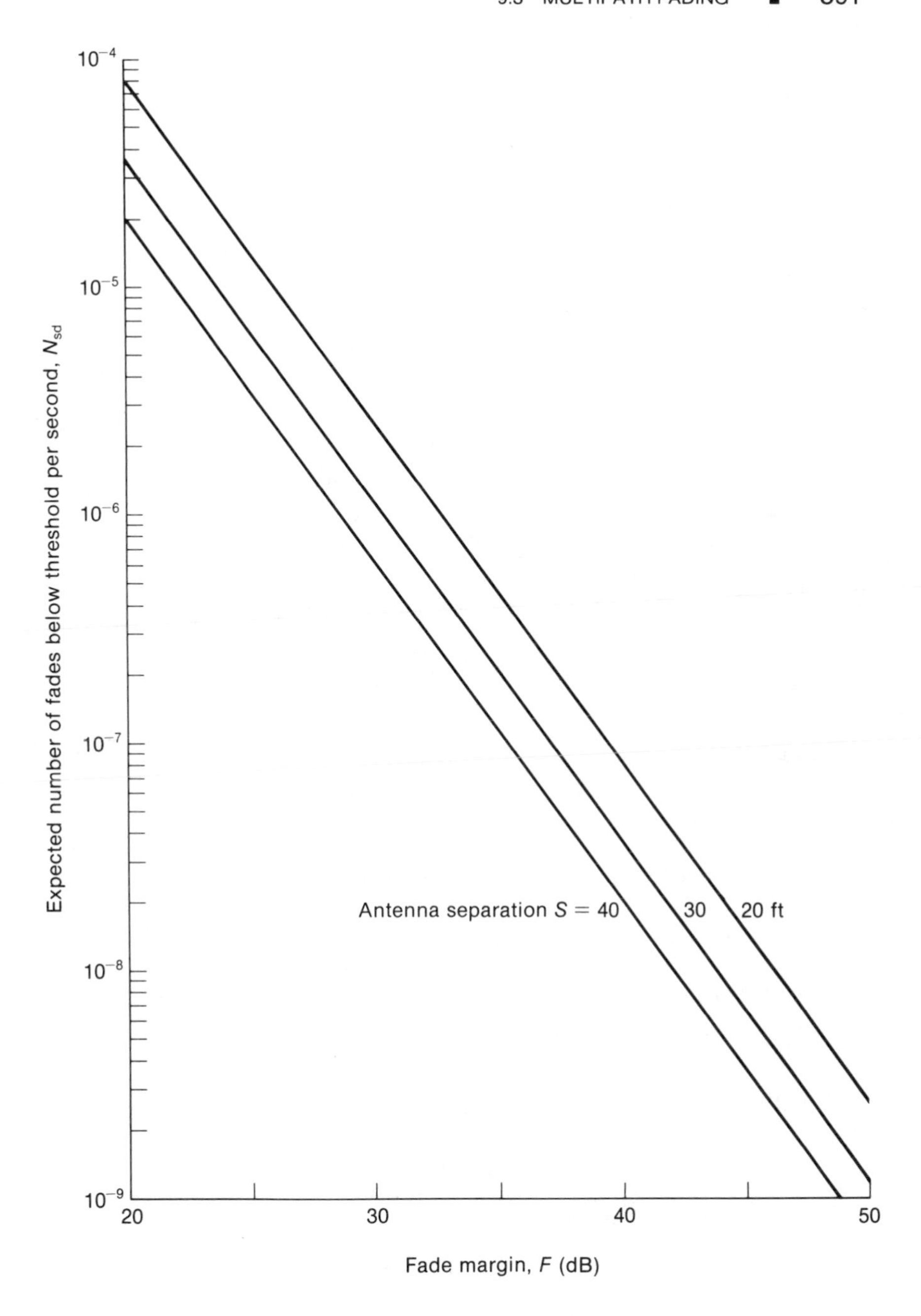

Figure 9.19 Expected Number of Fades Below Threshold Per Second vs. Fade Margin for LOS Space Diversity Link (30-mi Path with Average Terrain and Climate)

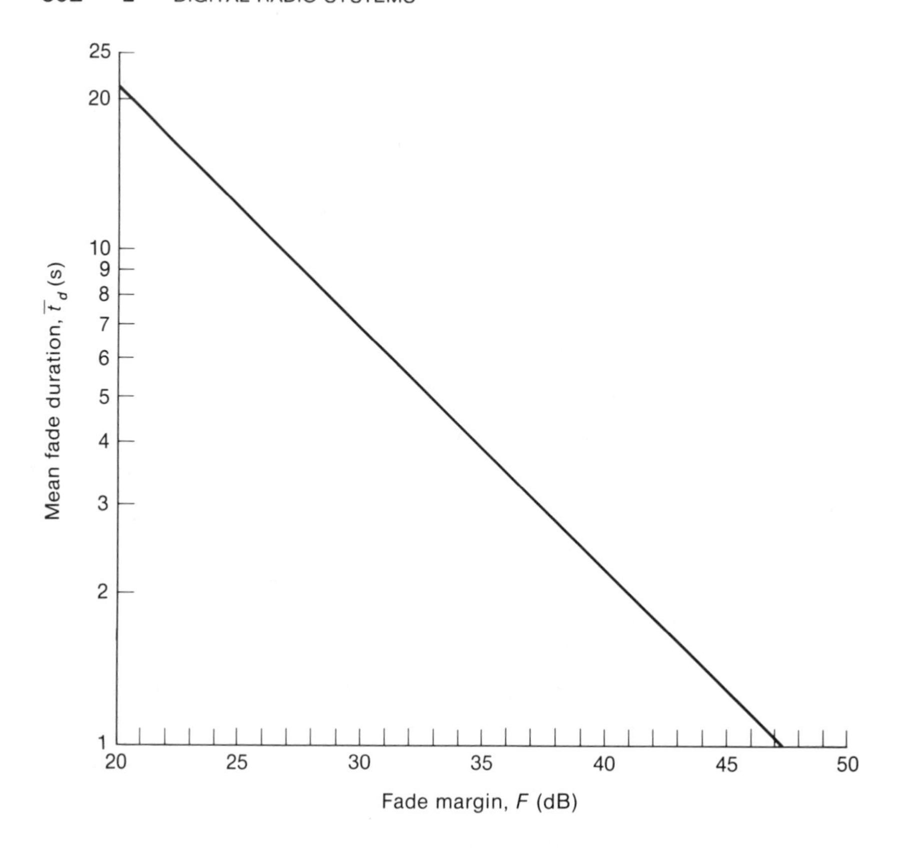

Figure 9.20 Mean Fade Duration on Diversity LOS Links

or, in terms of link availability (A),

$$A = 1 - P(o) = 0.9999865$$

(b) Using Equation (9.43) or Figure 9.18, we obtain

$$P_{\text{sd}}(o) = 8.0 \times 10^{-8}$$

or, in terms of link availability,

$$A = 0.99999992$$

(c) Using Equation (9.44) we obtain

$$P_{\text{fd}}(o) = 1.3 \times 10^{-7}$$

or, in terms of link availability,

$$A = 0.99999987$$

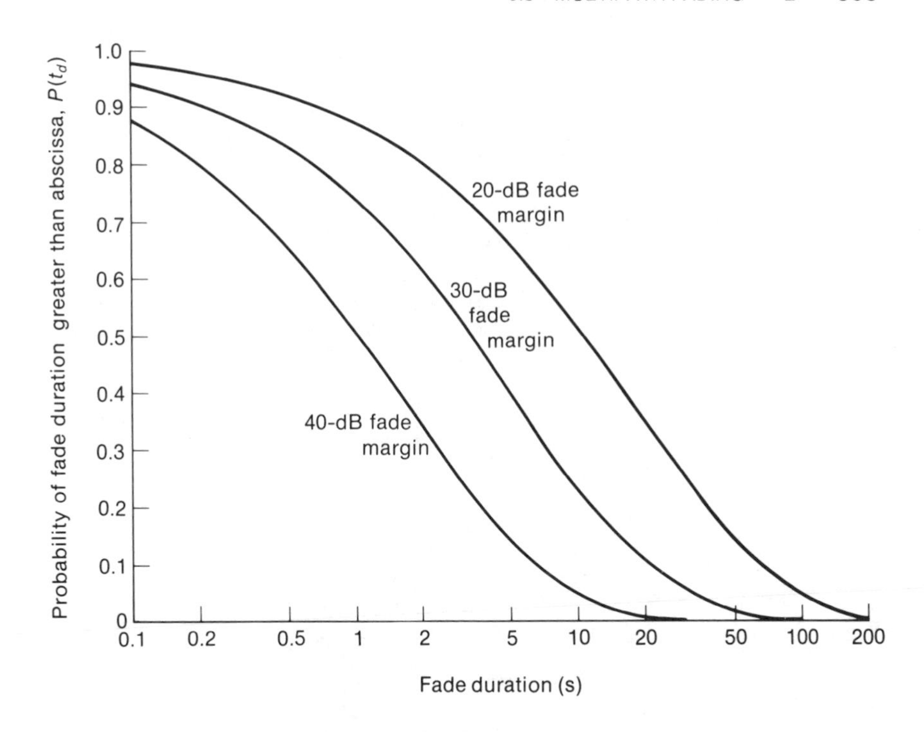

Figure 9.21 Probability Distribution of Fade Durations for LOS Diversity Links

For this example, we see that the improvement factor of space diversity is greater than that of frequency diversity.

9.3.3 Frequency-Selective Fading

The first experiences with wideband digital radios revealed that measured error performance fell far short of the performance predicted by the **flat fading** model assumed in our discussions so far. This result is due to the presence of **frequency-selective fading** during which the amplitude and group delay characteristics become distorted. For digital signals, this distortion leads to intersymbol interference that in turn degrades the system error rate. This degradation is directly proportional to system bit rate, since higher bit rates mean smaller pulse widths and greater susceptibility to intersymbol interference. Previously, in analog radio transmission, frequency-selective fading caused intermodulation distortion, but this effect was always secondary when compared to the received signal power. For

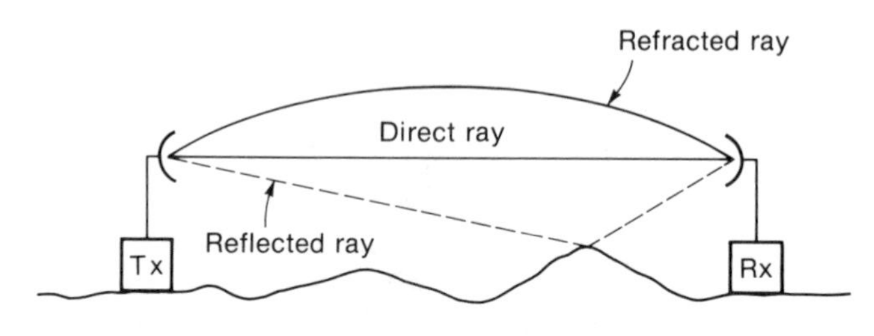

(a) Refraction or reflection geometry resulting in multipath propagation

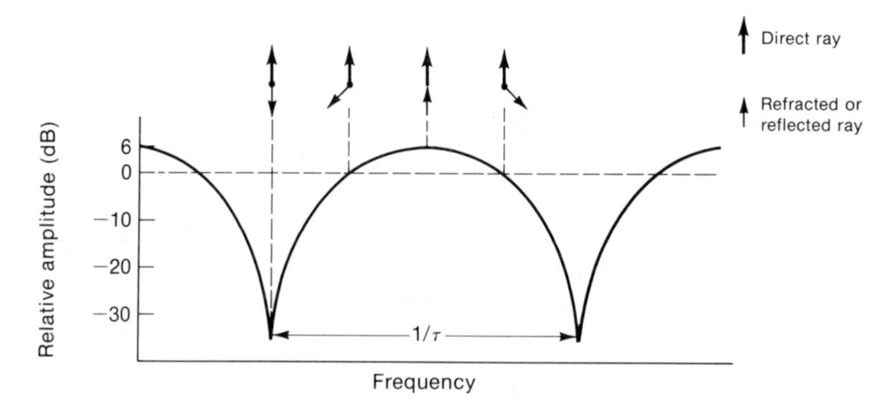

(b) Amplitude characteristics in multipath

Figure 9.22 Multipath Fading Effect

digital radio systems, however, the traditional fade depth is found to be a poor indicator of error rate.

Multipath geometry and its frequency-selective nature is illustrated in Figure 9.22. When the amplitude of the received signal is plotted versus frequency, deep amplitude notches appear when the direct ray is out of phase with the indirect rays. These notches are separated in frequency by $1/\tau$, where τ is the time delay between the direct and indirect rays. The notch depth is determined by the relative amplitude of the direct and indirect rays. When an amplitude notch or slope appears in the band of a radio channel, degradation in the error rate can be expected. This variation of amplitude with frequency is known as **amplitude dispersion** and is often the main source of degradation in digital radio systems.

Amplitude dispersion can be measured by recording the amplitudes across the RF or IF band of the received radio signal. Usually dispersion is calculated by taking the difference of spectral amplitudes at the two band edges and dividing that difference by the corresponding RF or IF band-

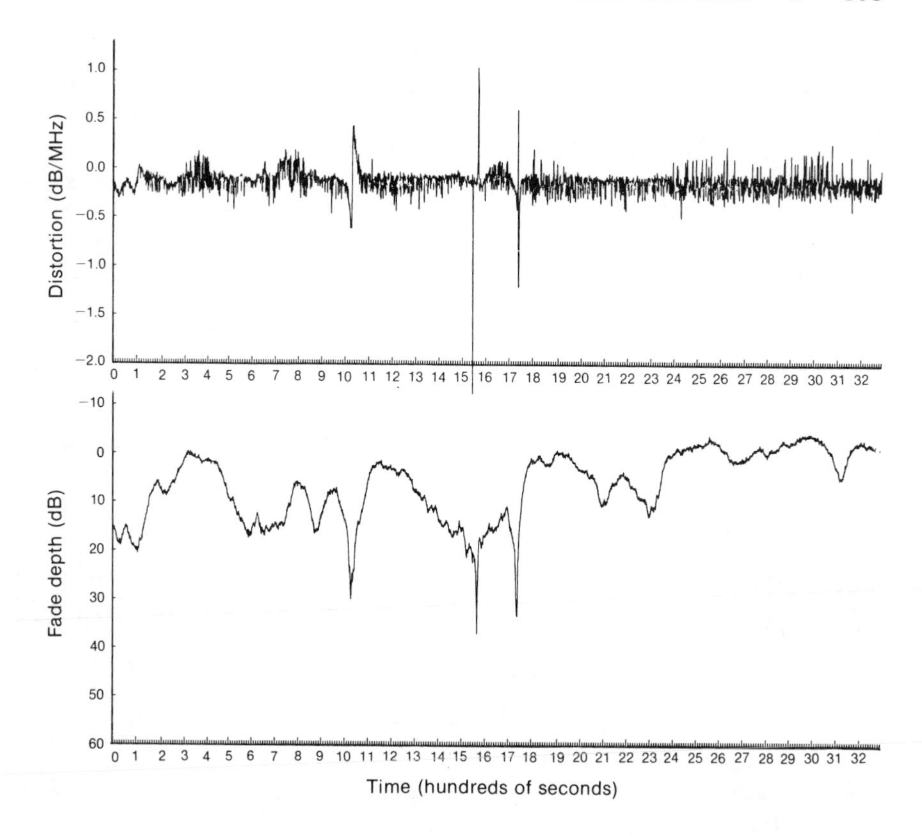

Figure 9.23 Recording of Received Signal Level and Amplitude Distortion for LOS Link

width. Figure 9.23 is such a recording of amplitude dispersion together with received signal level for an 8-GHz, 56-mi path. These recordings indicate the expected correlation of dispersion and signal level. Note that during the three deepest fades, the slope of the dispersion changes signs, which indicates that an amplitude notch has passed through the radio channel.

Channel Models

Both low-order power series [17] and multipath transfer functions [18] have been used to model the effects of frequency-selective fading. Several multipath transfer function models have been developed, usually based on the presence of two [19] or three [18] rays. In general, the multipath channel transfer function can be written as

$$H(\omega) = 1 + \sum_{i=1}^{n} \beta_i e^{j\omega\tau_i} \tag{9.50}$$

where the direct ray has been normalized to unity and the β_i and τ_i are amplitude and delay of the interfering rays relative to the direct ray. The two-ray model can thus be characterized by two parameters, β and τ. In this case, the amplitude of the resultant signal is

$$R = \left(1 + \beta^2 + 2\beta \cos \omega\tau\right)^{1/2} \tag{9.51}$$

and the phase of the resultant is

$$\phi = \arctan\left(\frac{\beta \sin \omega\tau}{1 + \beta \cos \omega\tau}\right) \tag{9.52}$$

The group delay is then

$$T(\omega) = -\frac{d\phi}{d\omega} = -\beta\tau\left(\frac{\beta + \cos \omega\tau}{1 + 2\beta \cos \omega\tau + \beta^2}\right) \tag{9.53}$$

The deepest fade occurs with

$$\omega_d \tau = \pi(2n - 1) \qquad (n = 1, 2, 3, \dots) \tag{9.54}$$

where both R and T are at a minimum, with

$$R_{\min} = 1 - \beta \tag{9.55a}$$

$$T_{\min} = \frac{\beta\tau}{1 - \beta} \tag{9.55b}$$

The frequency defined by ω_d is known as the *notch frequency* and is related to the carrier frequency ω_c by

$$\omega_d = \omega_c + \omega_o \tag{9.56}$$

where ω_o is referred to as the *offset frequency*.

Typical amplitude and group delay curves are shown in Figure 9.24 for 5-dB and 20-dB fades—that is for two different ratios of direct to interfering rays. Note that the delay peaks and amplitude nulls repeat for a frequency separation of $1/\tau$. When the amplitude of the direct ray is stronger than the interfering ray, the sign of the group delay is the same as the amplitude response. This case is illustrated in Figure 9.25*a* and is known as *minimum phase*. If the interfering ray is stronger than the direct ray, the sign of the group delay is opposite from the amplitude response and is known as the *nonminimum phase* case (Figure 9.25*b*). The significance of the group delay's sign is that amplitude equalizers actually increase the envelope delay distortion for nonminimum phase. Measurements of amplitude and group delay by Martin [20] indicate the presence of nonminimum fades for approximately 50 percent of fades greater than 20 dB but less than 16 percent of fades smaller than 20 dB.

Although the two-ray model described here is easy to understand and apply, most multipath propagation research points toward the presence of three (or more) rays during fading conditions. Out of this research, Rummler's three-ray model [18] is the most widely accepted.

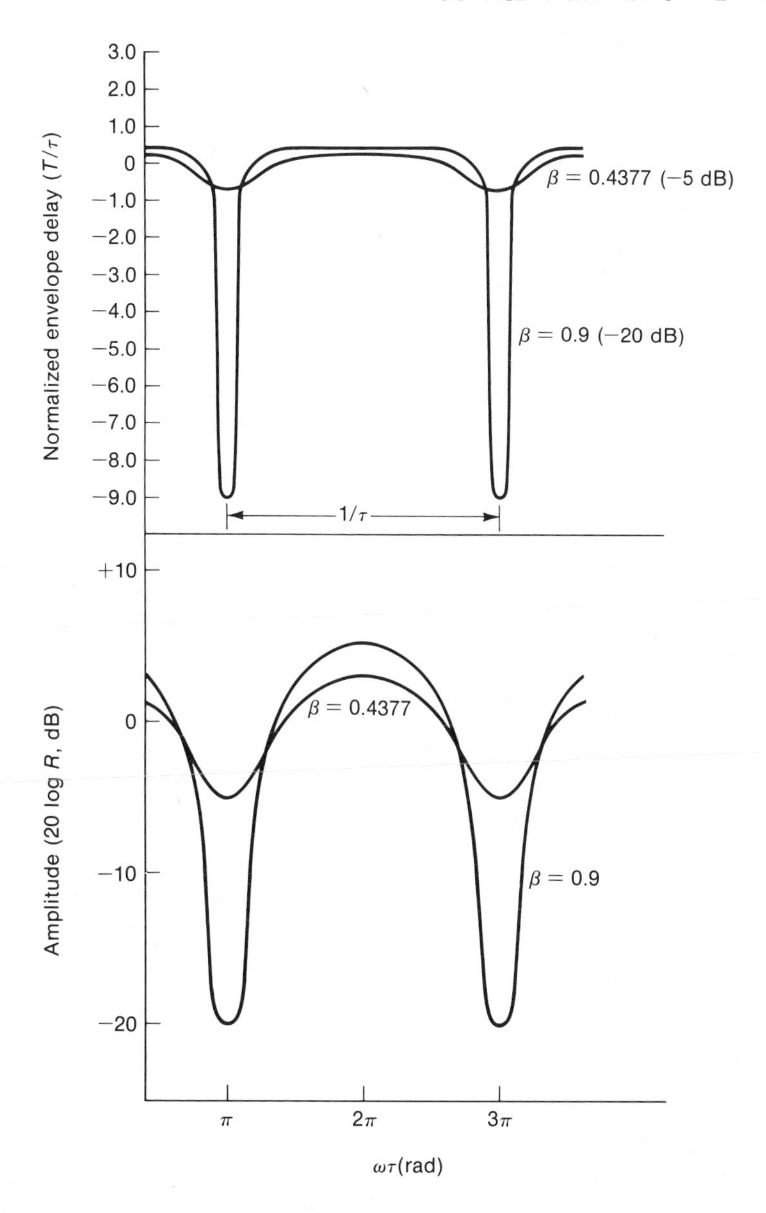

Figure 9.24 Two-Ray Envelope Delay and Amplitude Distortion for Fades of 5 and 20 dB

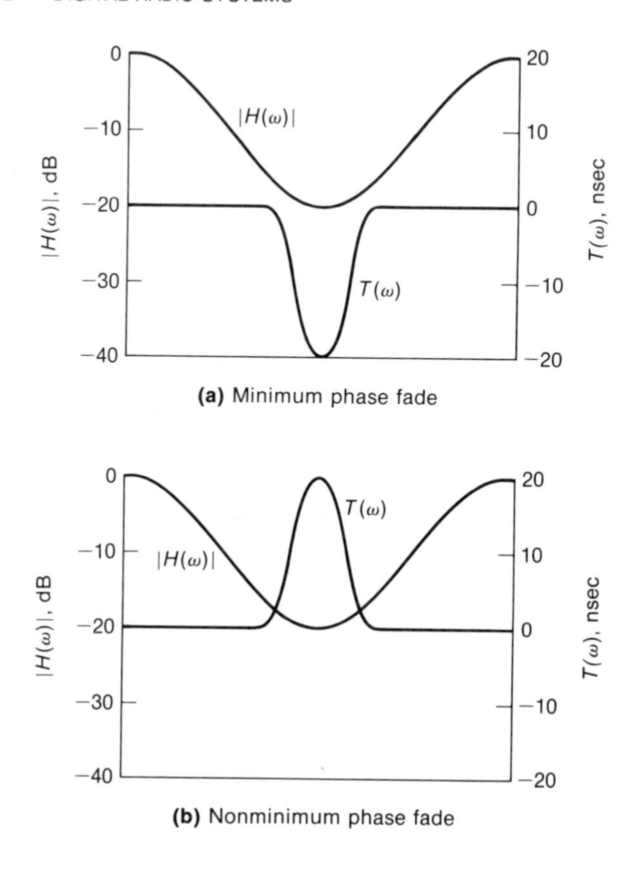

Figure 9.25 Fading Event Showing Minimum vs. Nonminimum Phase

Probability of Outage

Characterization of digital radio performance in the presence of selective fading can be accomplished by over-the-air testing or by separate testing of the link's dispersion characteristic and the radio performance for given dispersion. With regard to radio performance, a *signature* method of evaluation has been developed to determine radio sensitivity to multipath dispersion using analysis [21], computer simulation, or laboratory simulation [22]. The signature approach is based on an assumed fading model (for example, two ray or three ray). The parameters of the fading model are varied over their expected ranges, and the radio performance is analyzed or recorded for each setting. To develop the signature, the parameters are adjusted to provide a threshold error rate (say, 1×10^{-3}) and a plot of the parameter settings is made to delineate the outage area. In the two-ray model, for example, only two parameters are needed: notch depth $(1 - \beta)$

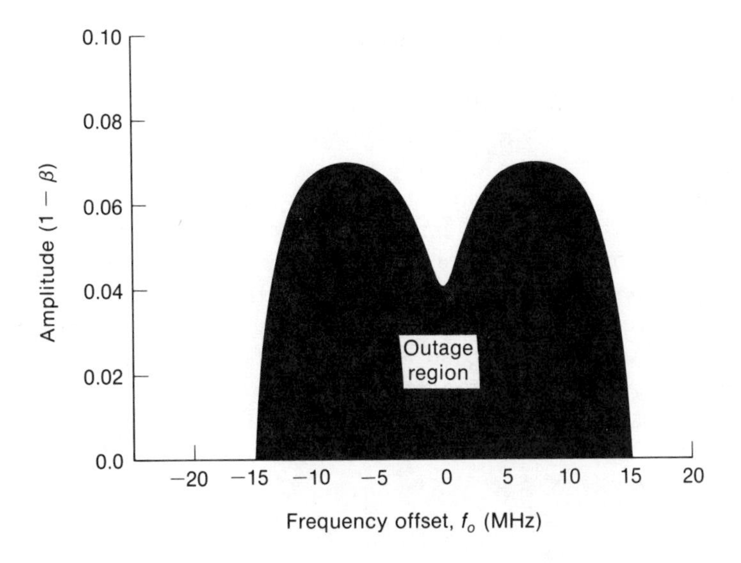

Figure 9.26 Typical Radio Signature *M* Curve

and offset frequency (f_o). A typical signature developed by using the two-ray model is shown in Figure 9.26. The area under the signature curve corresponds to conditions for which the bit error rate exceeds the threshold error rate; the area outside the signature corresponds to a BER that is less than the threshold value. The M shape of the curve (hence the term *M curve*) indicates that the radio is less susceptible to fades in the center of the band than fades off center. As the notch moves toward the band edges, greater notch depth is required to produce the threshold error rate. If the notch lies well outside the radio band, no amount of fading will cause an outage.

A method for predicting digital link performance in the presence of frequency-selective fading has not yet emerged that is as simple and well established as that for flat fading channels. Most of the outage prediction techniques assume a two-ray or three-ray multipath model and then measure or infer statistics of each parameter in the model. The probability of outage can then be calculated by integrating the probability density function (pdf) of the channel parameters over the radio outage region as established by the M curve. Emshwiller [21], for example, assumed a two-ray model with a joint pdf of $p(\beta, \tau)$ and a probability of outage given by

$$P(o) = \int_\tau \int_\beta p(\beta, \tau)\, d\beta\, d\tau \tag{9.57}$$

This integral was evaluated by assuming $p(\beta, \tau) = p(\beta)p(\tau)$, assuming a form for $p(\beta)$ and $p(\tau)$, and replacing the M curve by a rectangle with the same area. Rummler and Lundgren [22] assumed a three-ray model, developed statistics of the joint pdf based on experimental data, and evaluated the probability of outage using an integral similar in form to Equation (9.57).

Another approach to predicting digital radio system performance is based on the assumption that amplitude dispersion is the principal cause of outages. The probability of outage can then be expressed as the product of a dispersion occurrence factor and the multipath occurrence factor. The dispersion occurrence factor is conveniently expressed as a probability distribution function of in-band amplitude dispersion. This approach has the advantage of simplicity and independence from the ray model assumed. Amplitude dispersion is defined here as the difference in amplitude, ΔA, for two in-band frequencies, usually located at the two band edges; thus, in terms of decibels,

$$\Delta A = |20 \log \alpha| \tag{9.58}$$

where $\alpha = R_1/R_2$ and R_1 and R_2 are the amplitudes of frequencies located at the band edges. Assuming that R_1 and R_2 are correlated Rayleigh-distributed signals, Vigants [23] has shown the probability that α exceeds some value α_0 to be

$$P(\alpha > \alpha_0) = \frac{100\Delta f \alpha_0^2}{f^2 D\left(\alpha_0^2 - 1\right)^2} \qquad (\alpha_0 \gg 1) \tag{9.59}$$

where Δf = frequency separation of two signals and also the transmission bandwidth (GHz)
f = radio frequency (GHz)
D = path length (mi)

The probability of outage can now be written

$$P(o) = P(\alpha > \alpha_0) \cdot d \tag{9.60}$$

where α_0 now represents the threshold of in-band amplitude dispersion for a specified BER and where the multipath occurrence factor d is defined by (9.29). Substituting (9.29) and (9.59) into (9.60) we obtain the final form as

$$P(o) = (2.5 \times 10^{-4})\left(\frac{abD^2\Delta f}{f}\right)\left[\frac{\alpha_0^2}{\left(\alpha_0^2 - 1\right)^2}\right] \qquad (\text{for } \alpha_0 \gg 1) \tag{9.61}$$

The theoretical predictions of (9.61) show good agreement with measured results of outages for 8-PSK and 16-QAM radio systems as well as measured data of amplitude dispersion distributions [24].

Example 9.4

Consider an 8-PSK radio operating at a radio frequency of 6 GHz over a path of average terrain, and length 26.4 mi. The threshold of in-band amplitude dispersion for a BER of 10^{-3} is known to be 8 dB over a measured bandwidth of 23.1 MHz. Find the probability of outage (BER $\geq 10^{-3}$), for a heavy fading month (climate factor $b = 1$).

Solution

From (9.61) we obtain

$$P(o) = \frac{(2.5 \times 10^{-4})(0.0231)(26.4)^2(2.5)^2}{6(5.25)^2}$$

$$= 1.5 \times 10^{-4}$$

This result compares favorably with the measured result of 4.1×10^{-4} found by Barnett [25].

Improvements Due to Diversity and Equalization

Both diversity and adaptive equalization can be used, separately or together, to improve digital radio performance in the presence of frequency-selective fading. Diversity reduces the probability of in-band dispersion. Adaptive equalization reduces the in-band difference between the minimum and maximum amplitude values and, depending on the type of equalizer, reduces the in-band difference between group delay values also. In many instances, both diversity and equalization have been necessary to meet performance objectives. Interestingly, the combined improvement obtained by simultaneous use of diversity and equalization has been found to be larger than the product of the individual improvements. This synergistic effect has been reported in several experiments [26, 27], where the added improvement has resulted from the diversity combiner's ability to replace in-band notches with slopes that are easier to equalize. Giger and Barnett [28] have derived a formula for this improvement, given by

$$I_t = I_d \cdot I_e^2 \tag{9.62}$$

where I_t = total improvement factor

I_d = diversity improvement factor

I_e = equalization improvement factor

These improvement factors are measured or calculated as the ratio of the nondiversity, unequalized system outage probability to the total, diversity, or equalized system outage probability.

Field measurements of BER distributions for a 6-GHz, 90-Mb/s, 8-PSK radio on a 37.3-mi link are shown in Figure 9.27 [29]. This system was tested in four configurations: unprotected; with adaptive equalization; with space diversity; and with equalization plus diversity. These results are tabulated in Table 9.1 and compared to other results; all sets of data indicate a synergistic effect, and large improvement is observed with the combination of equalization and diversity.

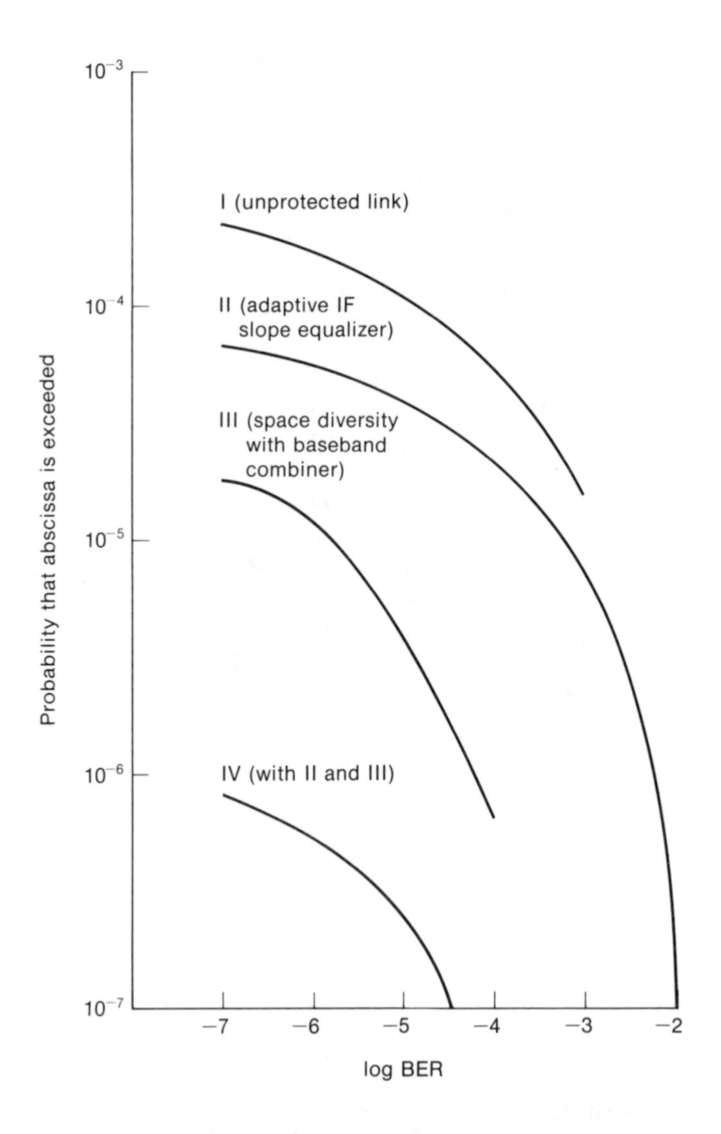

Figure 9.27 BER Distributions for 6-GHz, 90-Mb / s, 8-PSK, 37.3-mi Link Showing Effects of Equalization and Diversity [29]

Table 9.1 Summary of Space Diversity and Adaptive Equalizer Performance for Digital Radio Links

	Improvement Factors[a]		
Radio System	*Space Diversity*	*Adaptive Equalizer*	*Combined*
1[b]	38	2	770
2[c]	6	3–5	175
3[d]	14.2	2.6	320

[a]Improvement factors are based on fraction of time that BER ≥ threshold, with threshold = 10^{-4} for radio system 1, 10^{-3} for radio system 2, and 10^{-5} for radio system 3.
[b]8-GHz, 91-Mb/s, QPR, 32-mi link with IF combining and linear adaptive equalizer [26].
[c]6-GHz, 78-Mb/s, 8-PSK, 26.4-mi link with IF combining and IF amplitude slope equalizer [25, 27].
[d]6-GHz, 90-Mb/s, 8-PSK, 37.3-mi link with baseband combining and IF amplitude slope and notch equalizer (Figure 9.27).

9.4 FREQUENCY ALLOCATIONS AND INTERFERENCE EFFECTS

The design of a radio system must include a frequency allocation plan, which is subject to approval by the local frequency regulatory authority. In the United States, radio channel assignments are controlled by the Federal Communications Commission (FCC) for commercial carriers, by the National Telecommunications and Information Administration (NTIA) for government systems, and by the Military Communications and Electronics Board (MCEB) for military systems. FCC regulations pertaining to digital microwave systems are contained in FCC Docket 19311 [30]. This docket establishes standards for efficient use of the radio spectrum and controls interference between adjacent radio channels. A standard has also been developed for the U.S. military, MIL-STD-188-322, for the application of digital microwave to military frequency bands [31]. Table 9.2 lists the microwave frequency bands that are authorized for digital microwave in the United States together with the allowable bandwidths and minimum voice channel capacity. Note that in the common carrier bands, governed by FCC Docket 19311, a spectral efficiency of 2 bps/Hz or better is required to meet the minimum capacity requirement.

The International Radio Consultative Committee (CCIR) issues recommendations on radio channel assignments for use by national frequency allocation agencies. Although the CCIR itself has no regulatory power, it is important to realize that with the exception of the United States, CCIR recommendations are usually adopted on a worldwide basis. With regard to

Table 9.2 Authorized Frequency Bands in the U.S. for Digital Microwave Systems [30, 31]

Use	*Frequency Band (GHz)*	*Allowable Bandwidth (MHz)*	*Minimum Capacity of 64-kb/s PCM Voice Channels*
Common carrier	2.110 – 2.130	3.5	96
	2.160 – 2.180	3.5	96
	3.700 – 4.200	20	1152
Military	4.400 – 5.000	3.5	48
		7.0	96
		10.5	144
		14	192
Common carrier	5.925 – 6.425	30	1152
Military	7.125 – 8.400	3.5	48
		7.0	96
		10.5	144
		14	192
		20	288
Common carrier	10.700 – 11.700	40	1152
Military	14.400 – 15.250	20	288
		28	384
		40	576

digital microwave systems, some CCIR recommendations were issued with the 1982 Plenary Assembly, but other issues were deferred for further study. Table 9.3 lists reports and recommendations of CCIR Study Group 9 pertinent to the planning of frequency allocations for digital microwave systems [32].

The usual practice in frequency channel assignments for a particular frequency band is to separate transmit ("go") and receive ("return") frequencies by placing all go channels in one half of the band and all return channels in the other half. With this approach, all transmitters on a given station are in either the upper or lower half of the band, with receivers in the remaining half. Within each half-band, adjacent channels must be spaced far enough apart to avoid energy spillover between channels. A common scheme used to increase adjacent channel discrimination is to alternate between vertical and horizontal polarization. The isolation provided by cross-polarizing adjacent channels is on the order of 20 dB or more. At the edges of the band, a guard spacing is necessary to protect against interference into and from adjacent bands.

As an illustration of the frequency planning procedures described above, Figure 9.28 shows a suggested RF channel arrangement for a QPSK, 140-Mb/s digital radio system operating in the 10.7 to 11.7 GHz band. The band is first separated into halves to accommodate transmitting and receiving frequencies. Recommended spacings are shown for the band edge

Table 9.3 CCIR Frequency Allocations Applicable to Digital Radio

Frequency Band (GHz)	*Frequency Range (GHz)*	*Channel Spacing (MHz)*	*Modulation Technique*	*PCM Channel Capacity*	*Bit Rate (Mb/s)*	*CCIR Rec. or Report*
2	1.9 – 2.3	29	8-PSK	960	70	Report 934
			16-QAM	1344	90	and Rec. 382-3
	2.3 – 2.5	2		24		Report 933
		2		30		Report 933
		4		48		Report 933
		4		60		Report 933
4	3.7 – 4.2	29	8-PSK	960	70	Report 934
			16-QAM	1344	90	and Rec. 382-3
		40	16-QAM		200	Report 935
6	5.925 – 6.425	29.65	8-PSK	960	70	Report 934
			16-QAM	1344	90	and Rec. 383-2
		40	16-QAM		140	Rec. 384-3
8	7.725 – 8.275	29.65	8-PSK	960	70	Report 934
			16-QAM	1344	90	Report 934
11	10.7 – 11.7	67	4-PSK		140	Report 782-1
		60	8-PSK		140	Report 782-1
		48	16-QAM		140	Report 782-1
13	12.75 – 13.25	28		960	70	Rec. 497-2
		14		240	2×8.448	Rec. 497-2
		35		960	70	Rec. 497-2

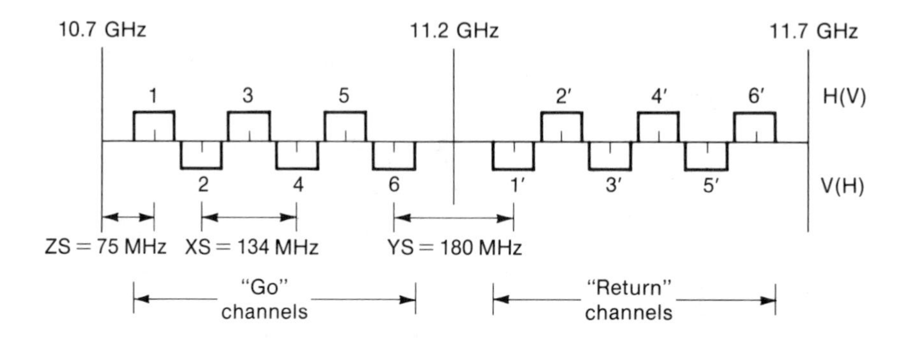

Figure 9.28 Suggested Radio-Frequency Channeling Arrangement for a High-Capacity Digital Radio-Relay System in the 10.7 to 11.7 GHz Band [32]

(ZS), adjacent channels (XS), and adjacent transmitter and receiver (YS). Note that adjacent channels also have opposite polarization, alternating between vertical (V) and horizontal (H).

Another important consideration in radio link design is RF interference, which may occur from sources internal or external to the radio system. The system designer should be aware of these interference sources in the area of each radio link, including their frequency, power, and directivity.

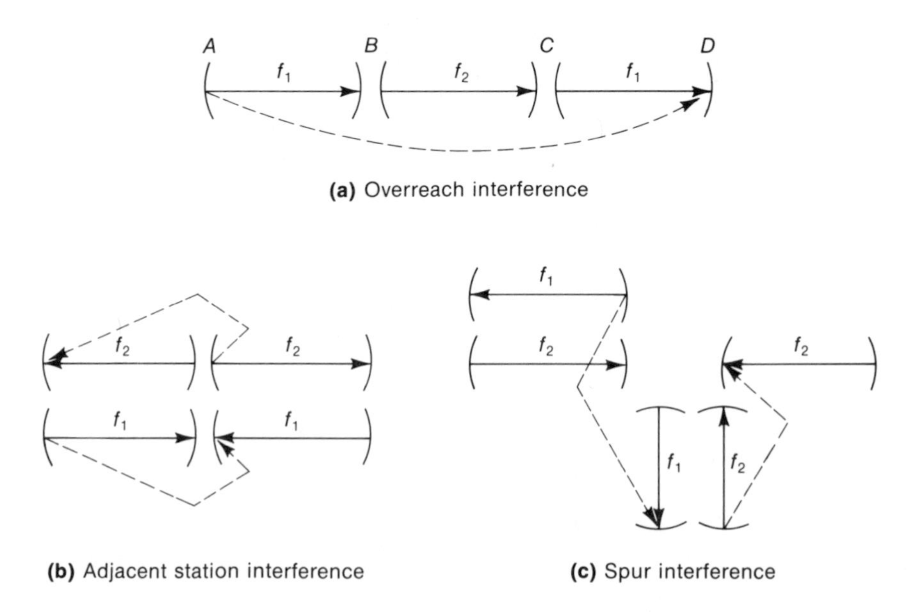

Figure 9.29 Intrasystem Sources of RF Interference [8]

Certain steps can be taken to minimize the effects of interference: good site selection, use of properly designed antennas and radios to reject interfering signals, and use of a properly designed frequency plan.

Figure 9.29 illustrates internal sources of RF interference, which are classified as overreach, adjacent station, and spur interference. (The solid lines in the figure signify the desired signal; the dashed lines represent the interfering signal.) Overreach may occur when radio links in tandem are positioned along a straight line. In Figure 9.29*a*, overreach interference becomes significant when energy transmitted from site *A* arrives at site

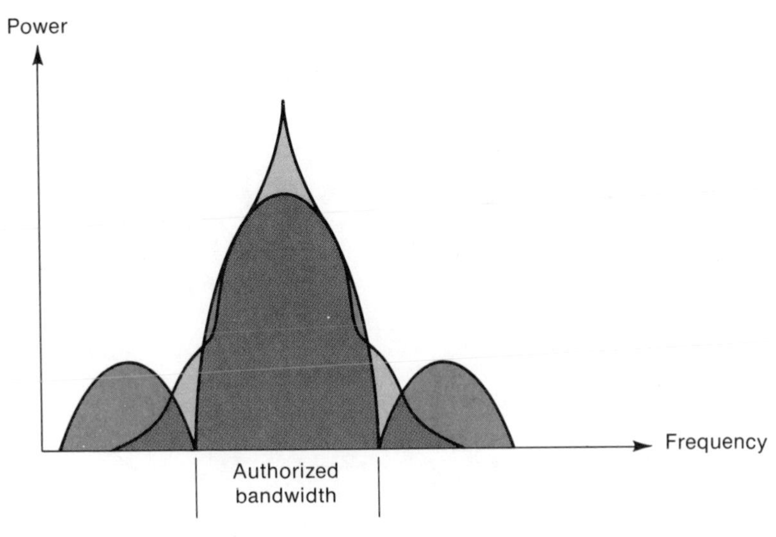

(a) Cochannel interference

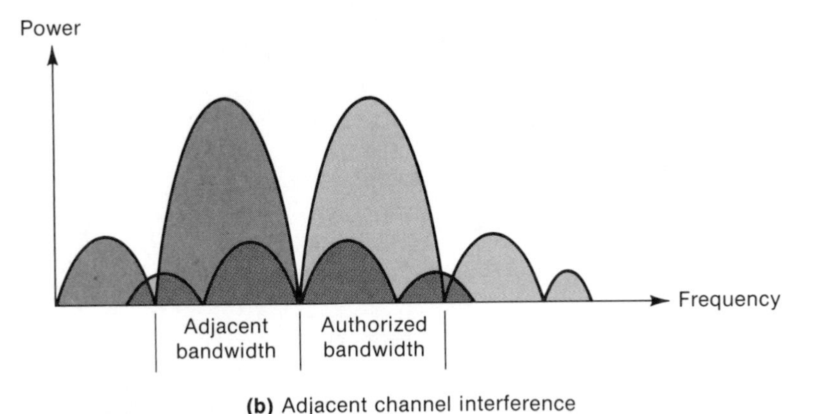

(b) Adjacent channel interference

Figure 9.30 Types of RF Interference

D—which may occur during superrefractive conditions—while there is fading present on the signal received from *C*. This problem can be reduced by staggering links to avoid a straight-line sequence of paths or by using earth blockage on the overreach path. Adjacent station and spur interference is more complex to analyze but is a function of antenna performance parameters, such as front-to-back ratio and side lobes.

Other sources of interference that may arise outside the radio system include radar, troposcatter, satellite, and other LOS systems. Radar systems often propagate energy at high levels and in a 360° arc and are therefore a

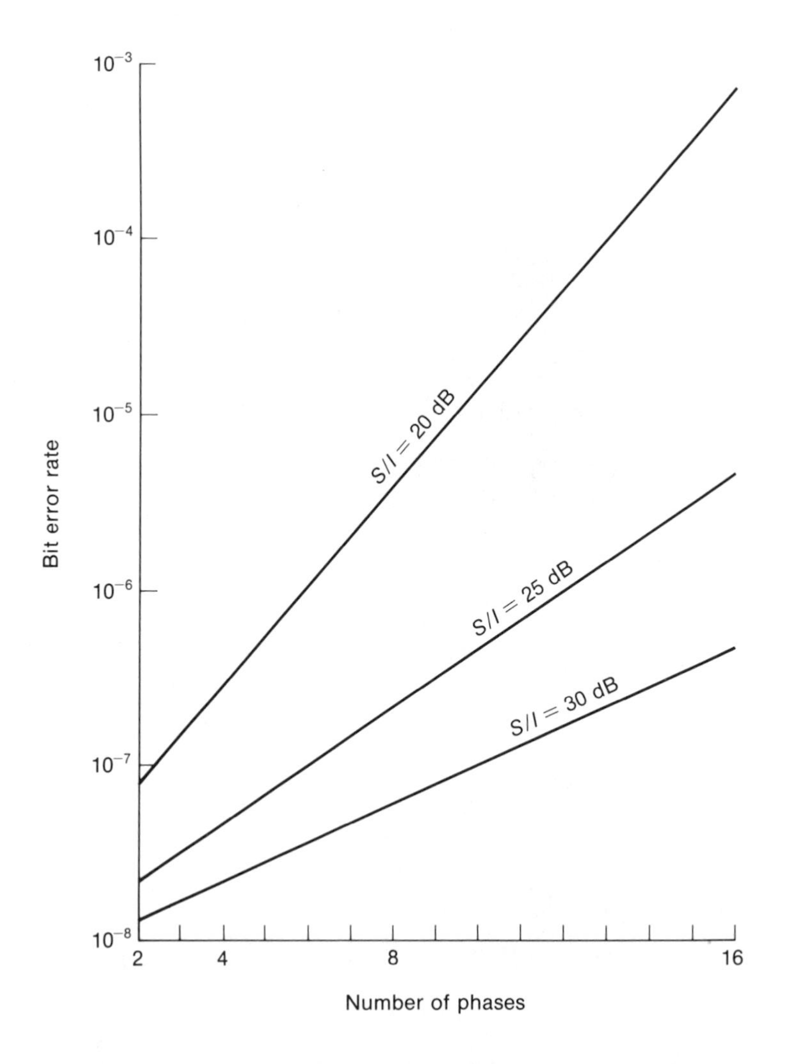

Figure 9.31 Effects of Cochannel Interference on BER Performance of a Coherent Detection PSK System as a Function of Number of Phases (Adapted from [34])

significant source of interference for microwave receivers. Earth blocking or filtering of the interfering signals is recommended to minimize interference from radar. Certain frequency bands are shared among line-of-sight, troposcatter, and satellite systems. Because of the large power used by troposcatter and satellite transmitters, LOS receivers are susceptible to interference from these transmitters even at distances well beyond the horizon. High-power transmitters for satellite and troposcatter transmission are usually located in isolated areas, however, which reduces this interference problem. Existing LOS systems, either analog or digital, may parallel or intersect the new link. Information needed of existing systems includes transmit power, frequency, distance, and antenna discrimination.

The effect of RF interference on a radio system depends on the level of the interfering signal and whether the interference is in an *adjacent* channel or is *cochannel*. A cochannel interferer has the same nominal radio frequency as that of the desired channel. Cochannel interference arises from multiple use of the same frequency without proper isolation between links. Adjacent channel interference results from the overlapping components of the transmitted spectrum in adjacent channels. Protection against this type of interference requires control of the transmitted spectrum, proper filtering

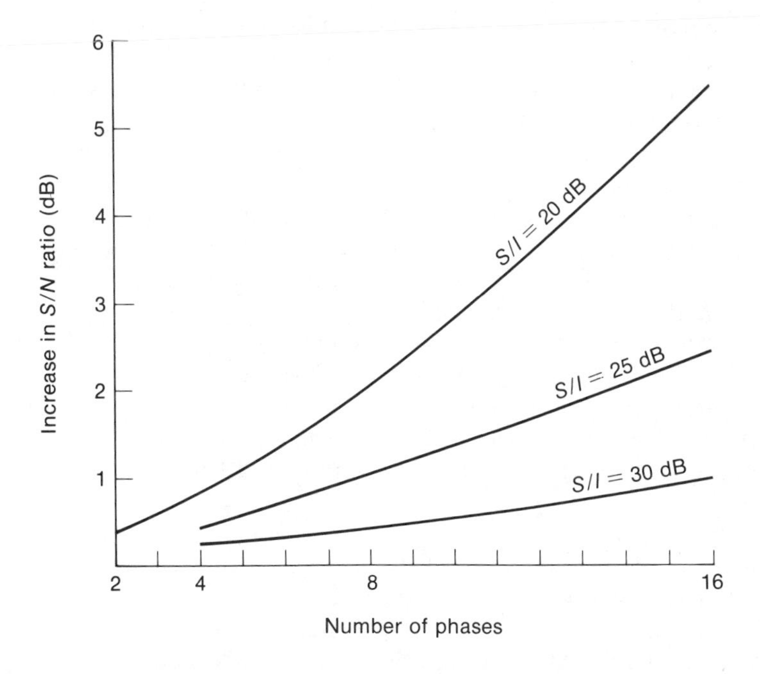

Figure 9.32 Required Increase in *S / N* Ratio to Maintain a 1 × 10⁻⁸ BER Performance of a Coherent Detection PSK System as a Function of Number of Phases (Adapted from [34])

within the receiver, and orthogonal polarization of adjacent channels. Both types of interference are illustrated in Figure 9.30.

The performance criteria for digital radio systems in the presence of interference are usually expressed in one of two ways: allowed degradation of the S/N threshold or allowed BER. Both criteria are stated for a given signal-to-interference ratio (S/I). For example, AT&T requirements for 6-GHz digital radio are a maximum allowed BER of 10^{-3} with an S/I of 24 dB for cochannel interference and the same BER with an S/I of 10 dB for adjacent channel interference [33]. As a further illustration, the effect of various cochannel interference levels, with S/I of 20, 25, and 30 dB, on the BER and S/N of coherent phase-shift-keyed (PSK) systems is shown in Figures 9.31 and 9.32 [34]. In Figure 9.31, the S/N ratio has been adjusted to give a BER of 1×10^{-8} with $S/I = \infty$.

9.5 DIGITAL RADIO DESIGN

A block diagram of a digital radio transmitter and receiver is shown in Figure 9.33. The traffic data streams at the input to the transmitter are usually in coded form, for example bipolar, and therefore require conversion to an NRZ signal with an associated timing signal. The multiplexer combines the traffic NRZ streams and any auxiliary channels used for orderwires into an aggregate data stream. This step is accomplished either by using pulse stuffing, which allows the radio clock rate to be independent of the traffic data, or by using a synchronous interface, which requires the radio and traffic data to be controlled by the same clock. The aggregate signal is scrambled to obtain a smooth radio spectrum and ensure recovery of the timing signal at the receiver. For phase modulation, some form of differential encoding is often employed to map the data into a change of phase from one signaling interval to the next. The modulator converts the digital baseband signals into a modulated intermediate frequency (IF), which is typically at 70 MHz when the final frequency is in the microwave band. The RF carrier is generated by a local oscillator, which is mixed with the IF modulated signal to produce the microwave signal. The RF power amplification is accomplished by a traveling-wave tube (TWT) or by a solid-state amplifier such as the gallium arsenide field-effect transistor (GaAs FET) amplifier. The final component of the transmitter is the RF filter, which shapes the transmitted spectrum and helps control the signal bandwidth.

At the receiver, the RF signal is filtered and then mixed with the local oscillator to produce an IF signal. The IF signal is filtered and amplified to provide a constant output level to the demodulator. Automatic gain control (AGC) in the IF amplifier provides variable gain to compensate for signal fading. Because the AGC voltage is a convenient indicator of received signal level, it is often used for performance monitoring or diversity combining. Fixed equalization is required to compensate for static amplitude or delay distortion from radio components, such as a TWT or filter, or to build out differential delay between RF channels in diversity operation. Adaptive

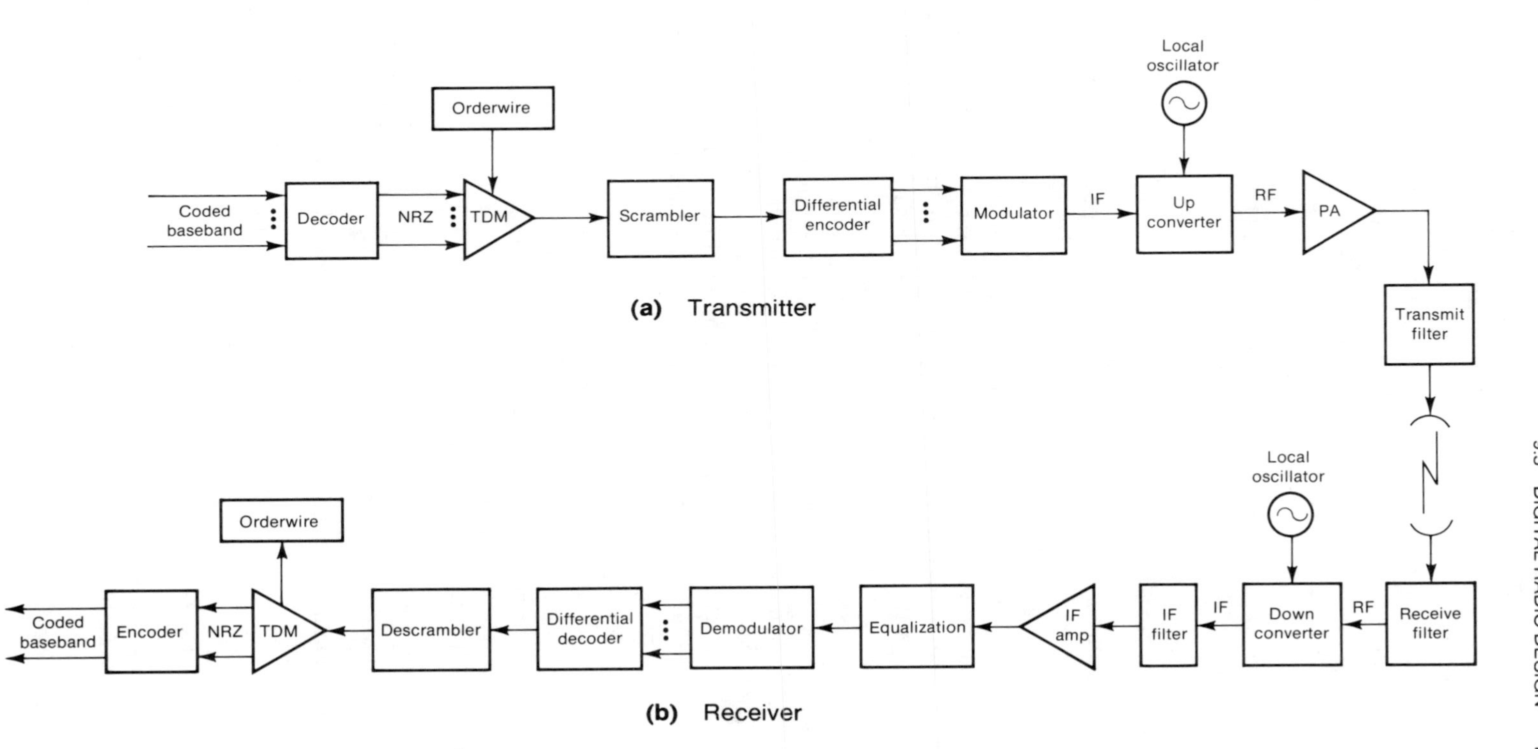

Figure 9.33 Block Diagram of a Digital Radio

equalization may also be required to deal with frequency-selective fading on the transmission path. Using the amplified and equalized IF signal, the demodulator recovers data and corresponding timing signals. Some type of performance monitoring is also commonly found in the demodulator, often based on eye pattern opening or pseudoerror techniques. The recovered baseband signal is next decoded and descrambled to reconstruct the aggregate data stream. The demultiplexer recovers the traffic data streams and auxiliary channels. Finally, in the baseband encoder the standard data interface is generated.

9.5.1 Effects of Nonlinear Amplifiers

In order to maximize efficiency (DC to RF) and RF power output, the microwave power amplifier (PA) shown in Figure 9.33*a* is often operated near saturation. An amplifier output is referred to as saturated when the output power is no longer increased by an increase in input level. For small input level, the output of a linear amplifier increases linearly with increasing input power. Near saturation, however, the input/output relationship becomes nonlinear. Although FM analog radios operate satisfactorily with saturated amplifiers, with digital modulation schemes, particularly those having an AM component, amplifiers operating in a nonlinear region are frequently a source of performance degradation and spectral spreading.

In order to examine these effects, we will consider the amplitude and phase characteristics of a typical TWT [35]. In Figure 9.34*a*, the power output increases linearly with increasing power input until saturation begins to occur where the output becomes nonlinear. This amplitude characteristic also produces intermodulation (IM) products that can be harmful if they fall within the radio channel. The most harmful IM product is third-order, since it falls in-band and contains more power than higher-ordered products. For Figure 9.34*b*, zero phase shift is observed with increasing power input until saturation begins to occur where the phase shift increases in a nonlinear fashion. These amplitude and phase characteristics of Figure 9.34 give rise to two forms of nonlinear distortion:

- Amplitude modulation to amplitude modulation (AM/AM) conversion
- Amplitude modulation to phase modulation (AM/PM) conversion

To model these nonlinear effects, let the TWT input signal be of the form

$$x(t) = A(t)\cos[\omega_0 t + \theta(t)] \tag{9.63}$$

Then the TWT output signal may be expressed by

$$y(t) = G[A(t)]\cos\{\omega_0 t + \theta(t) + F[A(t)]\} \tag{9.64}$$

where $G(\cdot)$ and $F(\cdot)$ are the AM/AM and AM/PM conversion characteristics, respectively. Considerable efforts have been made to derive analytic expressions that characterize TWT nonlinearity [36]. Experimental investigations [37] and computer simulations have also been used to examine the

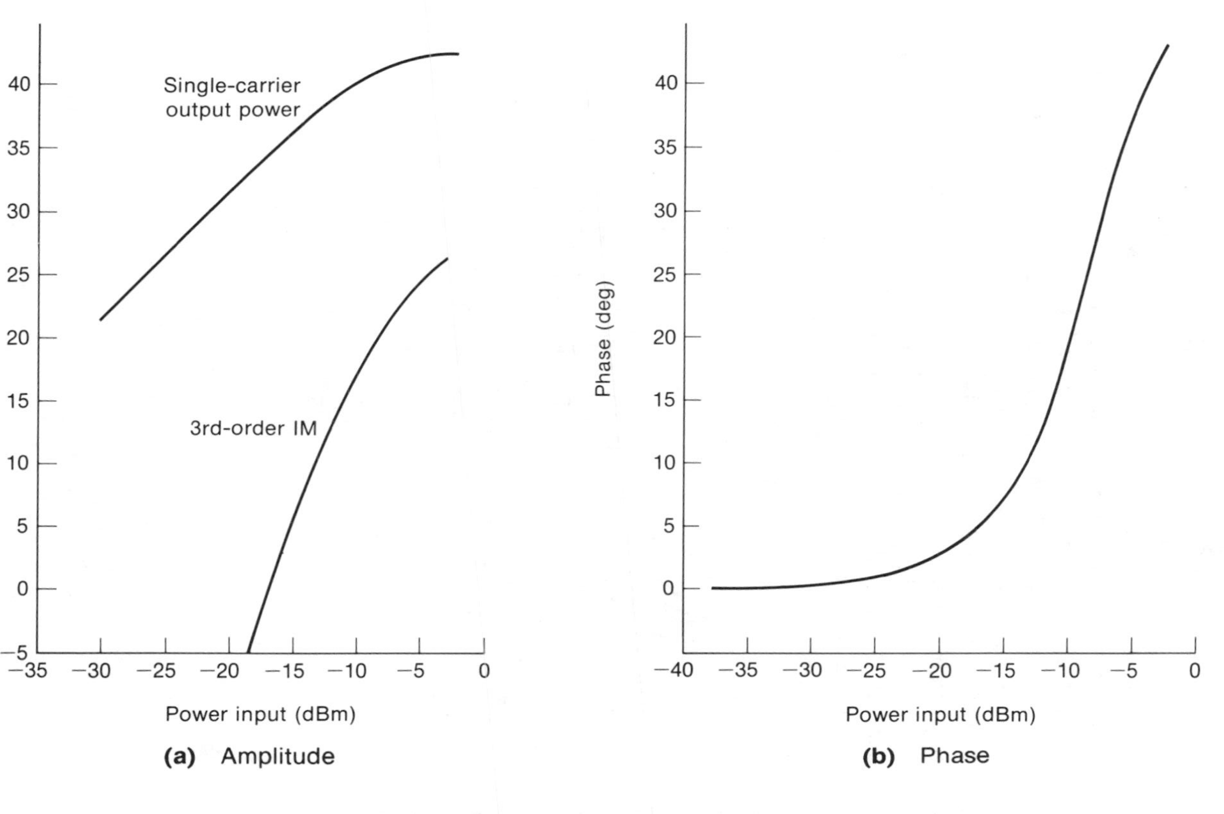

Figure 9.34 Amplitude and Phase Characteristics of Traveling Wave Tube (English Electric Valve Model N10022) [35]

effects of traveling-wave tube AM/AM and AM/PM conversion on system performance. The effects of TWT nonlinearity have been found to be twofold: degradation of the bit error rate and spectral spreading.

The BER degradation of a QPR modulated signal, caused by a nonlinear TWT, is illustrated in Figure 9.35. In this example, the effect of amplifier nonlinearity becomes apparent for any power output above 2.5 watts, where an irreducible error rate is observed. The effect of TWT nonlinearity on the transmitted spectrum of a QPR signal is illustrated in Figure 9.36. Digital modulation techniques such as QPR have spectra that decay slowly compared with FM spectra. Filtering to remove higher-order side lobes and reduce the width of the main lobe is usually done at baseband or IF, but nonlinearity in the RF amplifier can cause the restoration of out-of-band

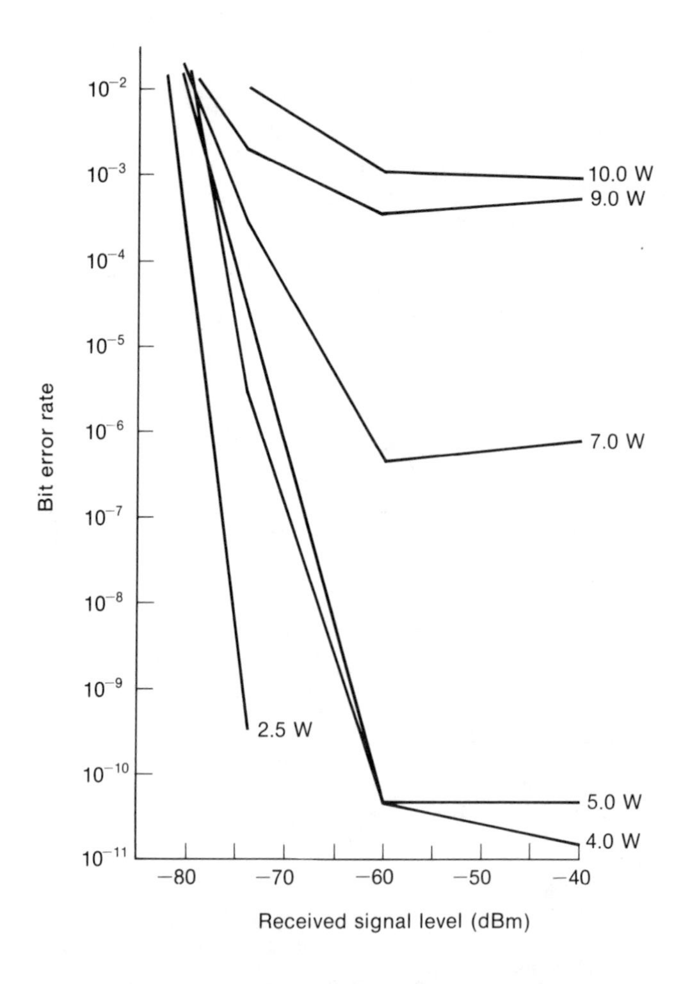

Figure 9.35 Effect of TWT Nonlinearity on BER for QPR Modulation and Different Output Power [38]

side lobes. This effect may cause interference with adjacent radio channels and violation of bandwidth allocations.

Figures 9.35 and 9.36 also illustrate one remedy for the nonlinear effects of a TWT—that is, to reduce, or back off, the output power so that the amplifier operates in a more linear region. To maintain desired output power, an alternative solution is to provide any spectrum control filtering at RF, after the nonlinear amplifier. This approach permits nearly saturated operation of the amplifier but at the expense of the insertion loss characteristic of an RF filter. Further, this approach means more difficult design and higher cost for RF filtering compared to IF or baseband filtering. As a third

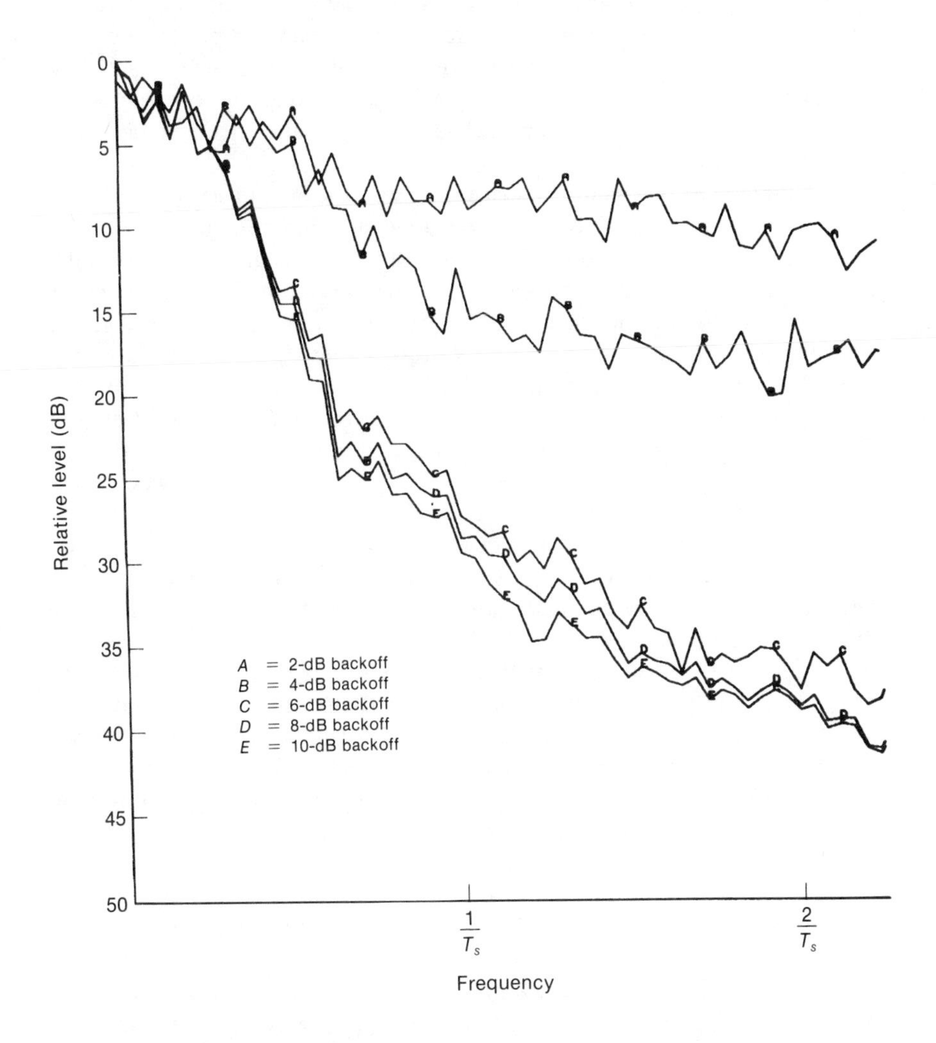

Figure 9.36 Spectrum of QPR Modulation for Different TWT Backoffs Using TWT Characteristics of Figure 9.34

alternative, the amplifier characteristics can be compensated by IF predistortion so that the overall characteristic approaches that of a linear amplifier. With this approach the amplitude and phase of the IF signal are adjusted to cancel the nonlinearity of the amplifier. Postdistortion can be used in the same manner as predistortion but is performed at the receiver IF rather than the transmitter IF.

9.5.2 Space and Frequency Diversity Design

Diversity in LOS links is used to increase link availability by reducing the effects of multipath fading, improving the combined output S/N ratio, and protecting against equipment failure. The most common forms of diversity use two parallel paths, separated in frequency or space, to provide one-for-one (1:1) protection on each link. The improvement afforded each link depends on the degree of correlation in fading between the two paths and the ability of a combiner to recognize and mitigate the effects of fading or equipment failure.

A typical arrangement for space diversity is shown in Figure 9.37. The two transmitters operate on the same frequency and can be switched for output to a common antenna. One transmitter can operate in a hot standby mode while the other is on-line, as shown in Figure 9.37; but as an alternative configuration, they can be combined to provide a 6-dB increase over the power available from a single transmitter. In this latter case, the failure of one transmitter power amplifier causes a 6-dB drop in output power but the link remains operational. The two receivers are connected to different antennas that are physically separated to provide the desired space diversity effect. The receiver outputs are fed to the combiner, which combines the two received signals. Space diversity, unlike frequency diversity, does not require an additional frequency assignment and is therefore more efficient in the use of spectrum. Its disadvantage is that additional antennas and waveguide are required, making it more expensive than frequency diversity arrangements.

Figure 9.38 illustrates a typical arrangement for frequency diversity. The two transmitters operate continuously on different frequencies but carry identical traffic. The receivers are connected to the same antenna but are tuned to separate frequencies. The combiner function is identical to that of the space diversity configuration. The use of frequency diversity doubles the spectrum amount required—a significant disadvantage in congested frequency bands. Unlike space diversity, however, frequency diversity provides two complete, independent paths, allowing testing of one path without interrupting service, while requiring only a single antenna per link end.

Variations of these protection arrangements include hot standby, hybrid diversity, and $M:N$ protection. Hot standby arrangements apply to those cases where a second RF path is not deemed necessary, as with short paths where fading is not a significant problem. Here both pairs of transmitters and receivers are operated in a hot standby configuration to provide

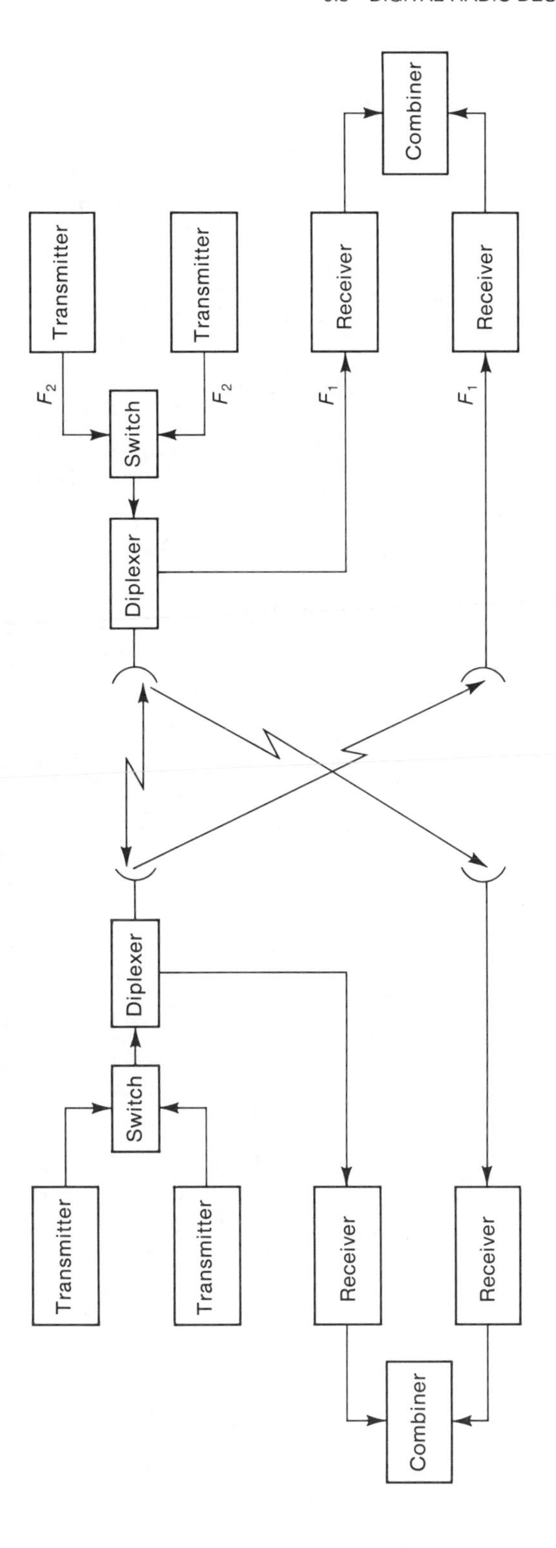

Figure 9.37 Space Diversity System with Hot Standby Transmitter

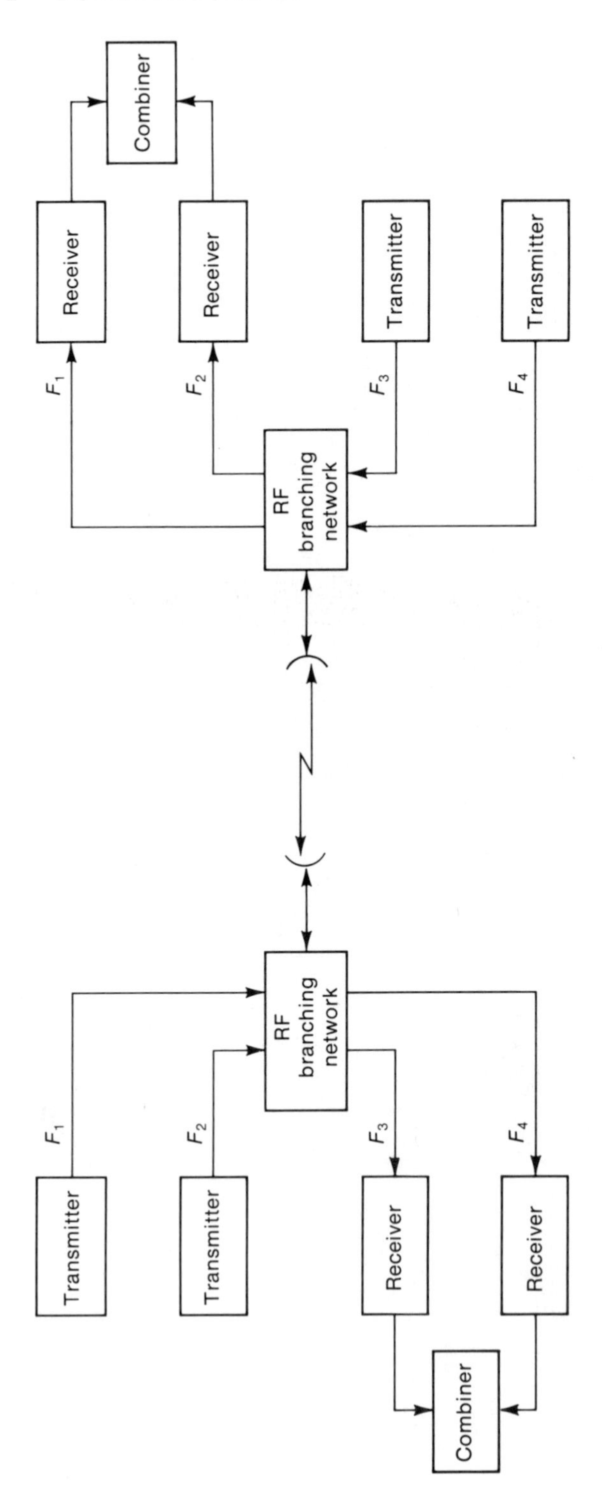

Figure 9.38 Frequency Diversity System (Single Antenna)

protection against equipment failure. The transmitter configuration is identical to that of space diversity. The receiving antenna feeds both receivers, tuned to the same frequency, through a power splitter. Hybrid diversity is provided by using frequency diversity but with the receivers connected to separate antennas that are spaced apart. This arrangement, which combines space and frequency diversity, improves the link availability beyond that realized with only one of these schemes.

For more efficient use of equipment and spectrum, diversity techniques are sometimes applied to a section of one or more links. In its simplest form, frequency diversity is used per section, with one protection channel used for N operational channels. This method can be extended to provide $M:N$ protection, where M protection channels are shared by N operational channels. Further protection can be provided by using space diversity on a per hop basis and frequency diversity on a section basis.

A diversity combiner performs the combining or selection of diversity signals. This function can be performed at RF [39], IF [40], or baseband [35]. Combiner techniques used in analog radio transmission [9] are generally applicable to digital radio. Phase alignment of the diversity signals becomes more important in digital radio systems, however, because of the potential occurrence of error bursts or loss of timing synchronization when combining misaligned diversity signals. Since delay equalization is simpler at baseband than at IF or RF, baseband "hitless" switching is a popular choice in digital radio combiners. This selection combiner uses some form of in-service performance monitor in each receiver to select the output signal after demodulation and data detection, as illustrated in Figure 9.39.

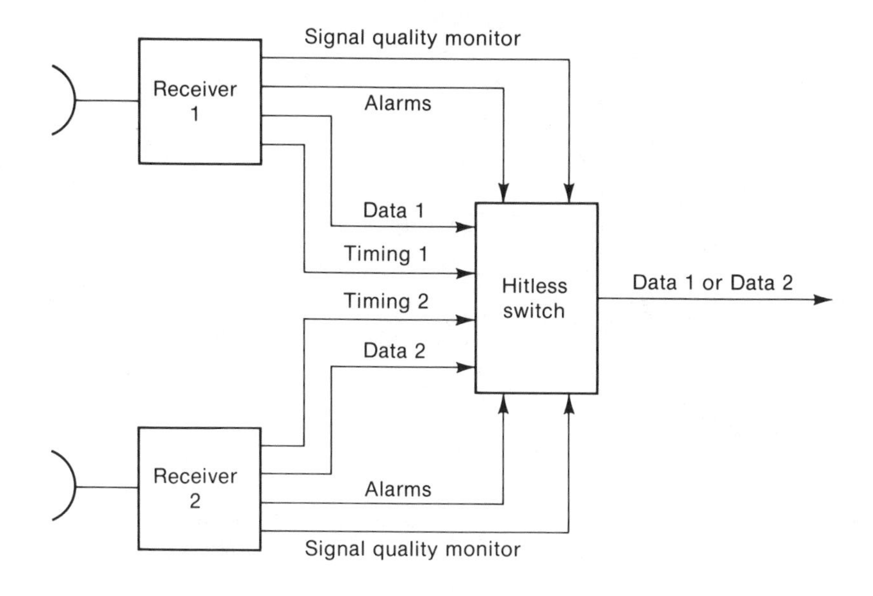

Figure 9.39 Baseband Selection Combiner

The performance of diversity combiners is greatly affected by the performance monitor used to sense signal quality and by delayed combiner action due to slow sensing and switching times. Performance monitors used in analog radio diversity combiners, such as receiver AGC, pilot tone, or out-of-band noise detection, are not sufficient in digital radio applications. Because of the presence of frequency-selective fading, the performance monitor must be responsive to the dispersion of the received signal and not just the total power received. Moreover, the performance monitor must be able to respond to other forms of signal degradation, including flat fading, interference, and additive noise, before the onset of an outage. To meet these requirements the BER of each receiver must be accurately estimated by the monitoring scheme; such estimation is possible by monitoring overhead (frame) bits or using the pseudoerror techniques described in Chapter 11.

The time required to assess each diversity channel and effect diversity selection combining can also adversely affect diversity performance. Multipath fading is a relatively slow process, however, so that a switching action can be initiated and completed before the channel has faded to threshold. Nevertheless, the response time of the performance monitor must match the speed of fading on LOS links.

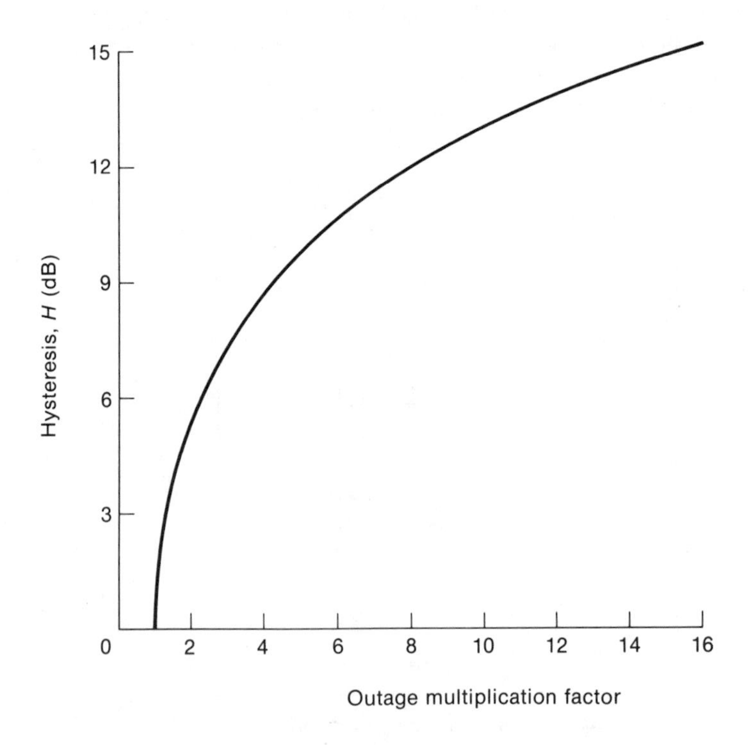

Figure 9.40 Outage Multiplication Due to Hysteresis in LOS Diversity Combiners

Ideally, the selection combiner will always instantaneously select the received signal with the highest S/N ratio. Some inaccuracy in the performance monitor is inevitable, however, and leads to the use of *hysteresis* [41], defined as the minimum ratio (h^2) of received signal levels required to initiate a switch from the signal with lower S/N to the signal with higher S/N. The hysteresis (H) in decibels is related to the ratio h^2 by

$$H = 10 \log h^2 \tag{9.65}$$

The additional outage caused by hysteresis is given by the multiplicative factor $0.5(h^2 + h^{-2})$. A plot of the hysteresis outage factor is given in Figure 9.40. This factor modifies the expression for probability of fade outage for space diversity (9.43) and for frequency diversity (9.44) to account for the effect of hysteresis on total outage. The hysteresis should be small near threshold to minimize outage extension. Typical values are in the range of 3 to 6 dB near threshold.

The control of diversity combining is usually based on a multitude of alarms and monitors ordered according to some priority. Switching criteria are based first on signal continuity and second on signal quality. Alarms resulting from equipment failure or loss of frame synchronization are assigned the highest priority in switchover control. Signal quality monitors, such as parity errors, frame errors, pseudoerrors, or eye pattern closure, are then secondary although essential during periods of fading.

9.5.3 Adaptive Equalizer Design

Initial applications of wideband digital radios revealed that dispersion due to frequency-selective fading was the dominant source of multipath outages. These experiences led to the development and use of adaptive equalization so that outage requirements could be met. Since the introduction of the first adaptive equalizers to digital radio, these devices have undergone a rapid evolution. The degree of sophistication required of the equalizer depends primarily on the bit rate and path length. The types of equalizers in use today range from simple amplitude slope equalizers to complex transversal filter equalizers.

The slope equalizer was designed to detect the presence of amplitude slope in the spectrum and perform slope correction. This objective is achieved by comparing the amplitude at the high end of the channel with the amplitude at the low end. A linear slope correction circuit then provides the desired slope across the passband. This type of equalizer performs well when the notch is outside the passband but is severely limited in the vicinity of amplitude notches. The next improvement was the addition of a notch detector that controls a notch correction circuit. The notch detection circuit operates by comparing the energy at midband with the total energy in the passband. Typical implementations of such an equalizer can correct up to ± 12 dB of amplitude slope and 12 dB of amplitude notch [42]. This performance is demonstrated in Figure 9.41, which shows the effects on the radio signature of adaptive equalization for simulated two-ray multipath.

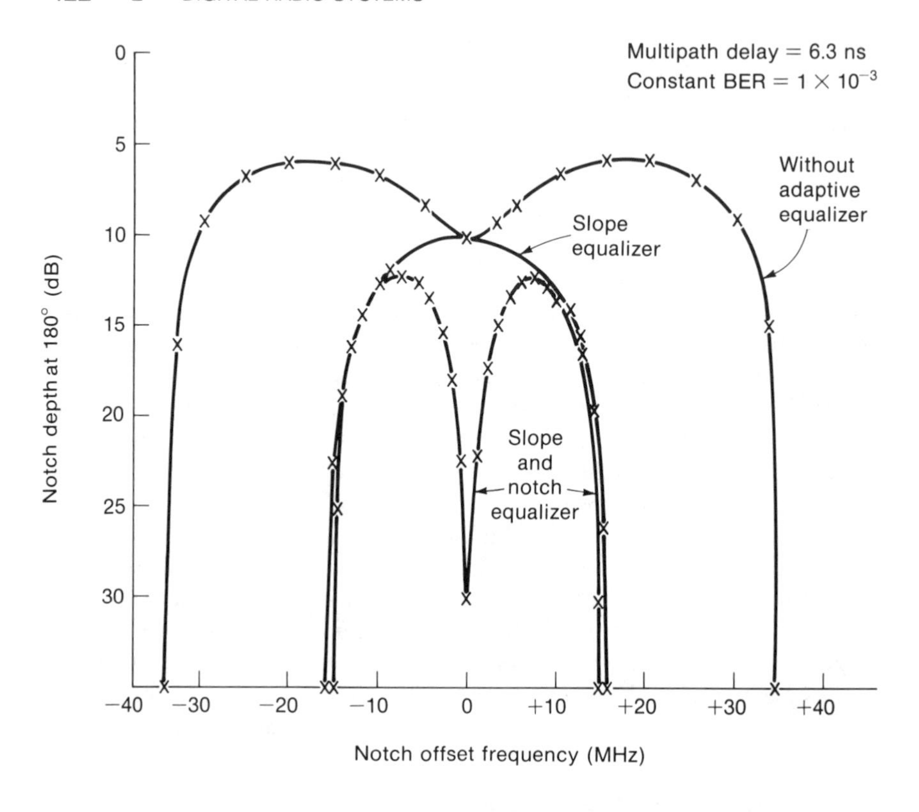

Figure 9.41 Radio Signature for Collins MDR-6 at 10^{-3} BER Showing Effects of Adaptive Equalizer [43] (Reprinted by Permission of Collins Transmission Systems Division, Rockwell International)

Here the delay of the interfering signal is fixed at 6.3 ns, the amplitude is adjusted to yield a BER of 1×10^{-3}, and the offset frequency is adjusted to sweep the notch across the passband [43].

The slope plus notch equalizer corrects linear and parabolic distortion, but it does not correct higher-order distortion. Moreover, it is ineffective for notches as they move away from midband, and it actually degrades group delay for nonminimum phase notches. A means of correcting multiple notches or a single notch as it moves through the passband is to place bump equalizers at selected frequencies across the passband. Each bump equalizer consists of a filter tuned to a selected frequency and matched to the inverse shape (*bump*) of the notch. The transfer function of a bump equalizer may be expressed as

$$H(\omega) = \frac{1}{1 + \beta e^{\pm j\omega\tau}} \tag{9.66}$$

where a two-ray model has been assumed as the source of multipath fading.

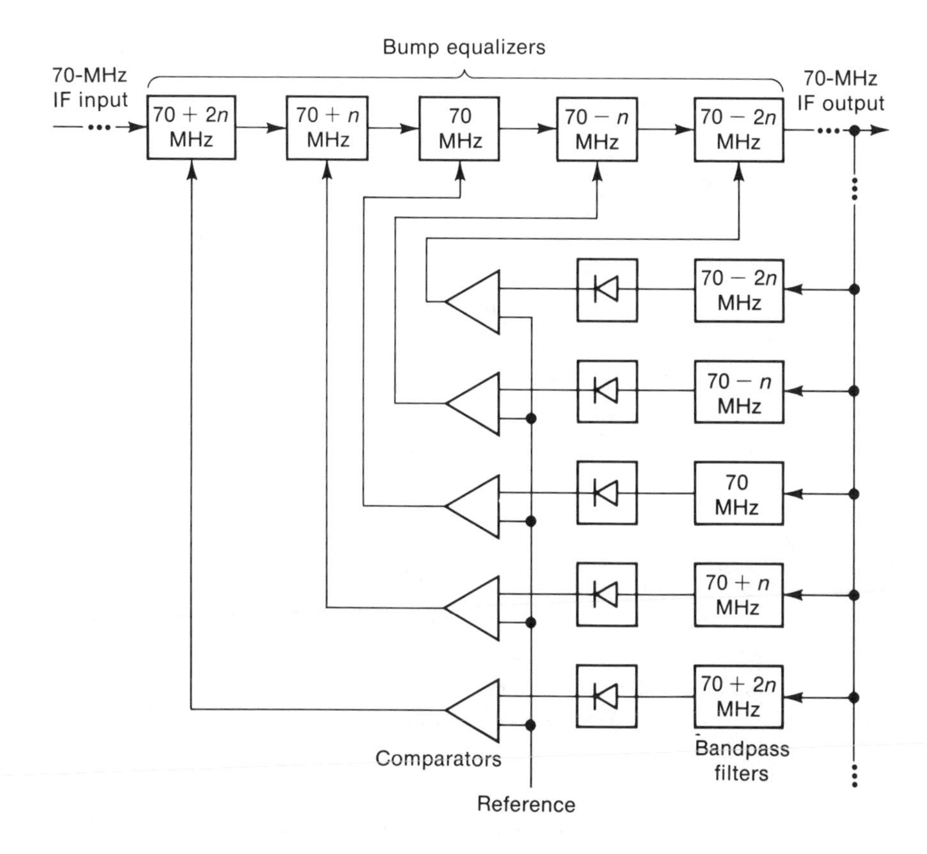

Figure 9.42 Block Diagram of IF Bump Equalizer

The sign of the term $j\omega\tau$ in (9.66) is set to yield either minimum phase (minus sign) or nonminimum phase (plus sign). Because detection of the group delay sign is difficult, the settings are usually preset and left fixed. A minimum phase bump equalization would then completely correct minimum phase notches whose frequency is the same as a bump frequency, but it would increase the group delay due to a nonminimum phase notch. Figure 9.42 is a block diagram of a generic N-bump equalizer for an IF passband with bandwidth B centered at 70 MHz [44, 45]. The IF signal is equalized by a bank of N bump circuits whose frequencies are spaced at B/N intervals across the passband. The gain of each bump is controlled by a feedback loop in which the IF signal is sampled at the bump frequencies and compared with an undistorted reference signal. The error signal from each comparator then controls the gain of the corresponding equalizer section.

To obtain better equalizer performance, particularly for group delay distortion, transversal equalization techniques have been recently applied to digital radio [46]. Figure 9.43 shows a block diagram of a QAM demodula-

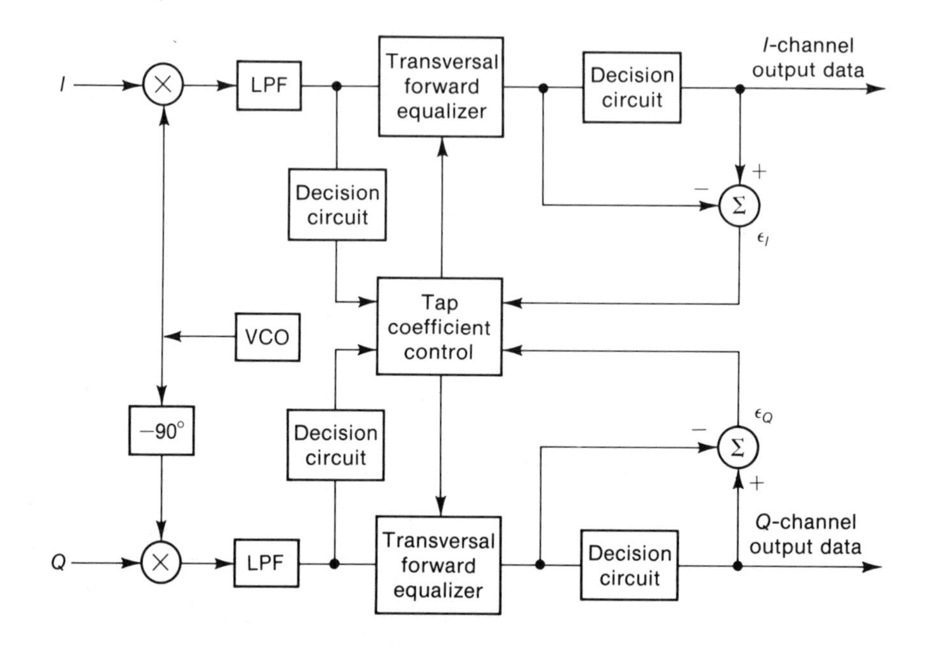

Figure 9.43 QAM Demodulator Equipped with Decision-Directed Transversal Filter Equalizer

tor equipped with a transversal filter equalizer. The demodulator outputs are equalized by a forward baseband equalizer whose tap weights are controlled by decision feedback. The tap weights are continuously adjusted to correct channel distortion and provide the desired pulse response. The number of taps and the tap spacing in the transversal filters are design parameters that are chosen according to the system bit rate, channel fading characteristics, and radio outage requirements. Field tests of a baseband adaptive transversal filter applied to a 90-Mb/s, 16-QAM radio show outage improvement by more than a factor of 3 compared with the same radio equipped with an IF slope equalizer only [47].

9.6 RADIO LINK CALCULATIONS

Procedures for allocating radio link performance based on end-to-end system requirements have already been discussed in Chapter 2. There it was noted that outages occur due to both equipment failure and propagation effects. It was noted, moreover, that the allocations for equipment and propagation outage are usually separated because of their different effect on the user and different remedy by the system designer. The subject of equipment reliability has been treated in Chapter 2. Here we will consider procedures for calculating values of key radio link parameters, such as transmitter power, antenna size, and diversity design, based on propagation outage requirements. These procedures include the calculation of intermediate parameters such as system gain and fade margin.

9.6.1 System Gain

System gain (G_s) is defined as the difference, in decibels, between the transmitter output power (P_t) and the minimum receiver signal level required to meet a given bit error rate objective (RSL_m):

$$G_s = P_t - \mathrm{RSL}_m \tag{9.67}$$

The minimum required RSL, also called receiver threshold, is determined by the receiver noise level and the signal-to-noise ratio required to meet the given BER. Noise power in a receiver is determined by the noise power spectral density (N_0), the amplification of the noise introduced by the receiver itself (noise figure N_f), and the receiver bandwidth (B). The total noise power is then given by

$$P_N = N_0 B N_f \tag{9.68}$$

The source of noise power is thermal noise, which is determined solely by the temperature of the device. The thermal noise density is given by

$$N_0 = kT_0 \tag{9.69}$$

where k = Boltzman's constant (1.38×10^{-23} joule/°K)

T_0 = absolute temperature in degrees Kelvin

The reference for T_0 is normally assumed to be room temperature, 290°K, for which $kT_0 = -174$ dBm/Hz. The minimum required RSL may now be written as

$$\begin{aligned} \mathrm{RSL}_m &= P_N + \frac{S}{N} \\ &= kT_0 B N_f + \frac{S}{N} \end{aligned} \tag{9.70}$$

It is often more convenient to express (9.70) as a function of data rate R and E_b/N_0. Using (6.4) we can rewrite (9.70) as

$$\mathrm{RSL}_m = kT_0 R N_f + \frac{E_b}{N_0} \tag{9.71}$$

The system gain may also be stated in terms of the gains and losses of the radio link:

$$G_s = L_p + F + L_t + L_m + L_b - G_t - G_r \tag{9.72}$$

where G_s = system gain in decibels

L_p = free space path loss in dB, given by (9.5)

F = fade margin in dB

L_t = transmission line loss from waveguide or coaxials used to connect radio to antenna, in dB (see Table 9.4)

L_m = miscellaneous losses such as minor antenna misalignment, waveguide corrosion, and increase in receiver noise figure due to aging, in dB

L_b = branching loss due to filter and circulator used to combine or split transmitter and receiver signals in a single antenna

G_t = gain of transmitting antenna (see Table 9.5)

G_r = gain of receiving antenna (see Table 9.5)

The system gain is a useful figure of merit in comparing digital radio equipment. High system gain is desirable since it facilitates link design—for example, by easing the size of antennas required. Conversely, low system gain places constraints on link design—for example, by limiting path length.

Table 9.4 Transmission Line Loss Factors (decibels/meter) [8, 48]

Transmission Line Type	*Frequency Band (GHz)* 2	4	6	8
Rectangular waveguide (WR)	—	0.027 (WR 229)[a]	0.068 (WR 137)[a]	0.087 (WR 112)[a]
Elliptical waveguide (EW)	—	0.028 (EW 37)[a]	0.039 (EW 52)[a]	0.058 (EW 77)[a]
Circular waveguide (WC)	—	0.013 (WC 269)[a]	0.030 (WC 166)[a]	0.022 (WC 166)[a]
Coaxial ($\frac{7}{8}''$, air dielectric)	0.062 (HJ 5)[b]	—	—	—

[a]Designates type of waveguide.
[b]Designates type of coaxial cable.

Table 9.5 Gain of Parabolic Antennas (decibels)

Parabolic Antenna Diameter (ft)	Frequency Band 2 GHz G_A	2 GHz $2G_A$	4 GHz G_A	4 GHz $2G_A$	6 GHz G_A	6 GHz $2G_A$	8 GHz G_A	8 GHz $2G_A$
4	25.3	50.6	31.3	62.6	34.8	69.6	37.3	74.6
6	28.8	57.6	34.8	69.6	38.3	76.6	40.8	81.6
8	31.3	62.6	37.3	74.6	40.8	81.6	43.3	86.6
10	33.3	66.6	39.3	78.6	42.8	85.6	45.3	90.6
12	34.8	69.6	40.8	81.6	44.3	88.6	46.8	93.6
15	36.8	73.6	42.8	85.6	46.3	92.6	48.8	97.6

Note: $G_A = 20 \log f + 20 \log d - 52.75$, where f = frequency in megahertz and d = antenna diameter in feet.

Example 9.5

Determine the system gain of an 8-GHz, 45-Mb/s, 16-QAM digital radio with a transmitter output power of 30 dBm and a receiver noise figure of 7 dB. Assume that the desired bit error rate is 1×10^{-6}.

Solution

To find the minimum required RSL, the required value of E_b/N_0 for 16-QAM is first found from Figure 6.21. Note that this figure plots probability of *symbol* error $P(e)$ versus E_b/N_0, so that we must convert $P(e)$ to BER according to (6.52):

$$P(e) = (\log_2 4)(\text{BER}) = 2\ \text{BER}$$

The E_b/N_0 for $P(e) = 2 \times 10^{-6}$ is 14.5 dB. The required RSL is found from (9.71):

$$\text{RSL}_m = -174 + 10 \log(45 \times 10^6) + 7 + 14.5 = -76\ \text{dBm}$$

The system gain then follows from (9.67):

$$G_s = 30 - (-76) = 106\ \text{dB}$$

9.6.2 Fade Margin

The traditional definition of fade margin is the difference, in decibels, between the nominal RSL and the threshold RSL as illustrated in Figure 9.14. An expression for the fade margin F required to meet allowed outage probability $P(o)$ may be derived from (9.30) for an unprotected link, from

(9.43) for a space diversity link, and from (9.44) for a frequency diversity link, with the following results:

$$F = 30 \log D + 10 \log(abf) - 56 - 10 \log P(o) \qquad \text{(unprotected link)} \tag{9.73}$$

$$F = 20 \log D - 10 \log S + 5 \log\left(\frac{h^2 + h^{-2}}{2}\right) + 5 \log(ab) - 7.2 - 5 \log P(o) \qquad \text{(space diversity link)} \tag{9.74}$$

$$F = 20 \log D + 15 \log f + 5 \log\left(\frac{h^2 + h^{-2}}{2}\right) + 5 \log(ab) - 5 \log \Delta f - 36.5 - 5 \log P(o) \qquad \text{(frequency diversity link)} \tag{9.75}$$

with combiner hysteresis effects included for space and frequency diversity. This definition of fade margin has traditionally been used to describe the effects of fading at a single frequency for radio systems that are unaffected by frequency-selective fading or to radio links during periods of flat fading. For wideband digital radio without adaptive equalization, however, dispersive fading is a significant contributor to outages so that the flat fade margins do not provide a good estimate of outage and are therefore insufficient for digital link design. Several authors have introduced the concept of effective [26], net [49], or composite [50] fade margin to account for dispersive fading. The effective (or net or composite) fade margin is defined as that fade depth which has the same probability as the observed probability of outage. The difference between the effective fade margin measured on the radio link and the flat fade margin measured with an attenuator is then an indication of the effects of dispersive fading. Since digital radio outage is usually referenced to a threshold error rate, BER_t, the effective fade margin (EFM) can be obtained from the relationship

$$P(A \geq \text{EFM}) = P(\text{BER} \geq \text{BER}_t) \tag{9.76}$$

where A is the fade depth of the carrier. The results of Equations (9.73) to (9.75) can now be interpreted as yielding the effective fade margin for a probability of outage given by the right-hand side of (9.76).

As an example, Figure 9.44 gives results for the same 90-Mb/s, 8-PSK system shown in Figure 9.27 [29]. Here observed fade outage probabilities have been plotted along with observed error rate probabilities for an unprotected link and the same link with adaptive equalization. The effective fade margin for a BER threshold of 10^{-6} is shown by the dashed lines indicating an EFM of 33 dB for the unprotected link and 38.5 dB for the equalized link. Figure 9.44 also plots the theoretical performance for flat fading, which indicates a flat fade margin (FFM) of 39 dB for the 10^{-6} BER threshold. The large difference between the unprotected-link EFM and the theoretical FFM indicates that multipath outage is dominated by dispersive effects. Use of equalization virtually eliminates the effects of dispersive fading, however, improving the EFM to within 0.5 dB of the theoretical FFM.

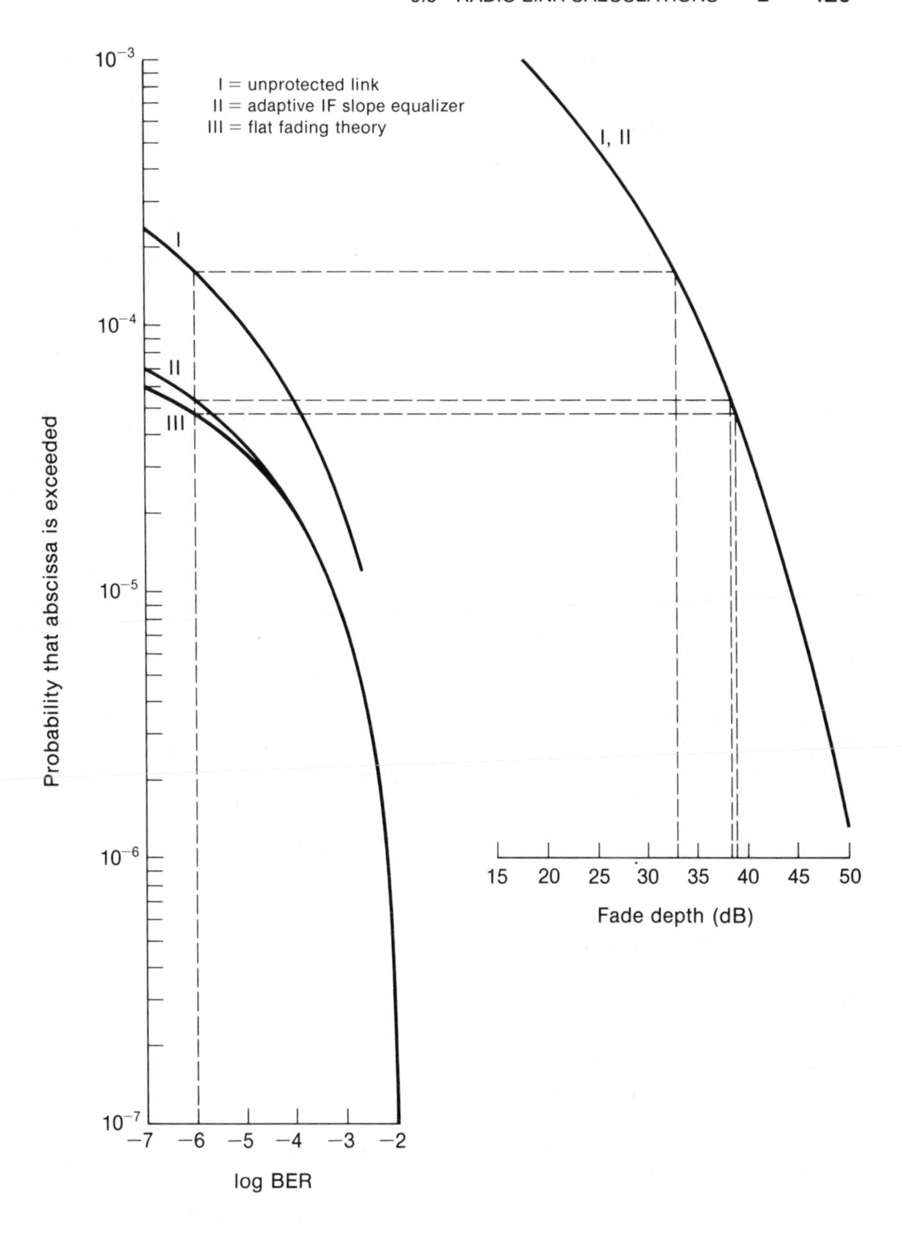

Figure 9.44 Fading and BER Distributions for 6-GHz, 90 Mb/s, 8 PSK System on 37.3 mi Path

The effective fade margin is derived from the addition of up to three individual fade margins that correspond to the effects of flat fading, dispersion, and interference [50]:

$$\text{EFM} = -10\log(10^{-\text{FFM}/10} + 10^{-\text{DFM}/10} + 10^{-\text{IFM}/10}) \quad (9.77)$$

Here the flat fade margin (FFM) is given as the difference in decibels between the unfaded signal-to-noise ratio (S/N_u) and the minimum signal-to-noise ratio (S/N_m) to meet the error rate objective, or

$$\text{FFM} = \left(\frac{S}{N_u}\right) - \left(\frac{S}{N_m}\right) \quad (9.78)$$

Similarly, we define the dispersive fade margin (DFM) and interference fade margin (IFM) as

$$\text{DFM} = \left(\frac{S}{N_d}\right) - \left(\frac{S}{N_m}\right) \quad (9.79)$$

and

$$\text{IFM} = \left(\frac{S}{I}\right) - \left(\frac{S}{N_m}\right) \quad (9.80)$$

where N_d represents the effective noise due to dispersion and I represents the effect of all interference. Each of the individual fade margins can also be calculated as a function of the other fade margins by using (9.77), as for example

$$\text{FFM} = -10\log(10^{-\text{EFM}/10} - 10^{-\text{DFM}/10} - 10^{-\text{IFM}/10}) \quad (9.81)$$

9.6.3 Link Calculation Procedure

Step-by-step procedures for calculating values for link design parameters are given in Table 9.6. For each parameter listed, a source of information is given for each entry made in the path calculation process. The following assumptions are made in Table 9.6:

1. An adequate path clearance is provided on the path under consideration (see Section 9.2.4).
2. Rain attenuation and atmospheric absorption have been ignored, which is appropriate for frequencies below 10 GHz. These two effects are easily added for use of higher frequencies (see Section 9.2.3).
3. As is often the case, antenna size is the final design parameter. Hence antenna size is calculated from other design parameters whose values are already selected.

The procedures shown can be rearranged to make any design parameter the last value to be determined. Often iteration is required in order to determine the best combination of design values. The following design changes are commonly made to meet performance objectives:

- Increase transmitter power.
- Use lower loss transmission line.
- Use lower noise receiver.
- Add diversity paths to the unprotected path.
- Increase antenna spacing for existing space diversity or frequency spacing for existing frequency diversity.
- Increase antenna size.
- Add an amplitude equalizer or, for greater improvement, add an adaptive transversal equalizer.

Table 9.6 also illustrates the use of these design procedures by giving values for a representative radio system. This example is the 6-GHz, 90-Mb/s, 8-PSK radio system whose performance characteristics were shown earlier in Figures 9.27 and 9.44. This radio system employs space diversity with baseband selection combining and IF adaptive amplitude slope and notch equalization.

9.7 SUMMARY

The design of digital radio systems for line-of-sight links involves engineering of the path to provide proper clearance, evaluation of the effects of propagation on performance, development of a frequency allocation plan, and proper selection of radio and link components. This design process must ensure that outage requirements are met on a per link and system basis.

Propagation losses occur primarily due to free-space loss, which is proportional to the square of both path length and frequency. Terrain and climate also have an effect, causing losses to vary from free-space predictions. Reflection or diffraction from obstacles along the path can cause secondary rays that may interfere with the direct ray. The use of Fresnel zones from optics facilitates the design of links by determining the clearances required over obstacles to minimize interference. Varying index of refraction caused by atmospheric structural changes can cause decoupling of the signal from the receiving antenna or fading due to interference from multiple rays (multipath). At frequencies above 10 GHz, atmospheric absorption and rain attenuation become dominant sources of loss, limiting the path length or availability that can be achieved with millimeter-wave frequencies.

Because of its dynamic nature, multipath fading is characterized statistically—usually by its probability of occurrence, average duration, and rate of fading. Work done primarily by Bell Telephone Laboratories has provided empirical models for multipath fading statistics based on analog radio links. Space or frequency diversity is used to protect radio links against multipath fading by providing a separate (diversity) path whose fading is

Table 9.6 Design Procedures for Digital Radio Links

Parameter	*Source*	*Sample Value for 6-GHz, 90-Mb/s, 8-PSK System*
1. Frequency, f	System plan	6 GHz
2. Path length, D	Map or site records	37.3 mi
3. Transmitter line losses		
a. Antenna height above ground	Site records or path profile	190 ft
b. Horizontal transmission line length	Site records	55 ft
c. Total transmission line length	Parameters 3a + 3b	245 ft
d. Transmission line type	System design: rectangular waveguide elliptical waveguide circular waveguide coaxial cable	Circular waveguide
e. Transmission line loss factor	Manufacturer's data (see Table 9.4)	0.01 dB/ft
f. Total transmitter transmission line loss	Parameters 3c × 3e	2.45 dB
4. Receiver line losses		
a. Antenna height above ground (upper antenna if space diversity is used)	Site records or path profile	390 ft
b. Horizontal transmission line length	Site records	55 ft
c. Total transmission line length	Parameters 4a + 4b	445 ft
d. Transmission line type	Same as parameter 3d	Circular waveguide
e. Transmission line loss factor	Manufacturer's data (see Table 9.4)	0.01 dB/ft
f. Total receiver transmission line loss	Parameters 4c × 4e	4.45 dB
5. Free space path loss, L_p	$20 \log f + 20 \log D + 96.6$ (f in GHz, D in miles)	143.6 dB
6. Branching loss, L_b	System design	0.5 dB
7. Miscellaneous losses, L_m	System design	3 dB
8. Total losses, L	Parameters 3f + 4f + 5 + 6 + 7	154 dB
9. Transmitter power, P_t	System design	39 dBm

Table 9.6 continued

Parameter	Source	Sample Value for 6-GHz, 90-Mb / s, 8-PSK System
10. Received signal level threshold, RSL_m	Equipment specification or following procedure	
a. Thermal noise density, N_0	kT_0	−174 dBm/Hz
b. Receiver noise figure, N_f	Equipment specification	8 dB
c. Bit rate, R	System plan	90 Mb/s
d. Required E_b/N_0 for BER $= 10^{-6}$	Equipment specification or theory with appropriate implementation factor	22.5 dB
e. Threshold, RSL_m	$kT_0 + N_f + 10 \log R + E_b/N_0$	−64 dBm
11. System gain, G_s	$P_t - RSL_m$	103 dB
12. Effective fade margin		
a. Terrain factor, a	Path profile	1
b. Climate factor, b	Weather map	1/4
c. Antenna spacing, S (space diversity)	System design	40 ft
d. Frequency spacing, Δf (frequency diversity)	System design	N.A.
e. Combiner hysteresis, H	Equipment specification	3 dB
f. Outage allowed	System design	0.00001
g. Effective fade margin, EFM	See equations (9.73), (9.74), or (9.75)	30.7 dB
13. Flat fade margin required		
a. Dispersive fade margin, DFM	Equipment specification or field measurement	40.3 dB
b. Interference fade margin, IFM	Equipment specification or field measurement	59 dB
c. Flat fade margin, FFM	See Equation (9.81)	31.2 dB
14. Total antenna gain required, $2G_A$	FFM $+ L - G_s$	82.2 dB
15. Required antenna diameter	Table 9.5	10 ft

somewhat uncorrelated from the main path. The recent introduction and testing of wideband digital radio has led researchers to conclude that outages in digital radio are due principally to dispersion caused by frequency-selective fading. Models for multipath channels have been developed for digital radio application, and radio performance is now characterized by a signature that plots the outage area for a given set of multipath channel model parameters. The use of diversity and adaptive equalization has been found to greatly mitigate the effects of dispersion, allowing performance objectives to be met for digital radio.

The frequency allocation plan is based on four elements: the local frequency regulatory authority requirements, selected radio transmitter and receiver characteristics, antenna characteristics, and potential intrasystem and intersystem RF interference.

Key components of a digital radio include the modulator/demodulator (already described in Chapter 6), power amplifier, diversity arrangement, and adaptive equalizer. When choosing a traveling-wave tube as the RF power amplifier, the system designer must carefully examine the effects of saturated operation on the error performance and spectral spreading due to nonlinearities. The type of diversity is a function of the path and system availability requirements. The choice of combiner type can significantly affect the performance expected from diversity systems, depending on the performance monitors used in controlling the combining functions. Adaptive equalizers are generally necessary in digital radio design, although simple amplitude slope equalizers have been found to be quite effective. More elaborate designs involving transversal filters have evolved to meet stringent performance objectives.

Finally, a step-by-step procedure has been presented for calculating values of digital radio link parameters. This procedure is easily adapted to calculate any given parametric value as a function of given values for other link parameters.

REFERENCES

1. J. D. Kraus, *Antennas* (New York: McGraw-Hill, 1950).
2. Henry R. Reed and Carl M. Rusell, *Ultra High Frequency Propagation* (London: Chapman & Hall, 1966).
3. Kenneth Bullington, "Radio Propagation Fundamentals," *Bell System Technical Journal* 36(3)(May 1957):593–626.
4. J. H. Van Vleck, *Radiation Lab. Report 664*, MIT, 1945.
5. D. E. Setzer, "Computed Transmission Through Rain at Microwave and Visible Frequencies," *Bell System Technical Journal* 49(8)(October 1970):1873–1892.
6. H. E. Bussey, "Microwave Attenuation Statistics Estimated from Rainfall and Water Vapor Statistics," *Proc. IRE* 38(July 1950):781–785.
7. CCIR XVth Plenary Assembly, vol. V, *Propagation in Non-Ionized Media* (Geneva: ITU, 1982).
8. *Engineering Considerations for Microwave Communications Systems* (San Carlos, Calif.: GTE Lenkurt Inc., 1981).
9. M. Schwartz, W. R. Bennett, and S. Stein, *Communication Systems and Techniques* (New York: McGraw-Hill, 1966).
10. W. T. Barnett, "Multipath Propagation at 4, 6, and 11 GHz," *Bell System Technical Journal* 51(2)(February 1972):321–361.
11. A. Vigants, "The Number of Fades and Their Durations on Microwave Line-of-Sight Links With and Without Space Diversity," *1969 International Conference on Communications*, pp. 28-25 to 28-31.

12. A. Vigants, "Number and Duration of Fades at 6 and 4 GHz," *Bell System Technical Journal* 50(3)(March 1971):815–841.

13. D. J. Kennedy, "Digital Error Statistics for a Fading Channel," *1973 International Conference on Communications*, pp. 18-17 to 18-23.

14. A. Vigants, "Space-Diversity Performance as a Function of Antenna Separation," *IEEE Trans. on Comm. Tech.*, vol. COM-16, no. 6, December 1968, pp. 831–836.

15. A. Vigants and M. V. Pursley, "Transmission Unavailability of Frequency Diversity Protected Microwave FM Radio Systems Caused by Multipath Fading," *Bell System Technical Journal* 58(8)(October 1979):1779–1796.

16. S. H. Lin, "Statistical Behavior of a Fading Signal," *Bell System Technical Journal* 50(10)(December 1971):3211–3270.

17. L. C. Greenstein and B. A. Czekaj, "A Polynomial Model for Multipath Fading Channel Responses," *Bell System Technical Journal* 59(7)(September 1980):1197–1205.

18. W. D. Rummler, "A New Selective Fading Model: Application to Propagation Data," *Bell System Technical Journal* 58(5)(May–June 1979):1037–1071.

19. W.C. Jakes, Jr., "An Approximate Method to Estimate an Upper Bound on the Effect of Multipath Delay Distortion on Digital Transmission," *1978 International Conference on Communications*, pp. 47.1.1 to 47.1.5.

20. L. Martin, "Phase Distortions of Multipath Transfer Functions," *1984 International Conference on Communications*, pp. 1437–1441.

21. M. Emshwiller, "Characterization of the Performance of PSK Digital Radio Transmission in the Presence of Multipath Fading," *1978 International Conference on Communications*, pp. 47.3.1 to 47.3.6.

22. C. W. Lundgren and W. D. Rummler, "Digital Radio Outage Due to Selective Fading—Observation vs. Prediction from Laboratory Simulation," *Bell System Technical Journal* 58(5)(May–June 1979):1073–1100.

23. A. Vigants, "Distance Variation of Two-Tone Amplitude Dispersion in Line-of-Sight Microwave Propagation," *1981 International Conference on Communications*, pp. 68.3.1 to 68.3.5.

24. Y. Serizawa and S. Takeshita, "A Simplified Method for Prediction of Multipath Fading Outage of Digital Radio," *IEEE Trans. on Comm.*, vol. COM-31, no. 8, August 1983, pp. 1017–1021.

25. W. T. Barnett, "Multipath Fading Effects on Digital Radio," *IEEE Trans. on Comm.*, vol. COM-27, no. 12, December 1979, pp. 1842–1848.

26. C. W. Anderson, S. G. Barber, and R. N. Patel, "The Effect of Selective Fading on Digital Radio," *IEEE Trans. on Comm.*, vol. COM-27, no. 12, December 1979, pp. 1870–1876.

27. T. S. Giuffrida, "Measurements of the Effects of Propagation on Digital Radio Systems Equipped with Space Diversity and Adaptive Equalization," *1979 International Conference on Communications*, pp. 48.1.1 to 48.1.6.

28. A. J. Giger and W. T. Barnett, "Effects of Multipath Propagation on Digital Radio," *IEEE Trans. on Comm.*, vol. COM-29, no. 9, September 1981, pp. 1345–1352.

29. D. R. Smith and J. J. Cormack, "Improvement in Digital Radio Due to Space Diversity and Adaptive Equalization," 1984 Global Telecommunications Conference, pp. 45.6.1 to 45.6.6.

30. FCC Docket 19311, FCC 74-985, adopted 19 September 1974, released 27 September 1974, revised 29 January 1975.
31. MIL-STD-188-322, "Subsystem Design/Engineering and Equipment Technical Design Standards for Long Haul Line-of-Sight (LOS) Digital Microwave Radio Transmission," U.S. Department of Defense, 1 November 1976.
32. CCIR XVth Plenary Assembly, vol. IX, pt. 1, *Fixed Service Using Radio-Relay Systems* (Geneva: ITU, 1982).
33. Bell System Technical Reference, "6 GHz Digital Radio Requirements and Objectives," Pub. 43501, AT & T, December 1980.
34. V. K. Prabhu, "Error Rate Considerations for Coherent Phase-Shift Keyed Systems with Co-Channel Interference," *Bell System Technical Journal* 48(3)(March 1969):743–768.
35. C. M. Thomas, J. E. Alexander, and E. W. Rahneberg, "A New Generation of Digital Microwave Radios for U.S. Military Telephone Networks," *IEEE Trans. on Comm.*, vol. COM-27, no. 12, December 1979, pp. 1916–1928.
36. R. G. Lyons, "The Effect of a Bandpass Nonlinearity on Signal Detectability," *IEEE Trans. on Comm.*, vol. COM-21, no. 1, January 1973, pp. 51–60.
37. D. Chakraborty and L. S. Golding, "Wide-Band Digital Transmission Over Analog Radio Relay Links," *IEEE Trans. on Comm.*, vol. COM-23, no. 11, November 1975, pp. 1215–1228.
38. J. E. Hamant, O. P. Connell, and H. S. Walczyk, "Digital Transmission Evaluation Project, DR8A Test Final Report." Report Number CCC-CED-77-DTEP-012, U.S. Army Communications-Electronics Engineering Installation Agency, Ft. Huachuca, Arizona.
39. I. Horikawa, Y. Okamoto, and K. Morita, "Characteristics of a High Capacity 16 QAM Digital Radio System on a Multipath Fading Channel," *1979 International Conference on Communications*, pp. 48.4.1 to 48.4.6.
40. G. deWitte, "DRS-8: System Design of a Long Haul 91 Mb/s Digital Radio," *1978 National Telecommunications Conference*, pp. 38.1.1 to 38.1.6.
41. A. Vigants, "Space-Diversity Engineering," *Bell System Technical Journal* 54(1)(January 1975):103–142.
42. P. R. Hartmann and E. W. Allen, "An Adaptive Equalizer for Correction of Multipath Distortion in a 90 Mb/s 8 PSK System," *1979 International Conference on Communications*, pp. 5.6.1 to 5.6.4.
43. P. R. Hartmann and E. W. Allen, "Transmission Engineering Considerations," Collins Transmission Engineering Symposium, Dallas, November 1982, pp. 3-1 to 3-48.
44. T. P. Murphy and others, "Practical Techniques for Improving Signal Robustness," *1981 National Telecommunications Conference*, pp. C3.3.1 to C3.3.7.
45. E. R. Johnson, "An Adaptive IF Equalizer for Digital Transmission," *1981 International Conference on Communications*, pp. 13.6.1 to 13.6.4.
46. C. A. Siller, Jr., "Multipath Propagation," *IEEE Comm. Mag.*, February 1984, pp. 6–15.
47. G. L. Fenderson, S. R. Shepard, and M. A. Skinner, "Adaptive Transversal Equalizer for 90 Mb/s 16-QAM Systems in the Presence of Multipath Propagation," *1983 International Conference on Communications*, pp. C8.7.1 to C8.7.6.

48. R. L. Freeman, *Telecommunication Transmission Handbook* (New York: Wiley, 1975).

49. P. Dupuis, M. Joindet, A. Leclert, and M. Rooryck, "Fade Margin of High Capacity Digital Radio System," *1979 International Conference on Communications*, pp. 48.6.1 to 48.6.5.

50. W. D. Rummler, "A Comparison of Calculated and Observed Performance of Digital Radio in the Presence of Interference," *IEEE Trans. on Comm.*, vol. COM-30, no. 7, July 1982, pp. 1693–1700.

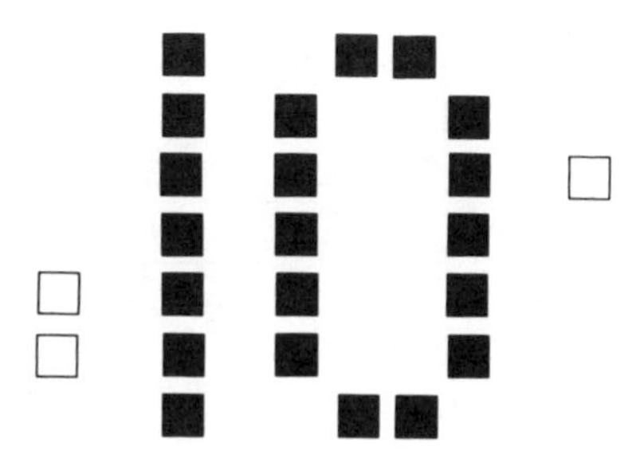

Network Timing and Synchronization

OBJECTIVES

- Develops the concept of time as the basic unit for timing and synchronization
- Discusses the types of clocks commonly found in timing systems
- Describes pertinent clock parameters with examples
- Considers candidate systems for accurate network timing and synchronization
- Examines time and frequency dissemination systems used by communication networks
- Cites examples of network synchronization schemes

10.1 INTRODUCTION

The evolution of digital transmission toward end-to-end digitally switched voice and data channels will lead to new timing and synchronization requirements. These timing requirements did not exist in analog transmission networks, and they differ greatly from most present-day digital transmission networks. In the past, store-and-forward data networks have used magnetic tape or punched paper tapes with large storage capabilities to eliminate the need for accurate network timing. Today's emerging data networks, however, rely on some form of network synchronization to provide greater efficiency and performance. Initial implementation of digital voice transmission using PCM employed point-to-point application of PCM channel banks in which all voice channels are converted to analog form for

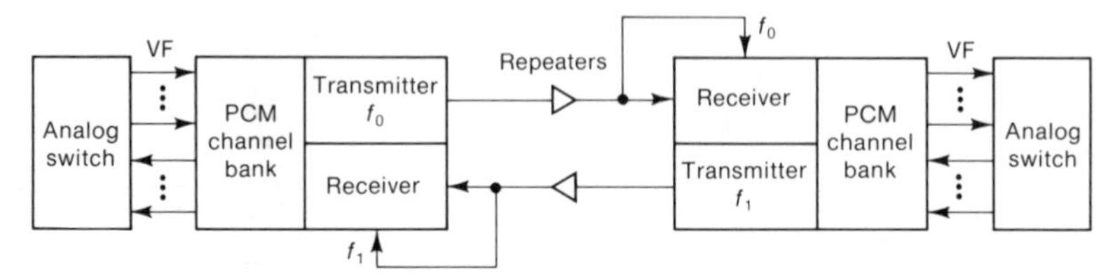

(a) PCM transmission and analog switching

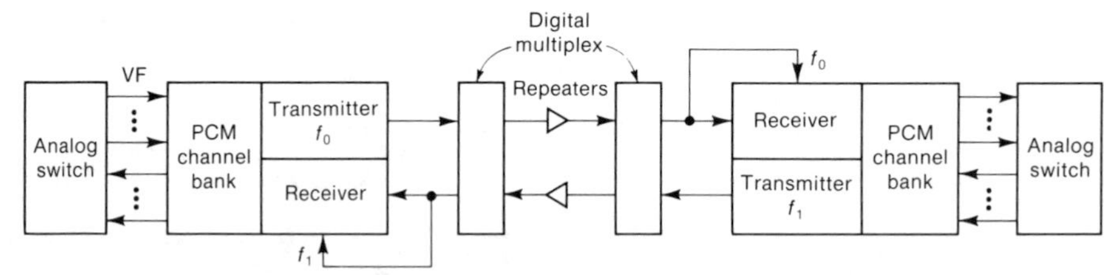

(b) Digital multiplexing and analog switching

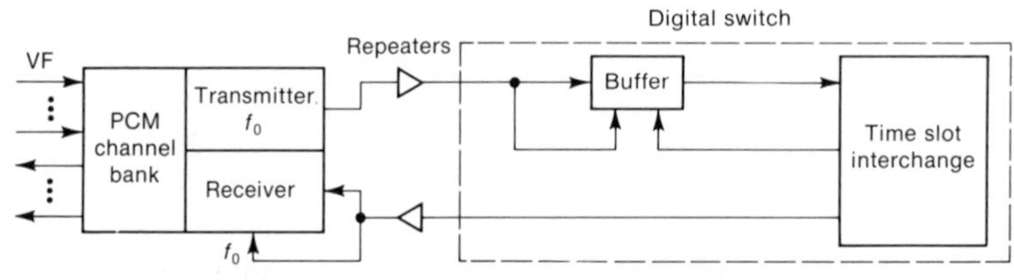

(c) Digital transmission and digital switching

Figure 10.1 Evolution of PCM Network Synchronization

switching, thus avoiding the need for accurate network synchronization. In this configuration, shown in Figure 10.1*a*, timing is supplied by clock circuits within each PCM channel bank, and received timing is slaved to the distant transmitter by use of clock recovery circuits. In the second phase of PCM applications, higher-level multiplexers are used to combine several lower-level PCM bit streams, as shown in Figure 10.1*b*. In this case pulse stuffing is conventionally used to convert each incoming PCM bit stream to a rate synchronous with the higher-level multiplexer. With the third phase of PCM applications, analog switches are replaced with digital switches that require each incoming PCM bit stream to be synchronized in frequency and in frame. Buffering is provided to compensate for small differences in clock frequencies throughout the network and to facilitate frame alignment, as indicated in Figure 10.1*c*.

Digital transmission inherently requires that each signal, no matter where it originates in the network, must be available at a precise time for multiplexing or switching functions. Because of variations in transmission times and timing inaccuracies and instabilities in equipment, there is some variation in the arrival time of individual bits. The velocity of propagation in coaxial cable varies with temperature. Terrestrial transmission paths of over-the-horizon tropospheric scatter, ionospheric scatter, or line-of-sight radio links vary with meteorological conditions. Path lengths through synchronous satellites vary because of cyclic position changes of the satellite relative to earth stations. Clocks used to time bits within transmission equipment vary in accuracy and stability from equipment to equipment. Therefore the number of bits in transit between two given nodes varies with time.

This variation in transmission paths and clock parameters is accommodated by storage buffers in which incoming bits are held until needed by the time-division multiplexer or switch. Some degree of timing precision is required for clocks in order to assure that buffers of finite size infrequently (or never) empty or overflow, thereby causing a loss of bit count integrity. Such losses of BCI that do not result in loss of alignment are known as **timing slips**. Acceptable slip rates depend on the services provided in the network, but in any case they should be accounted for in the development of BCI performance objectives as described in Chapter 2. The CCITT recommends that the end-to-end mean slip rate of an international digital connection should be less than one slip in 5 hr [1], an objective that has been adopted by AT&T [2] and is under consideration by other carriers [3].

10.2 TIME STANDARDS

Time is one of four independent standards or base units of measurement (along with length, mass, and temperature). The real quantity involved in this standard is the time *interval*. Today time is based on the definition of a second, given as "the duration of 9, 192, 631, 770 periods of the radiation corresponding to the transition between the two hyperfine levels of the ground state of the cesium atom 133" [4].

Time scales have been developed to keep track of dates or number of time intervals (seconds). Two basic measurement approaches are astronomical time and atomic time [5]. Methods for reconciling these two times also exist. A nationally or internationally agreed time scale then enables precise time and frequency dissemination to support a number of services including synchronization of communication networks.

10.2.1 Mean Solar Time

For many centuries, mean solar time was based on the rotation of the earth about its axis with respect to the sun. Because of irregularities in the earth's speed of rotation, however, the length of a solar day varies by as much as 16 min through the course of a year. Early astronomers understood the laws of motion and were able to correct the apparent solar time to obtain a more uniform *mean solar day*, computed as the average length of all the apparent days in the year. Mean solar time corrects for two effects:

- *Elliptical orbit*: The earth's orbit around the sun is elliptical. When nearer to the sun, the earth travels faster in orbit than when it is farther from the sun.
- *Tilted axis*: The axis of the earth's rotation is tilted at an angle of about $23\frac{1}{2}°$ with respect to the plane that contains the earth's orbit around the sun.

10.2.2 Universal Time

Universal Time (UT), like mean solar time, is based on the rotation of the earth about its axis. The time scale UT0 designates universal time derived from mean solar time. As better clocks were developed, astronomers began to notice additional irregularities in the earth's rotation, leading to subsequent universal time scales, UT1 and UT2. The UT1 time scale corrects UT0 for wobble in the earth's axis. The amount of wobble is about 15 m; if left uncorrected, it would produce a discrepancy as large as 30 ms from year to year. The UT2 time scale is UT1 with an additional correction for seasonal variations in the rotation of the earth. These variations are apparently due to seasonal shifts in the earth's moment of inertia—shifts that occur, for example, with changes in the polar ice caps as the sun moves from the southern to the northern hemisphere and back again in the course of a year.

10.2.3 Coordinated Universal Time

With the advent of atomic clocks [6, 7], improved timing accuracies led to the development of Coordinated Universal Time (UTC). The problem in using atomic clocks is that even with the refinements and corrections that have been made in UT (earth time), UT and atomic time will get out of step because of the irregular motion of the earth. Prior to 1972, UTC was maintained by making periodic adjustments to the atomic clock frequency that allowed the clock rate to be nearly coincident with UT2. Because a correction factor to UTC was necessary, this method of timekeeping was a problem for navigators who needed solar time. Since 1972, UTC has been generated not by offsetting the frequency of the atomic clock but by adding or subtracting "leap seconds" to bring atomic time into coincidence with UT1. Since the rotation of the earth is not uniform we cannot exactly predict when leap seconds will be added or subtracted, but to date these adjustments, if required, have been made at the last day, last second, of 31 December or 30 June. Positive leap seconds have been made at an average of about one per year since 1972, indicating that the earth is slowing down by approximately 1 s per year. To the user of time or frequency, UTC guarantees accuracy to within 1 s of UT1, sufficient for most users' requirements. Perhaps as important, since the frequency of the atomic clocks used for UTC is no longer periodically adjusted, UTC can also provide users a source of precise frequency limited only by intrinsic characteristics of the atomic clocks.

Numerous national and international organizations are responsible for maintaining standards of time, time interval, and frequency [8]. In the United States there are two organizations primarily responsible for providing time and frequency information: the National Bureau of Standards (NBS) and the U.S. Naval Observatory (USNO) [9]. The National Bureau of Standards develops and maintains the atomic frequency and time interval standards for the United States. The NBS also disseminates time and frequency via radio broadcasts. The USNO makes astronomical observations for determining UT1 and keeps accurate clocks running for use by the U.S. Department of Defense. The USNO also controls distribution of precise time and time interval (PTTI) from navy radio stations, satellites, and radio navigation (LORAN C) systems operated by the U.S. Coast Guard.

Time and frequency information from national organizations is collected by the Bureau International de l'Heure (BIH) in Paris, which is the agency responsible for international coordination. At the BIH this information is evaluated and used to determine corrections for each contributing clock. By international agreement, all UTC time scales must agree with the UTC time scale operated by the BIH to within ± 1 ms [10]. Relationships among the NBS, USNO, and BIH are illustrated in Figure 10.2.

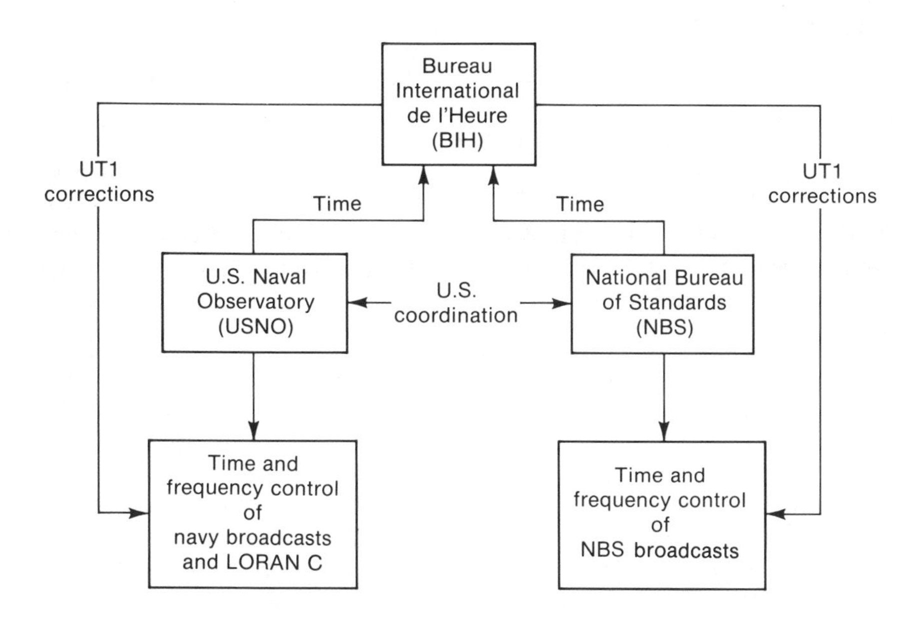

Figure 10.2 Relationships Among NBS, USNO, and BIH

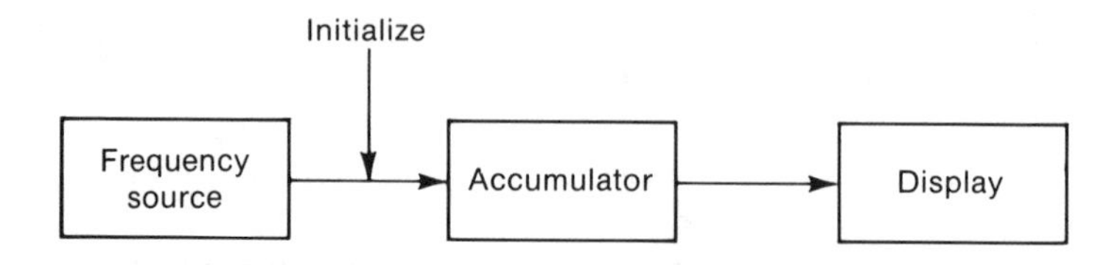

Figure 10.3 Basic Clock

10.3 FREQUENCY SOURCES AND CLOCKS

The term **frequency** can be defined as the number of events per unit of time. One definition of the unit of time is thus a specific number of periods of a well-defined event generator. For example, the second is today defined on the basis of a number of periods emitted by a certain resonant frequency of the element cesium. Thus a **clock** can be defined as a device that counts the number of seconds occurring from an arbitrary starting time. From this definition it appears that a clock needs three basic parts as shown in Figure 10.3: a source of events to be counted, a means of accumulating these events, and a means of displaying this accumulation of time. In a frequency source there are two major sources of error: accuracy and stability. In a clock there are these two major sources of error plus the accuracy of the initial time setting [11].

10.3.1 Definitions

Certain terms are of primary interest in network timing and synchronization.

Accuracy defines how close a frequency agrees with its designated value. A 1-MHz frequency source that has accuracy of 1 part in 10^6 can deviate ± 1 Hz from 1,000,000 Hz. The accuracy of a source is often specified with physical conditions such as temperature. **Fractional frequency difference** is the relative frequency departure of a frequency source, f_0, from its desired (nominal) value, f_D, defined as

$$\frac{\Delta f}{f} = \frac{f_0 - f_D}{f_D} = \frac{f_0}{f_D} - 1 \qquad (10.1)$$

and referred to variously as fractional, relative, or normalized frequency difference. The frequency difference Δf has units in hertz, so that $\Delta f/f$ is dimensionless. The abbreviated form $\Delta f/f$ is most commonly used in the literature and has been adopted here.

Settability is the degree to which frequency can be adjusted to correspond with a reference frequency.

Reproducibility is the degree to which independent devices of the same design can be adjusted to produce the same frequency or, alternatively, the degree to which a device produces the same frequency from one occasion to another.

Stability [12] specifies the rate at which a device changes from nominal frequency over a certain period of time. Stability is generally specified over several measurement periods, which are loosely divided into long-term and short-term stability. Usually the stability of a frequency source improves with longer sampling time, but there are exceptions to this rule. Random noise produces a jitter in the output frequency and is the dominant source of instability for short-term measurements. **Systematic drift** is the dominant source of long-term instability. Systematic drift results from slowly changing physical characteristics, which cause changes in frequency that over a long period of time tend to be in one direction. Quartz crystal oscillators, for example, exhibit a linear frequency drift, the slope of which reveals the quartz crystal's aging rate. A typical frequency stability curve is shown in Figure 10.4. Theoretically the units of drift are in hertz per second. In practice, however, **relative drift** is more commonly used, given in units as $\Delta f/f$ per unit time. Since $\Delta f/f$ is dimensionless, the unit of relative drift is time^{-1}.

Time base error, or **time interval error (TIE)**, is the error in time that accumulates when a frequency source is used as a clock. To calculate the time error, let the frequency at any time t be

$$f_t = f_0 + \alpha t \tag{10.2}$$

as illustrated in Figure 10.5 and where f_0 is the initial frequency at time $t = 0$ and where α is the drift rate in hertz per second. It should be noted that a general representation of f_t would include higher-order terms in time.

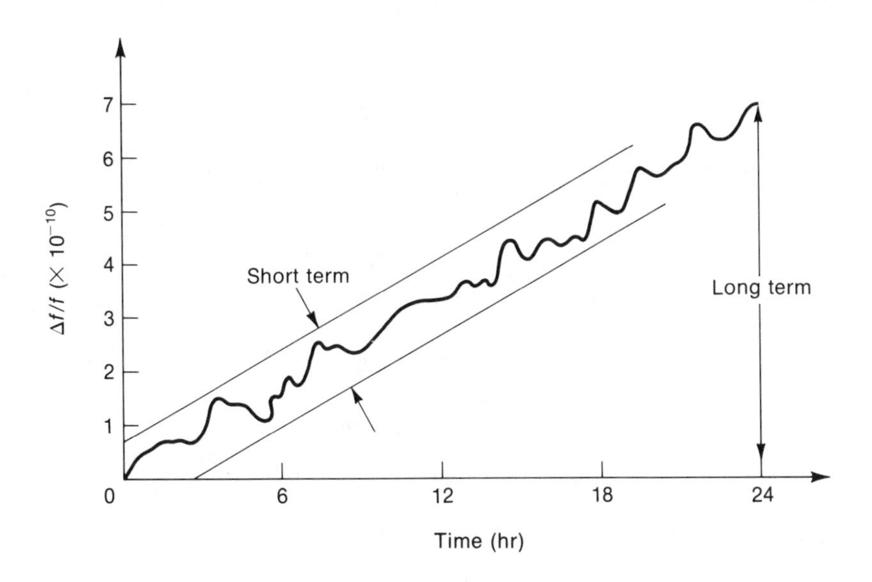

Figure 10.4 Typical Frequency Stability Curve

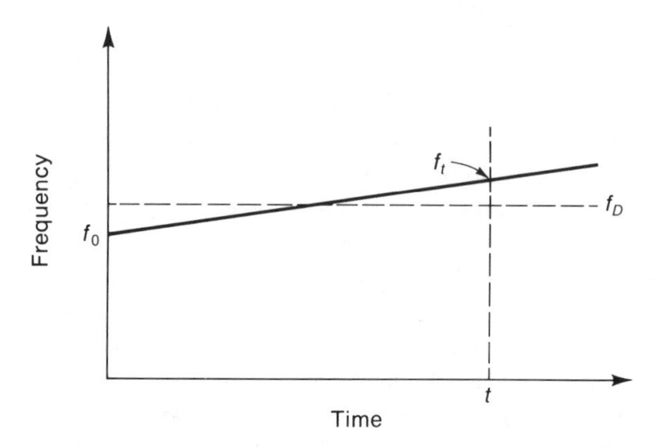

Figure 10.5 Oscillator Frequency vs. Time

When one is considering precision oscillators, however, the coefficients for these higher-order terms can be considered zero [13]. The phase accumulation in true time T is then

$$\phi(T) = \int_0^T f_t \, dt = f_0 T + \tfrac{1}{2}\alpha T^2 + \phi_0 \tag{10.3}$$

where true time is defined as that time determined by the international definition of the second. We assume the phase error at time $t = 0$ to be ϕ_0. The apparent clock time after true time T is

$$C(T) = \frac{\phi(T)}{f_D} \tag{10.4}$$

where f_D is the desired frequency. Thus the time difference between apparent clock time and true time is

$$\begin{aligned} \Delta_T &= C(T) - T = \frac{\phi(T)}{f_D} - T \\ &= \left(\frac{f_0}{f_D} T + \frac{1}{2} a T^2 + \epsilon_0 \right) - T \end{aligned} \tag{10.5}$$

where $a = \alpha / f_D$ = relative drift rate

$\epsilon_0 = \phi_0 / f_D$ = initial time error (setting error)

Simplifying (10.5), we get

$$\Delta_T = \epsilon_0 + \left(\frac{f_0}{f_D} - 1 \right) T + \frac{1}{2} a T^2 \tag{10.6}$$

or from (10.1)

$$\Delta_T = \epsilon_0 + \frac{\Delta f}{f_D} T + \frac{1}{2} a T^2 \tag{10.7}$$

This relationship of frequency error (due to setting error and drift) and time error is illustrated in Figure 10.6.

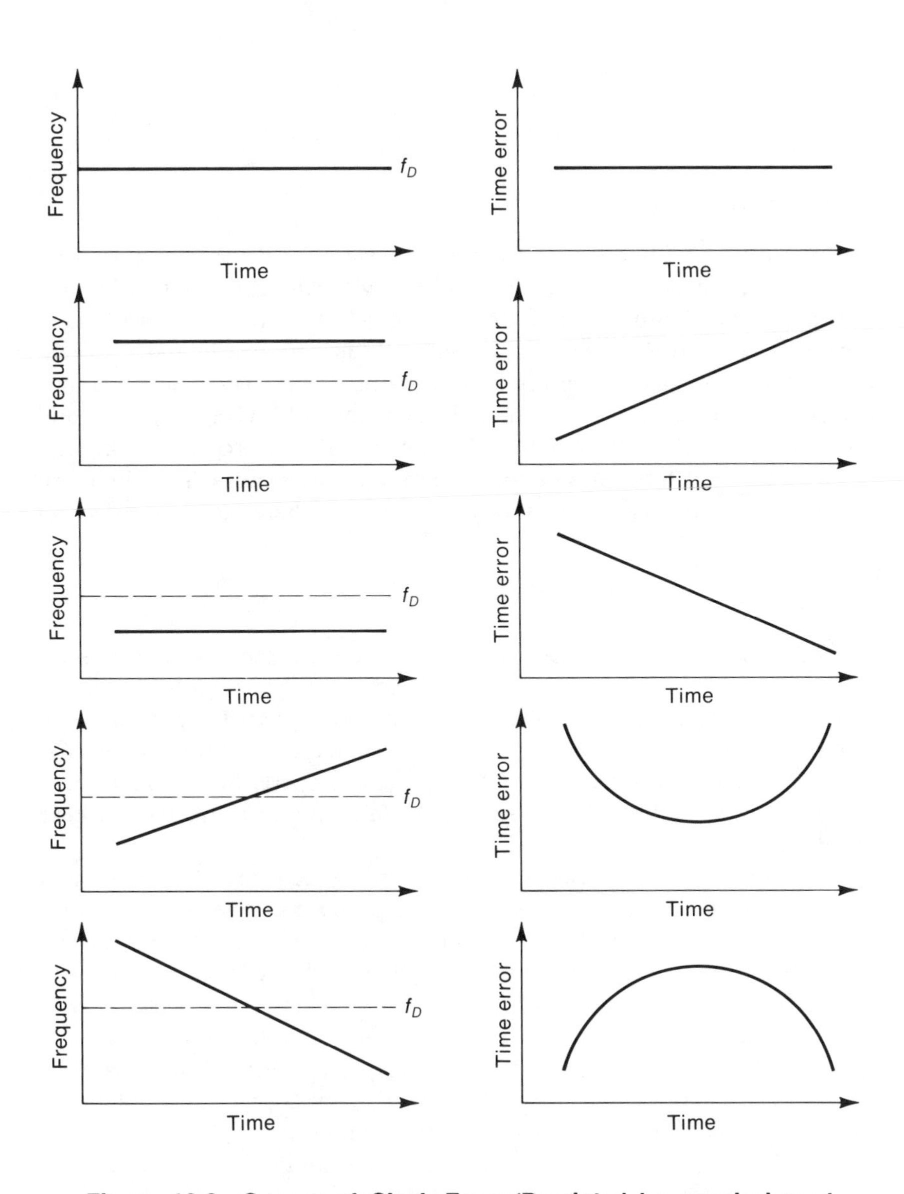

Figure 10.6 Causes of Clock Error (Reprinted by permission of Austron, Inc.)

Example 10.1

A 1-MHz crystal oscillator has a setting accuracy of 1×10^{-8} and a relative drift rate of 1×10^{-8} per day. Assuming the initial time error to be zero, the time base error in one 24-hr period is

$$\begin{aligned} \Delta_T &= (1 \times 10^{-8})(86{,}400 \text{ s}) + \frac{1}{2}\,\frac{(1 \times 10^{-8})(86{,}400)^2}{86{,}400} \\ &= (864 + 432) \times 10^{-6} \text{ s} \\ &= 1296\ \mu\text{s} \end{aligned}$$

10.3.2 Frequency Sources

The many kinds of frequency-determining devices can be grouped into three classes: mechanical, electronic, and atomic. Mechanical resonators like the pendulum and the tuning fork, and electronic resonators such as LC tank circuits and microwave cavities, are of little importance in today's high-performance frequency sources and will not be discussed here. Atomic frequency sources used in communication network synchronization and quartz crystal oscillators used as backup in network synchronization schemes or as an internal time base in communication equipment will be discussed extensively. Characteristics of the clocks described here are summarized in Table 10.1.

Quartz

Crystalline quartz has great mechanical and chemical stability, a most useful characteristic in a frequency source [14, 15]. The quartz crystal clock is actually a mechanical clock, since such crystals vibrate when excited by an electric potential. Conversely, if the crystal is made to vibrate, an electric potential is induced in nearby conductors. This property in which mechanical and electrical effects are linked in a crystal is known as the **piezoelectric effect** [16].

In building a quartz oscillator, the quartz crystal is first cut and ground to create the desired resonant frequency. To produce the piezoelectric effect, metal electrodes are attached to the crystal surfaces. Using the crystal, an oscillator is produced by adding an amplifier with feedback as shown in Figure 10.7. Its frequency in turn is determined by the physical dimensions of the crystal together with the properties of the crystalline quartz used.

Drift, or aging, and dependence on temperature are common traits of all crystal oscillators. When switched on, the crystal oscillator exhibits a "warm-up time" due to temperature stabilization. After this initial aging period of a few days to a month, the aging rate can be considered constant and usually remains in one direction. Thus periodic frequency checks and corrections are needed to maintain a quartz crystal frequency standard. The

Table 10.1 Comparison of Frequency Sources (Adapted from [5])

	Frequency Source				
Characteristic	*Quartz*	*Quartz (Temperature Compensated)*	*Quartz (Single Oven)*	*Cesium*	*Rubidium*
Basic resonator frequency		10 kHz to 100 MHz		9, 192, 631, 770 Hz	6, 834, 682, 613 Hz
Output frequencies provided		10 kHz to 100 MHz		1, 5, 10 MHz typical	1, 5, 10 MHz typical
Relative frequency drift, short term, 1 s	1×10^{-9} typical	1×10^{-9} typical	1×10^{-9} to 1×10^{-10}	5×10^{-11} to 5×10^{-13}	2×10^{-11} to 5×10^{-12}
Relative frequency drift, long term, 1 day	1×10^{-7} typical	1×10^{-8} typical	1×10^{-7} to 1×10^{-9}	1×10^{-13} to 1×10^{-14}	5×10^{-12} to 5×10^{-13}
Relative frequency drift, longer term	5×10^{-6} per year	1×10^{-8} to 5×10^{-7} per year	1×10^{-9} to 5×10^{-11} per year	$< 5 \times 10^{-13}$ per year	1×10^{-11} per month
Principal environmental effects		Motion, temperature, crystal drive level		Magnetic field, accelerations, temperature change	Magnetic field, temperature change, atmospheric pressure
Principal causes of long-term instability		Aging of crystal, aging of electronic components, environmental effects		Component aging	Light source aging, filter and gas cell aging, environmental effects

temperature dependence is caused by a slight change in the elastic properties of the crystal. If large temperature fluctuations are to be tolerated or if high-frequency precision is required, the crystals can be enclosed in electronically controlled ovens that maintain constant temperatures (Figure 10.7). An alternative solution to the temperature problem is the so-called temperature-compensated crystal oscillator (TCXO). A temperature sensor and varactor are added to cancel any changes in the resonant frequency due to temperature. The overall performance of the quartz oscillator is also affected by the line voltage, gravity, shock, vibration, and electromagnetic interference.

Performance of a crystal oscillator is determined primarily by the aging rate and temperature dependence. The most stable crystal oscillators utilize ovens and exhibit an aging characteristic as good as 10^{-11} per day. Crystal oscillators manufactured for use at room temperature exhibit aging rates of typically 5×10^{-7} per month. The good short-term stability characteristic and frequency control capability make crystal oscillators ideal for use as slave clocks with, for example, atomic clocks. Table 10.1 compares the various crystal oscillators and their typical specifications.

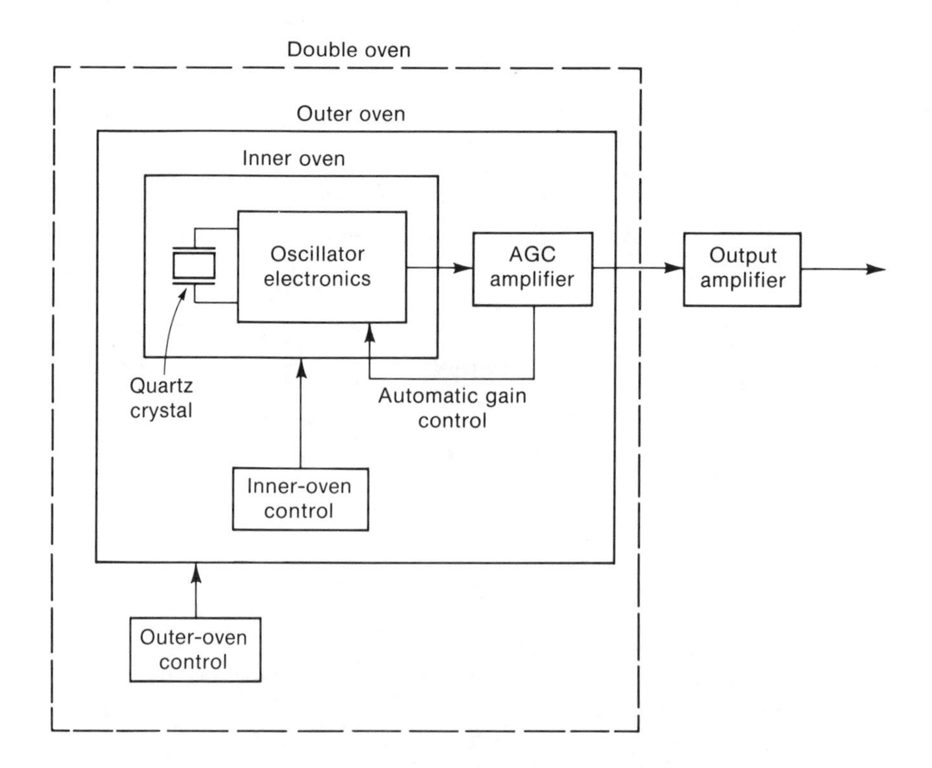

Figure 10.7 Typical Quartz Crystal Oscillator (Reprinted by permission of the Efratom Division of the Ball Corp. [17])

Cesium (Cs)

Use of atomic clocks such as the cesium standard is based on the quantum-mechanical phenomenon that atoms release energy in the form of radiation at a specific resonant frequency. The atom is a natural resonator whose natural frequency is immune to the temperature and frictional effects that beset mechanical clocks. In an atomic clock the oscillator is always a quartz oscillator, and the high − Q resonator is based on some natural frequency of an atomic source. The frequency-determining element is the atomic resonator, which consists of a cavity containing the atomic material and some means of detecting changes in the state of the atoms. The output of a voltage-controlled crystal oscillator (VCXO) is used to coherently generate a microwave frequency that is very close to resonance of the atomic resonator. A control signal, related to the number of atoms that changed their state due to the action of the microwave signal, is fed back to the crystal oscillator. Atomic resonances are at gigahertz frequencies whereas crystal oscillator frequencies are at megahertz frequencies, so that multiplication is required by a factor of about 1000. When the atomic standard is in lock, the VCXO output is exactly at its resonant frequency (typically 5 MHz) to within the accuracy tolerance provided by the atomic resonator [18].

For a cesium oscillator, atomic resonance is at 9,192,631,770 Hz. The so-called cesium-beam frequency standard is shown in Figure 10.8. The cesium metal is housed in an oven. When heated to about 100°C, cesium gas

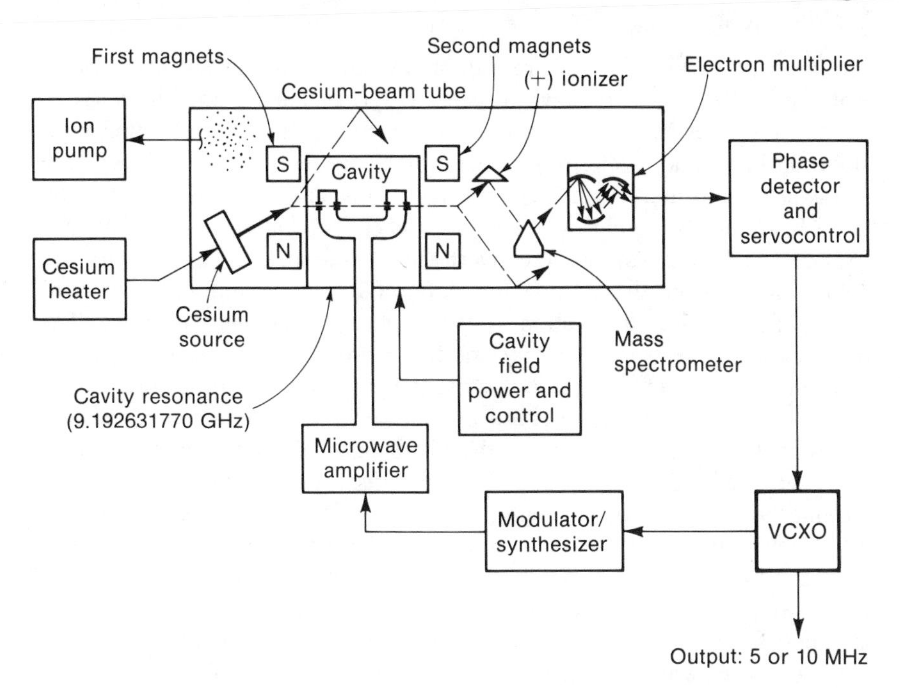

Figure 10.8 Cesium-Beam Frequency Standard (Reprinted by permission of the Efratom Division of the Ball Corp. [17])

forms an atomic beam that is channeled into the vacuum chamber. The magnetic field at the two ends of the tube separates the atoms according to their energy state. When passing through the microwave signal, which is set very near to the cesium resonant frequency, cesium atoms change energy state and are detected at the far end of the tube. The detector produces a control voltage that is related to the number of atoms reaching it, and this signal is fed back to control the microwave frequency through the crystal oscillator in order to maximize the number of atoms reaching the detector—which means that the radio signal is at the cesium atom's natural frequency.

Cesium frequency standards are used extensively where high reproducibility and long-term stability are needed. Cesium standards exhibit no systematic long-term drift. Frequency stability of parts in 10^{14} is possible at sampling times of less than 1 hr to days. The very short term stability of cesium standards is governed by the stability of the quartz oscillators within them. The time at which the longer-term stability properties of the atomic standard become dominant depends on the time constant of the quartz oscillator control loop, which is typically in the range 1 to 60 s. Used as a clock, cesium oscillators provide accuracy to a few microseconds per year. Cesium standards are the predominant source in today's precise time and frequency dissemination systems.

Rubidium (Rb)

The atomic resonator for the rubidium standard is a gas cell that houses rubidium (Rb^{87}) gas at low pressure [19]. As shown in Figure 10.9, the rubidium gas interacts with both a light source that is generated by a rubidium lamp and a microwave signal that is generated by a VCXO. One energy state is excited by the beam of light, while another state is excited by the RF signal. The microwave signal, when at the resonant frequency 6,834,682,613 Hz of rubidium, converts atoms into an energy state that will absorb energy from the light source. A photodetector monitors the amount of light absorbed as a function of the applied microwave frequency. The microwave signal is derived by multiplication of the quartz oscillator frequency. A servoloop controls the oscillator frequency so that the oscillator is locked to the atomic resonance.

Rubidium oscillators vary in their resonance frequency by as much as 10^{-9} because of differences in gas composition, temperature, and pressure and in the intensity of the light. Therefore rubidium oscillators require initial calibration and also recalibration because they exhibit a frequency drift or aging like crystal oscillators. This aging is due to such factors as drift in the light source and absorption of rubidium in the walls of the gas chamber. Nevertheless, stability performance of rubidium oscillators is quite good. As with cesium standards, the very short term stability of rubidium standards is governed by the stability of the slaved quartz oscillator. At 1-s sampling times, they display a stability of better than 10^{-11} and nearly 10^{-13} for sampling times of up to a day. For longer averaging times, the

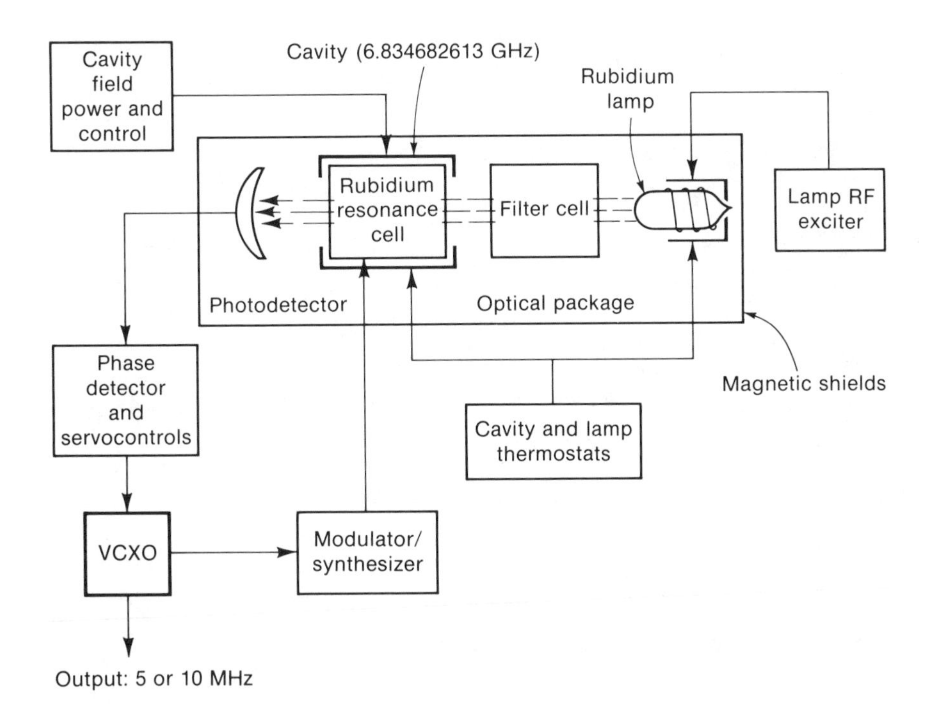

Figure 10.9 Rubidium Frequency Standard (Reprinted by permission of the Efratom Division of the Ball Corp. [17])

frequency stability is affected by long-term drift, which is typically 1 part in 10^{11} per month. Rubidium oscillators are used wherever excellent medium-term stability—minutes to a day—is required and where reduced cost and size, as compared with cesium oscillators, are important.

10.4 NETWORK SYNCHRONIZATION TECHNIQUES

Candidate systems for providing digital transmission network synchronization fall into two major categories: nondisciplined and disciplined techniques. All nondisciplined clocks are asynchronous since all clocks run independent of control from a reference. The slip rate is a function of accuracy between any two clocks and buffer size. For disciplined techniques, each nodal clock is synchronized to a reference signal. One of the principal differences between the disciplined techniques is the choice of reference. This difference has a direct impact on the ability of the technique to accommodate transmission delay variations and to adapt to a loss of reference signal. Other differences in disciplined techniques are the capabilities of providing precise time and clock error correction. Conceptual illustrations of both nondisciplined and disciplined network synchronization techniques are shown in Figures 10.10 and 10.11.

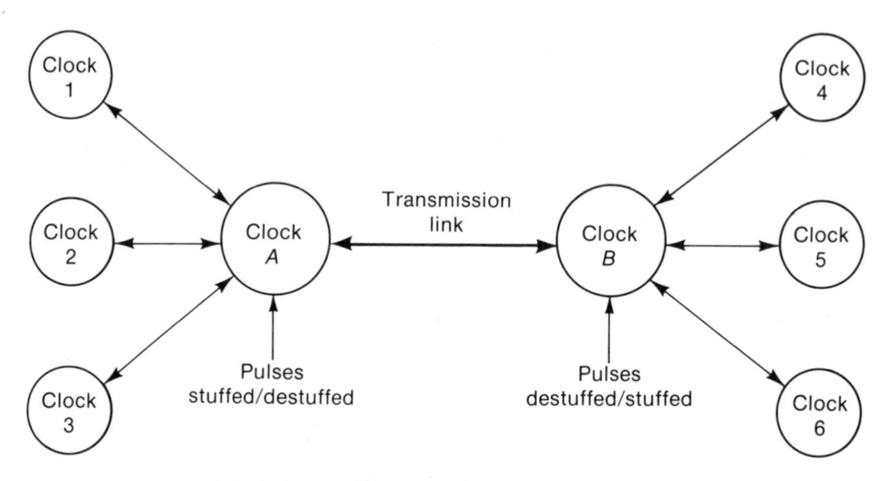

(a) Pulse stuffing synchronization concept

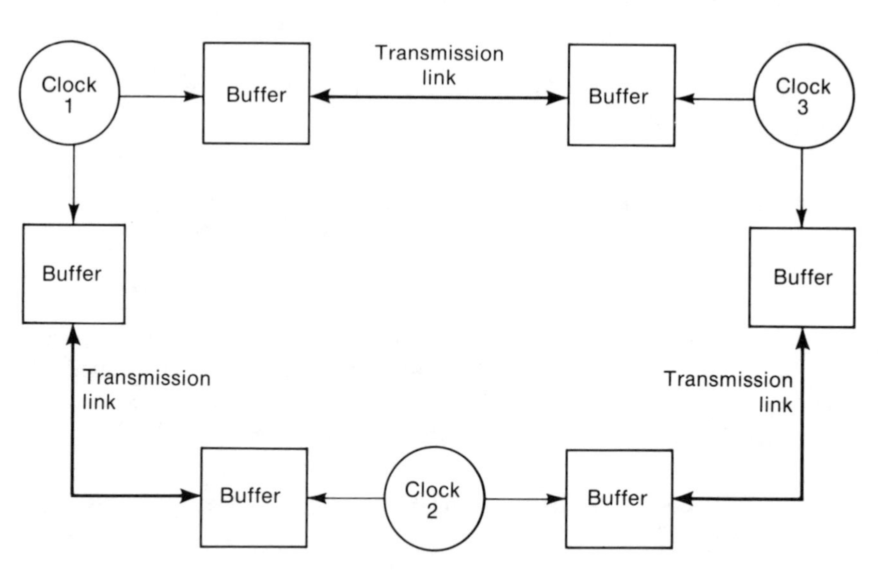

(b) Independent clock synchronization concept

Figure 10.10 Nondisciplined Network Synchronization Concepts

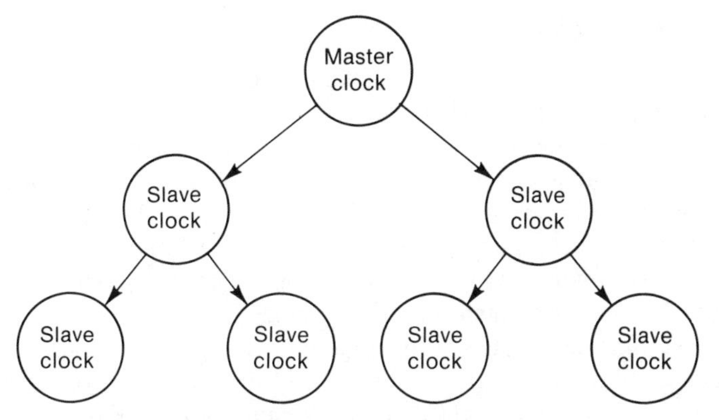

(a) Master-slave synchronization concept

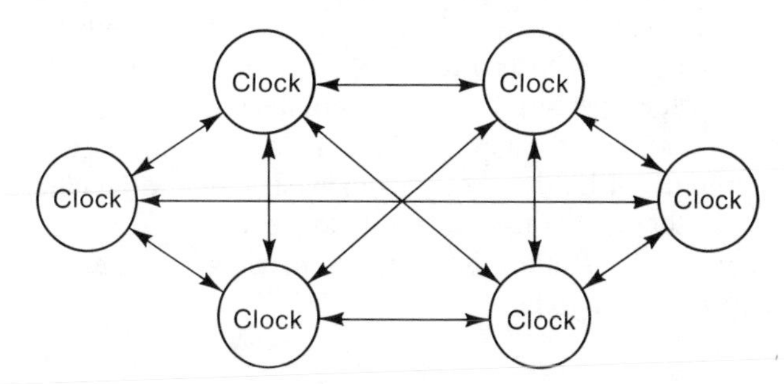

(b) Mutual synchronization concept

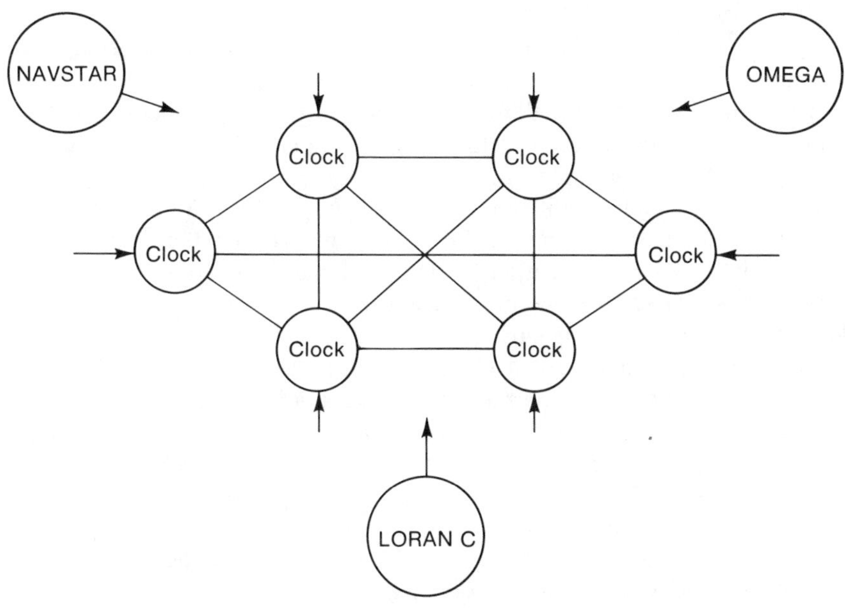

(c) External reference synchronization concept

Figure 10.11 Disciplined Network Synchronization Concepts

10.4.1 Pulse Stuffing

Pulse stuffing can be used as a form of network synchronization, allowing asynchronous interface at each node. Analog switching networks that use digital transmission generate TDM signals that are not normally synchronous with each other. The individual TDM signals (for example, 1.544 Mb/s for North American systems and 2.048 Mb/s for European systems) are synchronized to a common clock in a higher-level multiplexer via pulse stuffing. This technique has the advantage of not requiring stringent tolerance on the frequency of each multiplexer clock.

Pulse stuffing techniques were originally designed for operation over cable (such as T carrier) and later line-of-sight radio. When operated over links such as tropospheric scatter that are subject to bursts of errors or interruptions due to deep fades, however, errors may occur in the stuff control bits resulting in a loss of BCI on the individually derived bit streams. Moreover, timing jitter introduced by the pulse destuffing circuits can result in degraded performance in equipment that must extract timing from the jittered data or clock signals. Pulse stuffing also involves a considerable amount of hardware and somewhat reduces the efficiency of a multiplexer. For these reasons, and with the evolution toward all-digital switched systems, pulse stuffing has served chiefly as an interim means of network synchronization.

10.4.2 Independent Clock

The independent clock approach is based on a **plesiochronous** (nearly synchronous) concept of network synchronization in which data buffers are used to compensate for the differential phase in the system clocks. Highly stable references such as atomic clocks are used at each node of the network with a buffer large enough to maintain BCI over a predetermined period. Buffer length depends on differential clock drift, path length variation, link data rate, and buffer reset period. Since the length of the buffer is fixed, the buffer maintains synchronization (BCI) over some time period T. The rate of buffer reset $(1/T)$ can be made acceptably low with proper choice of clock stability and buffer length.

Conceptually, independent clocks constitute the simplest of all techniques for network synchronization. The chief advantage lies in system survivability, where link degradation or failure does not affect nodal timing and clock failure of one node does not affect other nodal clocks. There are two significant disadvantages, however, for the independent clock technique vis-à-vis any of the disciplined clock approaches. First, without some form of discipline, greater stability must be demanded from individual clocks, implying a higher cost penalty. Second, periodic slips due to buffer overflow or underflow must be tolerated along with attendant resynchronization periods.

Buffer Length

The buffer length (B) required for a single link under independent clock operation is given by

$$B = 2R(T_{\text{CE}} + T_{\text{PD}}) \qquad \text{bits} \tag{10.8}$$

where R = data rate in bits per second

T_{CE} = clock error ($\pm$) in seconds accumulated during buffer reset period

T_{PD} = path delay variation ($\pm$) in seconds accumulated during buffer reset period

The factor of 2 in (10.8) results from a requirement that the buffer be set initially at center position and be able to shift in either the positive or negative direction.

Calculation of Clock Error (T_{CE}). Time (phase) difference between two clocks can accumulate due to differences in initial frequency (accuracy) and stability. If $f_i(0)$ is the frequency of a clock at the ith node at time zero, then the frequency after time t is

$$f_i(t) = f_i(0) + \alpha_i t \tag{10.9}$$

where α_i is the drift rate in hertz per second of the ith nodal clock and where higher-order terms of time are assumed to have negligible contribution. The factor α_i is actually a function of time, but here α_i will be assumed constant and equal to the worst resulting offset due to instability. The difference in frequency between any two clocks is then

$$f_i(t) - f_j(t) = f_i(0) - f_j(0) + (\alpha_i - \alpha_j)t \tag{10.10}$$

Letting $f_i(t) - f_j(t) = \Delta f(t)$ and dividing by the desired (nominal) frequency f, we get

$$\frac{\Delta f(t)}{f} = \frac{\Delta f(0)}{f} + at \tag{10.11}$$

where $\Delta f(0)/f$ and $\Delta f(t)/f$ are the relative frequency differences between two clocks at times zero and t, respectively, and where

$$a = \frac{\alpha_i - \alpha_j}{f} \tag{10.12}$$

is the relative (difference) drift rate between the two clocks. To determine the accumulated clock error T_{CE}, the frequency difference given in (10.11) is integrated over some time T:

$$\begin{aligned} T_{\text{CE}} &= \int_0^T \left[\frac{\Delta f(0)}{f} + at \right] dt \\ &= \frac{\Delta f(0)}{f} T + \frac{1}{2} a T^2 \end{aligned} \tag{10.13}$$

Figure 10.12 is a family of plots of accumulated phase error in time versus operating time T for various initial frequency accuracies between two clocks. Figure 10.13 shows a similar family of plots for selected long-term (per day) drift rates; these drift rates are based on an assumed temperature range of 0 to 50°C. Total clock error, given by Equation (10.13), is determined by summing contributions from initial frequency inaccuracy (Figure 10.12) and from frequency drift (Figure 10.13). Figure 10.14 shows buffer fill in bits versus accumulated time error for standard data rates. To determine buffer lengths required to compensate for accumulated clock error, first select the buffer reset period T. For a given T, Figures 10.12, 10.13, and 10.14 can be used to yield the required buffer length. The factor of 2 in (10.8) doubles the buffer length calculated here to allow opposite direction (sign) of drift and initial inaccuracy. The underlying, worst-case assumption here is that both frequency inaccuracy and drift are biased in opposite directions between the two clocks. If two clocks are settable within ± 1 part in 10^{11}, for example, the maximum initial frequency offset is 2 parts in 10^{11}. In practice, accuracy and drift differences between two clocks are not always biased in opposite directions; rather, they differ from one pair of clocks to another.

Example 10.2

Assume a buffer reset period (T) of 1 week and the use of independent cesium frequency standards. Determine the required buffer length for a data rate of 1.544 Mb/s. From Figures 10.12 and 10.13 we have

$$\text{Phase error:} \quad \frac{\Delta f}{f} T \approx 6 \times 10^{-6} \text{ s}$$

$$\text{Phase error:} \quad \frac{1}{2} a T^2 \approx 2 \times 10^{-6} \text{ s}$$

$$\text{Total phase error} \quad \approx 8 \times 10^{-6} \text{ s}$$

Using Figure 10.14, we find the required buffer length to be approximately ± 15 (30) bits.

Determination of Path Delay Variation (T_{PD}). Fixed delay over a transmission path does not affect clock synchronization between nodes. Variations in transmission delay, however, can cause **doppler shift**, or frequency wander in timing derived from the received signal. The effect of path delay variations on the independent clock technique is to shift the "pointer" of the affected buffer by the amount of the phase change.

Delay variations on *line-of-sight microwave links* are due to meteorological parameters such as atmospheric pressure, temperature, relative humidity, wind velocity, and solar radiation. Measured data indicate that

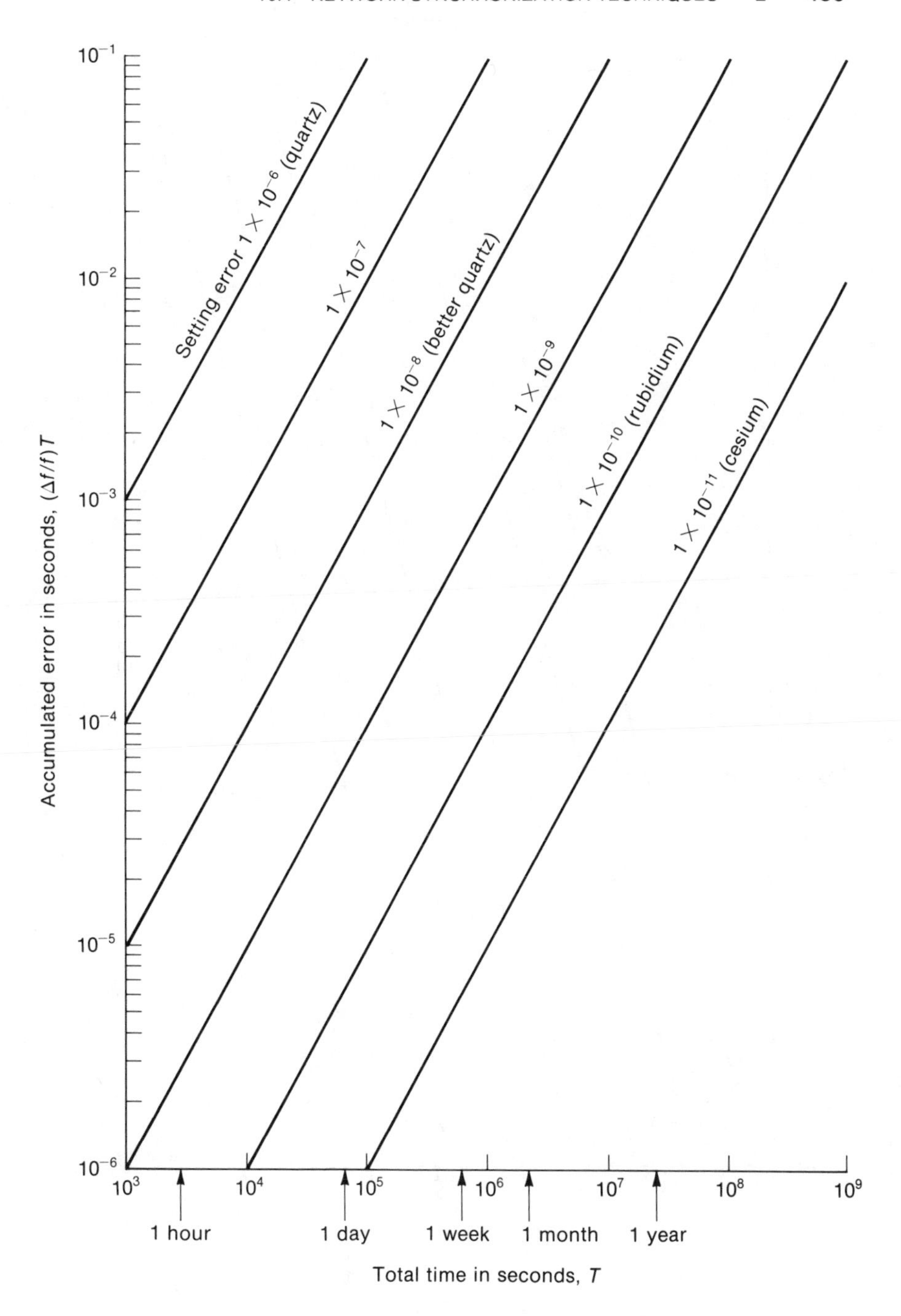

Figure 10.12 Accumulated Time Error vs. Time for Selected Frequency Offsets

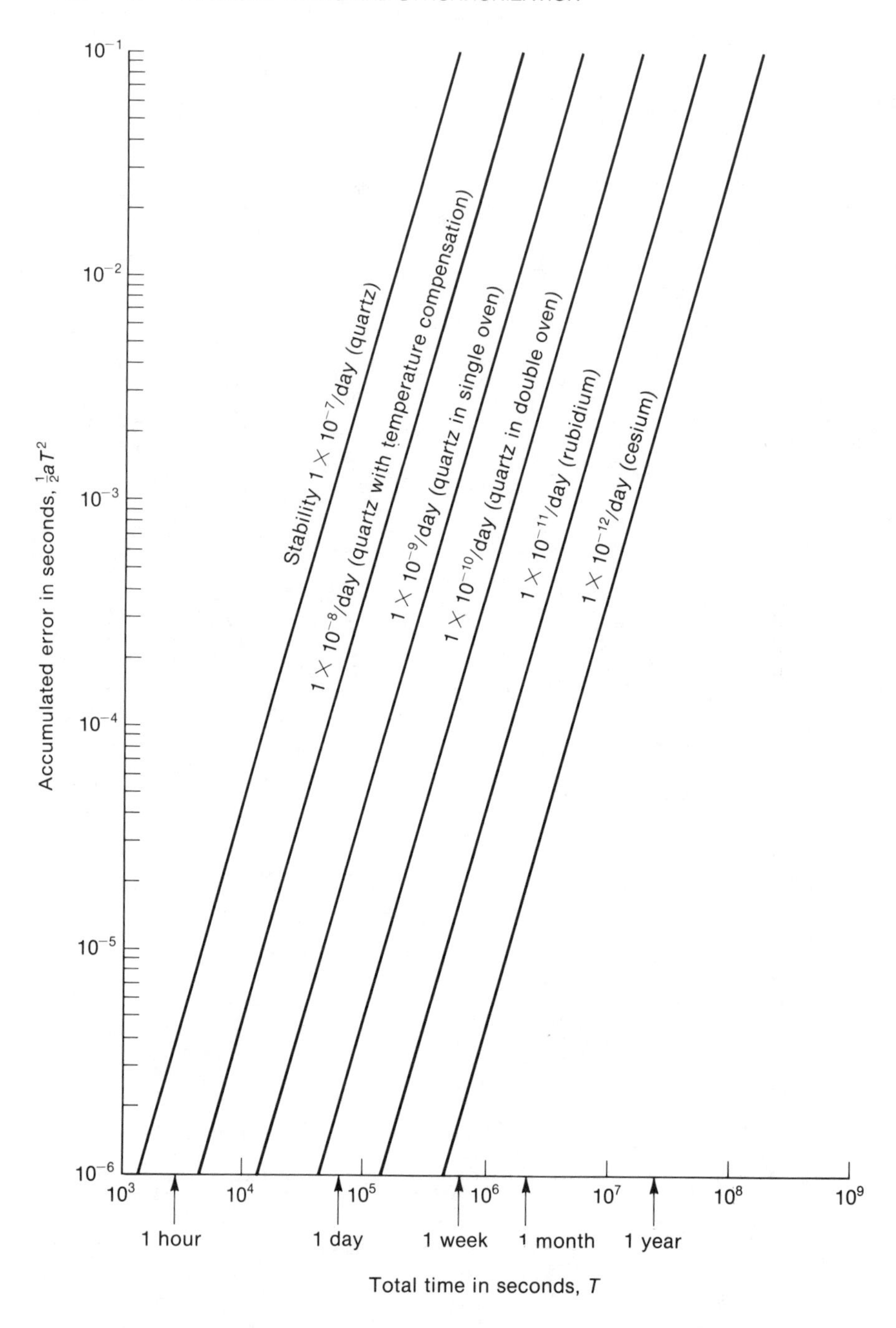

Figure 10.13 Accumulated Time Error vs. Time for Selected Drift Rates

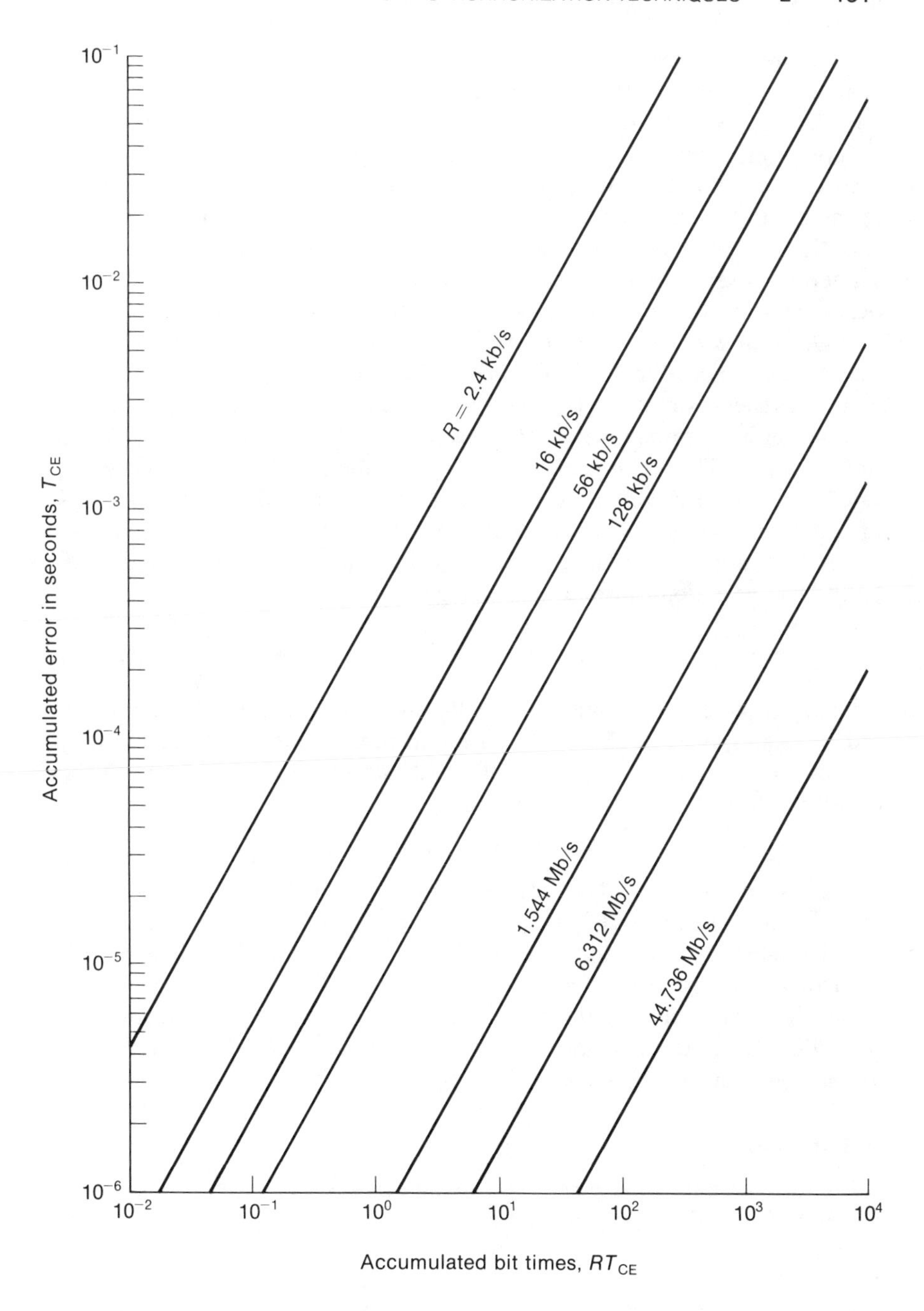

Figure 10.14 Accumulated Bit Times vs. Accumulated Error in Time for Selected Bit Rates

the long-term component of delay variation for LOS microwave transmission is typically 0.1 to 0.2 ns/km over a 24-hr period [20, 21].

Tropospheric scatter links are subject to daily variations that are several orders of magnitude larger than variations observed in line-of-sight. This increase in delay variation is due to a continuously varying index of refraction in the troposphere with respect to both space and time. The delay variations depend on the path length and the intersection angles of the scattering volume. For typical troposcatter links, the worst peak-to-peak delay variation that may be expected is on the order of 1.0 μs [21, 22].

Geostationary satellite paths experience path delay variation, primarily due to satellite orbit inclination, satellite orbit eccentricity, and atmospheric and ionospheric variations [23]. Unless corrected, the orbital plane changes inclination by approximately 0.86° per year due to lunar and solar gravitational effects. This orbital inclination causes the point directly below the satellite to move in a figure eight pattern, completing the pattern every 24 hr. The dimensions of the figure eight increase with the inclination. For an inclination angle I in radians, relative to the equatorial plane, the diurnal path delay variation causes a time change of

$$\Delta T \approx \frac{4Ir_e}{c} \qquad \text{for } I \ll 1 \tag{10.14}$$

where r_e is the earth's radius and c is the speed of light. Thus for $I = 0.015$ rad ($\approx 0.86°$), there is $\Delta T \approx 1.3$ ms peak-to-peak variation. The eccentricity e due to an elliptical orbit causes the path distance to an observer on the equator to vary by

$$\pm e(r_e + H)$$

for an altitude H. For a typical maximum of $e = 0.01$ and for $H = 22{,}300$ mi, the variation is ± 263 mi or ± 1.4 ms. Communication satellite eccentricities are typically on the order of 10^{-4}, however, and produce a small path delay change compared to the inclination. There are also the usual atmospheric and ionospheric mechanisms that cause change in the index of refraction. Resulting changes in delay are independent of orbital parameters and are smaller in magnitude; ± 1 ms is an upper bound.

Buffer Location

The choice of buffer placement in a demultiplexer hierarchy depends on a number of considerations, such as performance, network topology, cost, and reliability.

Performance. When buffering is carried out at one of the lower levels, all higher-level equipment must be clocked from the recovered clock, which exhibits degradation due to jitter and other media effects (such as diurnal variations for satellites). To avoid passing on this clock degradation, buffers can be placed at higher levels in the hierarchy. Another influence on performance is the resetting of buffers due to underflow or overflow; data are lost (or repeated), resulting in loss of BCI and the need to resynchronize downstream equipment. Here the choice of buffer location determines the number of users affected by a single buffer reset.

Network Topology. At some nodes, switching and multiplexing of data from several remote sources may be required. This is usually done with synchronously clocked data, implying that the data must be buffered prior to switching or multiplexing. The number of users (circuits) requiring synchronous operation also affects the choice of buffer location. If only a few low-speed users require synchronous operation, it may be less costly and more flexible to place buffers only at the low-data-rate location. Placement of the buffer at the highest level in the data hierarchy provides synchronous data to all users, independent of their needs, and facilitates transition to an all-synchronous switched system. However, one topological disadvantage of high-level buffer placement is that for many networks high-level multiplexers are used in greater proportion than lower-level multiplexers—implying the need for more frequent buffering at the higher level and hence a larger buffer reset period per link in order to meet end-to-end slip rate requirements.

Cost. Present technology suggests that the cost of buffers is somewhat insensitive to the size (length) of the buffer. A large buffer with high operating speed is probably cheaper than a number of smaller ones with similar total capacity, each operating at a slower speed, unless reliability requirements dictate a more expensive approach. Multiplex and switching equipment inherently requires buffers to perform their function, however, suggesting that such built-in buffers could be made sufficiently longer to provide a network synchronization function also. With this assumption, separate buffers can be avoided by using buffers built directly into transmission and switching equipment.

Reliability. A disadvantage of the high-level buffer placement is the addition of another series component that is critical to system reliability. If the high-level buffer fails, all downstream lower-level outputs are affected. The magnitude of such a failure forces the use of a high-reliability design or a standby buffer with automatic switchover. When the lower-level buffer location is used, the impact of a failed buffer is less and therefore a lower-reliability design could be used.

Example 10.3

Table 10.2 presents buffer length requirements for both terrestrial and satellite applications and both low-level and high-level buffer location. Buffer length is specified as plus and minus so that the buffer can initially be set at mid position and allowed to move in either direction to its specified maximum. The timing source is assumed to be a cesium clock. The low-level

Table 10.2 Typical Buffer Length Requirements for Independent Clock Synchronization

Buffer Location	Clock Accuracy	Long-Term Clock Stability	Propagation Delay Variations (T_{PD})	Buffer Reset Period	Clock Error (T_{CE})	Buffer Length Requirements: Bit Rate (kb/s)	Buffer Length Requirements: Buffer Length (bits)
Lower level (terrestrial link)	1×10^{-11}	1×10^{-13}	$\leq 1\ \mu s$	10 days	$9\ \mu s$	512	±6
						256	±3
						128	±2
						64	±1
Higher level (terrestrial link)	1×10^{-11}	1×10^{-13}	$\leq 1\ \mu s$	50 days	$54\ \mu s$	44,736	±2460
						6,312	±347
						1,544	±85
Lower level (satellite link)	1×10^{-11}	1×10^{-13}	10 ms	10 days	$9\ \mu s$	128	±1282
						64	±641
						32	±321
						16	±161

buffer placement corresponds to the lowest level in the multiplex hierarchy, or a user level, perhaps a single voice or data channel as indicated by the bit rates shown in Table 10.2. The higher-level placement corresponds to the highest multiplex level, equal to the total transmission rate. Buffer reset periods, here set at 10 and 50 days for lower and higher buffer placement respectively, are determined by the relative density of multiplex breakouts at low and high levels in the hierarchy. The conclusions that can be drawn from the examples in Table 10.2 are that phase error due to clock differences is much greater than phase error due to propagation delay variations in terrestrial systems but much less than phase error due to satellite doppler.

10.4.3 Master-Slave System

The simplest and most utilized [24] form of disciplined network synchronization is the **master-slave system**, which distributes a reference or master clock to all nodes via a tree-type structure. As with most disciplined systems, timing signals are distributed along with data over all transmission links. Timing is then recovered from the incoming data, either from bit transitions or from frame information. The most important feature is the maintenance of timing distribution in the event of link failure. According to this feature, several types of master-slave systems can be envisioned.

Fixed Hierarchy

With the fixed-hierarchy approach, when the normal clock distribution path has failed, restoration at each slave node is accomplished via a predetermined algorithm [25]. All nodal clocks are prioritized, and each node is capable of locking itself to the highest-rank clock in operation. If a clock distribution path fails, the node changes to the next priority source. In this system, there is no need to transfer control signals between nodes.

Self-Organizing

To implement the self-organizing synchronization scheme, control signals must be transferred between nodes to allow automatic rearrangement of the clock distribution network. This control information describes the performance of the clock signal being used at a given transmitting node, and this information is transmitted to all other connected nodes. Typically three kinds of control signals are transmitted [26, 27]:

- Designation of the node used as the master reference for the local clock
- The number and quality of links that have been encountered in the path from the reference clock to the local clock
- The rank of the local clock

The third type of control signal refers to a preassigned ranking to each node that depends, for example, on the quality of its clock.

Loosely Coupled

In the loosely coupled approach [28], each node maintains high-frequency accuracy even in the event of a fault. Thus no control is needed to rearrange the clock distribution network. This approach is made possible by use of a phase-locked oscillator that stores the previous input frequency in memory [29], by use of high-quality backup oscillators that allow plesiochronous operation [2], or by use of external reference, such as LORAN C, as a backup source of timing [30]. This version of the master-slave system may be viewed as the simplest from the viewpoint of network control and maintenance.

10.4.4 Mutual Synchronization

The technique known as **mutual synchronization**, or frequency averaging, is one in which each nodal clock is adjusted in frequency in order to reduce the timing error between itself and some weighted average of the rest of the network. The basic approach is illustrated in Figure 10.15, where a weighted sum of phase errors for all incoming clocks is used as a control signal to adjust the local clock. Figure 10.15 depicts the circuitry at one node where $f_1, f_2, \ldots, f_i$ represent clock signals regenerated through the use of a phase-locked loop from each incoming nodal bit stream, $r_1, r_2, \ldots, r_i$. The phase detector is modeled as a flip-flop with output

$$\begin{aligned} V_{1j} &= \phi_j - \phi_1 \\ V_{2j} &= \phi_j - \phi_2 \\ &\vdots \\ V_{ij} &= \phi_j - \phi_i \end{aligned} \tag{10.15}$$

where ϕ_j and $\phi_1, \phi_2, \ldots, \phi_i$ represent the nodal clock and incoming clock phases. The output of the summing network is

$$V_j = \sum_i W_{ij} V_{ij} \tag{10.16}$$

where Σ_i denotes summation over all nodes connected to node j. If the free-running clock of the jth node has frequency denoted as f_{0j}, then since the local frequency is controlled by voltage V_j through the filter with gain K_j, the nodal clock output is given as

$$f_j = f_{0j} - K_j V_j \tag{10.17}$$

where K_j is a constant with units of radians per volt.

The version of mutual synchronization described above is called *single-ended* because the error control signals at a given node are derived only from timing signals available at the node. In a *double-ended* mutual synchronization system, timing information from all neighboring nodes is transmitted to the given node and used in determining the local frequency. The advantage of this technique over the single-ended control technique is that the system frequency of the entire interconnected network is not a

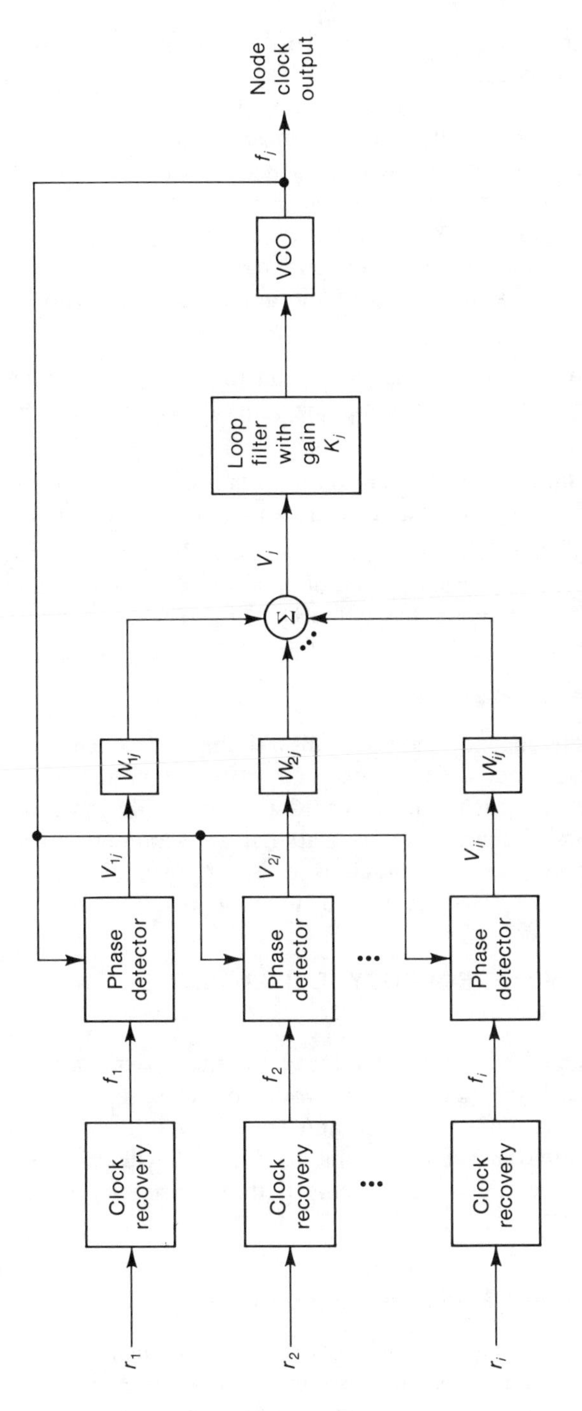

Figure 10.15 Implementation of Mutual Synchronization Technique

function of transmission delay but rather depends only on the parameters of the clock at each node.

Mutual synchronization is a widely studied technique [31 to 37], but it has been used operationally in few applications and then only outside the United States. One advantage is provided by the multinode source of timing, which assures that no single clock or transmission path is essential. Another basic advantage is its ability to adapt to system changes and clock drift. As a result, a less accurate reference clock can be used than with other approaches. Some disadvantages of the mutual synchronization technique are:

- The eventual system frequency is difficult to predict, since it is a function of VCO frequencies, network topology, link delays, and weighting coefficients.
- Changes in link delay or nodal dropout can cause significant perturbations in nodal frequencies and a permanent change in system frequency.
- Lack of a fixed reference results in offset with respect to any external network or other external source of time or frequency, making mutual synchronization incompatible with a UTC reference.

10.4.5 External Reference

The **external reference** technique obtains a timing or frequency reference for all nodal clocks from a source external to the communication network. There exist several such sources based on worldwide systems providing precise time and frequency to navigational and communication systems. Table 10.3 summarizes the characteristics of the major time and frequency dissemination systems that are described in more detail in the next section.

10.5 TIME AND FREQUENCY DISSEMINATION SYSTEMS

Numerous systems have been developed for the dissemination of precise time and frequency. These systems work on the basis of continuous or periodic radio broadcasts of time and frequency information, using terrestrial or satellite transmission techniques. Here we will examine those time and frequency dissemination systems that are used by communication networks.

10.5.1 LF and VLF Radio Transmission

Radio waves at low frequency (LF) and very low frequency (VLF) are used for precise time and frequency dissemination because of their excellent stability and potential for good earth coverage. Accuracies of 1 part in 10^{11} in frequency and 500 μs or better for time can be achieved using LF and VLF broadcasts. Some 20 individual LF–VLF stations exist worldwide that

Table 10.3 Characteristics of the Major Time and Frequency Dissemination Systems (Adapted from [5])

Dissemination Techniques	*Accuracy for Frequency Comparison or Calibration*	*Accuracy for Time Transfer*	*Ambiguity*	*Coverage*	*Reference*
VLF radio					
OMEGA	1×10^{-11}	Envelope: 1 – 10 ms	1 cycle	Nearly global	UTC
LF radio					
LORAN C	5×10^{-12}	1 μs (ground) 50 μs (sky)	30 – 50 ms	Most of northern hemisphere	UTC
HF / MF radio					
WWV and WWVH	1×10^{-6}	1000 μs	Code:year Voice:1 day Tick:1 s	Hemisphere	NBS master clock
TV (VHF / UHF radio)					
TV Line-10	1×10^{-11}	1 μs	N.A.	Network coverage	None
Color subcarrier	1×10^{-11}	N.A.	N.A.		
Satellite (SHF radio)					
GOES	3×10^{-10}	30 μs	1 year	Western hemisphere	UTC
TRANSIT	3×10^{-10}	30 μs	15 min	Nearly global	UTC
Global Positioning System	5×10^{-12}	1 μs	N.A.	Global	UTC
Defense Satellite Communications System	5×10^{-12}	1 μs	N.A.	Nearly global	UTC

provide dissemination of time and frequency information. Here, however, two systems will be described that utilize several stations collectively to provide good earth coverage; one system operates in the VLF band (OMEGA) and the other in the LF band (LORAN C).

OMEGA Navigation System

The OMEGA navigation system is operated by the United States (U.S. Coast Guard) and several other countries and is composed of eight VLF radio stations operating in the 10 to 14 kHz range [38, 39]. These stations radiate a nominal 10 kW of power sufficient for moderate accuracy navigation on a worldwide basis. All stations transmit according to a predetermined time sequence shown in Figure 10.16, where the asterisks signify the unique frequency for the station. During each transmission interval only three stations are radiating, each at a different frequency. The duration of each transmission varies from 0.9 to 1.2 s, depending on the station's assigned location within the signal pattern. With a silent interval of 0.2 s between each transmission, the entire cycle of the signal pattern repeats every 10 s. This signal format allows each station to be identified by its transmission of a particular frequency at a prescribed time.

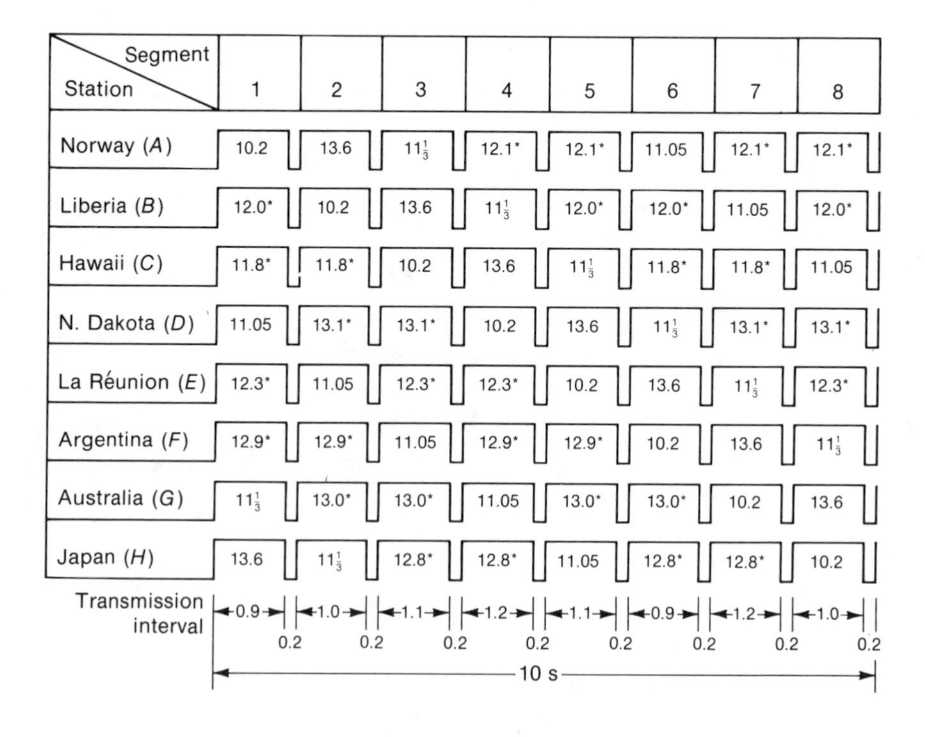

Station \ Segment	1	2	3	4	5	6	7	8
Norway (*A*)	10.2	13.6	$11\frac{1}{3}$	12.1*	12.1*	11.05	12.1*	12.1*
Liberia (*B*)	12.0*	10.2	13.6	$11\frac{1}{3}$	12.0*	12.0*	11.05	12.0*
Hawaii (*C*)	11.8*	11.8*	10.2	13.6	$11\frac{1}{3}$	11.8*	11.8*	11.05
N. Dakota (*D*)	11.05	13.1*	13.1*	10.2	13.6	$11\frac{1}{3}$	13.1*	13.1*
La Réunion (*E*)	12.3*	11.05	12.3*	12.3*	10.2	13.6	$11\frac{1}{3}$	12.3*
Argentina (*F*)	12.9*	12.9*	11.05	12.9*	12.9*	10.2	13.6	$11\frac{1}{3}$
Australia (*G*)	$11\frac{1}{3}$	13.0*	13.0*	11.05	13.0*	13.0*	10.2	13.6
Japan (*H*)	13.6	$11\frac{1}{3}$	12.8*	12.8*	11.05	12.8*	12.8*	10.2
Transmission interval	0.9	1.0	1.1	1.2	1.1	0.9	1.2	1.0

Figure 10.16 OMEGA Signal Format (frequencies in kHz)

Each OMEGA transmitting station employs four cesium-beam frequency standards for reliable timing and interstation synchronization. All eight stations are synchronized to each other to about 2 or 3 μs by means of interstation measurements made twice daily. OMEGA clocks are referenced to UTC to the extent that OMEGA system time "tracks" UTC but is not adjusted for yearly leap seconds. Time transfer via OMEGA can provide accuracies of 1 to 10 ms by using envelope detection. Frequency comparison can be made to 1×10^{-11} per day. Daily phase values referenced to UTC standards for OMEGA station transmissions are published weekly by the USNO.

LORAN C

LORAN C (long-range navigation) is a low-frequency radio navigational system operated by the U.S. Coast Guard [40, 41, 42]. Although primarily used for navigation, LORAN C transmission may also be used for time dissemination, frequency reference, and communication networks. LORAN C transmitting stations are grouped into chains of at least three stations. One transmitting station is designated master while the others are called secondaries. Each transmitting station derives its frequency from three cesium standards. Chain coverage is determined by the stations' transmitter power, their orientation and receiver sensitivity, and the distance between stations.

Low-frequency (LF) transmission produces both a groundwave propagation mode, which is extremely stable, and a skywave mode that results from ionospheric reflection and exhibits strong diurnal variation. Because of propagational delays in receiving the skywave signal, interference with the groundwave signal could occur without proper choice of LORAN C signal format. Using groundwave propagation, LORAN C can provide about 1 μs time transfer accuracy and about $\pm 5 \times 10^{-12}$ frequency stability control over a 24-hr period. Groundwave coverage with reliable performance extends to about 1500 mi for presently used transmitter power. Because of ionospheric perturbations, skywave propagation yields poorer time transfer than the groundwave mode and typically provides ± 50 μs accuracy. At distances greater than 1500 mi and up to 5000 mi, however, the skywave signal is usually stronger than the groundwave. Figure 10.17 shows groundwave and skywave coverage for the present LORAN C chains.

Signal Characteristics. The LORAN C carrier frequency is 100 kHz with a 20-kHz bandwidth. Phase and coding modulation is used to minimize the effects of skywave interference and to allow identification of a particular station being received. The 100-kHz LORAN C pulse, shown in Figure 10.18, has a fast rise time that achieves high power prior to the arrival of skywaves. The shape of the pulse also allows the receiver to identify one particular cycle of the 100 kHz carrier. In addition to transmitting pulses for separation of groundwave and skywave signals, the pulses are transmitted in

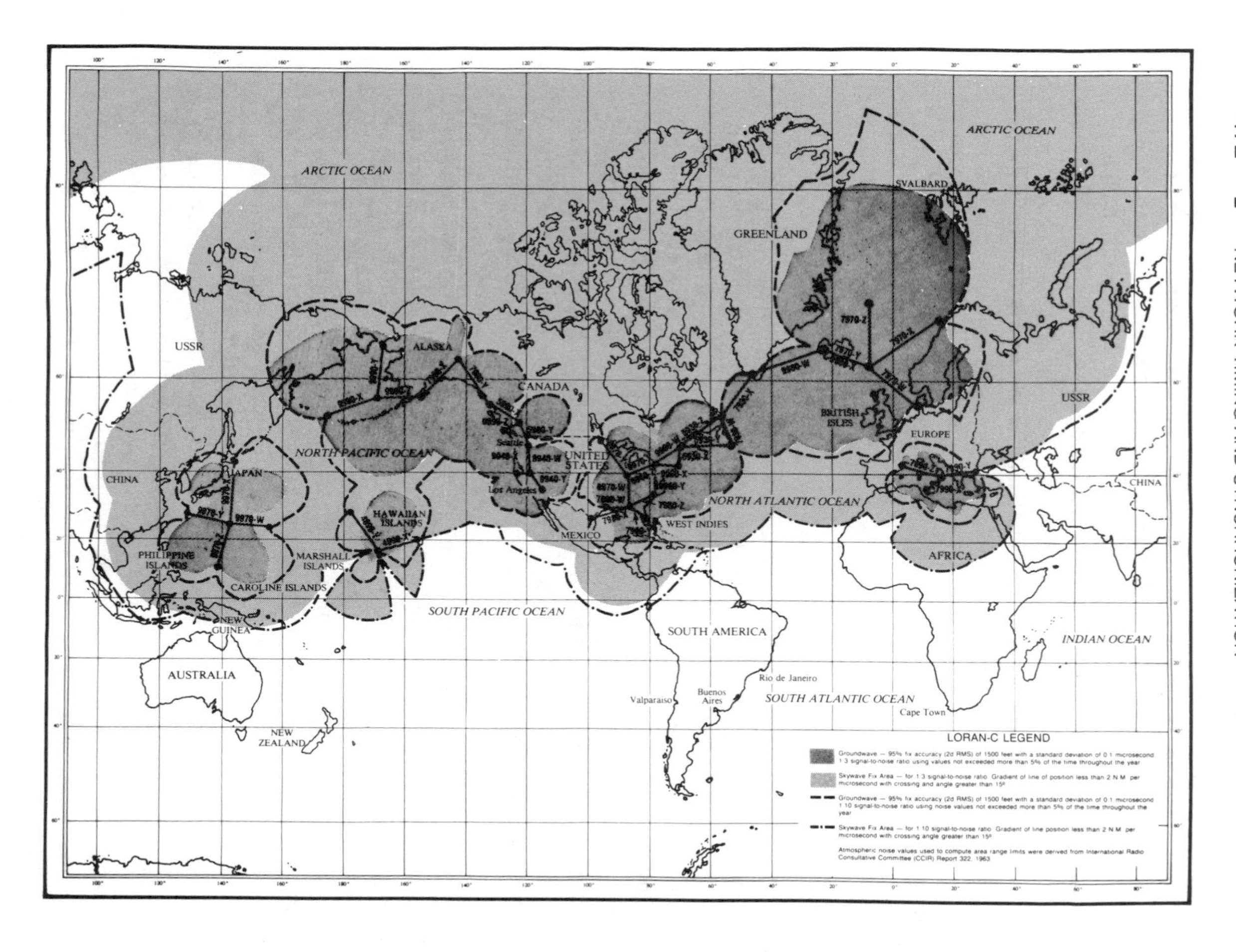

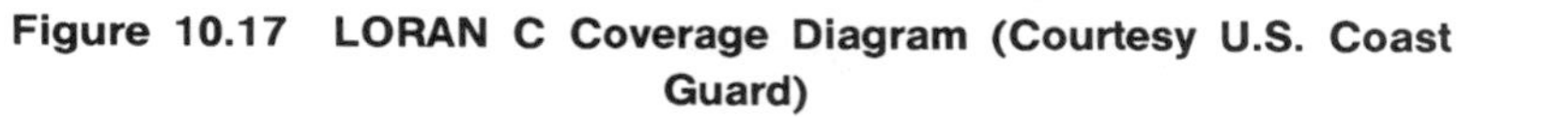

Figure 10.17 LORAN C Coverage Diagram (Courtesy U.S. Coast Guard)

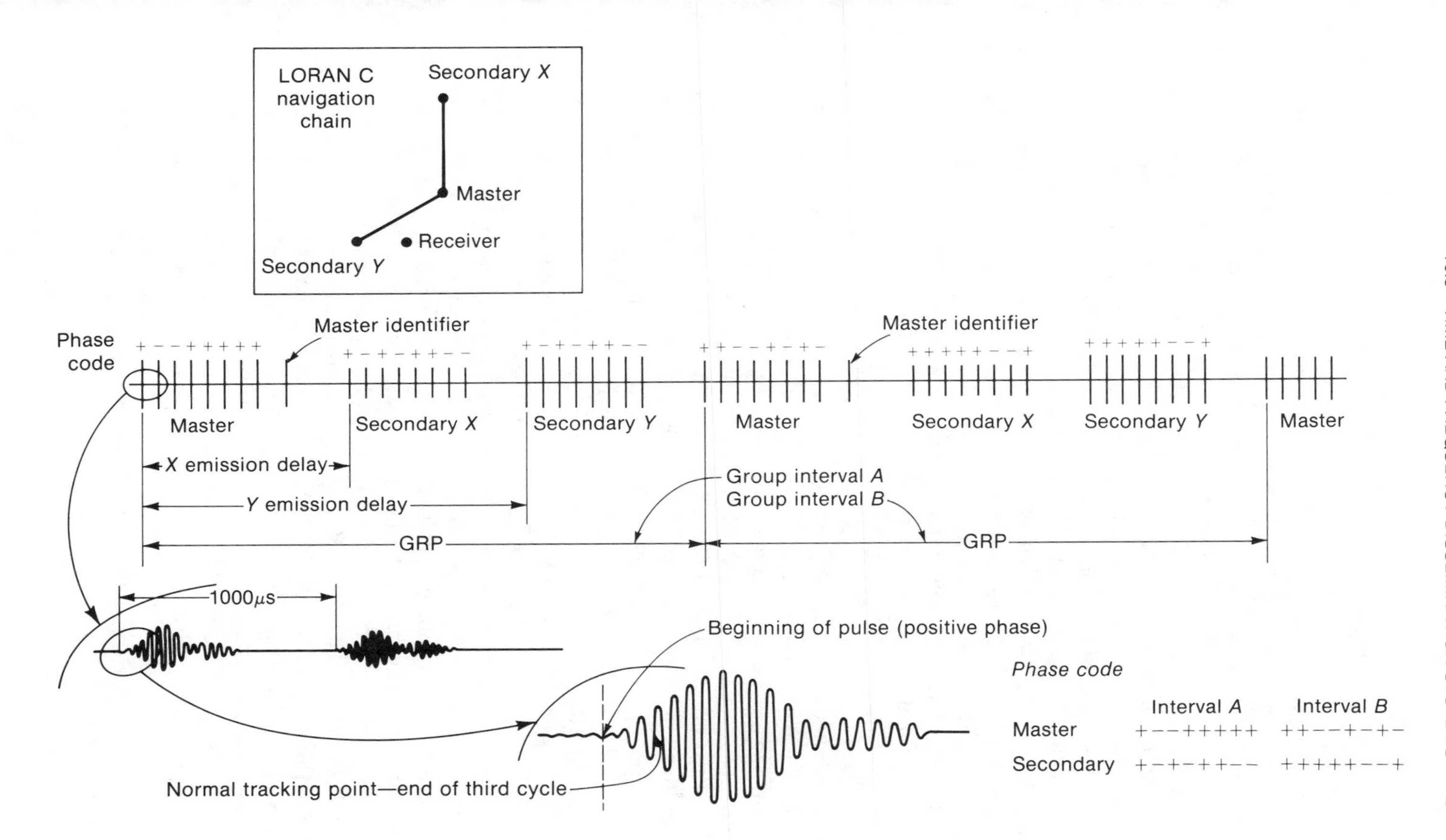

Figure 10.18 LORAN C Signal Format and Waveforms (Reprinted by permission of Austron, Inc.)

groups for station identification. The secondary stations in a LORAN C chain transmit eight pulses spaced 1 ms apart; the master station adds a ninth pulse, 2 ms after the eighth, for identification. To reduce skywave interference, binary phase shifting of the RF carrier of individual pulses in a group is used, according to an alternating code for the master and secondary stations. Figure 10.18 illustrates this signal format for a chain consisting of a master and two secondary stations labeled *X* and *Y*.

Since all LORAN C stations broadcast at 100 kHz, it is possible to receive many different LORAN chains at the same time. Therefore the pulses in a chain are transmitted at slightly different rates, called the *group repetition interval* (GRI) or *group repetition period* (GRP), so the receiver can distinguish between them. For each chain, a GRI is specified of sufficient length so that it contains time for transmission of the pulse group from each station plus a time delay between each pulse group. Thus the signals from two or more stations cannot overlap in time anywhere in the coverage area. The minimum GRI is then a function of the number of stations and the distance between them.

Time Determination. The LORAN C system does not broadcast a time code signal. With knowledge of the time of coincidence (TOC) of a particular GRP relative to a second, however, a timing signal can be extracted from a LORAN C signal. This procedure is illustrated in Figure 10.19 for the Gulf of Alaska LORAN C chain, which has a GRP of 79,600 μs. Since this GRP along with most others is not a submultiple of 1 s, there is only periodic coincidence between the group interval and a second. At the TOC interval shown in Figure 10.19, which is 3 min, 19 s, this chain's master station transmits a pulse that is coincident with the UTC second. A LORAN C receiver can then synchronize its internal time signal to UTC when the TOC occurs. Using LORAN C emission and propagation delay figures and the receiver's internal delay, the local clock time can be corrected to be on time. It should be noted that external coarse clock synchronization to within one-half of a repetition period is required to resolve the ambiguity inherent in a pulsed system. For LORAN C the half-period interval varies from 30 to 50 ms. Resolution to this degree is obtainable from standard HF or VLF time transmissions.

With the cooperation of the U.S. Coast Guard, the U.S. Naval Observatory maintains synchronization of LORAN C transmissions to the observatory's master clock. Since the master clock is on UTC, time derived from LORAN C is also on this scale. Reference to the master clock and corrections to LORAN transmitting stations are made by a combination of portable clock and satellite techniques. Tables showing the discrepancy between each LORAN chain and UTC are published periodically by the Naval Observatory.

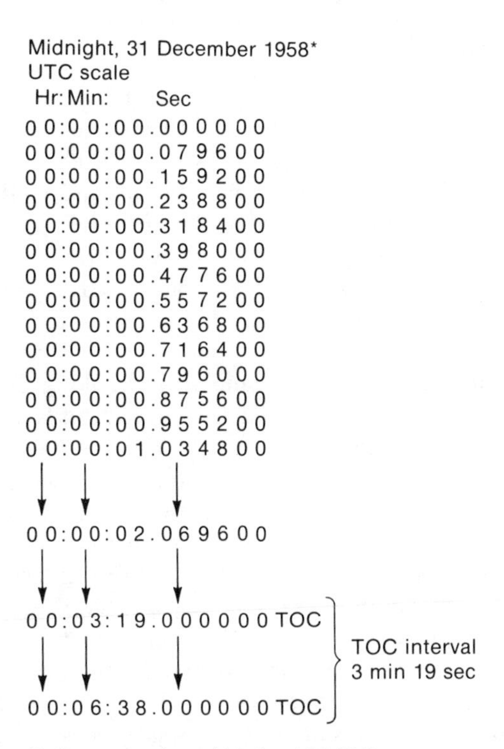

Figure 10.19 Time of Coincidence for Gulf of Alaska LORAN C Chain with GRP = 79,600 μs (Reprinted by permission of Austron, Inc.)

Frequency Determination. If the LORAN C receiver is continuously phase-locked to the received signals, it can provide a phase-locked 100 kHz or, more commonly, a 1-MHz output frequency. Since the frequency of LORAN C chains is controlled by a cesium-beam oscillator, the receiver output frequency exhibits the identical long-term stability of the cesium oscillator and can therefore be used as an excellent frequency reference. The short-term stability is determined by the choice of local oscillator (usually good-quality quartz crystal) and by receiver measurement fluctuations due to noise and interference. Since the frequency of each LORAN chain is traceable to UTC, LORAN C transmissions become a reliable frequency reference. The frequency of the radiated carrier at LORAN C stations is kept within $\pm 2 \times 10^{-12}$ with respect to UTC.

10.5.2 HF Radio Broadcasts

The long-range capability of high-frequency (HF) propagation makes HF one of the most commonly used sources of time and frequency dissemination [43, 44]. Western hemisphere HF stations such as WWV from Ft. Collins, Colorado, WWVH from Kauai, Hawaii, and CHU from Ottawa, Canada, provide essentially worldwide coverage.

WWV and WWVH stations broadcast on carrier frequencies of 2.5, 5, 10, and 15 MHz. WWV also broadcasts on 20 MHz. These two stations derive carrier frequencies, audio tones, and time of day from three cesium-beam frequency standards. The station references are controlled to within 1 part in 10^{12} of the NBS frequency standard in Boulder, Colorado. Time at these stations is controlled to within a few microseconds of NBS UTC.

Time and frequency are encoded on WWV and WWVH transmissions by means of UTC voice announcements, time ticks, and gated tones. The beginning of each hour is identified by an 800-ms tone burst at 1500 Hz. The beginning of each minute is identified by an 800-ms tone burst at 1000 Hz for WWV and 1200 Hz for WWVH. The remaining seconds in a minute are encoded by 5-ms audio ticks, at 1000 Hz for WWV and 1200 Hz for WWVH, which occur at the beginning of each second. The twenty-ninth and fifty-ninth second ticks are omitted to serve as identifiers for the half and full-minute ticks. Each of the second ticks is preceded by 10 ms of silence and followed by 25 ms of silence to make the ticks for seconds more discernible. The tones or ticks also preempt voice announcements, but this shortcoming causes only small audio distortion in the voice announcements. Voiced time-of-day announcements are broadcast once a minute.

For time calibration the voice announcements can be used if accuracy requirements are low (within 1 s). For higher accuracy, a user must measure the seconds' ticks or decode the WWV/WWVH time code. The time code indicates UTC by day of the year, hour, minute, and second, and it is broadcast once per second in binary coded decimal (BCD) format on a 100-Hz subcarrier. Use of this time code or seconds' ticks with appropriate electronics permits time accuracy to 1 ms. Frequency comparison with WWV or WWVH can usually be accomplished to about 1 part in 10^6 by a variety of methods, including time comparisons to compute the frequency offset according to the relationship*

$$\frac{\Delta f}{f} = -\frac{\Delta t}{T} \tag{10.18}$$

10.5.3 Television Transmission

The major television networks in the United States use atomic oscillators (cesium or rubidium) to generate their reference signals [45, 46]. A com-

*Easily derived by elementary calculus from the definition of frequency as $f = 1/t$.

munication station can calibrate a local precise oscillator by measuring the difference between the local oscillator and the received television signal. Accuracies obtainable by TV signal comparison range from 1 part in 10^9 to a few parts in 10^{11}. The two basic methods of calibrating a frequency source using TV transmission are color burst comparisons and TV Line 10 comparison. The TV Line 10 approach is also used for time transfer and time comparison. At one time, atomic frequency standards of the TV networks were referenced to NBS time. On 1 February 1983, however, the National Bureau of Standards discontinued its frequency calibration service using network TV.

TV Color Subcarrier Comparison

The major network color subcarrier frequency is approximately 3.58 MHz, derived from network atomic frequency standards. The most straightforward method for use of the color subcarrier frequency in calibration of a local oscillator is to compare, via a phase detector, a locally generated 3.58 MHz signal with a received sample of 3.58 MHz from a color TV set. The dc output of the phase detector, which represents differential frequency for large offsets and differential phase for small offsets, is amplified, plotted, and calibrated in phase change per unit time. Since relative frequency is related to the relative phase (or time) by Equation (10.18), we can derive the frequency error from the phase error. In most cases of TV reception, resolution is limited to about 10 ns in 15 min, which corresponds in frequency resolution to

$$\frac{\Delta f}{f} = \frac{\Delta t}{T} = \frac{10\text{ ns}}{15\text{ min}} = 1.1 \times 10^{-11} \tag{10.19}$$

Some knowledge of network scheduling is required in order to make use of the 3.58-MHz subcarrier for frequency calibration. During station breaks, the color signal often originates from the local station, which uses a lower accuracy source. Tape delay broadcasts also suffer from degraded subcarrier frequency and are therefore invalid as a precision reference.

TV Line 10

In the United States a television picture frame is made up of 525 lines that are scanned or traced to produce an image on the screen. First the odd lines are traced; then the even lines are traced in between the odd. Line 10 is one of the odd lines that make up each picture; it was chosen because it is easy to pick out from the rest of the lines with simple circuitry. To calibrate a local frequency source, a clock pulse from the source is compared to the trailing edge of the Line 10 synchronizing pulse at a certain time each day. The Line 10 technique can be used to determine the frequency stability of a local oscillator compared to another driven by Line 10 measurements. Likewise, frequency or time source calibration between two stations can be accomplished by simultaneously recording the times of arrival of the Line

10 pulse output and comparing them to local clock time. Frequency accuracy to 1 part in 10^{11} over a period of several years and time accuracy to several microseconds have been reported for Line 10.

10.5.4 Satellite Transmission

Time and frequency dissemination via satellite facilitates global coverage while avoiding the ionospheric effects that beset terrestrial systems. Three different modes of satellite time and frequency transfer have been developed and come into practice [47]. The first method uses geostationary satellites to broadcast time signals referenced to the United States' NBS; the second method uses navigation satellites carrying time standards; the third method uses communication satellites to transfer time. All three methods are discussed in the following paragraphs.

Geostationary Operational Environmental Satellites (GOES)

The GOES, under the management of the U.S. National Oceanic and Atmospheric Administration (NOAA), provides a PTTI capability via two

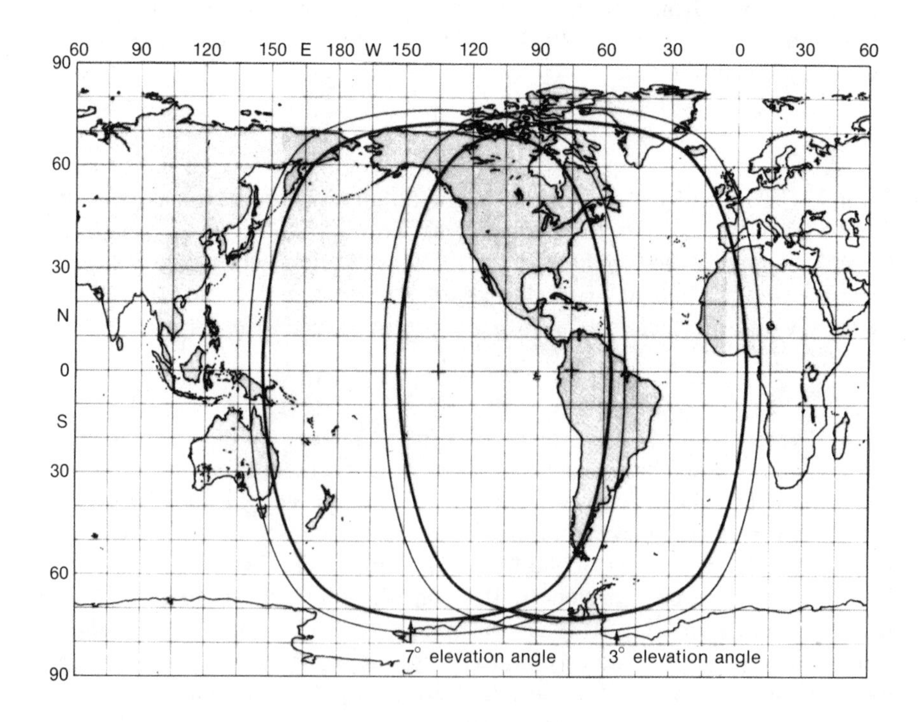

Figure 10.20 Coverage of the GOES Satellites

geostationary satellites. When both are operational,* the GOES satellites, designated GOES/East and GOES/West, are positioned over the equator at 75°W and 135°W longitude, respectively. From these locations they provide continuous coverage to most of the western hemisphere as indicated in Figure 10.20. Although their primary mission involves collection of environmental data, the GOES signal format transmitted from satellite to earth includes a digital time code generated and controlled by the NBS UTC time scale. The time code is transmitted twice a second and consists of a synchronization word, a time-of-year word (UTC), a UT1 correction, and the satellite's position in terms of latitude, longitude, and altitude [48, 49].

GOES time-code receivers decode the satellite position data, compute the satellite-to-user propagation delay, and adjust a one pulse per second (pps) output signal to be on time with respect to the master clock at the GOES satellite control facility. Time transfer accuracy is 10 to 30 μs. The time accuracy depends on the accuracy of the orbital data provided by NBS and accuracy of UTC provided by LORAN C and TV transmissions [5].

TRANSIT

TRANSIT is a satellite-based navigational system operated by the U.S. Navy and consisting of five satellites and associated ground stations [49, 50]. The satellites are in nearly circular, polar orbits at an altitude of about 1100 km; thus any user on the earth will see a TRANSIT satellite about every 90 min. TRANSIT provides worldwide coverage and is, in fact, the first time and frequency dissemination system to work anywhere in the world.

The satellites broadcast ephemeris data continuously on approximately 150 MHz and 400 MHz carriers. Navigational information is broadcast in 2 min intervals, including a time mark every even minute. Each satellite derives its clock time and carrier frequencies from a temperature-stabilized, crystal-controlled 5-MHz oscillator. Satellite clock time is monitored from earth-control stations, and clock corrections are periodically sent to the satellites.

TRANSIT allows a user to recover time signals that can be used as a clock or to steer an oscillator's frequency. Time to better than 30 μs and frequency accuracy to a few parts in 10^{10} per day can be achieved. The instrumentation needed by a user in general consists of a TRANSIT receiver, a frequency source, and a clock. Moreover, it is necessary to obtain the time of day from another source to within 15 min in order to resolve the time ambiguity of TRANSIT.

NAVSTAR Global Positioning System

The NAVSTAR Global Positioning System (GPS) is a satellite navigation system under development for the U.S. military that will provide highly accurate position and velocity information in three dimensions and precise

*The GOES/East satellite failed on 29 July 1984 and is not due for replacement until 1986.

time and time interval on a global basis continuously [51 to 54]. When fully deployed at the end of the 1980s, the space segment will consist of 18 satellites providing continuous global coverage with at least four satellites in view at a given location. The GPS concept is predicated on accurate and continuous knowledge of the spatial position of each satellite in the system. Each satellite will continuously broadcast data on its position. This information will be periodically updated by the master control station based upon data obtained from monitoring earth stations. The USNO will also monitor GPS and provide PTTI data to the GPS control station.

Data are transmitted from each satellite on two carrier frequencies designated L1 (1575.42 MHz) and L2 (1227.6 MHz). Using spread spectrum, the navigation message at a rate of 50 b/s is transmitted in about 20 MHz bandwidth, using an advanced cesium-beam standard onboard as a frequency reference. For time and frequency users, the receiver configuration is as shown in Figure 10.21. The time transfer unit (TTU) accepts stable reference time (1 pps) and frequency (5 MHz) signals from a local source and the L1 and L2 signals from any GPS satellite. Ephemeris data from the satellite are detected and processed to determine satellite position and to estimate time of arrival of the satellite signal epoch. The actual time of

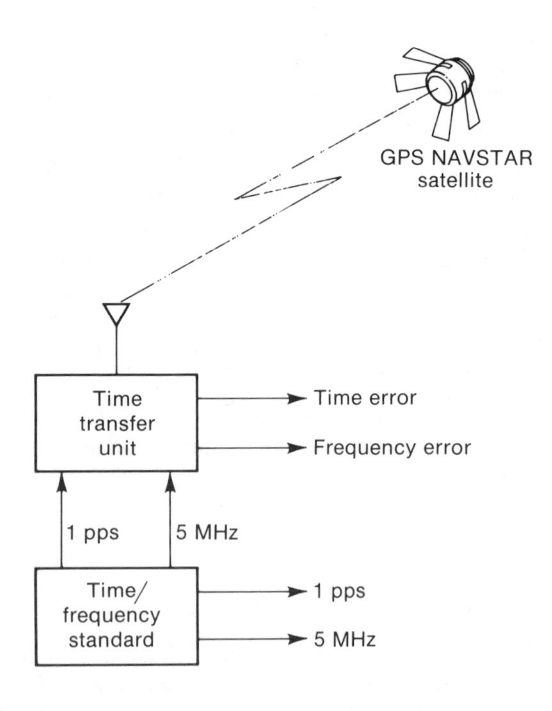

Figure 10.21 GPS Time Transfer Unit

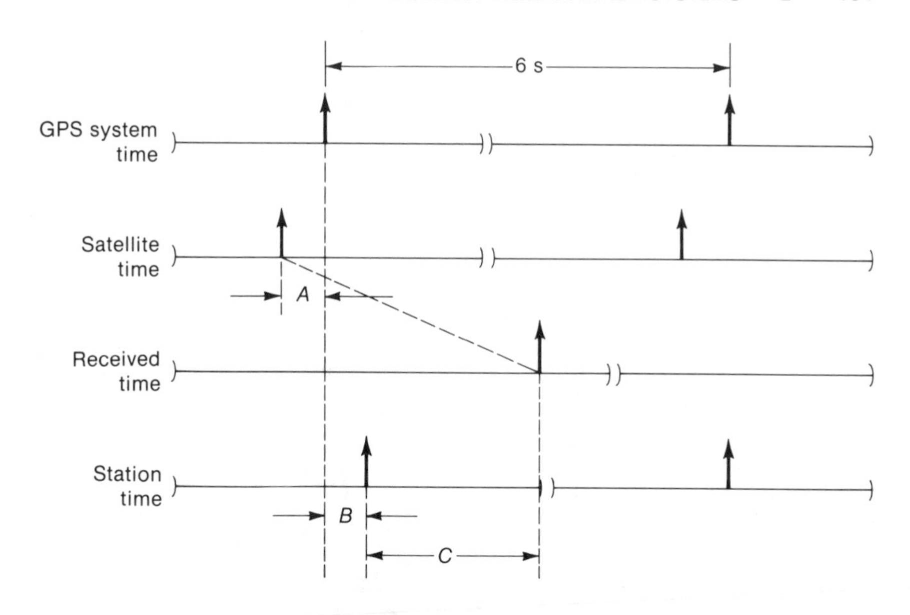

Figure 10.22 GPS Time Relationships

arrival is recorded and compared with the expected value after correction for atmospheric and relativistic errors. The difference between the calculated and actual time of arrival can be displayed as the local time error or used to correct the local time or frequency standard [52].

The GPS time-measuring concept is shown in Figure 10.22. The difference between GPS time and onboard satellite time is designated time A. The master control system knows A, and this is transmitted to the satellite via uplink telemetry for satellite broadcast. Time A is decoded by the TTU and stored. Time B, the difference between GPS time and the user station time, is an unknown parameter. Time C, the difference between the user station time and the time received from GPS, is measured by the TTU. The satellite-to-user transmit time (T_T) can be estimated from satellite ephemeris and atmospheric propagation correction data. The time error between station clock and the GPS clock (B) is then equal to

$$B = T_T - (A + C) \tag{10.20}$$

Accuracy of time transfer to better than 1 μs and accuracy of frequency determination better than 5 parts in 10^{12} have been reported [53].

Defense Satellite Communications System

The U.S. Defense Satellite Communications System (DSCS) provides time transfer via satellite relay as depicted in Figure 10.23. Each station measures

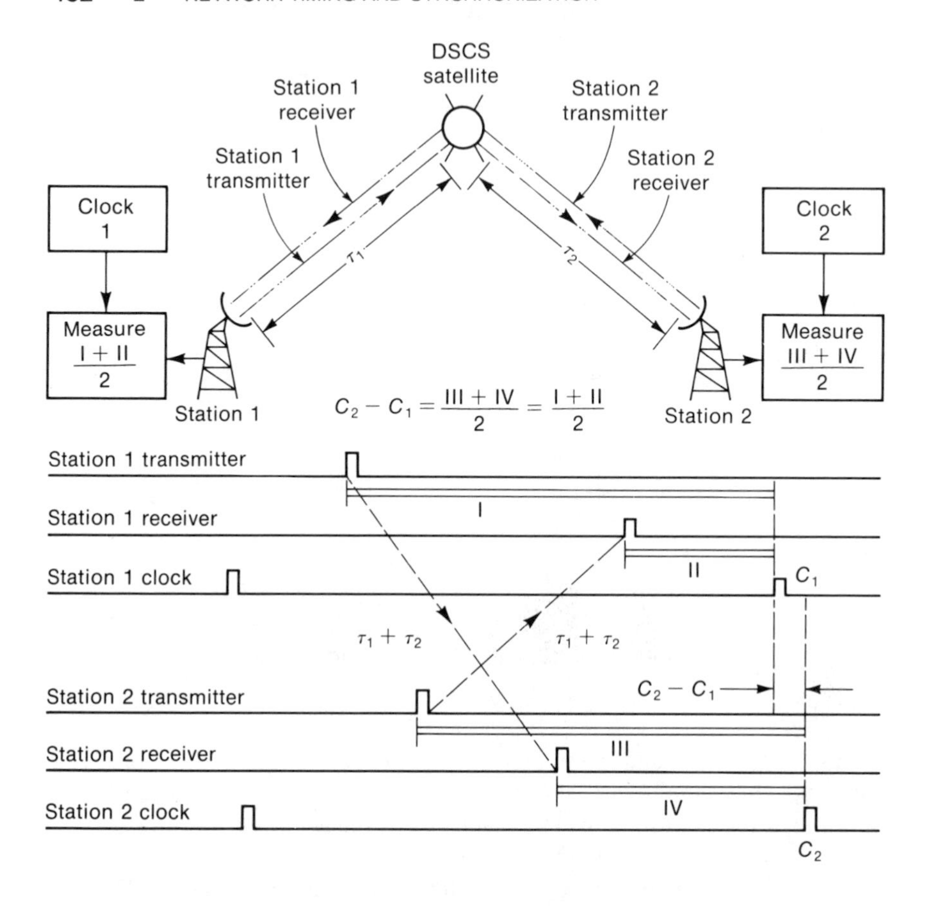

Figure 10.23 Time Transfer via Satellite (DSCS)

its own transmit time tick with respect to its clock and also measures the tick received from the other station with respect to the same clock. This approach assumes identical equipment and procedures at the two stations and reciprocity in the satellite paths. If the sum of the double measurements obtained at one station is subtracted from the sum of the double measurements taken at the other station, the result is twice the difference between the two clocks:

$$C_2 - C_1 = \frac{t_{III} + t_{IV}}{2} - \frac{t_I + t_{II}}{2} \tag{10.21}$$

The equipment configuration used for DSCS time transfer is depicted in Figure 10.24. The modem [55] makes use of spread-spectrum techniques and a pseudorandom code to transfer timing codes without interfering with communications. The time transfer unit compares the transmitted and

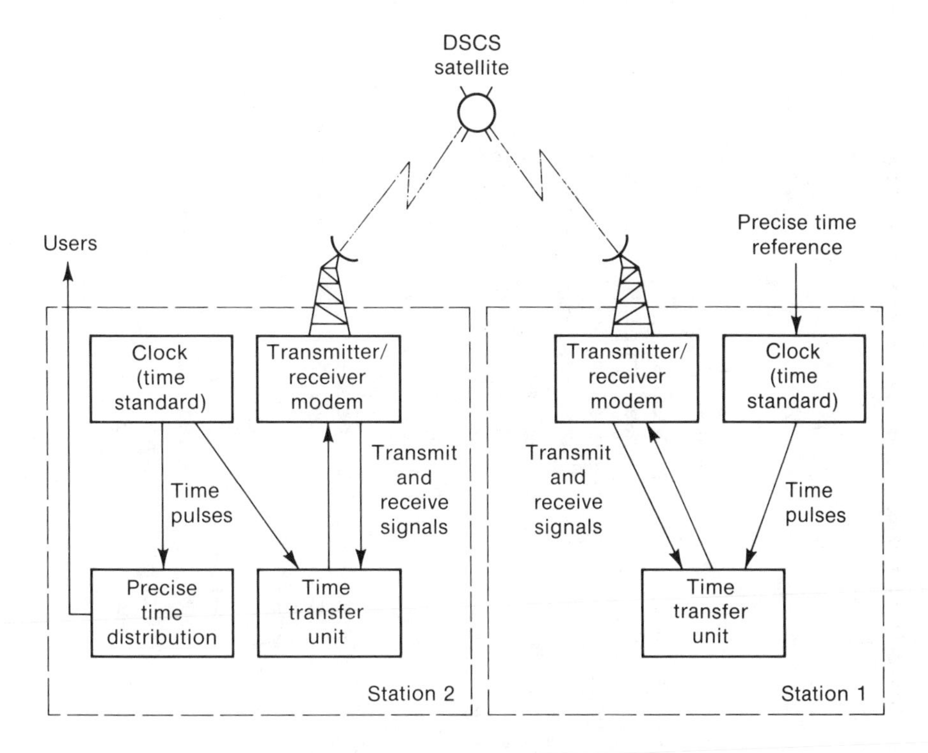

Figure 10.24 DSCS Time Transfer Equipment Configuration

received pseudorandom codes at each station with the local clock. The DSCS station having access to a precise time reference such as UTC acts as a master to the distant station where the local clock can be updated periodically by comparison with the master. Since clocks at DSCS terminals are driven by atomic (cesium) frequency standards, they need not be compared continuously. Time transfer to DSCS stations may be made daily, weekly, or monthly, depending on stability of the clock and required time accuracy. The DSCS method of PTTI transfer and calibration is accurate to within less than 1 μs. This DSCS time transfer capability is also used as a primary timing reference for certain LORAN C stations, which eliminates the need to send portable clocks for LORAN C calibration to UTC.

10.6 EXAMPLES OF NETWORK SYNCHRONIZATION SCHEMES

Table 10.4 lists salient characteristics of the network synchronization schemes employed by various national systems. The table shows that these network

Table 10.4 Examples of Existing Network Synchronization Systems (Adapted from [60])

Country	*Network*	*Service*	*Synchronization System*	*Master or Reference Clock*
Canada [27]	Dataroute	Data	Master-slave (self-organizing)	Rubidium
USA (AT&T) [28]	Digital data system	Data	Master-slave (loosely coupled)	Cesium
USA (AT&T) [2]	Switched digital network	Telephony	Master-slave (loosely coupled)	Cesium
USA (SPC) [29][a]	Data transmission (DATRAN)	Data	Master-slave (loosely coupled)	Rubidium
Japan (NTT) [25]	Digital data network	Data	Master-slave (fixed hierarchy)	Rubidium
Japan (NTT) [56]	Digital telephone system	Telephony etc.	Master-slave (loosely coupled)	Cesium
Italy [57]	Digital network	Telephony etc.	Master-slave (self-organizing)	Atomic standard
United Kingdom [58]	Digital network	Telephony etc.	Mutual (double-ended hierarchical)	Cesium
West Germany [59]	Digital network	Telephony etc.	Mutual (single-ended)	(unpublished)

[a]Southern Pacific Communications.

Table 10.5 CCITT Rec. G.811: Plesiochronous Operation of Digital Links [61]

Occurrence of Slips

Not greater than one controlled slip in every 70 days per digital international link for any 64-kb/s channel

Network Nodal Clocks

Long-term frequency inaccuracy not greater than 1 part in 10^{11}

Time Interval Error (TIE)[a]

Measurement Interval S (s)	Allowed TIE (ns)
$S < 5$	$(100S)$ ns + $\frac{1}{8}$ unit interval (UI)
$5 \leq S \leq 500$	$(5S + 500)$ ns
$S > 500$	$(10^{-2}S + 3000)$ ns

[a]See Figure 10.25 for plots of TIE for various data rates

synchronization systems are either master-slave or mutual, and no plesiochronous system is planned for use in a national network [60].

The CCITT [1] has specified plesiochronous operation of network synchronization for international connections, thus allowing national systems to choose their own form of network synchronization. Table 10.5 lists the salient performance characteristics of CCITT Rec. G.811 for plesiochronous operation of international digital links in terms of allowed slip rate, clock inaccuracy, and time interval error (TIE). These characteristics are based on ideal undisturbed (that is, theoretical) conditions although in practice performance may be degraded; for example, CCITT Rec. G.822 permits one slip per 5 hr on a 64-kb/s international connection versus one slip in 70 days specified by Rec. G.811. The restrictions on time interval error, shown in Table 10.5 and Figure 10.25, are imposed to facilitate the evolution from plesiochronous to synchronous operation.

10.7 SUMMARY

The development and widespread use of digital transmission, in combination with digital switching, has led to all-digital networks that require timing and synchronization. Sources of precise time and frequency begin with time standards, both atomic and astronomical. Today, Coordinated Universal Time (UTC) is a universally accepted time scale that is used by communication networks as a source of precise time.

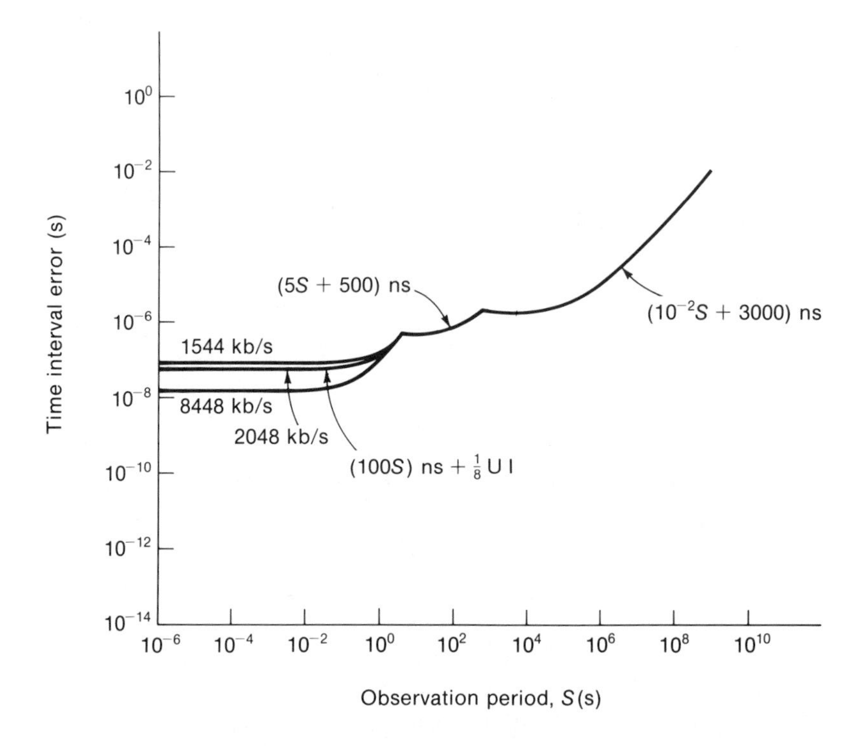

Figure 10.25 CCITT Rec. G.811 for Permissible Time Interval Error vs. Observation Time for a Reference Clock [61]

Frequency sources used in communication applications include quartz crystal oscillators for medium-quality requirements and atomic oscillators, such as cesium and rubidium, for high-quality requirements. These frequency sources are characterized by their accuracy, or degree to which the frequency agrees with its specified value, and by their stability, or the rate at which the frequency changes in time. Quartz oscillators are inexpensive and have good short-term stability but exhibit long-term instability. Cesium and rubidium standards both offer the short-term stability of the quartz oscillator but provide greater long-term stability that is important to communication applications.

Network synchronization schemes are divided into two categories: nondisciplined, wherein all nodal clocks are independent of a reference, and disciplined, wherein each nodal clock is synchronized to a reference. Both pulse stuffing (see Chapter 4) and plesiochronous operation are nondisciplined forms. Both schemes suffer from occasional losses of bit count integrity and a requirement for buffering of all signals. Of the disciplined forms of network synchronization, the master-slave system is the simplest and most popular. For this system a master frequency is distributed to all slave nodes via transmission paths. Secondary frequency sources and alternative transmission paths provide backup to the normally available master. Another disciplined scheme, mutual synchronization, uses a weighted average of clock signals derived from incoming transmission paths to obtain each nodal clock frequency. Although widely studied, this scheme has been implemented in very few applications. Another form of disciplined synchronization is available through external reference to one of several time and frequency dissemination systems.

Time and frequency dissemination systems useful in network synchronization are based on radio broadcasting via both terrestrial and satellite media. OMEGA uses eight VLF radio stations to broadcast a sequence of frequencies derived from cesium clocks and referenced to UTC. LORAN C is based on worldwide chains of stations that broadcast codes on a 100 kHz carrier, also derived from cesium clocks and referenced to UTC. Although both OMEGA and LORAN C are navigational systems (operated by the U.S. Coast Guard), both provide nearly global dissemination of time and frequency information. High-frequency radio broadcasts of time and frequency, encoded by means of tones and time ticks, also provide nearly global coverage. WWV (Colorado) and WWVH (Hawaii) HF stations provide full U.S. coverage. The major TV networks in the United States use atomic standards in broadcasting such that certain TV signals (color subcarrier and Line 10) can be used as a source of frequency calibration. Several satellite systems are currently used for dissemination of time and frequency, including the GOES environmental satellites, TRANSIT navigational system, and DSCS communication satellites. The Global Positioning System being implemented in the 1980s will provide position, time, and frequency information using 18 satellites for full earth coverage.

Network synchronization schemes for national systems are based principally on the master-slave concept. For international digital connections, the CCITT has prescribed the use of plesiochronous operation, thus allowing national systems to interoperate using separate network synchronization schemes.

REFERENCES

1. CCITT Yellow Book, vol. III.3, *Digital Networks—Transmission Systems and Multiplexing Equipment* (Geneva: ITU, 1981).
2. J. E. Abate, C. A. Cooper, J. W. Pan, and I. J. Shapiro, "Synchronization Consideration for the Switched Digital Network," *ICC '79 Conference Record*, June 1979, pp. 11.5.1 to 11.5.4.
3. J. C. Luetchford, J. Heynen, and J. W. Bowick, "Synchronization of a Digital Network," *ICC '79 Conference Record*, June 1979, pp. 11.3.1 to 11.3.7.
4. 13th Annual Conference on Weights and Measures, Paris, 1968.

5. G. Kamas and S. L. Howe, eds., "Time and Frequency Users' Manual," NBS Special Publication 559 (Washington: Government Printing Office, 1979).
6. D. W. Allan, J. E. Gray, and H. E. Machlan, "The National Bureau of Standards Atomic Time Scales: Generation, Dissemination, Stability, and Accuracy," *IEEE Trans. Instr. and Meas.*, vol. IM-21, no. 4, November 1972, pp. 388–391.
7. B. Guinot and M. Granveaud, "Atomic Time Scales," *IEEE Trans. Instr. and Meas.*, vol. IM-21, no. 4, November 1972, pp. 396–400.
8. J. T. Henderson, "The Foundation of Time and Frequency in Various Countries," *Proc. IEEE* 60(5)(May 1972):487–493.
9. J. A. Barnes and G. M. R. Winkler, "The Standards of Time and Frequency in the USA," *Proc. 26th Ann. Symp. on Frequency Control* (Ft. Monmouth, N.J.: Electronics Industries Assn., 1972).
10. H. M. Smith, "International Time and Frequency Coordination," *Proc. IEEE* 60(5)(May 1972):479–487.
11. H. Hellwig, "Frequency Standards and Clocks: A Tutorial Introduction," NBS Technical Note 616 (Washington: Government Printing Office, 1972).
12. J. A. Barnes and others, "Characterization of Frequency Stability," *IEEE Trans. Instr. and Meas.*, vol. IM-20, no. 2, May 1971, pp. 105–120.
13. J. A. Barnes, "Atomic Timekeeping and the Statistics of Precision Signal Generators," *Proc. IEEE* 54(2)(Feb. 1966):207–220.
14. E. A. Gerber and R. A. Sykes, "Quarter Century of Progress in the Theory and Development of Crystals for Frequency Control and Selection," *Proc. 25th Ann. Symp. on Frequency Control* (Ft. Monmouth, N.J.: Electronics Industries Assn., 1971).
15. "Fundamentals of Quartz Oscillators," Hewlett-Packard Application Note 200-2.
16. W. G. Cady, *Piezoelectricity: An Introduction to the Theory and Applications of Electromechanical Phenomena in Crystals* (New York: Dover, 1964).
17. H. Fruehauf, "Atomic Oscillators Stand Test of Time," *Microwaves and RF*, vol. 23, no. 6, June 1984, pp. 95–124.
18. A. S. Risley, "The Physical Basis of Atomic Frequency Standards," NBS Technical Note 399 (Washington: Government Printing Office, 1971).
19. D. H. Throne, "A Rubidium Vapor Frequency Standard for Systems Requiring Superior Frequency Stability," *Hewlett-Packard Journal* 19(11)(July 1968):8–16.
20. M. C. Thompson, Jr., and H. B. Jones, "Radio Path Length Stability of Ground-to-Ground Microwave Links," NBS Technical Note 219 (Washington: Government Printing Office, 1964).
21. Peter Alexander and J. W. Graham, "Time Transfer Experiments for DCS Digital Network Timing," *Proc. 9th Annual PTTI Applications and Planning Meeting*, March 1978, pp. 503–522.
22. P. A. Bello, "A Troposcatter Channel Model," *IEEE Trans. Comm. Tech.*, vol. COM-17, no. 2, April 1969, pp. 130–137.
23. J. J. Spilker, Jr., *Digital Communications by Satellite* (Englewood Cliffs, N.J.: Prentice-Hall, 1977).
24. R. L. Mitchell, "Survey of Timing/Synchronization of Operating Wideband Digital Communications Networks," *Proc. 10th Annual PTTI Applications and Planning Meeting*, November 1978, pp. 405–435.

25. N. Inoue, H. Fukinuki, T. Egawa, and N. Kuroyanagi, "Synchronization of the NTT Digital Network," *ICC '76 Conference Record*, June 1976, pp. 25.10 to 25.15.

26. G. P. Darwin and R. C. Prim, "Synchronization in a System of Interconnected Units," U.S. Patent 2,986,732, May 1961.

27. J. G. Baart, S. Harting, and P. K. Verma, "Network Synchronization and Alarm Remoting in the Dataroute," *IEEE Trans. on Comm.*, vol. COM-22, no. 11, November 1974, pp. 1873–1877.

28. B. R. Saltzberg and H. M. Zydney, "Digital Data Network Synchronization," *Bell System Technical Journal* 54(May–June 1975):879–892.

29. F. T. Chen, H. Goto, and O. G. Gabbard, "Timing Synchronization of the DATRAN Digital Data Network," *National Telecommunications Conference Record*, December 1975, pp. 15.12 to 15.16.

30. R. G. DeWitt, "Network Synchronization Plan for the Western Union All-Digital Network," *Telecommunications* 8(7)(July 1973):25–28.

31. M. B. Brilliant, "The Determination of Frequency in Systems of Mutually Synchronized Oscillators," *Bell System Technical Journal* 45(December 1966):1737–1748.

32. A. Gersho and B. J. Karafin, "Mutual Synchronization of Geographically Separated Oscillators," *Bell System Technical Journal* 45(December 1966):1689–1704.

33. M. B. Brilliant, "Dynamic Response of Systems of Mutually Synchronized Oscillators," *Bell System Technical Journal* 46(February 1967):319–356.

34. M. W. Williard, "Analysis of a System of Mutually Synchronized Oscillators," *IEEE Trans. on Comm.*, vol. COM-18, no. 5, October 1970, pp. 467–483.

35. M. W. Williard and H. R. Dean, "Dynamic Behavior of a System of Mutually Synchronized Oscillators," *IEEE Trans. on Comm.*, vol. COM-19, no. 4, August 1971, pp. 373–395.

36. J. Yamato, S. Nakajima, and K. Saito, "Dynamic Behavior of a Synchronization Control System for an Integrated Telephone Network," *IEEE Trans. on Comm.*, vol. COM-22, no. 6, June 1974, pp. 839–845.

37. J. Yamato, "Stability of a Synchronization Control System for an Integrated Telephone Network," *IEEE Trans. on Comm.*, vol. COM-22, no. 11, November 1974, pp. 1848–1853.

38. U.S. Coast Guard, U.S. Department of Transportation, "OMEGA Global Radionavigation—A Guide for Users," Washington, D.C., November 1983.

39. E. R. Swanson, "Omega," *Proc. IEEE* 71(10)(October 1983):1140–1155.

40. L. D. Shapiro, "Time Synchronization from LORAN-C," *IEEE Spectrum* 5(8)(August 1968):46–55.

41. C. E. Botts and B. Wiedner, "Precise Time and Frequency Dissemination via the LORAN-C System," *Proc. IEEE* 60(5)(May 1972):530–539.

42. R. L. Frank, "Current Developments in LORAN-C," *Proc. IEEE* 71(10)(October 1983):1127–1139.

43. N. Hironaka and C. Trembath, "The Use of National Bureau of Standards High Frequency Broadcasts for Time and Frequency Calibrations," NBS Technical Note 668 (Washington: Government Printing Office, 1975).

44. P. P. Viezbicke, "NBS Frequency-Time Broadcast Station WWV, Fort Collins, Colorado," NBS Technical Note 611 (Washington: Government Printing Office, 1971).
45. J. Tolman, V. Ptacek, A. Soucek, and R. Stecher, "Microsecond Clock Comparison by Means of TV Synchronizing Pulses," *IEEE Trans. Instr. and Meas.*, vol. IM-16, September 1967, pp. 247–254.
46. D. D. Davis, J. L. Jespersen, and G. Kamas, "The Use of Television Signals for Time and Frequency Dissemination," *Proc. IEEE* 58(6)(June 1970):931–933.
47. R. L. Easton and others, "Dissemination of Time and Frequency by Satellite," *Proc. IEEE* 64(10)(October 1976):1482–1493.
48. D. W. Hanson and others, "Time from NBS by Satellite (GOES)," *Proc. 8th Annual PTTI Applications and Planning Meeting*, November 1976.
49. R. E. Bechler and others, "Time Recovery Measurements Using Operational GOES and TRANSIT Satellites," *Proc. 11th Annual PTTI Applications and Planning Meeting*, November 1979.
50. T. D. Finsod, "Transit Satellite System Timing Capabilities," *Proc. 10th Annual PTTI Applications and Planning Meeting*, November 1978.
51. R. P. Denaro, "NAVSTAR: The All-Purpose Satellite," *IEEE Spectrum* 18(5)(May 1981):35–40.
52. J. T. Witherspoon and L. Schuchman, "A Time Transfer Unit for GPS," *Proc. 9th Annual PTTI Applications and Planning Meeting*, March 1978, pp. 167–176.
53. J. A. Buisson, R. L. Easton, and T. B. McCaskill, "Initial Results of the NAVSTAR GPS NTS-2 Satellite," *Proc. 9th Annual PTTI Applications and Planning Meeting*, March 1978, pp. 177–199.
54. G. W. Parkinson and S. W. Gilbert, "NAVSTAR: Global Positioning System—Ten Years Later," *Proc. IEEE* 71(10)(October 1983):1177–1186.
55. J. A. Murray, Jr., "Mini-Modem for PTTI Dissemination," *Proc. 3rd Annual PTTI Planning Meeting*, November 1971, pp. 16–18.
56. N. Inoue, H. Fukinuki, and R. Komiya, "Higher-Bit-Rate Terminals in a Synchronous Digital Network," *ICC '78 Conference Record*, pp. 11.5.1 to 11.5.6.
57. M. Decina, A. Pietromarchi, A. Bovo, and L. Musumeci, "Development of Network Synchronization Techniques in Italy," *ICC '76 Conference Record*, pp. 25.21 to 25.25.
58. P. A. Mitchell and R. A. Boulter, "Synchronization of the Digital Network in the United Kingdom," *ICC '79 Conference Record*, pp. 11.2.1. to 11.2.4.
59. W. R. Slabon, "Result of Investigations for the Clock Frequency Control and Distribution System in the Digital Telephony and Data Networks of the Deutsche Bundespost and Future Plans," *Proc. Frequency Control Symposium*, 1977, p. 445.
60. K. Okimi and H. Fukinuki, "Master-Slave Synchronization Techniques," *IEEE Comm. Mag.* 19(3)(May 1981):12–21.
61. "Part of Report of Working Party 4—Draft Revision of Recommendation G.811," CCITT Study Group XVIII, Question 3/XVIII, Geneva, 1 June 1984.

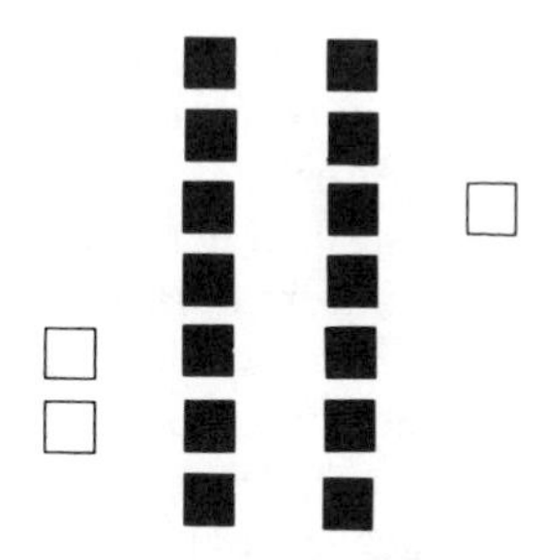

Transmission System Testing, Monitoring, and Control

OBJECTIVES

- Discusses techniques for testing the performance of digital transmission systems in terms of bit error rate, jitter, bit count integrity, and equipment reliability
- Describes techniques for continuous monitoring of system and equipment performance
- Covers the techniques for isolating faults in system performance
- Considers various schemes for reporting alarms, monitors, and restoral actions to other stations in the system

11.1 INTRODUCTION

During the process of installation, operation, and maintenance of a digital transmission system, the system operator must be able to test, monitor, and control the system on an end-to-end basis. This capability allows verification of design objectives as part of system installation. After system commission, components must be monitored and controlled to ensure that operation and maintenance (O&M) performance standards are met.

This chapter discusses techniques for testing digital transmission systems for performance-related parameters such as bit error rate, jitter, bit count integrity, and equipment reliability. These methods are suitable for factory or acceptance testing prior to system commission for operation. During system operation, maintenance personnel must be provided alarm and status indications that allow identification of failed equipment, its repair or replacement, and restoral of service. Fault isolation within a network requires performance monitors and alarm indications associated with each piece of equipment and each station. Performance monitors are described here that allow system and equipment margins to be continuously monitored, usually by estimation of the bit error rate. Fault isolation techniques identify the need for prompt maintenance when service has been interrupted or deferred maintenance when automatic switchover to standby units has been used to restore service or when standby units fail leaving no backup to the operational unit. Since maintenance actions taken at one station may affect other stations, it becomes necessary to report alarms, monitors, and restoral actions to other stations, often at a central location. In discussing various schemes used for reporting, the emphasis is on the unattended remote station where it may be necessary to remotely control redundant equipment.

It should be noted that the O&M performance standards discussed here differ from the design objectives discussed in Chapter 2. Various sources of degradation not always included in the design methodology—such as aging, acts of nature, and human error—result in performance limits during the lifetime of the operating system. Further, real circuits differ from the hypothetical reference circuit used in design methodology because of different length, media, equipment, and so forth. These differences must be taken into account when specifying and measuring O&M performance standards.

11.2 TESTING TECHNIQUES

Performance specifications or standards are verified by testing, whether in the factory or in an operational environment. To facilitate the testing process, these specifications and standards should include descriptions of the tests to be used: test equipment, configurations, procedures, and the like. This section describes commonly used techniques for testing the key performance parameters of digital transmission systems including bit error rate, jitter, and bit count integrity. The testing of individual transmission equipment and links for other performance specifications, such as reliability and radio fade margin, is also discussed.

11.2.1 Bit Error Rate Measurement

The most commonly used performance indicator in the testing of digital transmission systems is bit error rate (BER). As discussed in Chapter 2, error performance can be expressed in many forms, such as errored seconds (see the accompanying box), errored blocks, and average BER. Usually the error parameter used in measuring system performance is selected to match the error parameter used in the system design process to allocate performance. Whatever the error parameter, there are two general approaches to measurement: **out-of-service** and **in-service**. In the case of out-of-service measurement, operational traffic is replaced by a known test pattern. Here a pseudorandom binary sequence (PRBS) is used to simulate traffic. The received test pattern is compared bit by bit with a locally generated pattern to detect errors. The repetition period of the test pattern, given by $2^n - 1$ for a shift register of n bits, is selected to provide a sufficiently smooth spectrum for the system data rate. The most common patterns for standard data rates are shown in Table 11.1. Since out-of-service measurement eliminates traffic carrying capability, it is therefore best suited to production testing, installation testing, or experimental systems.

Figure 11.1 is a block diagram of a typical configuration for BER measurement. At the transmitter and receiver, the test pattern of the PRBS generator and detector is selected from available options. At the receiver, a form of error measurement is selected, which might include options for error pulses, average BER, errored seconds, or errored blocks, depending on the design of the bit error rate tester (BERT). The measured error performance is given by visual display and may also be available in printed form. Usually the BER test set includes some diagnostics, such as the ability to force errors at the transmitter and observe them at the receiver and an indication of loss of synchronization at the receiver.

Error-Free Second Measurements

Error-free seconds, a form of error measurement, is expressed as the percentage of error-free seconds using measurement periods of 1 s. There are two ways of measuring errored seconds: *synchronous* and *asynchronous*. In the synchronous mode an errored second is defined as a 1-s interval following the occurrence of the first error. The advantage of the synchronous mode is that measurements made with different instruments yield the same reading on the same link. Its disadvantage is that the measure of synchronous errored seconds does not directly yield error-free seconds but rather error-free time. In the asynchronous mode each discrete 1-s interval is checked for errors. The advantage of the asynchronous mode is that it directly yields error-free seconds. The disadvantage is that different measurements may be obtained with different equipment. CCITT Rec. G.821 implies the use of asynchronous errored seconds; the emerging industry standard in North America, however, is based on synchronous errored seconds.

In-service error measurement is possible when the traffic has an inherent repetitive pattern, the line format has inherent error detection, or the

Table 11.1 Pseudorandom Binary Sequences Recommended by the CCITT for the Measurement of Error Rate

Applicable Bit Rates	*Pattern Length*	*CCITT Recommendation [1, 2]*
Up to 20 kb/s	$2^9 - 1$	V.52
20 to 72 kb/s	$2^{20} - 1$	V.57
1.544 Mb/s	$2^{15} - 1$	O.151
2.048 Mb/s	$2^{15} - 1$	O.151
6.312 Mb/s	$2^{15} - 1$	O.151
8.448 Mb/s	$2^{15} - 1$	O.151
32.064 Mb/s	$2^{15} - 1$	O.151
34.368 Mb/s	$2^{23} - 1$	O.151
44.736 Mb/s	$2^{15} - 1$	O.151
139.264 Mb/s	$2^{23} - 1$	O.151

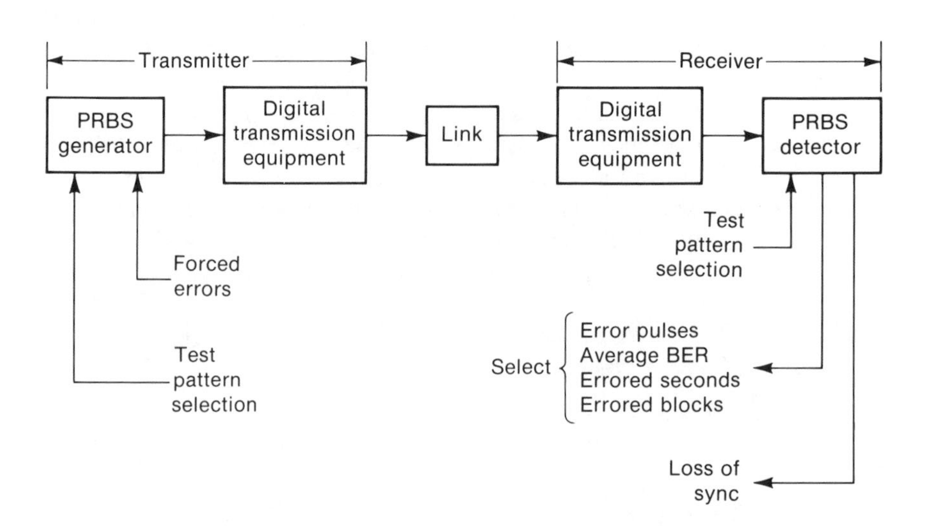

Figure 11.1 Block Diagram of BER Measurement

received signal is monitored for certain threshold crossings. Thus in-service techniques only estimate the error rate and do not yield a true measurement. These techniques are useful as performance monitors during live system operation. Further, if the error rate estimate can be made quickly enough, in-service schemes can be used to control redundancy or diversity protection switching. Categories of in-service error rate measurement will be described later in this chapter when we consider performance monitoring techniques.

Measured bit error rates have little significance unless a confidence level is stated. For bit error rate measurements, the *confidence level* is defined as the probability that the measured error rate is within an accuracy factor α of the true average BER or, in terms of number of errors,

$$\text{Confidence level} = P(np \leq \alpha k_1) \tag{11.1}$$

where k_1 = number of measured errors
p = probability of bit error
n = number of trial bits
np = expected number of errors

A mathematical model for the confidence level is well known for the case where errors are independently distributed. Here the Bernoulli trials model can be used to yield the probability that αk_1 or fewer errors occur in n independent trials:

$$P(\leq \alpha k_1 \text{ errors in } n \text{ bits}) = \sum_{k=0}^{\alpha k_1} \binom{n}{k} p^k q^{n-k} \tag{11.2}$$

where $q = 1 - p$. Expression (11.2) is difficult to evaluate for large n, but approximations exist for two common cases. If n is so large that the expected number of errors is also large ($np \gg 1$), the expression may be approximated by

$$P(\leq \alpha k_1 \text{ errors in } n \text{ bits}) = 1 - \text{erfc}\left[(\alpha - 1)\sqrt{k_1}\right] \qquad (np \gg 1) \tag{11.3}$$

where the complementary error function, erfc(x), was defined earlier in (5.28). In the case where p is small and n is large, so that the expected number of errors is small ($np \approx 1$), this same probability is more nearly Poisson as given by

$$P(np \leq \alpha k_1) = 1 - e^{-\alpha k_1} \sum_{k=0}^{k_1} \frac{(\alpha k_1)^k}{k!} \qquad (np \approx 1) \tag{11.4}$$

Expressions (11.3) and (11.4) can be solved for the probability (confidence level) as a function of the errors measured (k_1) and the actual bit error rate (np), as shown in Figure 11.2 for Expression (11.3). Alternatively, one can fix the probability and solve for the actual BER as a function of measured errors, as shown in Figure 11.3 for 99 percent confidence limits. In both Figures 11.2 and 11.3 the plotted curves indicate probability that

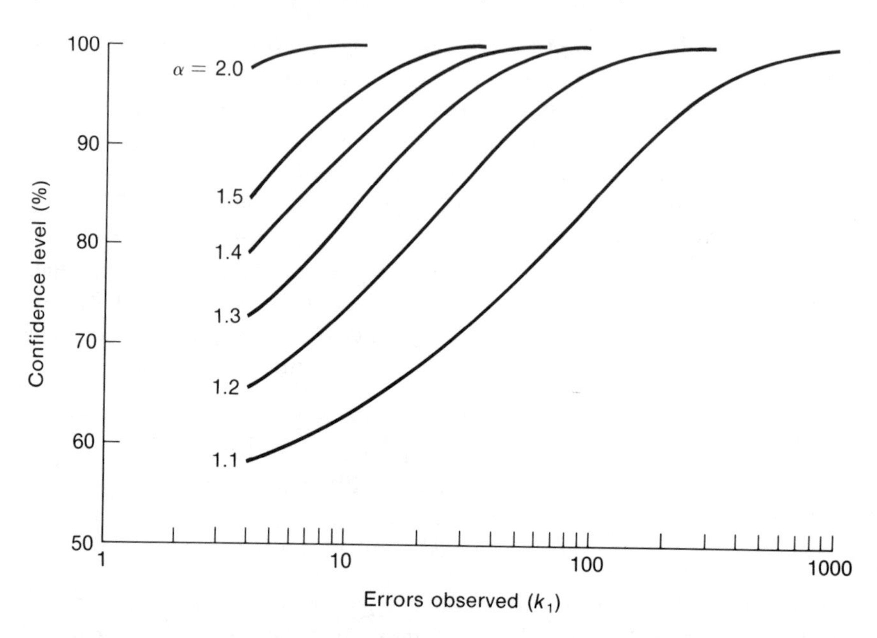

Figure 11.2 Confidence Level That Actual BER Is Less Than αk_1 (Adapted from [3])

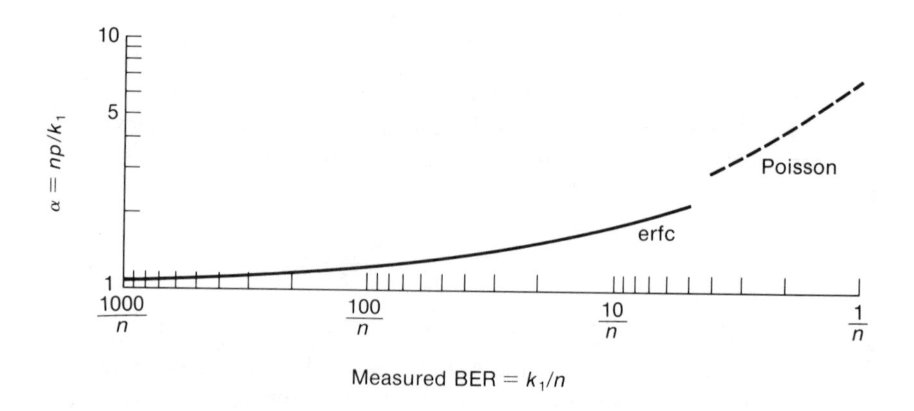

Figure 11.3 99 Percent Confidence Level That Actual BER Is Less Than α k_1

the actual BER is less than α times the measured BER, where $\alpha = np/k_1$. With 10 errors recorded, for example, the actual BER is within a factor of two times the measured BER with 99 percent confidence.

Example 11.1

Suppose that for acceptance of a 10-Mb/s system, the actual BER must be less than 1×10^{-9} with a 90 percent confidence. Assuming that the errors are independently distributed, what duration of test would be required and how many errors would be allowable?

Solution

From Figure 11.2 or Expression (11.3) we see that for seven measured errors there exists a 90 percent confidence that the actual BER is less than 1.5 times the measured BER. Therefore if we measure over $(1.5)(7)(10^9) = 1.05 \times 10^{10}$ bits, we are 90 percent confident that the actual BER is less than 10^{-9} if seven or fewer errors are recorded. The measurement period required is then

$$\frac{(1.5)(7)(10^9)\text{ bits}}{10\text{ Mb/s}} = 1050\text{ s}$$

11.2.2 Jitter Measurement

Recall from Chapter 2 that jitter performance is characterized in three ways:

- Amount of jitter tolerated at input (tolerable input jitter)
- Output jitter in the absence of input jitter (intrinsic jitter)
- Ratio of output to input jitter (jitter transfer function)

The instrumentation required to measure jitter for these three characteristics is shown in Figure 11.4. These measurements are all out-of-service and are conducted typically with factory or system acceptance tests.

The tolerance of digital transmission equipment to input jitter can be measured by the test setup of Figure 11.4*a*. The source of input jitter consists of a frequency synthesizer, jitter generator, and a pseudorandom test pattern generator. The clock signal produced by the frequency synthesizer is modulated by the jitter generator and then used to clock the pattern generator. The output of the pattern generator is passed through the unit under test from input port to output port. The pattern detector then accepts the output port signal and provides a detected error output to a counter. The amplitude of the induced jitter is incrementally increased until bit errors are detected. This test is repeated for a range of jitter frequencies, thus permitting a plot of maximum tolerable input jitter versus jitter frequency to be constructed for the equipment under test. Figure 2.19 is an

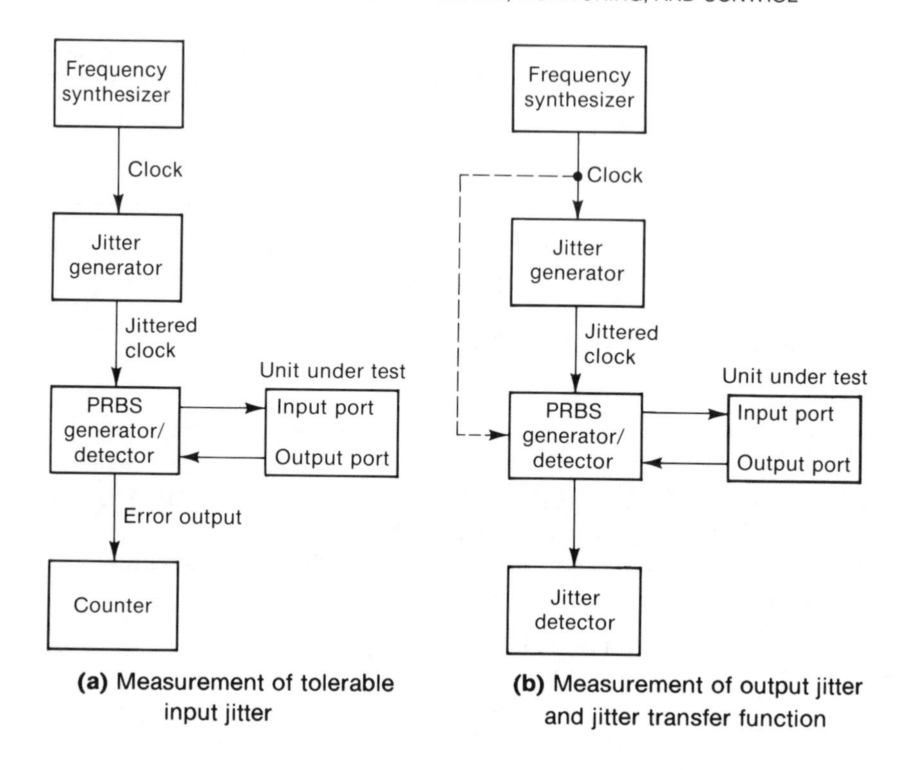

Figure 11.4 Block Diagram for Jitter Measurement

example of such a plot of maximum tolerable input jitter for several digital transmission hierarchical levels. The type of jitter modulation must also be specified; here sinusoidal jitter is often specified for convenience of testing, even though it is not representative of the type of jitter found in a network. For test purposes, a sufficiently long pseudorandom binary sequence is used as the test pattern, as specified for example in Table 11.1.

A test configuration for measuring intrinsic jitter and the jitter transfer function is shown in Figure 11.4*b*. This set of test equipment is similar to Figure 11.4*a* except that a jitter detector replaces the error detector. To measure output jitter in the absence of input jitter, the jitter generator is bypassed so that a jitter-free clock is used to generate the PRBS pattern; the received pattern is applied to the jitter detector, which measures jitter amplitude. The jitter transfer function is characterized by using a PRBS signal modulated by the jitter generator and by measuring the gain or attenuation of jitter in the detected PRBS signal. For both tests, the jitter amplitude, rms or peak-to-peak, is measured for selected bandwidths of interest. To accomplish this, the jitter measuring set may require bandpass, low-pass, or high-pass filters to limit the jitter bandwidth for the measurement of specified jitter spectra. The jitter measurement range is selected to be compatible with the equipment specifications of allowed output jitter or jitter transfer function. Test results are plotted in a manner similar to Figure 2.19. CCITT Rec. O.171 provides additional information on test configurations for jitter measurement [1].

11.2.3 Bit Count Integrity Testing

Bit count integrity (BCI) performance is determined by various parameters, such as signal-to-noise ratio in a clock recovery circuit, error rate in a multiplexer frame synchronization circuit, or clock accuracy and stability in a plesiochronous network. A test of BCI performance requires stressing the equipment or system to observe conditions under which BCI is lost and regained. Since BCI testing is done out-of-service, it is generally included as part of factory tests or system installation. As an example, the BCI performance of a digital multiplexer is usually specified for a certain error rate. To test this specification, the multiplexer aggregate data stream must be subjected to the specified error rate while measurement is made of the times between losses of BCI and the times to regain BCI. These measured times are compared to specified times in order to determine acceptance or rejection of the unit being tested.

11.2.4 Reliability Testing

Equipment reliability can be verified with a combination of factory testing, analysis, and field results. Nonredundant equipment can be tested directly to verify the mean time between failure (MTBF) specification. Procedures for such tests are well established by industry and government standards [4]. Using enough units under test at one time, it is possible to demonstrate reliability in relatively short periods of time through the use of sequential test plans such as that illustrated in Figure 11.5. Here the cumulative number of failures is plotted versus cumulative test time until an accept or reject decision is reached or until the maximum allowable test time (truncation) is reached.

For redundant equipment, direct testing is not a practical way of verifying mean time between outage (MTBO) because of the inordinate length of time that would be required. A recommended means of verifying the MTBO is to use analysis and MTBF testing. The analysis would produce a list of failure modes and effects, verify that performance sensors will detect failures, and verify that switching circuits will cause switchover from failed units to standby units. The MTBO can then be calculated from a demonstrated MTBF and predicted redundancy switching.

Reliability of equipment can also be monitored and recorded in actual field operation. This practice allows high-failure items to be identified and improved. For example, the CCIR has published field reliability results for a number of national radio systems that show equipment failure to be the

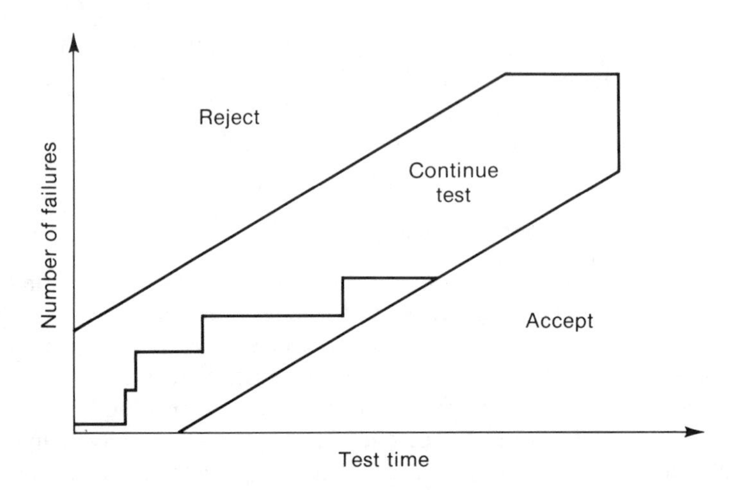

Figure 11.5 Reliability Test Plan Showing Accept / Reject / Continue Regions

dominant source of unavailability compared to outages due to human error, propagation, and primary power [5].

11.2.5 Link Testing

Before commission of a cable or radio link, testing is necessary to ensure that link installation has followed the design criteria. A first step is the verification of basic equipment interface compatibility and operation in a nonstressed environment. This type of test is facilitated by operating the equipment in a looped-back configuration rather than over the link. This test period is an appropriate time to identify cable mismatches, faulty grounds, and other defects in equipment installation.

The second phase of link acceptance testing requires operation of the transmission equipment over the link. Although it is impractical to measure and verify link performance objectives directly in a short test, certain design parameters related to link performance can be quickly measured. The signal-to-noise margin at a radio or cable receiver can be directly measured. Error rates can be measured for digital data; voice-channel measurements can be made for 4-kHz analog signals. During quiet periods, with no fading or failures, these design parameters under test should meet their specifications.

The link should also be artificially stressed to verify the design margins. In the case of radio links, for example, the fade margin is determined by the following procedure. First the nominal (nonfaded) received signal level

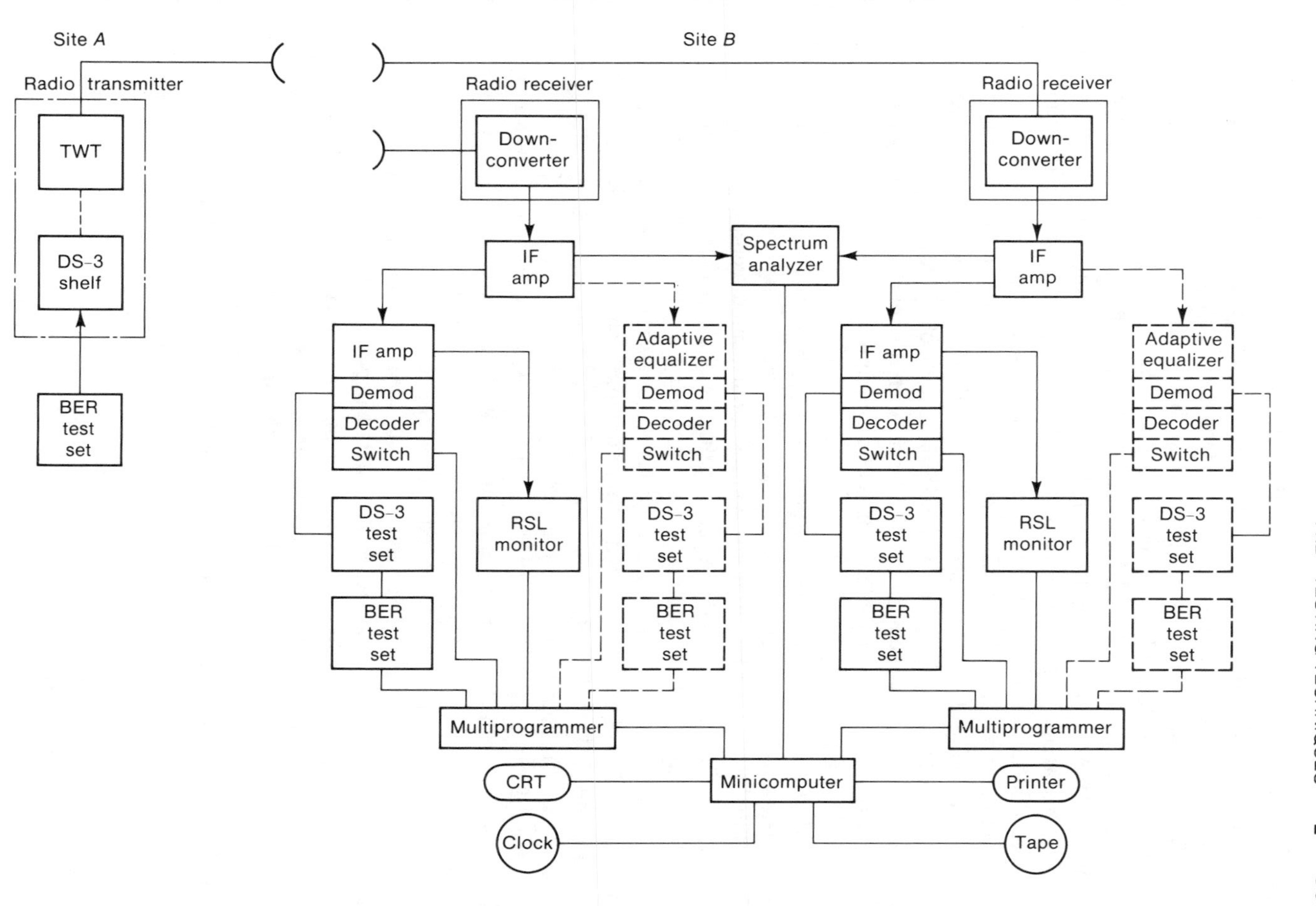

Figure 11.6 Configuration for Multipath Propagation Tests

(RSL) is determined by monitoring RSL during stable propagation time (normally around midday). A BER test is then conducted with a calibrated attenuator inserted in the received signal path and set to yield the threshold BER. The attenuator setting is then equal to the link fade margin. The fade margin should be determined separately for each received path of a diversity combiner output as well. Hysteresis of a selection combiner can be tested by fixing the attenuation of one diversity branch at a constant value and then varying the attenuation of the second diversity branch above and below the hysteresis design value. If the measured link fade margin and hysteresis are approximately equal to their design values, the link can be expected to meet performance objectives. Measured fade margins that fall short of design values indicate the possibility of inadequate path clearance or persistent anomalous propagation conditions. If the measured nominal RSL is significantly less than the calculated nominal RSL, antenna misalignment or waveguide losses should be suspected.

A direct measurement of link performance requires considerable test instrumentation and time. In the case of a radio link, the testing of anomalous propagation conditions requires sophisticated techniques to collect meteorological data and indices of refractivity along the full length of the propagation path. Significant test time may be required to collect sufficient data, since anomalous conditions are by definition rare. Testing of the effects of multipath fading in digital radio links requires instrumentation of both the RSL and BER. An example of a multipath test configuration, used to collect data presented in Chapter 9 (see Figures 9.27 and 9.44), is shown in Figure 11.6. This test configuration was designed for one-way transmission, with the transmitter at site *A* and the receiver at site *B* for a 90-Mb/s, 6-GHz, 8-PSK radio. The receiver site has two space diversity radio configurations—one equipped with adaptive equalizers and the other without equalizers. Both BER and RSL are measured simultaneously for both radios; a spectrum analyzer was shared between the two radios. The radio under test used selection combining, so that diversity switch position was also recorded with BER, RSL, and spectrum shape. Radio link testing as described here and shown in Figure 11.6 requires an extended test period to obtain statistically significant results. Both the CCITT and CCIR suggest a month of testing for measurement of bit error rate or unavailability.

11.3 PERFORMANCE MONITORING TECHNIQUES

Digital transmission systems are characterized by rugged performance when operating above their design thresholds and by abrupt degradation below threshold. Both of these characteristics of digital transmission are illustrated in Figure 11.7, which compares the performance of a PCM-derived VF

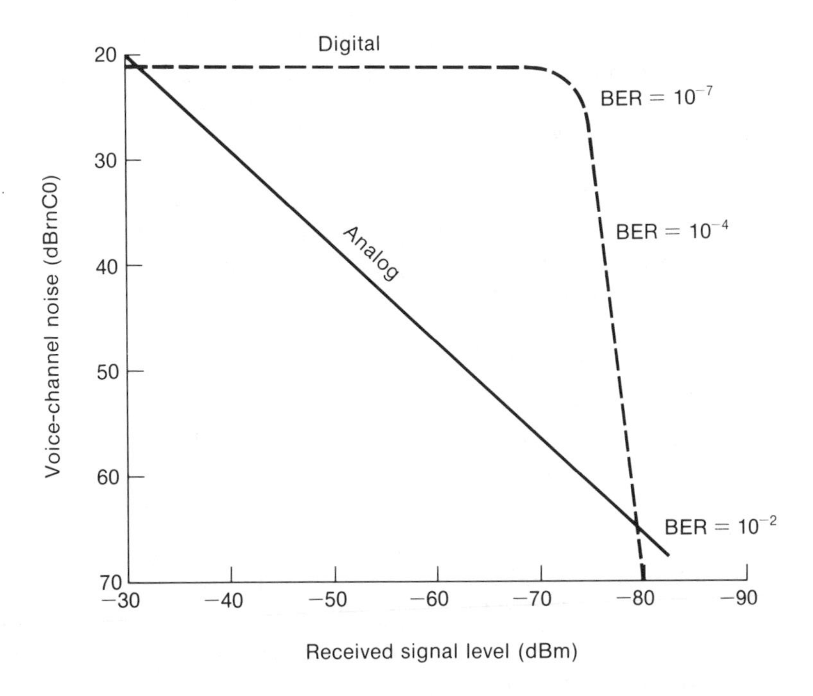

Figure 11.7 Performance of Analog vs. Digital Transmission

channel versus that of an FDM-derived VF channel for a typical radio link.* In comparing the performance of analog versus digital transmission, note that the FDM-derived VF channel suffers a gradual degradation in performance versus decreasing RSL while the PCM-derived VF channel maintains a high, constant level of performance versus decreasing RSL. Note also that below threshold the digital system performance rapidly degrades over a very small margin of RSL.

The inherent ruggedness and threshold effect of digital transmission create new problems in performance assessment when compared to analog transmission. The gradual degradation of an analog signal is translated directly into degraded user performance and can be easily monitored; however, gradual degradation of a digital signal has no discernable effect on system performance until errors are introduced. Once the digital signal has deteriorated to the point where errors result, only a small margin exists between initial errors and unacceptable performance. An effective performance monitor must be capable of constantly measuring the margin be-

*Voice-channel noise in Figure 11.7 is given in units of dBrnC0, which is defined as C-message weighted noise power in decibels referenced to 1 picowatt (−90 dBm) and measured at a zero transmission level point.

tween the state of the transmission channel and the threshold of that channel. A performance monitor indicating such a margin enables the system operator to implement diagnostic and maintenance procedures before user performance degrades.

In the design and use of performance monitors, several attributes should be considered. The following list of attributes applies equally to performance monitors used by system operators in monitoring system health and to those used within transmission equipment to control diversity or redundancy switching automatically:

- *In-service* techniques should be used to allow normal traffic and not affect the end user.
- *Simplicity* should be emphasized to maintain reliability and minimize cost.
- There should be *rapid* and *observable* response to any cause of system degradation.
- The *dynamic range* should extend from threshold to received signal levels that are significantly higher than threshold.
- Sufficient *resolution* should be provided to allow a continuous quantitative measure of the margin.
- Responses should be *stable* and *repeatable* to minimize or eliminate the need for calibration.
- A functional or empirical *relationship with BER* should be determinable.

The performance monitoring techniques described in the following paragraphs provide many of the attributes listed here. The suitability of each technique depends on the requirements and constraints of the system in question.

11.3.1 Test Sequence Transmission

Often spare channels exist in multiplex equipment or special test channels are provided in auxiliary service channels that allow transmission of a pseudorandom binary sequence at a low bit rate. This test sequence is interleaved with the traffic digital signal at a fraction $1/N$ of the system bit rate. At the receiver the BER is estimated by taking the ratio of errors detected to the number of transmitted test bits. Apart from the added overhead required of this technique, the time required to recognize a specified error rate is N times that required by counting errors of all the transmitted bits. The relatively long times required to estimate low error rates make this technique unsuitable for application to control of protection switching.

11.3.2 Frame Bit Monitoring

Digital multiplexers contain a repetitive frame pattern used for synchronization. Since this pattern is fixed and known beforehand, an estimate of BER can be obtained by measuring frame bit errors. For performance monitoring purposes, multiplex equipment often provides a digital pulse output for each frame bit error, so that only a counter is required to yield the frame bit error rate. In some cases the required counting and averaging are directly incorporated in the multiplexer, resulting in a front panel display of error rate, usually in decade steps—for example, 1×10^{-3}, 1×10^{-4}, 1×10^{-5}, and so forth. As an alternative, the framing signal can be monitored by an instrument separate from the multiplexer [1].

11.3.3 Parity Bit Monitoring

Check bits or parity bits are added to blocks of data or to digital multiplexer frames in order to provide error detection. At the transmitter a parity bit is added to each block of data to make the sum of all bits always odd or always even. At the receiver the sum of bits in each block or frame is compared to the value of the received parity bit. Disagreement indicates that an error has occurred. When used with digital radio, two types of parity bits are commonly added to each transmitted frame: a hop parity bit P_H and a section parity bit P_S. The P_H is examined at each radio regenerative repeater to indicate error occurrences on each hop. A new P_H that provides correct parity with the received data is then inserted into the frame for transmission over the next hop. In contrast, the value of P_S is not changed along a section, so that at any station along the radio section P_S can be checked to determine whether transmission errors have occurred up to that point [6].

Like many other in-service monitoring techniques, parity checking has limitations in its error detection capability. Only an odd number of errors in the block or frame will produce a parity error. Thus parity checking yields an accurate estimate of system error rate only for low error rates (below 10^{-N}, where N is the length of the block or frame).

11.3.4 Signal Format Violations

Certain multilevel coding techniques have inherent constraints on transitions from one level to another. A received signal can be monitored for violations of these level constraints to provide error detection. As mentioned earlier in Chapter 5, both bipolar and partial responses codes have intrinsic redundancy and are therefore candidates for this form of error monitoring. The level constraints vary for different codes. Some examples of level violations for standard codes are:

- Bipolar: two consecutive marks of the same polarity
- HDB3: two consecutive bipolar violations of the same polarity
- B6ZS: two consecutive marks of the same polarity excluding violations caused by the zero substitution code

11.3.5 Eye Pattern Monitoring

The monitoring of eye pattern opening can provide a qualitative measure of error performance. With signal degradation, the eye pattern becomes fuzzy and the opening between levels at the symbol sampling time becomes smaller. Noise and amplitude distortion close the vertical opening; jitter or timing changes close the horizontal opening. One approach in monitoring the eye pattern is to measure the eye closure and compare the measured values with those obtained for the system operating in an ideal environment. The eye closure can be expressed as system degradation in decibels by

$$\text{System degradation} = -20\log(1 - \text{eye closure}) \qquad (11.5)$$

Such a measure may be converted to a BER estimate by simple calibration of the eye pattern monitor. A more general technique for eye pattern monitoring is to perform analysis on selected bit patterns. This technique compares received eye patterns to the ideal eye pattern for a selected bit pattern, which may be useful in isolating the fault (noise, distortion, jitter) causing the eye closure.

11.3.6 Pseudoerror Detector

The pseudoerror detector is a technique that permits extrapolation of the error rate by modifying the decision regions in a secondary decision circuit to degrade the receiver margin intentionally and create a pseudoerror rate. The pseudoerror technique is implemented by offsetting the sampling instant [7], decision threshold [8], or both [9] from their optimum value and then counting the number of detected bits (pseudoerrors) that fall into the offset region. For a given type of degradation, there is a fixed relationship between the actual BER and pseudoerror rate (PER). This relationship depends on the amount of intentional offset selected for the pseudo decision circuit, but in general the PER is always greater than the BER. This multiplication factor is an important aspect of pseudoerror techniques, since error rate can be estimated in a shorter time, especially for low error rates, than by the other methods considered here, including both out-of-service and in-service. And since the pseudoerror technique detects channel impairment before actual errors are made, it can be used to control diversity switching in digital radio systems [10]. However, the multiplication factor may vary depending on the source of degradation (jitter, noise, multipath).

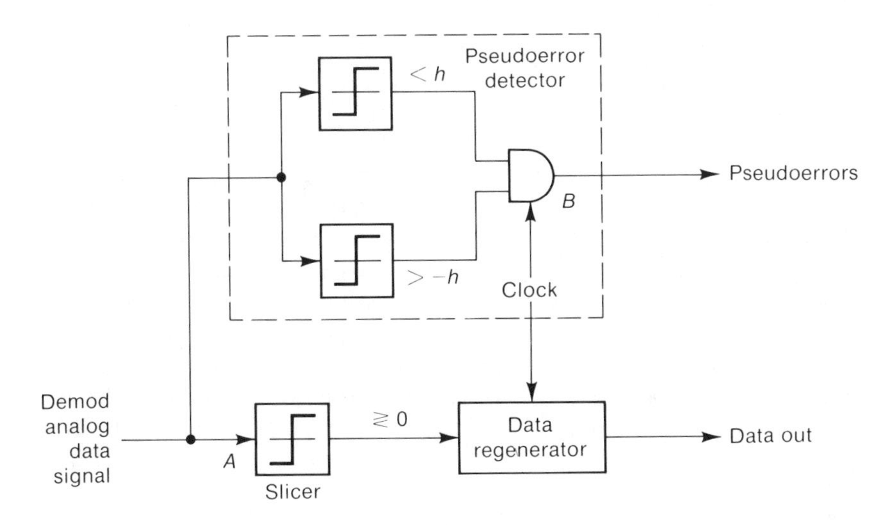

(a) Functional block diagram

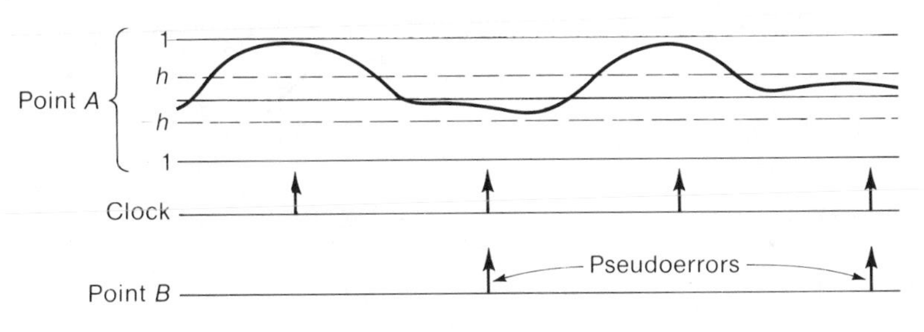

(b) Signal waveforms

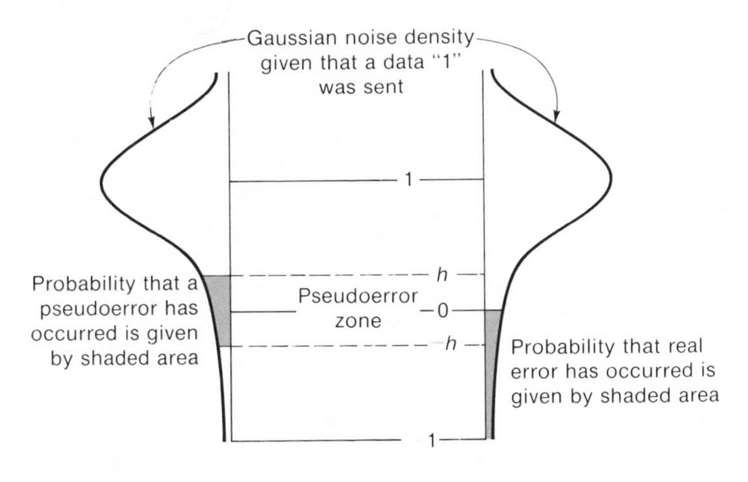

(c) Theory of operation

Figure 11.8 Pseudoerror Detection Technique

Hence the calibration of PER to BER should be related to the potential sources of degradation for the application.

A functional diagram of a pseudoerror detector based on decision threshold offset is shown in Figure 11.8. The pseudoerror zone is established by two slicers, one at h and the other at $-h$ relative to a full bit of height ± 1.0. The pseudoerror decision is made by using the same clock as the actual data decision. Note that the timing of the pseudoerror decision could be offset from the optimum sampling instant to further increase the probability of a pseudoerror. Likewise, the pseudoerror probability can be increased or decreased by adjustment of the pseudoerror zone height.

A model of the performance of the pseudoerror detector circuit in Figure 11.8*a* is shown in Figure 11.8*b* and *c*. The degradation is assumed to be additive gaussian noise with zero mean. For a transmitted 1, the noise

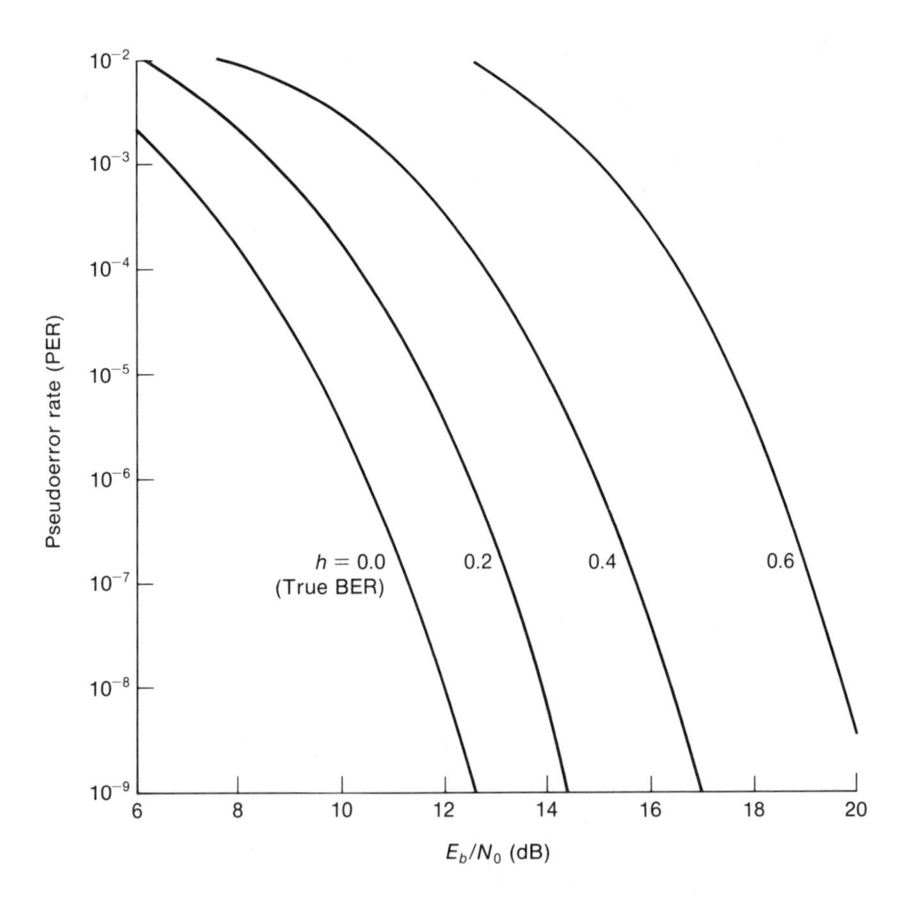

Figure 11.9 Pseudoerror Rate for BPSK or QPSK as a Function of Pseudoerror Zone Height *h*

density is centered about the received signal at a level 1. The probability of pseudoerror can be written mathematically by simple modification of the expressions for actual error probability. As an example, consider the case of coherent BPSK or QPSK, where the probability of pseudoerror is given by

$$\text{PER} = P[\text{pseudoerror}] = \text{erfc}\sqrt{\frac{2E_b}{N_0}(1-h)} - \text{erfc}\sqrt{\frac{2E_b}{N_0}(1+h)} \tag{11.6}$$

Figure 11.9 shows a comparison of the BER and PER as a function of the pseudoerror zone height. Note that over the range of zone heights the PER is roughly parallel to the BER. The pseudoerror rate for an offset $h = 0.2$ is approximately two orders of magnitude worse than the ideal error rate. With increasing values of h, the multiplication factor increases even further, meaning that the time required to estimate the true BER is reduced. As an example, for $h = 0.4$ and a true BER of 10^{-7} the PER will register 10^{-3} and allow the true BER to be estimated 10,000 times faster than the time that would be required to measure BER with conventional out-of-service methods.

11.4 FAULT ISOLATION

Digital transmission systems typically include built-in alarm and status indicators that allow operator recognition of faults and isolation to the failed equipment. Fault and status indicators are displayed on the equipment's front panel, usually by light-emitting diodes (LED), or are provided by teletypewriter output and are available for remote display through relay drivers. Ideally each major function (multiplexer, demultiplexer, timing circuits, and so forth) and each module (individual printed circuit board) have an associated alarm or status indicator. Often, built-in test equipment is used to recognize faults either automatically or with operator assistance. With the advent of the microprocessor, many transmission systems now contain embedded microprocessors that continuously monitor system performance by a sequence of tests designed to check signal paths and modules. The information provided by fault indicators is used to facilitate routine maintenance, to aid in troubleshooting and fault isolation, and to effect automatic equipment protection switching.

Alarm and status indicators are displayed and analyzed locally if repair personnel are available. For unmanned sites, however, these signals are transmitted to a manned facility where the decision is made whether to activate remote control or dispatch repair personnel. The following types of alarm and status points are provided either locally or remotely:

- Failure of transmission equipment (isolated to a module or function)
- Failure of primary power
- Failure of alarm circuitry

- Failure of protection switch circuitry
- Status (position) of automatic switching equipment
- Status of off-line equipment
- Status of various performance monitors

The information contained with these indications should be sufficient to isolate a failure to the site, equipment, and even module (circuit board). Further, an alarm should be classified as affecting service (major) or not affecting service (minor). For multiplex equipment, the level of traffic affected by a failure should also be indicated (for example, number of DS-1 signals affected, number of DS-2 signals affected, and so on).

As examples of the level of control, status, and monitoring points used in digital radio and multiplex equipment, consider the front panel displays illustrated in Figures 11.10 and 11.11. The radio [11] used for Figure 11.10 includes a TDM that combines two mission bit streams (MBS) with a service channel bit stream (SCBS). Moreover, both the transmitter and receiver are fully redundant with two independent units, *A* and *B*, that can be automatically or manually switched on-line. The alarm LED matrix and meter readings can be used for troubleshooting and preventive maintenance checks. The meter is used to monitor dc voltages from the power supply and significant RF signals. Controls include selection of manual versus automatic protection switching, manual selection of *A* versus *B* units, and loopback of the off-line transmitter and receiver.

The digital multiplexer used for Figure 11.11 combines up to eight 1.544 Mb/s channels (or ports) into a single mission bit stream (MBS). The port, multiplexer, and demultiplexer modules are all fully redundant; independent units *A* and *B* are controlled by a peripheral module. Alarm indications include loss of frame, individual frame errors, loss of input or output signals at the port or MBS level, and faulty modules. Each fault is isolated to a single module—either one of the port modules or the multiplex, demultiplex, peripheral, or power supply modules. Status displays include switchover control, whether automatic, manual or remote; indication of whether the *A* or *B* unit is on-line; and whether the frame search inhibit is enabled or disabled. Controls via the three toggle switches are for switchover from one unit to the other, loopback of the off-line transmitter and receiver, and lamp test.

In addition to the display of local alarms, major alarms should be transmitted to the remote end and to other levels of equipment affected by the alarm condition. Alarm indications may be conveniently transmitted to the remote end by changing prescribed bits of the frame synchronization pattern or other overhead patterns. In typical PCM multiplex equipment, major alarms result in the suppression of analog VF outputs. Similarly, for digital multiplex equipment an alarm indication signal (AIS) is transmitted in the direction affected (downstream). The AIS is typically an all 1's pattern or some other substitute for the normal signal, indicating to down-

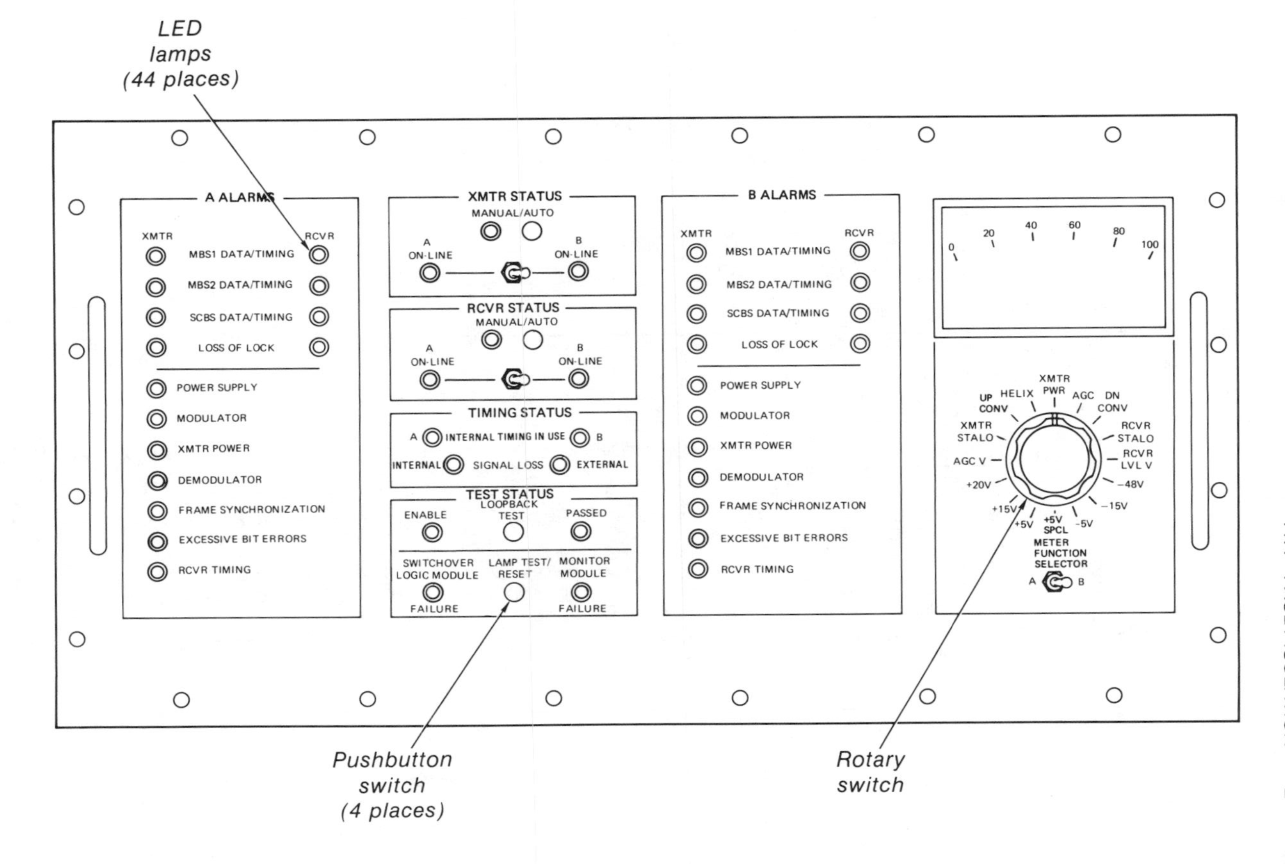

Figure 11.10 Display Panel Controls, Monitors, and Status Indicators for a Digital Radio (Reprinted by permission of TRW)

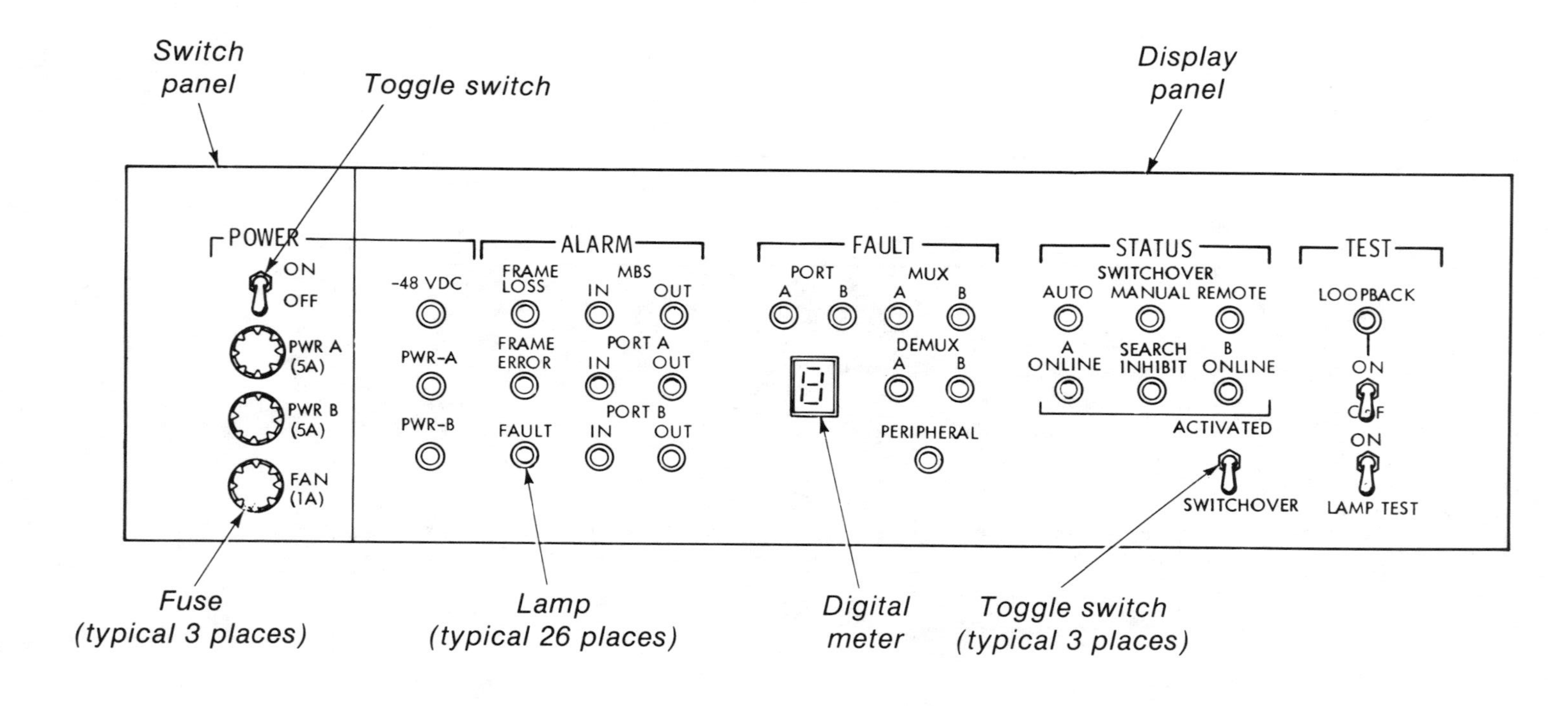

Figure 11.11 Display Panel Controls, Monitors, and Status Indicators for a Digital Multiplexer (Reprinted by permission of TRW)

stream equipment that a failure has been detected and that other maintenance alarms triggered by this failure should be ignored or inhibited in order to eliminate unnecessary actions. Detection of fault conditions and consequent actions have been recommended by the CCITT for PCM and digital multiplex equipment [12].

For redundant systems, a failed unit is automatically switched out and replaced with an operational standby unit, thus deferring the need for maintenance action. If the automatic protection switching circuitry itself fails, however, manual switching becomes necessary. Manual switching can be accomplished by local control and usually by remote control as well. Thus manual switching, whether local or remote, serves only as a backup to the preferred method of automatic protection switching.

For nonredundant systems, a failed unit requires manual maintenance action for restoral of service. Depending on the availability of properly trained personnel, test equipment, and spare parts, this restoral action may take considerable time. As an alternative, particularly for high-priority circuits, alternate transmission paths can be planned that are activated upon failure of equipment in the primary path. This approach will probably still require manual action, but limited to simple patching of circuits. In general, the maintenance approach for service restoral depends on the performance requirements of the service.

11.5 MONITORING AND CONTROL SYSTEM

A transmission monitoring and control system block diagram is shown in Figure 11.12 for a simple two-station network. The two types of stations illustrated in Figure 11.12, remote and master, are generic to transmission monitoring and control. The remote station must collect alarm and status signals for transmission to a master station and receive and execute control commands from a master station. The master station displays local alarm and status signals, receives and displays alarm and status signals for each remote site, and generates and transmits control commands to each remote station. In general, a master may interface and control several remote stations. Conversely, any remote station may report to two or more master stations, one of which is designated primary and the others backup. The primary master can request information or initiate control actions at remote stations. Backup masters listen to the primary master's requests and to remote responses and are capable of performing as primary masters on command from the system operator.

Monitoring and control signals are carried by a service channel that is usually separate from the mission traffic. The service channel is composed of voice orderwires and telemetry signals required for supervision and monitoring of the overall network and each individual link. As shown in Figure 11.12, a multiplexer is used to combine voice orderwires and alarm, status, and control signals for transmission via the service channel. The voice orderwire is used by maintenance personnel to coordinate work on the radio

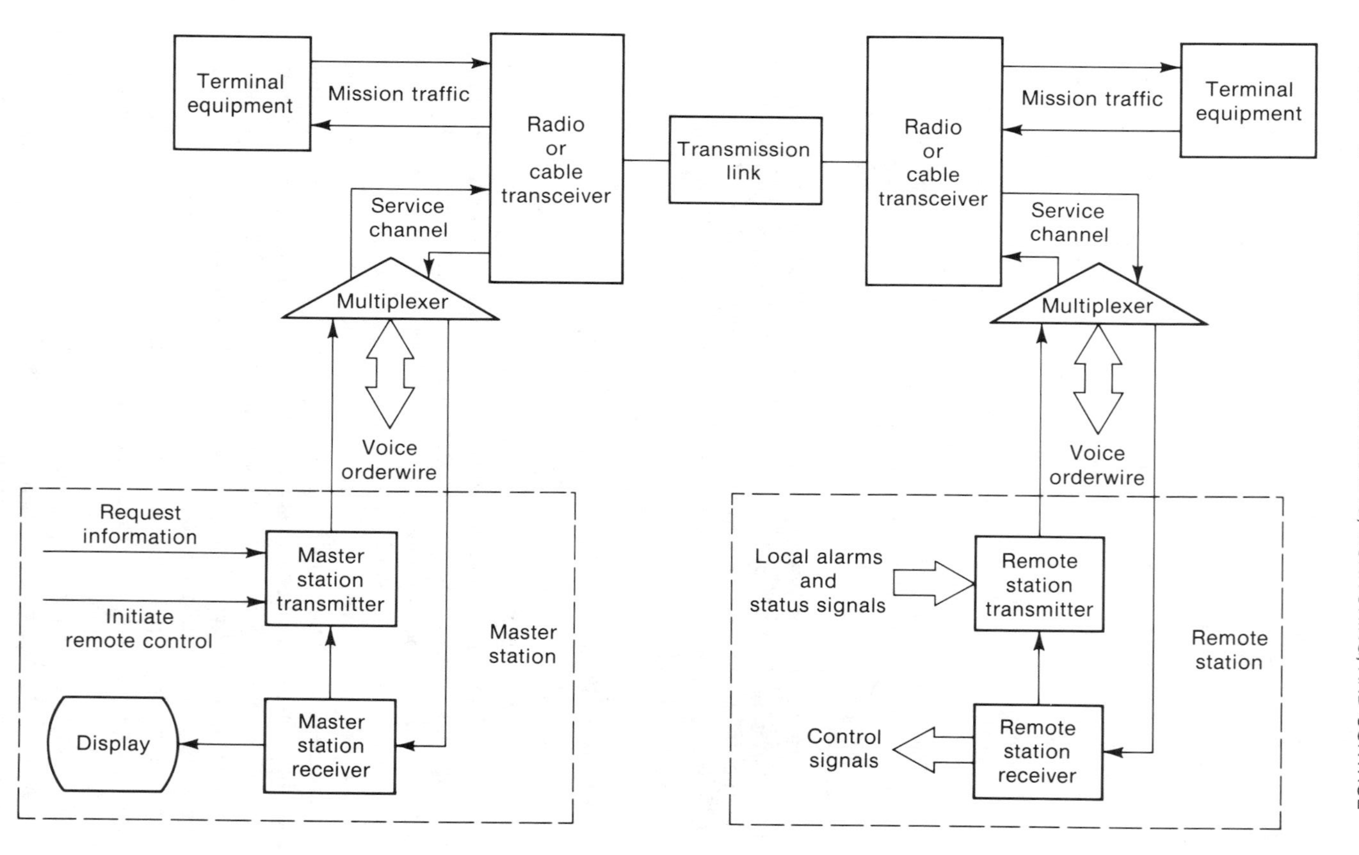

Figure 11.12 Block Diagram of Transmission Monitoring and Control System

or cable link. Equipment alarm and status signals take the form of contact closures, pulse outputs, and analog voltages, which are collected from the remote station's transmission equipment and encoded into a serial, digital form for interface with the service channel multiplexer. Control signals are also distributed via the service channel, emanating from the master station for interface with transmission equipment at the remote station. Other requirements and characteristics of a monitoring and control system are described in the following paragraphs.

11.5.1 Reporting System

Two general types of reporting can be used: continuous and polled. In a continuous reporting system, each remote station transmits its alarm and status signals continuously; a different transmission channel is required for each remote station. The master station receives these alarm and status signals and transmits control data to each remote station via separate transmission channels. As the number of remote stations reporting to a master increases, the number of transmission channels also increases—a disadvantage of the continuous reporting system. Most practical systems use the polling technique, whereby the remote stations all share a common transmission channel on a party-line basis. Each remote station has its own address and responds only when addressed, or *polled*, by the master. All other remote stations remain idle until addressed. When polled, the remote station's reply contains its address and the requested data.

11.5.2 Response Time

In the continuous reporting scheme, alarm and status signals are transmitted without delay so that response time is not of concern. In the polling scheme, however, each remote station must wait its turn to report. Thus information must be stored at each remote station until that station has been polled. The actual waiting time for reporting from a remote station is determined by the number of remote stations, the size of the remote stations, and the capacity of the telemetry channel used for reporting. The required response time is determined by the maintenance philosophy—that is, by what an operator at the master does when an alarm condition is detected. If the operator's response is to dispatch a maintenance team to the remote station, quick response should not be important. Quick response from the reporting system is important only if remote controls are available that will allow removal of the failed unit and restoral with a standby unit. System response time can be improved, if necessary, by increasing the telemetry channel data rate or reducing the number of remote stations reporting to a given master station.

11.5.3 System Integrity

The reporting system must also provide protection against transmission errors, short-term outages, and equipment failure. Some form of error detection is required in order to prevent erroneous reporting of alarms or accidental initiation of remote controls. Moreover, an acknowledge/not acknowledge exchange is required whenever control actions are initiated at a master station for execution at a remote station. Finally, the reporting system must be fail-safe, so that equipment failure does not cause extraneous alarms or control actions.

11.5.4 Transmission Techniques

Techniques for transmission of service channels in digital transmission systems fall into three categories: (1) transmission within the main digital bit stream, (2) transmission in an auxiliary channel that is combined with the main digital bit stream using multiplexing or modulation techniques, and (3) transmission that is separate from the media used for the main bit stream. The first technique simply takes channels of the mission equipment, say the multiplexer, for use as the service channel. This approach has the advantage of not requiring any auxiliary equipment or media, but it does require demodulation and demultiplexing of the mission traffic at each station in order to gain access to the service channels. The third technique requires a separate medium or network, such as the public telephone network, to provide service channel connections among the various stations. This approach provides the desired accessibility and independence from the main traffic signal, but it has the distinct disadvantage of higher cost and potentially less reliability. Only the second technique meets the requirement to drop and insert the service channel at each station easily with only a modest increase in transmission cost and complexity.

Thus the most commonly used technique for service channel transmission is the auxiliary channel that is multiplexed or modulated into the same passband as the mission traffic. This technique can be implemented in two ways: by time-division multiplexing the service channel in digital form with the mission digital bit stream or by modulation of the service channel, in analog or digital form, using a secondary modulation technique (AM, FM, and so on). The TDM approach requires multiplexing circuitry that combines the service channel and mission channels at the transmitter and provides the inverse operation at the receiver. This additional circuitry and transmission overhead is minimal and thus has a negligible impact on system cost and performance. The second implementation of auxiliary channel transmission is generally applicable to radio systems, in which the auxiliary signal is frequency-modulated onto the carrier, amplitude-modulated with the RF signal, or placed on a subcarrier that is inserted above or below the mission spectrum. These service channel transmission techniques for radio must utilize low levels of modulation and small bandwidths in order to avoid degradation of the mission traffic and exceeding of frequency allocation limitations. Because of these restrictions, the TDM approach is more commonly used in radio systems.

If an analog radio is to be used for digital transmission, the TDM approach to service channel transmission often is impractical, particularly when the service channels are analog. Figure 11.13 shows two approaches for interface of an analog FDM service channel multiplexer with a TDM bit stream for transmission over an FM radio. In Figure 11.13a the mission signal is passed through a partial response filter to allow insertion of an FM subcarrier above the digital baseband. After summing, the composite signal is applied to the radio baseband. At the receiver, the subcarrier is demod-

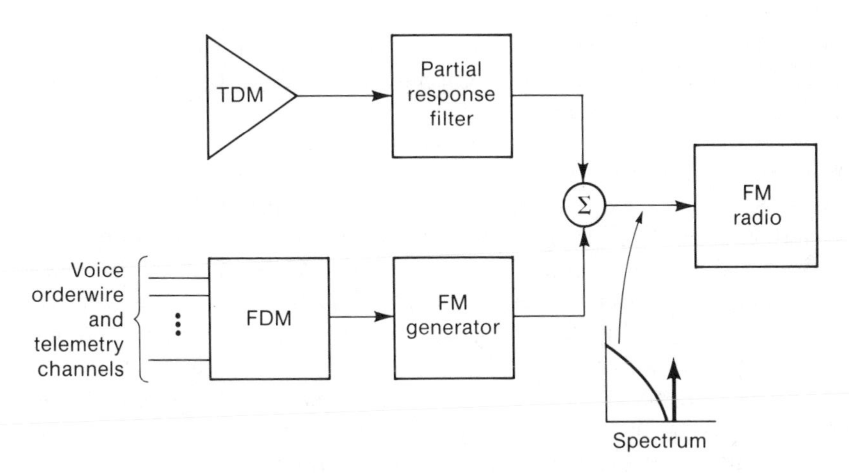

(a) Service channel above digital baseband

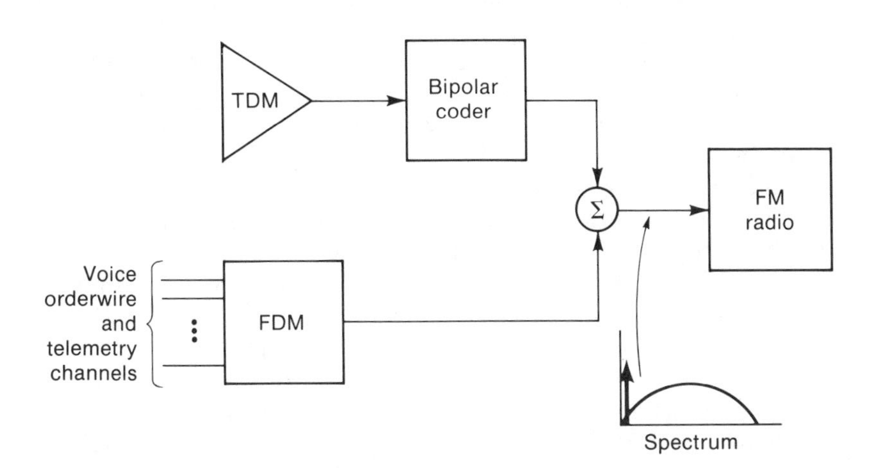

(b) Service channel below digital baseband

Figure 11.13 Service Channel Transmission Techniques for Digital Transmission via FM Radio

ulated and demultiplexed to derive the individual service channels. A similar configuration, shown in Figure 11.13*b*, uses bipolar coding of the mission signal to permit insertion of the service channel in the lower part of the baseband. This technique avoids the necessity of the subcarrier circuitry at the expense of additional system performance degradation due to the use of bipolar rather than partial response coding.

11.6 SUMMARY

During installation and operation of a transmission system, performance must be verified and continuously monitored to ensure that design and operating standards are met. This requirement has led to the development of testing, monitoring, and control techniques to support the introduction of digital transmission systems.

Testing techniques can be classified as in-service, when the mission traffic is unaffected by the test, or out-of-service, when the mission traffic must be removed to conduct the test. During installation and prior to system commission, out-of-service tests allow complete characterization of various digital performance parameters, such as bit error rate, jitter, and bit count integrity. Testing of other performance parameters, such as equipment reliability and multipath fading, requires longer test periods and more sophisticated techniques.

The robustness of digital transmission systems leads to new problems when one is trying to monitor its health during system operation. User performance is unaffected by loss of margin until the "last few decibels," where performance degrades rapidly. Performance monitors must provide a precursive indication of system degradation to allow operator intervention before user service becomes unacceptable. A number of in-service performance monitors are described here—including the pseudoerror monitor, which possesses the attributes necessary for monitoring of digital transmission systems.

Fault isolation is accomplished largely by alarms and status indicators that are built into the transmission equipment. For redundant equipment, most failures are automatically recognized and removed by protection switching. If redundancy is not used or protection switching itself fails, the operator must have adequate alarm and status indications to isolate the fault to the site and equipment. After fault isolation, service is restored by local or remote control of protection switches, by use of spare equipment, or by dispatching a maintenance team for equipment repair.

The collection and distribution of monitor, alarm, and status information requires a service channel that consists of telemetry and voice signals. The service channel is usually combined with the mission traffic using TDM or a subcarrier. At each site, the service channel is accessed to allow operator coordination between sites over a voice orderwire or exchange of alarm, monitor, and control signals. The typical monitor and control system employs a master station that periodically polls each of a number of remote stations to collect alarm and status signals. The master station then displays that information to assist the system operator in fault isolation and restoral of service.

REFERENCES

1. CCITT Yellow Book, vol. IV.4, *Specifications of Measuring Equipment* (Geneva: ITU, 1981).
2. CCITT Yellow Book, vol. VIII.1, *Data Communication Over the Telephone Network* (Geneva: ITU, 1981).
3. W. M. Rollins, "Confidence Level in Bit Error Rate Measurement," *Telecommunications* 11(12)(December 1977):67–68.
4. U. S. MIL-STD-781C, *Reliability Design Qualification and Production Acceptance Tests: Exponential Distribution*, U.S. Department of Defense, 21 October 1977.
5. CCIR XVth Plenary Assembly, vol. IX, pt. 1, *Fixed Service Using Radio-Relay Systems* (Geneva: ITU, 1982).
6. T. L. Osborne and others, "In-Service Performance Monitoring for Digital Radio Systems," *1981 International Conference on Communications*, pp. 35.2.1 to 35.2.5.
7. D. R. Smith, "A Performance Monitoring Technique for Partial Response Transmission Systems," *1973 International Conference on Communications*, pp. 40.14 to 40.19.
8. B. J. Leon and others, "A Bit Error Rate Monitor for Digital PSK Links," *IEEE Trans. on Comm.*, vol. COM-23, no. 5, May 1975, pp. 518–525.
9. J. A. Crossett, "Monitor and Control of Digital Transmission Systems," *1981 International Conference on Communications*, pp. 35.6.1 to 35.6.5.
10. J. L. Osterholz, "Selective Diversity Combiner Design for Digital LOS Radios," *IEEE Trans. on Comm.*, vol. COM-27, no. 1, January 1979, pp. 229–233.
11. C. M. Thomas, J. E. Alexander, and E. W. Rahneberg, "A New Generation of Digital Microwave Radios for U.S. Military Telephone Networks," *IEEE Trans. on Comm.*, vol. COM-27, no. 12, December 1979, pp. 1916–1928.
12. CCITT Yellow Book, vol. III.3, *Digital Networks—Transmission Systems and Multiplexing Equipment* (Geneva: ITU, 1981).

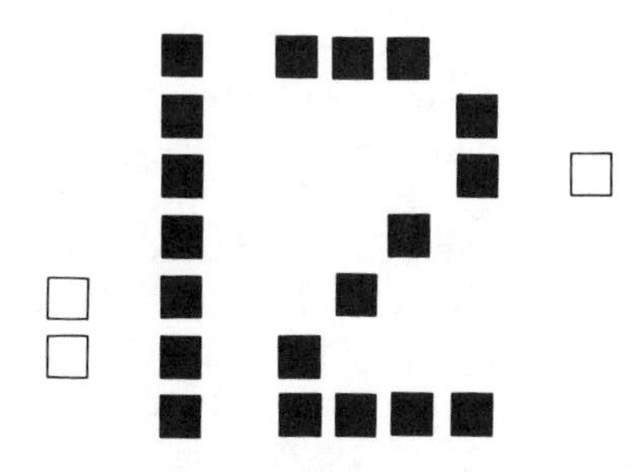

The Future of Digital Transmission

OBJECTIVES

- Discusses the meaning of the breakup of the Bell System
- Describes new digital services that have become a reality: electronic mail, videotext, teletex, and others
- Explains how more efficient coding schemes and the proliferation of fiber optics have changed the face of digital transmission
- Covers the use of digital transmission in local area networks (LAN) and the integrated services digital network (ISDN)

12.1 PROSPECT

With the invention of the transistor, the world entered a new Information Age that will mature in the next two decades. The global telecommunication network today forms the infrastructure for this information-based society. Not surprisingly, telecommunications is the world's fastest-growing industry, owing to the demands of society, the rapid growth of technology, and the declining cost of telecommunication services.

The telephone network is now and will continue to be the cornerstone of the telecommunication industry. In 1982, the global telephone network encompassed 550 million telephones and handled roughly 2 billion telephone calls a day. However, the demand for data, video, and other telecommunication services will cause major changes in today's telephone networks. The development of microelectronics has resulted in low-cost memory, greater computing power, the microprocessor, and other technological advances that have created a requirement for efficient, low-cost communications. Simultaneously, the computer and microelectronics industries have penetrated the telecommunication industry in the form of digital communication systems. In the future, the distinction between the telephone and computer, and other home or office terminals, will blur. Tomorrow's telephones will be "smart terminals" that integrate features now unavailable or provided by separate terminals. This integration of services is made possible by the pervasive use of "1's and 0's" in telecommunication networks. Digital technology will be extended all the way to the customer's premises. This anticipated availability of a completely digital network has led to the development of a concept for an **integrated services digital network (ISDN)**, now being planned by the CCITT and its member nations.

Digital transmission will continue to play an important role in the evolution of telecommunication networks. Historically, the transmission plant of telecommunication networks has been first to be digitized, while switches and terminals have remained analog. Breakthroughs in digital transmission technology and declining transmission costs allow the creation of new networks and service. Already the introduction of digital switches has led to integration of switches and transmission systems, which has been termed the **integrated digital network (IDN)** by the CCITT. New technologies such as local area networks and packet radio networks blend transmission with switching and networking. In the future, we can expect transmission systems to merge even further with switches, terminals, and networks, so that transmission systems will be inseparable from telecommunication systems.

In addition to technological advances in telecommunications, positive changes have also occurred in government regulation of this industry. In the United States technological advances made in the 1960s and 1970s were often at odds with public policy. The Bell System's technical innovations and entry into the data processing market were limited by U.S. regulatory structure, for example, while other common or specialized carriers were at a disadvantage due to AT & T's monopolistic structure. However, deregulation of the telecommunication industry and divestiture of the Bell System in the 1980s promise to bring advances in the U.S. telecommunication network in the 1980s. Outside the United States, most national telecommunication systems are owned and operated by a ministry or administration of Post, Telegraph, and Telephone (PTT). Compared to the multiplicity of public and private networks found in the United States, most countries have only one or a few networks. The United States, Canada, and the United Kingdom are among the few countries that permit competition in their telecommunication networks, although several countries including Japan, West Germany, and France are reviewing their policies for possible deregulation. Europe, in particular, has advantages over the United States in future planning because of standardization through the CCITT and CEPT. For example, the various European PTTs are playing a major role through their representation in the CCITT and CEPT in the definition and planning of the ISDN.

Despite the bright prospect of high technology, greater competition, and less regulation, there are still some concerns about the future of telecommunication networks. The major concern is the development of standards in the United States and throughout the world. With increased competition and deregulation, the need for standards will be even greater in the future but may be more difficult to attain than in the past. In the United States, AT&T has traditionally determined communication standards, which were followed by other telecommunication carriers and manufac-

Table 12.1 Organization of the T1 Telecommunications Standards Committee

Technical Subcommittee	*Working Groups*
T1C1: Customer premise equipment to carrier interfaces	T1C1.1: Analog interfaces T1C1.2: Digital interfaces T1C1.3: Special interfaces T1C1.4: Editing
T1D1: Integrated services digital networks (ISDN)	T1D1.1: ISDN architecture and services T1D1.2: ISDN switching and signaling protocols T1D1.3: ISDN physical layer
T1M1: Internetwork operations, administration, maintenance, and provisioning	T1M1.1: Internetwork planning and engineering T1M1.2: Internetwork operations T1M1.3: Testing and operations support systems and equipment T1M1.4: Administrative systems
T1Q1: Performance objectives	T1Q1.1: 4-kHz Voice T1Q1.2: Voice-band data T1Q1.3: Digital circuit T1Q1.4: Digital packet T1Q1.5: Wideband program T1Q1.6: Wideband analog
T1X1: Carrier to carrier interfaces	T1X1.1: Common channel signaling T1X1.2: Carrier interfaces T1X1.3: Digital network synchronization T1X1.4: Hierarchical rates and formats
T1Y1: Specialized subjects	T1Y1.1: Specialized video and audio services T1Y1.2: Specialized voice and data processing T1Y1.3: Advanced technologies and services T1Y1.4: Environmental standards for central office equipment

turers. With the divestiture of the Bell System and the entry of new carriers and networks into the U.S. market, standards will be required in order to avoid incompatibility and fragmentation in telecommunication systems. The planning of new international networks such as the ISDN will have to accommodate multiple providers of service as found in countries such as the United States. The breakup of the Bell System has also meant loss of a single point of contact for planning and providing end-to-end service to customers. From the customer's perspective, this change has created restrictions on the provision of end-to-end service, potential increases in cost, more complex contracting procedures, and a general uncertainty about future standards and regulation. (For a brief survey of the Bell System's breakup, see the accompanying box.)

Divestiture of the Bell System

In 1974, the U.S. Justice Department filed an antitrust suit against the Bell System, which went to trial in January 1981. In January 1982, AT & T and the Justice Department agreed upon a settlement in which the Bell System would be dismembered and the antitrust suit would be dropped. This settlement, which was effective 1 January 1984, required AT & T to divest itself of the local telephone network found in the 22 Bell operating companies. In return for the divestiture, AT & T is now free of all restrictions to pursue new markets such as data processing and electronic publishing. AT & T has reorganized its remaining assets into AT & T Communications (interstate long-distance network) and AT& T Technologies (research, development, manufacturing, customer premises equipment, and enhanced services). AT & T is also free to enter the local loop business in competition with the local Bell operating companies. The divested local operating companies must provide equal access to all long-distance carriers and information service providers. Moreover, local Bell companies are barred from providing long-distance service, other telecommunication services, and customer premises equipment. Further, local Bell cannot discriminate between AT & T and other equipment suppliers in the purchase of transmission and switching equipment. Finally, the local companies are required to form a centralized organization for coordination in matters of national security and emergency preparedness.

Divestiture of the Bell System and the proliferation of new carriers, networks, and services has, however, stimulated new attempts at standardization. Under the auspices of the American National Standards Institute (ANSI), the T1 Telecommunications Standards Committee has been established. This committee, whose organization is shown in Table 12.1, is chartered to develop standards and technical reports related to interfaces for U.S. networks. The six technical subcommittees of the T1 Committee are responsible for services, switching, signaling, transmission, performance, operation, administration, and maintenance aspects of telecommunication networks.

12.2 NEW DIGITAL SERVICES

Apart from the traditional services of voice, data, and video, new services are emerging from the fields of data processing, broadcast video, and interactive video. The marriage of telecommunications with data processing has resulted in "telematic" services such as facsimile, teletex, videotext, and electronic mail, which are most conveniently carried with digital transmission. Video services such as teleconferencing, interactive CATV, and direct broadcast by satellite are also entering the market, although with the exception of teleconferencing these services are now being provided via analog transmission. Introduction of these new services has led to the concept of "automated offices" for improved office efficiency and productivity and "wired households" for improved home environment [1].

12.2.1 Facsimile

Facsimile provides transmission of documents through public telephone or data networks. Three categories of facsimile terminals have been standardized by the CCITT, all designed to operate over public telephone networks: Group 1 takes 6 min to transmit a standard page; Group 2 takes 3 min per page; Group 3 takes 1 min per page. Group 3 terminals are designed to operate with 2.4-kb/s or 4.8-kb/s modems (CCITT Rec. V.27) or optionally with 7.2-kb/s or 9.6-kb/s modems (CCITT Rec. V.29). A Group 4 terminal is also under study by the CCITT for use in public data networks [2]. Because of the early standardization of facsimile terminals, facsimile services have enjoyed more rapid growth than other telematic services.

The trend in the future will be toward convergence of facsimile with other forms of telematics. Mixed-mode terminals have been developed, for example, which use an optical character reader (OCR) to read characters on the page and facsimile to transmit any noncharacter information on the page. A standard on this mixed-mode terminal is expected sometime after the Group 4 standard is completed by the CCITT. In Japan, recognized as the world's leader in facsimile technology, NTT has inaugurated a public facsimile network that provides multiple communication services [3].

12.2.2 Teletex

Teletex is a service providing communication of character-coded text that can be used between offices for word processing and editing or in the home for magazine or newspaper display. A teletex terminal includes a typewriter function, ability to send and receive a basic repertoire of characters, a display device, storage for characters being transmitted or received, and a hard copy capability. Teletex transmission can be provided by data or telephone switched networks, with a transmission rate of typically 2.4 kb/s or greater, or by unused lines in broadcast television.

The CCITT is proceeding with standardization of teletex service with the Series S recommendations. The basic teletex functions prescribed by the CCITT are the character repertoire, text dimensions and positioning, control procedures, and network-independent data transport service [2].

12.2.3 Videotext

Videotext is an interactive system of information retrieval offering the display of alphanumeric text and basic graphics on a display terminal. Using a telephone line and keyboard, the subscriber may access and interrogate a data bank. Selected data can then be displayed on the user's standard television receiver. A huge range of services can be offered on videotext systems, including news, games, advertising, shopping and banking at home, and electronic publishing (such as newspapers and magazines). Numerous public and private videotext offerings are already available, notably in the United States, Canada, Europe, Japan, and Australia [1].

With the transmission rate often limited by use of the switched telephone network, emphasis has been put on development of bandwidth-efficient picture coding techniques. Several digital coding techniques—alphamosaics, alphageometrics, alphaphotographics—are competing for international standardization. In the future, the limitations of the presently analog telephone network should be eliminated by the use of a 56 or 64-kb/s digital channel as part of the integrated services digital network (ISDN).

12.2.4 Electronic Mail

Historically, postal service has been separated from the electronic communication services provided by telephone and data networks. Even in Europe, where post office, telephone, and telegraph services are provided mainly by a single government agency, these services have been provided separately. However, the evolution of public voice and data networks has stimulated the growth of interactive electronic mail service for both the home and office. This service allows users to exchange text messages by means of terminals linked together by a data communication network. Messages are stored until the addressee logs on and requests his or her messages.

The U.S. Postal Service has developed Electronic Computer-Originated Mail (E-COM), a form of electronic mail that allows a customer to send a text message over telephone lines to specially equipped post offices. The receiving office makes a hard copy and delivers it by first-class mail.

Competition to E-COM exists from several U.S. telecommunication networks—notably GTE Telenet with its Telemail service and Tymnet, both of which employ packet-switching technology. Canada has also introduced an electronic mail service, called EnvoyPost, which uses the postal service and TransCanada Telephone System.

12.2.5 Video Teleconferencing

Video teleconferencing permits conferees at two or more facilities to have real-time, interactive, visual contact. Bandwidths used for broadcast television are too costly and unnecessary for teleconferencing. Typically the picture to be transmitted consists of graphics, slides, or conferees whose movement is very limited. Due to this limited motion, bandwidth compression techniques can be applied while maintaining sufficiently good video resolution. Video teleconferencing services now commercially available use digital encoding, bit rate reduction, and digital transmission to achieve lower costs. Standard bit rates for this service include 6.3 Mb/s [4], 3 Mb/s [5], 1.544 Mb/s [6], and 2.048 Mb/s [7]. In July 1982, AT&T introduced its Picturephone Meeting Service (PMS), initially with a transmission rate of 3 Mb/s that was reduced to 1.544 Mb/s in 1983. The PMS is part of a new high-speed switched digital service (HSSDS) offered by AT&T between major cities [8]. Most of the world's major telecommunication administrations and carriers offer similar video teleconferencing services, both by satellite and by terrestrial networks.

Standards for video teleconferencing have been based on video broadcasting standards such as those established by the U.S. National Television Systems Committee. As the video teleconferencing industry matures, however, it is expected to separate from the broadcasting industry and establish its own standards. The trends in video teleconferencing are toward lower bit rates, a single international standard, and the integration of video with voice and data signals.

12.2.6 Cable TV and Satellite Broadcasting

Cable TV (CATV) and direct broadcast by satellite (DBS) are now receiving wide publicity in the United States, Europe, and elsewhere. Although the transmission of video signals for TV is expected to remain analog for some time, two-way cable TV is becoming popular. Here the subscriber is given a low-speed digital connection for interactive services in addition to the one-way video channels.

12.3 NEW DIGITAL TRANSMISSION TECHNOLOGY

The penetration of digital transmission in future telecommunication networks will depend largely on cost considerations. Digital transmission for voice and video currently requires greater bandwidth and often greater cost than analog transmission techniques. This situation is expected to change, however, with the introduction of more efficient coding schemes for analog signals and the proliferation of fiber optics.

In local area and short-haul environments, digital radio and fiber optics will compete with the existing metallic cable plant. In long-haul applications, fiber optics and in a few cases digital radio will compete with satellite transmission. Costs are independent of distance with satellite transmission, so this medium may prove to be the least expensive for very long haul transmission. The use of digital transmission techniques in satellite transmission, including digital speech interpolation (DSI) and time-division multiple access (TDMA), is now cost-competitive with analog transmission techniques in present-day satellite systems. Terrestrial forms of transmission are characterized by costs that have a linear dependence on distance, longer system life, and greater transmission capacity than satellites, but a requirement for right-of-way. In submarine cables, fiber optics is planned for coming applications that may prove more cost-effective than satellite transmission.

12.3.1 Digital Speech

PCM at 64 kb/s is the present standard for digital speech in telephone networks throughout the world. For certain applications, however, there are advantages to reducing the bit rate for speech down to 32 kb/s, 16 kb/s, or even lower rates. Where transmission costs are high or bandwidth is limited as in long-haul terrestrial networks or satellite links, reduced-rate coding for speech offers more channels, by a factor of 2 or 4 or more, than conventional PCM channels. The local loop is another candidate facility for low-bit-rate speech, because of limited bandwidth available in existing metallic loop plant. The local loop bandwidth constraint is further compounded when considering the addition of a multiservice capability, such as digital speech and data provided simultaneously as envisioned with ISDN. A third application already in use involves encrypted voice at rates of 16 kb/s and less, operating via modems over switched telephone networks.

To be considered as a replacement for 64-kb/s PCM in public switched networks, a candidate scheme must provide speech quality comparable to PCM, the ability to sustain speech quality with tandem conversions, and the capability of handling nonspeech signals such as voice-band data and in-band signaling. Various carriers, manufacturers, and the CCITT have settled on 32 kb/s as a suitable rate for a new standard. Candidate schemes include adaptive delta modulation (ADM), adaptive differential PCM (ADPCM), and nearly instantaneous companding (NIC) [9]. The CCITT has already selected ADPCM as its recommended 32 kb/s standard (see the accompanying box). AT & T has also selected 32 kb/s ADPCM to double the voice channel capacity of their digital facilities [10].

Even lower rates, such as 16 kb/s, are envisioned for more limited speech networks, especially private networks that are digital end to end.

Speech quality comparable to 64-kb/s PCM has been demonstrated with several 16-kb/s coders, but transmission of voice-band data is not considered feasible. Tandem encodings would be accomplished on a digital-to-digital basis, which is also termed *synchronous* tandeming to avoid the degradation associated with *asynchronous* tandeming that provides analog representation with each encoding. Candidates for a 16-kb/s coding scheme include adaptive predictive coding (APC), subband coding (SBC), and adaptive transform coding (ATC) [11]. However, the CCITT has not yet developed recommendations on 16-kb/s speech coders.

32-kb / s Adaptive Differential PCM Standard

At the June 1982 meeting of CCITT Working Group 18, an ad hoc group was established to assess 32-kb/s ADPCM algorithms and prepare a draft recommendation for a single worldwide standard. This ad hoc group studied the performance and complexity of ADPCM coders from France, Japan, and the United States. By November 1983, a compromise algorithm had been developed, characterized by tests, and described in draft recommendation G.7ZZ, which was then adopted at the 1984 CCITT Plenary Assembly. Characteristics of this 32-kb/s ADPCM coder include:

- Transcoding is used for digital-to-digital conversion from 64-kb/s A-law or μ-law PCM to 32-kb/s ADPCM.
- Using subjective tests, ADPCM was judged equivalent to three PCM asynchronous tandem encodings.
- Performance of ADPCM improves over that of PCM for transmission line error rates of 10^{-4} and higher.
- Voice-band data performance at rates up to 2.4 kb/s (CCITT Rec. V.26 modems) was comparable for PCM and ADPCM for up to two asynchronous encodings.
- Voice-band data performance at 4.8 kb/s (CCITT Rec. V.27 modems) was acceptable but experienced degradation compared to 64-kb/s PCM.
- Voice-band data performance at 9.6 kb/s (CCITT Rec. V.29 modems) was unacceptable.
- Voice frequency telegraphy with 24 channels cannot be satisfactorily conveyed over 32-kb/s ADPCM.
- Dual-tone multifrequency (DTMF) signaling performed satisfactorily over 32-kb/s ADPCM.
- Group 2 facsimile performed satisfactorily over 32-kb/s ADPCM.

12.3.2 Digital Multiplexing

For applications to telephone networks, reduced-rate coding can be used with multiplexers and existing digital transmission facilities, including T carrier, fiber optics, and digital radio, to increase channel capacity by 2:1 or greater. Using 32-kb/s coding, several multiplexers are commercially available that provide up to 48 voice channels in 1.544 Mb/s or 60 voice channels in 2.048 Mb/s. These multiplexers typically are designed to interface directly with PCM multiplexers and use **transcoding** to digitally convert from PCM to the 32-kb/s code. Other features include the ability to accommodate voice-band data and signaling tones, transmit digital data and 64-kb/s PCM at their normal rates, and provide 64-kb/s digital data channels.

Another form of multiplexing used to compress the bit rate for speech transmission is digital speech interpolation (DSI), which was briefly discussed in Chapter 4. Compression ratios of 2:1 are common for DSI applications, in which two PCM multiplexed signals are compressed into a single 1.544-Mb/s or 2.048-Mb/s transmission signal. Even greater speech compression is possible by combining DSI with reduced-rate coding. Digital speech interpolation with variable-rate ADPCM has been developed for the U.S. military [12] and is being introduced by U.S. common carriers. The variable coding rate limits the degradation that normally occurs under heavy loading. That is, when speech activity is high or there is a high percentage of full period signals as in voice-band data or digital data, the ADPCM coder is adapted by reducing the number of bits used to encode each speech sample.

In the United States, AT & T is introducing further improvements to its 1.544-Mb/s terminals to expand their digital network and become compatible with ISDN standards. First, 64-kb/s "clear channel capability" is replacing the present 56-kb/s capability. The present use of bipolar coding on T carrier systems imposes constraints on 1's density (at least a single 1 required in every 8 bits) for adequate clock recovery performance, which then places a constraint on the data rate available to the customer. By using bipolar with 8 zero substitution (B8ZS), as discussed in Chapter 5, constraints on signal transitions are eliminated and the customer can be provided a full, clear 64 kb/s as envisioned with ISDN. Coincident with the new 64-kb/s clear channel, AT & T has introduced the extended framing format (F_e), which is a new DS-1 framing format [13]. (See Table 12.2 and the accompanying box.) This F_e feature improves performance and facilitates maintenance. Together with 64-kb/s clear channels, F_e is built into all new designs of DS-1 level equipment, starting with the D5 channel bank announced in 1982. However, this new terminal equipment is also capable of using the DS-1 compatible format whenever an F_e-compatible format is not available at both the transmit and receive terminals. On the basis of contributions from AT & T, the CCITT is studying F_e and 64-kb/s clear channel capability as a possible standard.

Table 12.2 Comparison of Extended Framing Format with DS-1 Format

Frame Number	*Extended Frame Format, F_e (Framing Bits)*			*Signaling Channel*	*DS-1 Format (Framing Bits)*		*Signaling Channel*
	Framing	*Data*	*Error Detection*		*Frame Synchronization*	*Multiframe Synchronization*	
1	—	D	—		—	1	
2	—	—	CB_1		0	—	
3	—	D	—		—	0	
4	0	—	—		0	—	
5	—	D	—		—	1	
6	—	—	CB_2	*A*	1	—	*A*
7	—	D	—		—	0	
8	0	—	—		1	—	
9	—	D	—		—	1	
10	—	—	CB_3		1	—	
11	—	D	—		—	0	
12	1	—	—	*B*	0	—	*B*
13	—	D	—		—	1	
14	—	—	CB_4		0	—	
15	—	D	—		—	0	
16	0	—	—		0	—	
17	—	D	—		—	1	
18	—	—	CB_5	*C*	1	—	*A*
19	—	D	—		—	0	
20	1	—	—		1	—	
21	—	D	—		—	1	
22	—	—	CB_6		1	—	
23	—	D	—		—	0	
24	1	—	—	*D*	0	—	*B*

Extended Framing Format (F_e)

The extended framing format (F_e) extends the multiframe structure from 12 to 24 frames and redefines the 8-kb/s framing overhead in 1.544-Mb/s PCM multiplex equipment (Bell DS-1 and CCITT Rec. G.733). Table 12.2 compares the DS-1 framing format and the extended framing format. The 8-kb/s overhead channel in F_e consists of three separate patterns:

- *2-kb/s Framing*: Beginning with frame 4, the framing bit of every fourth frame forms the 2-kb/s pattern 001011...001011, which is used to obtain frame and multiframe synchronization. Frame synchronization is used to identify each 64-kb/s PCM channel of each frame. Multiframe synchronization is used to locate each frame within the multiframe in order to extract the CRC and data link information and identify the frames that contain signaling information. Signaling bits carried by frames 6, 12, 18, and 24 are called respectively *A*, *B*, *C*, and *D* signaling bits.
- *2-kb/s Cyclic Redundancy Check (CRC)*: This is contained within the framing bit position of frames 2, 6, 10, 14, 18, and 22 of every multiframe. The check bits (CB) that make up the CRC are used to detect errors within the extended multiframe. The CRC can be monitored at any DS-1 access point and can be used for end-to-end performance monitoring, false framing protection, protection switching, automatic restoral after alarms, line verification in maintenance actions, and the measurement of error-free seconds.
- *4-kb/s Data Link*: Beginning with frame 1 of every multiframe, every other 193rd bit is part of the 4-kb/s data link. Using X.25 level 2 protocol, this data link has several possible applications, including transmission of alarms, status signals, supervisory signaling, network configuration, performance indicators, and maintenance information.

In the early 1980s AT & T introduced a new element, the Digital Access and Cross-Connect System (DACS), for use in digital telephone networks. A DACS is capable of terminating up to 128 DS-1 signals and providing digital cross-connection of any 64-kb/s channel. The cross-connection feature thus eliminates the need for D/A and A/D conversion of channels, drop and insert multiplexing, and manual patch panels in favor of direct digital rerouting and dropping and inserting under electronic control. This control may be accomplished locally by means of a standard (TTY) terminal or remotely by means of a standard data link. A test feature is also provided in which a DS-1 signal is reserved for test access and any channel can be tested by connecting it to the test access DS-1 termination. Thus the DACS combines multiplexing, network control, and testing functions in a way that should significantly simplify digital network operation and maintenance. AT & T and the Bell operating companies already have DACS installed within their networks. Future plans include national networking of DACS and the addition of a 2.048-Mb/s capability for international gateways.

12.3.3 Digital Radio

In the early 1980s, most digital radio applications were characterized by spectral efficiencies of up to 3 bps/Hz, transmission rates of 90 Mb/s or less, and routes limited to a few hundred miles. Long-haul applications of digital radio are still constrained by high costs compared to analog radio. Techniques such as single sideband make analog radio more bandwidth-efficient and therefore less expensive. However, digital terminal equipment, including channel banks and multiplexers, is so economical that the overall cost of a digital transmission system is usually less than that of an analog transmission system, at least for short routes. This economic advantage with digital transmission has been proved in the United States for intercity applications and in densely populated countries such as Japan [14]. Moreover, with the advent of digital switches, such as AT & T's no. 4 ESS, digital radio can be used to advantage in interconnecting these digital switches. Another important factor that will contribute to the growth of digital radio is the rapid spread of digital cable systems, both T-carrier types and fiber optics, which can be easily interconnected via digital radio.

The future challenges in digital radio technology include increased spectral efficiency, greater protection against multipath fading for higher transmission rates, and improved reliability. Already a spectral efficiency of 4.5 bps/Hz has been realized using 64-QAM, leading to digital radio upgrades in the United States [15] and Canada [16]. Other 64-QAM digital radio applications are expected to follow quickly. Combinations of coding and modulation have also been proposed, such as the Ungerboeck codes [17], as a means of achieving higher bandwidth efficiencies. However, the complexity of the modulation or coding scheme must be governed by its ability to perform in the presence of multipath fading and other forms of degradation. The trend in adaptive equalizer design is toward greater sophistication, such as the use of transversal equalization, in order to allow transmission rates in excess of 100 Mb/s. Improved reliability and availability can be realized through improved components, particularly with the replacement of frequently unreliable TWTs with solid-state amplifiers (such as the GaAs FET).

12.3.4 Fiber Optics

Advances in fiber optics have come rapidly in the last 10 years and are expected to continue, making fiber optics the predominant medium for digital transmission in the 1990s. In the short span of a decade, the fiber optics industry has seen three generations of technology. First-generation systems used multimode fibers and short-wavelength sources (850 nm) and have been in place since the 1970s. Second-generation systems, first fielded in the early 1980s, typically employ graded-index fibers at longer wavelengths (1300 nm). Third-generation systems based on single-mode fiber and long wavelength technology (1300 and 1500 nm) will be used for high data rates and long transmission distances. Applications are planned for both terrestrial and transoceanic cable.

The market for fiber optics by geographical area and by component as of 1983 and projected for 1989 is shown in Figure 12.1. The leading users are the United States, Canada, Western Europe, and Japan, while telephony is the dominant service. Commencing with the field trials conducted by the Bell System in Atlanta in 1976 and Chicago in 1977 (both systems at 45 Mb/s), AT & T has become the world's largest user of fiber optics. The first long-distance application of fiber optics was initiated in 1983 by AT & T between New York and Washington, D.C., using the AT & T Technologies' FT3C 90-Mb/s system. This route was the first part of the Northeast Corridor system, which was completed in 1984 to link Boston, Massachusetts, and Richmond, Virginia, for a total length of 776 mi. In California, Pacific Telesis inaugurated service in 1983 on the first part of an intrastate fiber optic system. By 1986 this system will provide metropolitan networks in Los Angeles and San Francisco with intercity links connecting Sacramento, San Francisco, Los Angeles, and San Diego. AT & T also plans five other intercity fiber optic routes—to be completed in 1986 and totaling about 1200 mi—that will use third-generation technology and have a capacity of 432 Mb/s. By 1990, most metro areas in the United States will have fiber optic connections provided by AT & T [18].

Other telecommunication carriers in the United States are installing fiber optic systems, particularly along railroad rights-of-way for long-distance transmission. MCI Communications Corp. is installing fiber optic cable in the eastern United States along 4250 mi of railroad right-of-way acquired from Amtrak and CSX Corp. The cable to be used by MCI will carry 44 single-mode fibers, each operating at a bit rate of 405 Mb/s. With complete installation planned by 1987, MCI would become the country's largest user of single-mode fiber optics. Southern New England Telephone has formed a partnership with CSX Corp. to build a 5000-mi (expandable to 27,000 mi) fiber optic network in the eastern United States. This network, called Lightnet, will initially operate at 45, 90, or 180 Mb/s over each of 60 fibers, which are single-mode cable to support planned expansion to 1 Gb/s transmission rates.

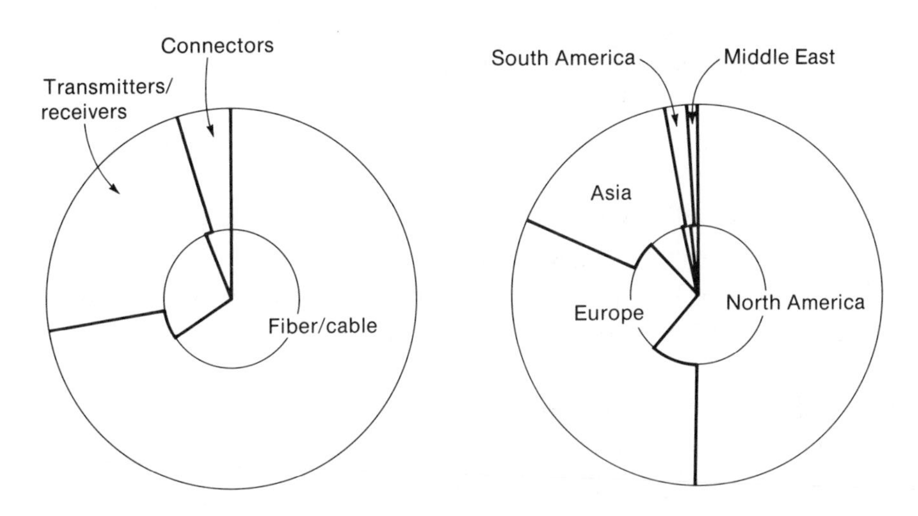

Figure 12.1 Worldwide Fiber Optic Markets by Component and Geographical Segment (Reprinted by permission of Kessler Marketing Intelligence)

In other countries, Nippon Telegraph and Telephone (NTT) has completed a field trial of its F-400M network—a 1300-nm, single-mode, 400-Mb/s fiber optic system—and in 1983 began installation of a 2500-km F-400M network along the length of Japan [19]. In the United Kingdom, British Telecom has developed a 1300-nm, single-mode fiber optic cable capable of bit rates up to 565 Mb/s with 30-km repeater spacing [20]. Mercury Communications Ltd., in competition with British Telecom, plans a similar single-mode fiber optic system along the British Rail right-of-way. European telephone administrations, including those in the United Kingdom, Belgium, France, West Germany, Italy, and Holland, have developed fiber optic systems, mainly for the bit rates of 8, 34, and 140 Mb/s, using graded-index fibers. Future systems are likely to use single-mode fibers operating at rates of 565 Mb/s and higher [21].

The coming generation of transoceanic cable will use optical fibers rather than coaxial cable. The next transatlantic system, called TAT-8, will link North America with Europe and is scheduled for service in 1988. TAT-8 will consist of six single-mode fibers, four active and two backup, with automatic protection switching. With a transmission rate of 274 Mb/s (T4) per fiber, TAT-8 will have four times the capacity of its predecessor the TAT-7, which used copper coaxial cable. Fiber optic submarine cable systems are also planned in the late 1980s to link Japan and the United States beneath the Pacific. By contracts and agreements already made, AT & T will be the major supplier in both the Atlantic and Pacific submarine cable ventures.

Several technological advances are expected to spur the growth of fiber optic communications in the near future:

- Single-mode fibers for very high data rates and very long transmission distances
- Longer wavelength source, at 1500 nm or higher, for lower loss
- Wavelength-division multiplex to increase the information transfer
- Integrated optical circuits that combine transmitter and receiver functions
- Transmission of power over optical fibers, possibly in the form of infrared radiation
- New glasses and new processes for manufacturing optical fibers
- Processes to improve fiber strength for military applications

12.4 DIGITAL TRANSMISSION IN LOCAL AREAS

Digital transmission is now being introduced into local areas for networking and distribution. A wide variety of media are under consideration for this application, including microwave and millimeter-wave radio, broadcast and packet radio, free-space optical transmission using lasers, fiber optic cable, and coaxial cable.

12.4.1 Digital Radio in Local Loops

The **local loop** is the portion of a communication network that serves a local community of users. In the commercial sector, telephone companies have traditionally supplied local loops, and any specialized or value-added carrier had to rely on the local telephone company to furnish the local loop. Radio systems offer an alternative in which the user is provided local area distribution by radio, thus bypassing the telephone network entirely unless long-haul service is required. In military applications, the need for mobility and other special features often dictates the choice of radio for local area distribution. Two technologies developed for the use of digital radio in local loops are discussed here: digital termination systems and packet radio.

In 1978, the Xerox Corporation applied to the U.S. Federal Communications Commission for approval of a proposed Xerox Telecommunications Network (XTEN). The FCC eventually granted this service by allocating the microwave band from 10.55 to 10.68 GHz for **digital termination systems** (DTS). The basis for XTEN, which provides insight to DTS, was the use of omnidirectional microwave transmission from a local node for data communications with a set of users within a geographic cell. Each user would receive the broadcast (TDM) signal and transmit back to the local node using time-division multiple access (TDMA). Local nodes were connected to each other and to long-haul facilities by point-to-point microwave [22]. Although Xerox abandoned XTEN in 1981, several other specialized carriers have used DTS in the 10 GHz band, and further frequency allocations for DTS are expected in the 18 GHz band.

Military requirements for mobile subscribers in data communication networks have led to the development of **packet radio**, a technology that combines the advantages of both packet switching and radio broadcast. Each radio node is accessed by multiple users who share a high-speed common channel. Transmission is half-duplex, with users transmitting their traffic in bursts during scheduled transmission times (that is, TDMA). Packet radio uses computers at each node to control access to the radio channel and provide efficient sharing of the common channel among all users. The design of a packet radio system involves a complex interaction of many design parameters: network topology, control of access to the channel, modulation technique, network protocols, and network management [23]. During the 1970s, two prototype packet radio systems were developed. First came the ALOHA project at the University of Hawaii, which then led to the Packet Radio Network (PRNET) under the sponsorship of the U.S. Department of Defense. With these demonstrations, packet radio technology is well established and is expected to play a major role in local distribution, particularly when the subscribers are mobile.

12.4.2 Local Area Networks

Local area networks (LAN) are used to interconnect computers, terminals, word processors, facsimile, and other office machines within a building, campus, or metropolitan area. A single definition of a local area network is difficult to achieve because of the divergent applications and design alternatives, but generally the characteristics include [24]:

- Geographically local with distances limited to a few miles
- Multiple services often possible, including voice, data, and video
- High-speed transmission, with typical data rates in the range of 50 kb/s to 150 Mb/s
- Use of cable transmission media, including twisted pair, coaxial cable, and fiber optics
- Some type of switching technology, typically packet switching
- Some form of network topology and network control
- Owned by a single company

Thus various technologies are necessary in local area networks, including transmission, switching, and networking. Here we will examine the alternative choices for transmission media.

Twisted pairs are limited in bandwidth and therefore have application principally in a voice network using a private automatic branch exchange (PABX) or for low-speed data possibly for personal computer networking. Coaxial cable is presently the most commonly used medium for LANs, although use of optical fiber is expected to grow rapidly. LAN transmission on coaxial cable falls into two categories: baseband and broadband. In the baseband mode, data rates to 50 Mb/s have been transmitted by using baseband coding schemes such as Manchester coding. The most prominent of the baseband LANs is Xerox's Ethernet, which transmits data at 10 Mb/s. The broadband mode of transmission uses a RF modem, which results in greater noise immunity and the ability to provide multiple FDM channels for simultaneous voice, data, and video services. Wangnet from Wang Laboratories is an example of a broadband system capable of providing all three services.

The development of standards for local area networks has been a difficult process because of competition and rapid advances in LAN technology. Nevertheless the trend is clearly toward integration of voice, data, and video to permit a single LAN for all services. The wide bandwidths required with integrated LANs also suggest a bigger role for fiber optics in future LANs.

12.5 INTEGRATED SERVICES DIGITAL NETWORK

The natural evolution of the public switched telephone network and the emergence of digital communication technology has led to the development of a concept known as the integrated services digital network (ISDN). The principle behind the ISDN concept is end-to-end digital connectivity that will support a wide range of voice and nonvoice services. The ISDN will be based on current integrated digital networks (IDN) that use 64-kb/s switched digital connections. Initial services under consideration include voice, data, facsimile, teletex, and videotext; full-motion video is likely to be one of the last services implemented. The CCITT has been the international focal point in defining and standardizing the ISDN concept. According to the CCITT in 1980, the transition period from existing networks to a comprehensive ISDN may require one or two decades [25].

As shown in Figure 12.2, two basic types of communication channels are defined for ISDN subscriber access: the B channel at 64 kb/s and the D channel at 16 kb/s. The B channels can carry digital voice or data. The D channel's primary function is to provide out-of-band signaling for B channel services, but it may also be accessed for telemetry and packet-switched data. Subscriber requirements for all but video services can be satisfied by providing two B channels and a D channel, or $2B + D$, for a total of 144

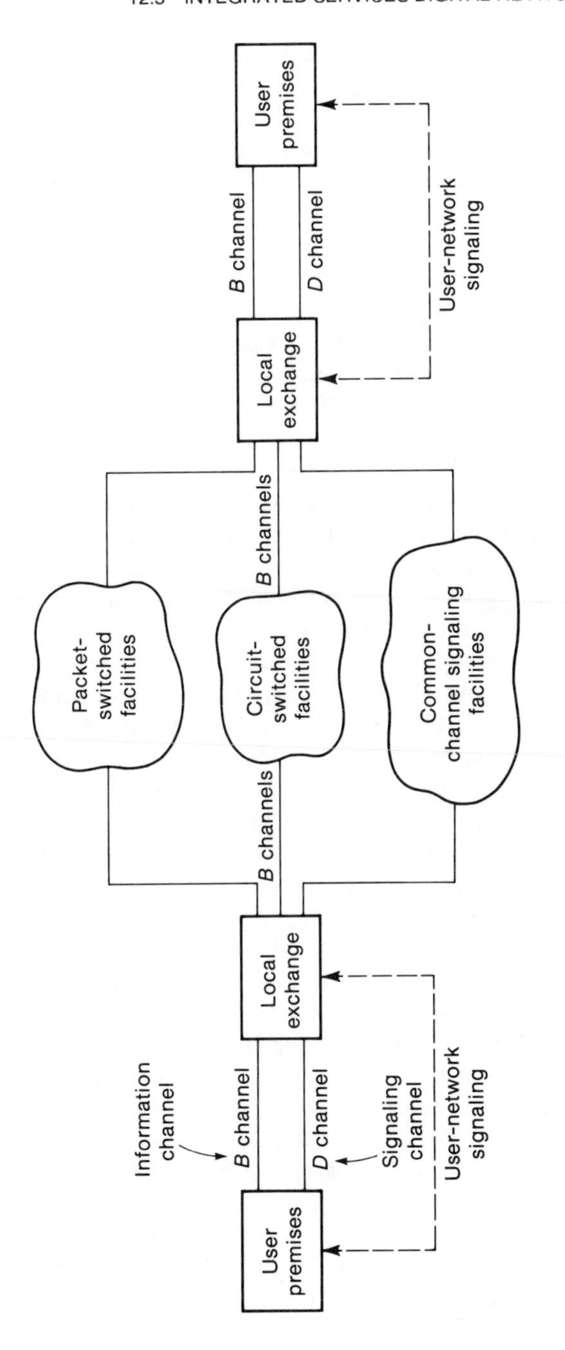

Figure 12.2 Functional Model of ISDN

kb/s. Subscribers requiring higher-speed connections can be provided multiple *B* channels, in which case the *D* channel may have to operate at 64 kb/s.

The ability of the transmission plant to support these ISDN channels is most limited in the local loop, which consists primarily of metallic two-wire loops. Various transmission techniques have been studied for full-duplex digital transmission over two-wire loops. A leading candidate is time-compressed multiplexing (TCM). The TCM scheme shares the loop between the two directions of transmission by transmitting bursts of data in each direction at a rate slightly higher than twice the nominal rate [26]. In the United States, the Bell operating companies have introduced a 56-kb/s circuit-switched digital capability (CSDC) that uses TCM in the local loop and IDN in the long-haul plant [27].

Techniques such as TCM can provide sufficient bandwidth on existing subscriber loops for all services except video. However, CATV networks using coaxial cable are already offering video services independent of ISDN planning. In the future, subscriber demand for switched video service or two-way video, plus the availability of low-cost, high-bandwidth fiber optics, should introduce video into the ISDN arena.

In the United States, the first significant planning for ISDN was undertaken by AT & T and the Bell operating companies with the introduction of 56-kb/s CSDC. However, ISDN in the United States will be influenced by the competitive market and government deregulation of telecommunication services. As a result, several ISDNs could emerge from competing telecommunication networks. Standards for interface between these networks will be necessary for the realization of a national ISDN in this country. In other parts of the world, national telecommunication administrations and various organizations have developed detailed plans and standards for ISDNs. In Europe, the CEPT has pursued plans for ISDN based on the use of digital transmission and digital switching equipment now being implemented to modernize European telephone networks. Standardization in Europe is facilitated since telecommunication networks are controlled mainly by PTT administrations.

The key to success in international and perhaps even national ISDNs rests with the efforts of the CCITT. The new I series recommendations published from the 1984 CCITT Plenary Assembly deal exclusively with the ISDN. Moreover, Study Group XVIII is considering ISDN performance standards, first based on a 64-kb/s hypothetical reference circuit. Among the transmission parameters that have been addressed are availability and error performance (Rec. G.821) and slip rate performance (Rec. G.822). (The details of these CCITT recommendations are discussed in Chapter 2 and shown in Tables 2.3 and 2.7 of this book.) These parameters apply to voice and data services independent of whether circuit switching or packet switching is used. The challenges ahead lie with the difficulties of prescribing transmission performance allocations for a wide variety of services and bit rates envisioned with ISDN.

12.6 SUMMARY

The prospect for digital transmission in telecommunication networks of the future is bright due to the pervasive use of digital technology. The merging of digital transmission, switching, and terminal equipment has led to the concept and development of integrated services digital networks (ISDN). In the United States, the breakup of the Bell System and government deregulation has stimulated competition but raised new questions about standardization for interconnection of telecommunication networks. Outside the United States, particularly in Europe, planning and implementation of standardized digital networks are proceeding at a faster pace.

Most new telecommunication services now emerging will be transmitted by digital networks. The combination of telecommunications with data processing has led to the introduction of telematic services, such as facsimile, teletex, videotext, and electronic mail. Although these telematic services are provided primarily by digital transmission, new video services for the home and office will be carried mostly by analog transmission with the exception of video teleconferencing.

New digital transmission technology is now reducing both cost and bandwidth. Reduced-rate coding and speech interpolation applied to voice transmission have allowed bandwidth-efficient use of digital transmission for both satellite and terrestrial networks. The emerging 32-kb/s ADPCM standard has promise for widespread application in voice networks. In the United States, the 1.544-Mb/s DS-1 standard is being revised with an extended framing format to allow introduction of 64-kb/s channels for compatibility with ISDN. Advances in modulation, adaptive equalization, and reliability have made digital radio cost-competitive with analog radio. The newest fiber optic systems, with single-mode fiber and long-wavelength components, will make fiber optics the predominant transmission media in the 1990s.

The evolution of digital transmission has led to other diversified applications. Digital radio has been used for local area distribution in order to bypass the local telephone company. Digital cable transmission is used in local area networks, where fiber optics is expected to dominate. Finally, digital transmission will play a key role in the introduction of ISDN, which will provide end-to-end digital connections for a wide range of voice and nonvoice services.

REFERENCES

1. E. B. Carne, "New Dimensions in Telecommunications," *IEEE Comm. Mag.* 20(1)(January 1982):17–25.
2. CCITT Yellow Book, vol. VII.2., *Telegraph and Telematic Services Terminal Equipment* (Geneva: ITU, 1981).
3. T. Kamae, "Public Facsimile Communication Network," *IEEE Comm. Mag.* 20(2)(March 1982):47–51.

4. K. Takikawa, "Simplified 6.3 Mbit/s Codec for Video Conferencing," *IEEE Trans. on Comm.*, vol. COM-29, no. 12, December 1981, pp. 1877–1882.
5. H. S. London and D. B. Menist, "A Description of the AT & T Video Teleconferencing System," *1981 National Telecomm. Conf.*, pp. F5.4.1 to F5.4.5.
6. T. Ishiguro and K. Iinuma, "Television Bandwidth Compression by Motion-Compensated Interframe Coding," *IEEE Comm. Mag.* 20(6)(November 1982):24–30.
7. J. E. Thompson, "European Collaboration on Picture Coding Research for 2 Mbit/s Transmission," *IEEE Trans. on Comm.*, vol. COM-29, no. 12, December 1981, pp. 2003–2004.
8. H. S. London and T. S. Giuffrida, "High Speed Switched Digital Service," *IEEE Comm. Mag.* 21(2)(March 1983):25–29.
9. D. L. Duttweiler and D. G. Messerschmitt, "Nearly Instantaneous Companding for Nonuniformly Quantized PCM," *IEEE Trans. on Comm.*, vol. COM-24, no. 8, August 1976, pp. 864–873.
10. D. W. Petr, "32 kbps ADPCM-DLQ Coding for Network Applications," *1982 Global Telecommunications Conference*, pp. A8.3.1 to A8.3.5.
11. R. E. Crochiere and J. L. Flanagan, "Current Perspectives in Digital Speech," *IEEE Comm. Mag.* 21(1)(January 1983):32–40.
12. R. P. Gooch, "DCEM: A DSI System for North America and Europe," *1982 International Conference on Communications*, pp. 4.G.5.1 to 4.G.5.5.
13. AT & T Compatibility Bulletin No. 142, "The Extended Framing Format Interface Specification" (Parsippamy, N.J.: AT & T, 1981).
14. S. Katayama, M. Iwamoto, and J. Segawa, "Digital Radio-Relay Systems in NTT," *1984 International Communications Conference*, pp. 978–983.
15. D. C. Riker, "Digital Transmission in the MCI Network," *IEEE Comm. Mag.* 22(10)(October 1984): 22–25.
16. J. McNichol, S. Barber, and F. Rivest, "Design and Application of the RD-4 and RD-6 Digital Radio Systems," *1984 International Conference on Communications*, pp. 646–652.
17. G. Ungerboeck, "Channel Coding with Multilevel/Phase Signals," *IEEE Trans. Info. Theory*, vol. IT-28, no. 1, January 1982, pp. 55–67.
18. D. C. Gloge and others, "Characteristics and Operation of the FT4E-432 Mb/s Repeater Line," *1984 International Conference on Communications*, pp. 775–778.
19. T. Ito, K. Nakagawa, and Y. Hakamada, "Design and Performances of the F-400M Trunk Transmission System Using a Single-Mode Fiber Cable," *1982 International Conference on Communications*, pp. 6.D.1.1 to 6.D.1.5.
20. J. E. Midwinter, "Development of High Bit Rate Monomode Fibre Systems in the UK," *1982 International Conference on Communications*, pp. 6.D.6.1 to 6.D.6.4.
21. K. Mouthaan, "Long-Wavelength Optical Transmission Systems in Europe," *1982 International Conference on Communications*, pp. 5.D.5.1 to 5.D.5.2.
22. D. Nielson, "The Role of Radio in Local Area Data Distribution," *Journal of Telecomm. Networks* 1(1)(Spring 1982):39–47.
23. R. E. Kahn and others, "Advances in Packet Radio Technology," *Proc. IEEE* 66(11)(November 1978):1468–1496.

24. T. Lissack, B. Maglaris, and I. T. Frisch, "Digital Switching in Local Area Networks," *IEEE Comm. Mag.* 21(3)(May 1983):26–37.
25. CCITT Yellow Book, vol. III.3, *Digital Networks—Transmission Systems and Multiplexing Equipment* (Geneva: ITU, 1981).
26. B. S. Bosik and S. V. Kartalopoulos, "A Time Compression Multiplex System for a Circuit Switched Digital Capability," *IEEE Trans. on Comm.*, vol. COM-30, no. 9, September 1982, pp. 2046–2052.
27. F. C. Kelcourse and E. H. Siegel, "Switched Digital Capability: An Overview," *1982 International Conference on Communications*, pp. 3.D.1.1 to 3.D.1.4.

Glossary

Many of these definitions have been taken from the following sources: CCITT 1981, vol. X.1, *Terms and Definitions*; CCIR 1982, vol. XIII, *Vocabulary*; and U.S. Federal Standard 1037, *Glossary of Telecommunication Terms*, 1980.

Adaptive equalization. A process in a receiver in which received symbols are used to continuously update the settings of a filter used for equalization.

Aliasing distortion. The error in reconstructing a sampled analog signal when the signal is sampled at a rate that is less than twice the signal bandwidth. Also called *foldover distortion*.

Amplitude dispersion. See *dispersion*.

Amplitude distortion. Distortion in a transmission system caused by non-uniform gain or attenuation with respect to frequency. Also called *attenuation distortion*.

Amplitude-shift keying (ASK). A form of digital modulation in which discrete amplitudes of the carrier are used to represent a digital signal.

Anomalous propagation. A propagation condition that results from abnormal changes with altitude in the atmospheric index of refraction.

Asynchronous transmission. A transmission mode in which each character is individually synchronized by use of start and stop elements. The time intervals between transmitted characters may be of unequal length.

Availability. The probability or fraction of time that a system is not in a state of failure.

Bandwidth-distance product. The product of bandwidth and distance (usually in MHz · km), which specifies a limit on an optical fiber's bandwidth and repeater spacing.

Baud. The unit of symbol rate in modulation. Corresponds to a rate of one symbol or unit interval per second.

Bipolar. A pseudoternary signal, conveying binary digits, in which successive "marks" are of alternate polarity (positive and negative) but equal in amplitude and in which "spaces" are of zero amplitude. Also called *alternate mark inversion* (AMI). [FED-STD-1037]

Bipolar *N*-zero substitution (BNZS). A form of bipolar transmission in which all strings of N zeros are replaced with a special N bit sequence containing bipolar violations.

Bipolar violation. A mark that has the same polarity as the previous mark in the transmission of bipolar signals.

Bit count integrity. Preservation of the precise number of bits (or characters or frames) that are originated in a message or unit of time.

Bit error rate. The ratio of the number of bits incorrectly received to the total number of bits transmitted.

Bit interleaving. A method of time-division multiplexing in which each channel is assigned a time slot corresponding to a single bit.

Bit stuffing. See *pulse stuffing*.

Buffered timing. A form of timing used with independent clocks in which buffers compensate for small differences in clock rates between equipments.

Bulk encryption. The process whereby two or more channels of a telecommunications system are encrypted by a single piece of crypto-equipment. [FED-STD-1037]

Bunched frame structure. A TDM frame structure in which the frame alignment signal occupies contiguous bit positions. Also called *burst frame structure*.

Burst frame structure. See *bunched frame structure*.

Carrier. A wave suitable for modulation by an information-bearing signal to be transmitted over a communication system. [FED-STD-1037]

Cesium clock. An atomic clock using the cesium element to produce a stable, accurate timing source.

Channel. A single path provided from a transmission medium either by physical separation (e.g., multipair cable) or by electrical separation (e.g., frequency or time-division multiplexing). [FED-STD-1037]

Channel vocoder. A vocoder in which the vocal tract filter is estimated by samples taken in contiguous frequency bands within the speech signal.

Character interleaving. A method of time-division multiplexing in which each channel is assigned a time slot corresponding to a single character. Also called *word interleaving*.

Circuit. The electrical path between two endpoints over which one-way or two-way communications may be provided.

Clipping. In speech interpolation, the loss of speech segments due to overload or delay in speech detection. Also see *overloading*.

Clock. Equipment providing a time base used in a transmission system to control the timing of certain functions such as the duration of signal elements or the sampling. [CCITT]

Coaxial cable. A metallic communications cable consisting of an inner conductor surrounded by and insulated from a concentric outer conductor.

Coded mark inversion. A two-level code in which binary 0 is coded so that both amplitude levels are attained within the bit duration, each for half the interval. Binary 1 is coded by either of the amplitude levels for the full bit duration in such a way that the levels alternate with each occurrence of a 1.

Codirectional timing. A form of timing for synchronous transmission in which timing is provided along with data by the data terminal equipment.

Coherent detection. Detection using a reference signal that is synchronized in frequency and phase to the transmitted signal. Also called *synchronous detection*.

Companding. The process of compressing a signal at the transmitter and expanding it at the receiver to allow signals with a large dynamic range to be sent through a device with a more limited range.

Conditioned diphase. A two-level code in which diphase is applied to a conditioned NRZ signal.

Conditioned NRZ. A two-level nonreturn-to-zero code in which the presence or absence of a level change is used to signal a binary number.

Conditioning. Special treatment of voice-band data circuits provided by telephone companies. As provided by AT & T, for example, C-type conditioning controls phase and amplitude distortion whereas D-type conditioning limits noise and harmonic distortion.

Continuously variable slope delta modulation (CVSD). A type of delta modulation in which the size of the steps of the approximated signal is progressively increased or decreased to make the approximated signal closely match the input analog wave. [FED-STD-1037]

Contradirectional timing. A form of timing for synchronous transmission in which timing of the data terminal equipment is slaved to timing from the data circuit terminating equipment.

Correlative coding. See *partial response*.

Crosstalk. The phenomenon in which a signal transmitted on one circuit or channel of a transmission system creates an undesired effect in another circuit or channel. [FED-STD-1037]

Cryptographic equipment. A scrambling device based on digital logic used to encrypt communications.

Data above voice. A hybrid transmission scheme that places a data signal above the voice spectrum in a cable or radio system. Also called *data over voice*.

Data circuit terminating equipment (DCE). The interfacing equipment sometimes required to couple the data terminal equipment (DTE) into a transmission circuit or channel and from a transmission circuit or channel into the DTE. [FED-STD-1037]

Data in voice. A hybrid transmission scheme that places a data signal in the middle of the voice spectrum in a cable or radio system.

Data over voice. See *data above voice*.

Data terminal equipment (DTE). Equipment consisting of digital instruments that convert the user information into data signals for transmission or reconvert the received data signals into user information. [FED-STD-1037]

Data under voice. A hybrid transmission scheme that places a data signal under the voice spectrum in a cable or radio system.

dbm. Power level in decibels relative to 1 milliwatt.

dBm0. Power level in dBm referred to or measured at a point of zero relative level.

dBrnC0. C-message weighted noise power in decibels referenced to 1 picowatt (−90 dBm) and measured at a zero transmission level point.

Decision feedback equalizer. An equalizer that uses a feedback loop operating on the detector outputs to cancel intersymbol interference due to previous symbols.

Dedicated circuit. A circuit designated for exclusive use by two users. [FED-STD-1037]

Delay distortion. The distortion of a waveform made up of two or more different frequencies that is caused by the difference in arrival time of each frequency at the output of a transmission system. [FED-STD-1037]

Delta modulation. A form of DPCM in which the magnitude of the difference between the predicted value and the actual value is encoded by one bit only—that is, where only the sign of that difference is detected and transmitted. [CCITT]

Differential phase-shift keying (DPSK). A method of phase-shift keying in which each symbol is a change in the phase of the carrier with respect to its previous phase angle.

Differential pulse code modulation (DPCM). A process in which a signal is sampled, and the difference between the actual value of each sample and its predicted value derived from previous samples is quantized and converted by encoding to a digital signal. [CCITT]

Digital channel bank. PCM multiplexer equipment used in the North American digital hierarchy.

Digital speech interpolation (DSI). A form of digital multiplexing for voice that uses speech interpolation.

Digital termination systems. A form of local loop or local distribution in which digital radio is used to connect users to a long-haul communication network.

Diphase. A two-level code generated by the modulo 2 addition of an NRZ signal with its associated clock signal.

Dispersion. The spreading, separation, or scattering of a waveform during transmission. Also called *amplitude dispersion*.

Distortion. Any departure from a specified input/output relationship over a range of frequencies, amplitudes, or phase shifts during a time interval. [FED-STD-1037]

Distributed frame structure. A TDM frame structure in which the frame alignment signal occupies noncontiguous bit positions.

Doppler. A shift in the observed frequency of a signal caused by variation in the path length between the source and the point of observation.

Drift. An undesirable progressive change in frequency with time. Also called *frequency drift* or *wander*. [CCIR]

Ducting. The propagation of radio waves within an atmospheric layer whose refractivity gradients are such that radio waves are guided or focused within the duct.

Duobinary. A three-level coding scheme that uses controlled amounts of intersymbol interference to achieve transmission at the Nyquist rate.

Duplex. A type of transmission that affords simultaneous operation in both directions.

Effective earth radius. The radius of a hypothetical earth for which the distance to the radio horizon, assuming rectilinear propagation, is the same as that for the actual earth with an assumed uniform vertical gradient of refractive index. [FED-STD-1037]

Elastic buffer. A storage device using digital logic designed to accept data timed by one clock and deliver it timed by another clock.

Encoder. A device that generates a code, frequently one consisting of binary numbers, to represent individual characters or signal samples.

Envelope delay distortion. In a given passband of a device or a transmission facility, the maximum difference of the group delay time between any two specified frequencies. Also called *group delay distortion*. [FED-STD-1037]

Envelope detection. A form of demodulation in which detection is based on the presence or absence of the signal envelope.

Equalization. The process of reducing frequency distortion or phase distortion, or both, of a circuit by the introduction of networks to compensate for the difference in attenuation, time delay, or both, at the various frequencies in the transmission band. [FED-STD-1037]

Error-free blocks. A measure of error performance based on the percentage or probability of data blocks that are error free.

Error-free seconds. A measure of error performance based on the percentage or probability of seconds that are error free.

Exponential delta modulator. A type of delta modulator that uses an RC circuit for an integrator that provides an exponential response to a constant input.

Extended framing format. A framing format for 1.544-Mb/s digital channel banks introduced by AT & T with the D5 channel bank.

External reference synchronization. A form of network synchronization in which a timing or frequency reference is obtained from a source external to the communication network.

Eye pattern. An oscilloscope display of a digital signal used to examine performance.

Fade margin. The amount by which a received signal level may be reduced without causing the system (or channel) output to fall below a specified threshold. [FED-STD-1037]

Fading. The variation, with time, of the intensity or relative phase, or both, of any frequency component of a received signal due to changes in the characteristics of the propagation path with time. [FED-STD-1037]

Failure. The termination of the ability of an item to perform a required function. [CCITT]

Failure rate. The average rate at which failures can be expected to occur throughout the useful life of an item.

Far-end crosstalk (FEXT). Crosstalk that is propagated in a disturbed channel in the same direction as the propagation of a signal in the disturbing channel. The receiving terminals of the disturbed channel and the energized terminals of the disturbing channel are usually remote from each other. [FED-STD-1037]

Flat fading. Fading in which all frequency components of the received radio signal vary in the same propagation simultaneously. [FED-STD-1037]

Foldover distortion. See *aliasing distortion*.

Formant vocoder. A vocoder in which the vocal tract filter is estimated by the frequencies and amplitude of the spectral peaks, or formants.

Fractional frequency difference. The algebraic difference between two normalized frequencies. Also called *relative frequency difference* or *normalized frequency difference*.

Frame. A set of consecutive digit time slots in which the position of each slot can be identified by reference to a frame alignment signal. The frame alignment signal does not necessarily occur, in whole or in part, in each frame. [CCITT]

Frame acquisition mode. A mode of operation in TDM frame synchronization in which the frame alignment signal is detected. Also called *frame search mode*.

Frame acquisition time. The time that elapses between a valid frame alignment signal being available at the receiver terminal equipment and frame alignment being established. Also called *frame alignment recovery time*. [CCITT]

Frame alignment. The state in which the frame of the receiving equipment is correctly phased with respect to that of the received signal. [CCITT]

Frame maintenance mode. A mode of operation in TDM frame synchronization in which the frame alignment signal, after detection, is continuously monitored to ensure that frame alignment is maintained.

Frame reacquisition time. The time that elapses between a loss of frame synchronization and the recovery of the frame alignment signal.

Frame search mode. See *frame acquisition mode.*

Free-space loss. The signal attenuation that would result if all obstructing, scattering, or reflecting influences were sufficiently removed that they have no effect on propagation. [FED-STD-1037]

Frequency. The number of cycles or events per unit of time. When the unit of time is one second, the measurement unit is the hertz (Hz). [FED-STD-1037]

Frequency accuracy. The degree of conformity to a specified value of a frequency. [FED-STD-1037]

Frequency deviation. In frequency modulation, the peak difference between the instantaneous frequency of the modulated wave and the carrier frequency. [FED-STD-1037]

Frequency diversity. A method of radio transmission used to minimize the effects of fading wherein the same information signal is transmitted and received simultaneously on two or more independent carrier frequencies.

Frequency drift. See *drift.*

Frequency-selective fading. Fading in which not all frequency components of the received radio signal vary simultaneously.

Frequency-shift keying (FSK). A form of frequency modulation in which discrete frequencies are used to represent a digital signal.

Fresnel zones. A cigar-shaped shell of circular cross section surrounding the direct path between a transmitter and a receiver. For the first Fresnel zone, the distance from the transmitter to any point on this shell and on to the receiver is one half-wavelength longer than the direct path; for the second Fresnel zone, two half-wavelengths, etc. [FED-STD-1037]

Graded-index fiber. An optical fiber which has a refractive index that gets progressively higher toward the center, causing light rays continually to be refocused inward.

Group. In frequency-division multiplexing, a number of voice channels (generally 12) occupying a 48-kHz frequency band.

Group delay distortion. See *envelope delay distortion*.

Half duplex. A type of transmission that affords communication in either direction but only in one direction at a time.

High-density bipolar *N* (HDBN). A form of bipolar transmission that limits the number of consecutive zeros to N by replacing the $(N + 1)$th zero with a bipolar violation.

High-information delta modulation. A form of delta modulation in which the step size is doubled for consecutively identical bits at the coder output and halved for consecutively opposite bits at the coder output.

Hybrid transmission. A type of transmission that carries FDM voice and digital data on the same medium, such as cable or radio.

Independent clock synchronization. A form of network synchronization based on the use of clocks that are independently timed but constrained within specified limits of accuracy.

In-service testing. A form of testing in which operational traffic is not interrupted.

Instantaneous companding. A form of companding used in delta modulation in which changes in step size are made at a rate equal to the sampling rate.

Integrated digital network. A network in which connections established by digital switching are used for the transmission of digital signals. [CCITT]

Integrated services digital network (ISDN). An integrated digital network in which the same digital switches and digital paths are used to establish connection for different services—for example, telephony and data. [CCITT]

Intermodal dispersion. See *multimode dispersion*.

Intersymbol interference. Extraneous energy from the signal in one or more signaling intervals that tends to interfere with the reception of the signal in another signaling interval.

Intramodal dispersion. See *material dispersion*.

Jitter. Short-term variations of the significant instants of a digital signal from their ideal positions in time. [CCITT]

Jumbogroup. In frequency-division multiplexing, a number of voice channels (generally 3600) composed of six mastergroups.

Justification. See *pulse stuffing*.

***k* factor.** The ratio of effective to true earth radius.

Linear predictive coding. A voice coder based on prediction of speech samples from a linear weighted sum of previously measured samples.

Line-of-sight. Radio propagation in the atmosphere along a path that is unobstructed by the earth or any other opaque object.

Loading factor. In a quantizer, the ratio of peak to rms amplitude for the input signal.

Local area network. A network using principally cable transmission to connect multiple services within a geographically local area.

Local loop. A single connection from a local central office to the end terminal.

Mastergroup. In frequency-division multiplexing, a number of voice channels (generally 300 or 600) consisting of five or ten supergroups.

Master-slave synchronization. A form of network synchronization in which a master clock is distributed and used to control other clocks, equipment, or nodes.

Material dispersion. In an optical fiber, the dispersion caused by the variation in propagation velocity with the wavelength of light. Also called *intramodal dispersion*.

Mean time between failure (MTBF). For a specific interval the ratio of total operating time to the number of failures in the same interval.

Mean time between outages (MTBO). For a specific interval the ratio of total operating time to the number of outages in the same interval.

Mean time to repair (MTTR). The total corrective maintenance time divided by the total number of corrective maintenance actions during a given period of time. [FED-STD-1037]

Mean time to service restoral (MTSR). The mean time to restore service following system failures that result in a service outage. The time to restore includes all time from the occurrence of the failure until the restoral of service. [FED-STD-1037]

Medium. A substance regarded as the means of signal transmission.

Microbending. Minute curvatures in an optical fiber caused by external forces. Since the principle of total internal reflection is not fully satisfied at such bends, additional loss in the fiber is introduced.

Midpoint equalization. A combination of postequalization and preequalization used at intermediate points in a circuit.

Minimum-shift keying (MSK). A form of digital modulation that uses offset QPSK with sinusoidal pulse shaping. Also a special case of FSK in which continuous phase is maintained at symbol transitions using a minimum difference in signaling frequencies.

Modem. A contraction of "modulator/demodulator." The term may be used when the modulator and the demodulator are associated in the same signal-conversion equipment. [CCITT]

Modified duobinary. A form of duobinary coding whose spectral shape has no dc component.

Modulation. The process of varying certain characteristics of a carrier in accordance with a message signal.

Modulation index. In FSK, the ratio of the frequency deviation of the modulated signal to the bandwidth of the modulating signal.

Modulator. A device that performs modulation.

Multiframe. A set of consecutive frames in which the position of each frame can be identified by reference to a multiframe alignment signal.

Multimode dispersion. In an optical fiber, the dispersion resulting from the different arrival times of optical rays that follow different paths. Also called *intermodal dispersion*.

Multimode fiber. A fiber that supports propagation of more than one mode (optical ray) at a given wavelength.

Multipath fading. Fading that results when radio signals reach the receiving antenna by two or more paths.

Mutual synchronization. A network synchronizing arrangement in which each clock in the network exerts a degree of control on all others. [FED-STD-1037]

Narrowband coder. A voice coder with a transmission rate that can be accommodated by 3-kHz telephone channels using modems.

Narrowband FSK. A form of FSK in which the modulation index is much less than 1.

Near-end crosstalk (NEXT). Crosstalk that is propagated in a disturbed channel in the direction opposite to the direction of propagation of the circuit in the disturbing channel. The receiving terminals of the disturbed channel and the energized terminals of the disturbing channel are usually near each other. [FED-STD-1037]

Negative pulse stuffing. In time-division multiplexing, the controlled delection of bits from channel inputs so that the rates of the individual channel inputs correspond to a rate determined by the multiplex equipment. The deleted information is transmitted via a separate overhead channel.

Network. An organization of stations capable of intercommunication but not necessarily on the same channel. [FED-STD-1037]

NEXT coupling loss. For near-end crosstalk, the ratio of power in the disturbing circuit to the induced power in the disturbed circuit.

Noncoherent detection. Any form of detection that does not require a phase reference.

Nonreturn-to-zero (NRZ). A two-level code in which each level is held for the duration of a signal interval.

Normalized frequency. The ratio between the actual frequency and its nominal value. [CCIR]

Numerical aperture (NA). A number that expresses the light-gathering characteristics of an optical fiber, expressed by the sine of the maximum angle (with respect to the fiber axis) at which an entering ray will experience total internal reflection.

Nyquist bandwidth. As first described by Nyquist, the minimum channel bandwidth required for transmission without intersymbol interference.

Nyquist rate. As described by Nyquist, the sampling rate (equal to twice the signal bandwidth) that permits the samples of a signal to completely determine the signal waveform.

Offset quadrature phase-shift keying (OQPSK). A form of QPSK in which the in-phase and quadrature bit streams are offset in time by one bit period.

Optical fiber. A long, thin cylinder having a central core of one transparent material of high refractive index surrounded with a cladding of another transparent material of lower refractive index.

Outage. A condition wherein a user is deprived of service due to failure in the communication system.

Out-of-service testing. A form of testing in which operational traffic is interrupted and replaced by a test pattern.

Overloading. The error introduced in quantization when the input signal amplitude exceeds the allowed amplitude range of the quantizer. Also called *clipping*.

Packet radio. A digital radio designed to handle multiple users who share a common high-speed channel and transmit data in bursts (packets) during scheduled transmission times.

Partial response. A multilevel coding scheme that uses prescribed amounts of intersymbol interference to increase the transmission rate in a given bandwidth.

Path profile. A graphic representation of a propagation path showing the surface features of the earth, such as trees, buildings, and other features that may cause obstruction or reflection, in the vertical plane containing the path. [FED-STD-1037]

Phase delay. In the transfer of a single-frequency wave from one point to another in a system, the time delay of the part of the wave that identifies its phase. [FED-STD-1037]

Phase-shift keying (PSK). A form of digital modulation in which discrete phases of the carrier are used to represent a digital signal.

Piezoelectric effect. An effect in certain materials such as quartz in which electrical energy causes vibration and, conversely, mechanical vibration induces electrical energy.

Plesiochronous. The relationship between two signals such that their corresponding significant instants (transitions) occur at nominally the same rate, any variation in rate being constrained within a specified limit.

Polar. A two-level signal using balanced (symmetric) positive and negative levels.

Positive-negative pulse stuffing. A combination of positive and negative pulse stuffing in which the two stuffing states are indicated by uniquely coded signals.

Positive pulse stuffing. In time-division multiplexing, the provision of a fixed number of dedicated time slots used to transmit either information from the individual channel inputs or no information, according to the relative bit rates of the channel input and the TDM channel output signals.

Positive-zero-negative pulse stuffing. A combination of positive and negative pulse stuffing in which the two stuffing states are indicated by uniquely coded signals and the state of no stuffing is indicated by a third signal.

Postequalization. Equalizition of a circuit performed at the receiver.

Power fading. Fading caused by anomalous propagation conditions and characterized as slowly varying in time and causing long outages.

Preequalization. Equalization of a circuit performed at the transmitter using a form of predistortion.

Preset equalization. A form of automatic equalization in which a training sequence is used to fix the settings of the equalizer.

Pseudoerror detector. A BER estimation technique that uses modified decision regions in a separate decision circuit to degrade the receiver margin intentionally and reduce the time required to estimate the BER.

Pseudorandom sequence. A sequence of numbers or bits that exhibits properties of a random signal. Although it seems to lack a definite pattern, there is a sequence that repeats after a long time interval.

Pulse code modulation. A process in which a signal is sampled, and the magnitude of each sample is quantized independently of other samples and converted by encoding to a digital signal. [CCITT]

Pulse stuffing. A process of changing the rate of a digital signal in a controlled manner so that it can accord with a rate different from its own inherent rate, usually without loss of information. Also called *bit stuffing* and *justification*. [CCITT]

Pulse stuffing jitter. Jitter caused by the removal of stuffed bits at the demultiplexer and the inability of a clock recovery circuit to eliminate completely the resulting gaps in the information signal.

Quadrature amplitude modulation (QAM). Independent amplitude modulation of two orthogonal channels using the same carrier frequency.

Quadrature partial response (QPR). The use of partial response filtering on the two orthogonal channels of a QAM system to increase the bandwidth efficiency of QAM.

Quadrature phase-shift keying (QPSK). A form of PSK in which four discrete phases, separated by 90°, are used to represent two information bits per signaling interval.

Quantization. A process in which the continuous range of values of a signal is divided into nonoverlapping but not necessarily equal subranges and to each subrange a discrete value of the output is uniquely assigned. Whenever the signal value falls within a given subrange, the output has the corresponding discrete value. [FED-STD-1037]

Quantization noise. An undesirable random signal caused by the error of approximation in a quantizing process. [FED-STD-1037]

Quantizing distortion. The distortion resulting from the process of quantizing. [FED-STD-1037]

Quartz clock. A clock made of quartz crystal.

Radio refractivity (N). One million times the amount by which the refractive index of the atmosphere exceeds unity.

Raised-cosine filter. A filter with a specific characteristic that produces no intersymbol interference at the sample times of adjacent signaling intervals.

Receiver. A device that converts electrical or optical signals used for transmission back to information signals.

Relative (frequency) drift. The frequency drift divided by the nominal frequency value. Also called *normalized frequency drift*. [CCIR]

Reliability. The probability that an item will perform its intended function for a specified interval under stated conditions. [FED-STD-1037]

Reproducibility. The degree to which independent devices of the same design can be adjusted to produce the same frequency, or the degree to which a device will produce the same frequency from one occasion to another.

Return-to-zero (RZ). A two-level code in which each level is held for a fraction (usually half) of the signaling interval and then returned to a reference level for the remainder of the signaling interval.

Rubidium clock. An atomic clock using the element rubidium to produce a stable, accurate timing source.

Scintillation. In radio propagation, a random fluctuation of the received field about its mean value; the deviations usually are relatively small. [FED-STD-1037]

Secure voice. Telephone communications that are protected against compromise through the use of an encryption system.

Settability. The degree to which frequency can be adjusted to correspond with a reference frequency.

Signal constellation. The signal magnitudes and phases of a digital modulation scheme when displayed by a phasor diagram.

Signal-to-distortion ratio. The ratio of the amplitude of the desired signal to the amplitude of the distortion.

Simplex. Permitting the transmission of signals in either direction, but not simultaneously. [CCITT]

Single-mode fiber. Optical fiber with a core radius so close to a wavelength of light that only one mode can be propagated.

Slip. The irretrievable loss or gain of a set of consecutive bits without loss of alignment. Also called *timing slip*.

Slope overload distortion. In quantization, the distortion that results when the slope of the input signal exceeds the slope of the quantizer.

Space diversity. A method of radio transmission employed to minimize the effects of fading by the simultaneous use of two or more antennas spaced apart.

Speech interpolation. A voice multiplexing scheme in which the gaps and pauses occurring in one voice channel are filled with speech bursts from other voice channels to reduce the bandwidth or transmission rate.

Stability. The rate at which a device changes from its nominal frequency over a selected period of time.

Standard refraction. The refraction resulting from a gradient of radio refractivity equal to -40 N/km. Also corresponds to atmospheric refraction with $k = 4/3$.

Statistical multiplexing. A form of concentration in data multiplexing in which time slots are allocated to whichever terminals are active at any one time.

Step-index fiber. An optical fiber that has an abrupt change in refractive index at the boundary of the core and cladding.

Subrefraction. Refraction for which the refractivity gradient is greater than that for standard refraction. Also corresponds to atmospheric refraction with $k < 1$.

Supergroup. In frequency-division multiplexing, a number of voice channels (generally 60) occupying a 240-kHz bandwidth and composed of five groups.

Supermastergroup. In frequency-division multiplexing, a number of voice channels (generally 900) occupying a 3872-kHz bandwidth and composed of three 300-channel mastergroups.

Superrefraction. Refraction for which the refractivity gradient is less than that for standard refraction. Also corresponds to atmospheric refraction with $2 < k < \infty$.

Switched circuit. A circuit that may be temporarily established at the request of one or more of the connected stations. [FED-STD-1037]

Syllabic companding. A form of companding in which the step sizes are controlled by the syllabic rate of human speech.

Synchronous detection. See *coherent detection*.

Synchronous signals. Signals are synchronous if their corresponding significant instants have a desired constant phase relationship with each other. [CCITT]

Synchronous transmission. A transmission process such that between any two significant instants in the overall bit stream, there is always an integral number of unit intervals. [FED-STD-1037]

Systematic drift. Frequency drift in an oscillator resulting from slowly changing physical characteristics and causing frequency changes that over a long period of time tend to be in one direction.

System gain. The difference, in decibels, between the transmitter output power and the minimum receiver signal level required to meet a given performance objective.

System rise time. In an optical fiber system, the time for the system response to a step function to rise from 10 percent to 90 percent of maximum amplitude.

Time accuracy. The degree to which a clock agrees to a specified standard time.

Time assignment speech interpolation (TASI). A form of speech interpolation generally used with analog voice transmission.

Time base error. See *time interval error*.

Time-division multiplexing (TDM). Multiplexing in which two or more channels are interleaved in time for transmission over a common channel. [CCITT]

Time interval error (TIE). The error in time that accumulates when a frequency source is used as a clock.

Time to loss of frame alignment. After proper frame alignment, the time to loss of frame synchronization, often specified as a function of BER.

Timing slip. See *slip*.

Transcoder. A device that performs direct digital-to-digital conversion between two different voice encoding schemes without returning the signals to analog form.

Transitional coding. A coding scheme used to convert asynchronous data to synchronous transmission by transmitting the data values and data transition times.

Transmitter. A device that converts information signals to electrical or optical signals for transmission purposes.

Transmultiplexer. Equipment that transforms frequency-division multiplexed signals (such as group or supergroup) into corresponding time-division multiplexed signals that have the same structure as those derived from PCM multiplex equipment. The equipment also carries out the inverse function. [CCITT]

Transversal filter. An equalizer that uses a tapped delay line with weighting coefficients at each tap to eliminate intersymbol interference.

Twisted-pair cable. A communication cable consisting of two insulated conductors twisted together.

Unavailability. The probability or fraction of time that a system is in a state of failure.

Unipolar. A two-level code using a single polarity for one level and zero voltage for the other level.

Vocoder. A type of voice coder that is based on a model of the human speech mechanism, in which the source of sound is separated from the vocal tract. A speech analyzer determines the source of sound and vocal tract characteristic and formats this information into a bandwidth that is much smaller than the speech signal itself. A speech synthesizer converts the formatted information back into artificial speech sounds.

Waiting time jitter. In a pulse stuffing multiplexer, the jitter occurring because of the delay between the time a stuff is needed and the prescribed times that stuffs are allowed.

Wander. See *drift*.

White noise. Noise whose frequency spectrum is continuous and uniform over a wide frequency range. [FED-STD-1037]

Wideband coder. A voice coder with a transmission rate that exceeds the rate that can be accommodated by 3-kHz telephone channels using modems.

Wideband FSK. A form of FSK in which the modulation index is much greater than 1.

Word interleaving. See *character interleaving.*

Zero code suppression. In PCM transmission, the replacement of the all-zero code with a code that contains a single 1 and the remainder zeros. It is used to maintain clock recovery performance when using bipolar coding.

Zero-forcing equalizer. An equalizer in which the weighting coefficients of a transversal filter are selected to force the equalizer output to zero at sampling instants on either side of the desired signal.

INDEX